SCALEUP OF
CHEMICAL PROCESSES

SCALEUP OF CHEMICAL PROCESSES

Conversion from Laboratory Scale Tests to Successful Commercial Size Design

ATTILIO BISIO
Exxon Research and Engineering Co.

ROBERT L. KABEL
Pennsylvania State University

A Wiley-Interscience Publication
JOHN WILEY & SONS
New York · Chichester · Brisbane · Toronto · Singapore

Copyright © 1985 by John Wiley & Sons, Inc.

All rights reserved. Published simultaneously in Canada.

Reproduction or translation of any part of this work beyond that permitted by Section 107 or 108 of the 1976 United States Copyright Act without the permission of the copyright owner is unlawful. Requests for permission or further information should be addressed to the Permissions Department, John Wiley & Sons, Inc.

Library of Congress Cataloging in Publication Data:

Bisio, Attilio.
   Scaleup of chemical processes.

   "A Wiley-Interscience publication."
   Includes bibliographies and index.
     1. Chemical plants—Design and construction.
2. Chemical engineering.   I. Kabel, Robert L.   II. Title.
TP155.5.B57   1985   660.2'8073   84-25767
ISBN 0-471-05747-9

Printed in the United States of America

10 9 8 7 6 5 4 3 2

CONTRIBUTORS

F. G. Aerstin, Dow Chemical (USA), 633 Building, Midland, Michigan 48640

G. Astarita, University of Delaware, Newark, Delaware 19711

A. Bisio, Exxon Research and Engineering Company, Route 22 East, Annandale, New Jersey 08801

W.-D. Deckwer, Fachbereich Chemie Universitat Oldenburg, Oldenburg, D-2900, Federal Republic, Germany

J. R. Fair, Department of Chemical Engineering, University of Texas, Austin, Texas 78711

G. F. Froment, Laboratorium voor Petrochemische Techniek, University of Gent, Rijksuniversiteit, Krijgslaan 271, 9000 Gent, Belgium

D. M. Himmelblau, Department of Chemical Engineering, University of Texas, Austin, Texas 78712

R. L. Kabel, Pennsylvania State University, 164 Fenske Laboratory, University Park, Pennsylvania 16802

P. E. Krystow, Exxon Chemical Company, Central Engineering Division, P.O. Box 271, Florham Park, New Jersey 07932

P. B. Lederman, Roy F. Weston, Inc., 1 Weston Way, West Chester, Pennsylvania 19038

J. M. Matsen, Exxon Research and Engineering Company, Exxon Engineering Technology Department, P.O. Box 101, Florham Park, New Jersey 07932

E. B. Nauman, Rensselaer Polytechnic Institute, Department of Chemical Engineering, Troy, New York 12181

J. Y. Oldshue, Mixing Equipment Company, 135 Mt. Road Blvd., Rochester, New York 14603

L. A. Robbins, Dow Chemical (USA), 845 Building, Midland, Michigan 48640

Y. T. Shah, Department of Chemical and Petroleum Engineering, University of Pittsburgh, Pittsburgh, Pennsylvania 15261

L. Svarovsky, Postgraduate School of Studies in Power Technology, University of Bradford, Bradford, West Yorkshire BD7 1DP, England

A. J. Vogel, Dow Chemical (USA), 1604 Building, Midland, Michigan 48640

To students who wonder about the real world
To practitioners who know about it and still wonder
To our wives, Rosemary and Barbara, who wonder about us

PREFACE

Scaleup is inherent in all industrial activity. No plant is built or product made without supporting calculations, studies, and demonstrations under conditions that are not the ones that will be practiced commercially at a future date. Scaleup is the process or group of activities by which one moves from the calculations, studies, and demonstrations to a successful commercial operating facility.

Scaleup, therefore, includes many facets of industrial activities. In its broadest sense it includes such activities as marketing, product design, product testing, plant design, and plant construction. However, our focus in this book is on the technical factors that are critical to the design and startup of a commercial manufacturing facility. Obviously, the separation of technical factors from all the other factors that are critical to a successful commercial enterprise is somewhat artificial, but it is both convenient and necessary.

Scaleup is constantly on the minds of chemists and engineers concerned with the development of new processes or expansion of existing ones. Scaleup, in the sense of the definition given in Chapter 1, involves answering the technical question, "How will this play on a larger scale?" When batch processing was the rule (and it still is in some industries), scaleup meant doing what you were already doing, only with a larger volume of reagents. As the efficiency of continuous processing became evident, scaleup often implied conversion from batch to continuous operation. Current emphasis on pharmaceutical and other speciality chemicals is resulting in a resurgence of commercial batch processing. Moreover, many processes that will ultimately operate continuously are still explored in the batch mode. Today, there is the added attempt of aiming for an optimum operation, for example, optimum from the point of view of energy or optimum from the point of view of utilization of feeds.

There is a great urgency about scaleup: the earlier the plant startup date, the longer the useful life of the plant, the greater the economic benefit. Thus there is a premium for scaling up directly from the lab bench. This places the "fundamentals" approach to scaleup in conflict with the traditional "evolutionary" approach. The virtues of both approaches are identified and exploited in this book, although the "fundamentals" approach is championed by most authors.

Many technological tools (theories, models, experimental methods, and computer software) are readily available today. However, even with these tools we are always involved in making an important judgment: how much calculation, how much demonstration, how large a unit is necessary before sufficient confidence is gained? Scaleup is not separable from the organization practicing it and the constraints that have been imposed upon that organization by its management or the marketplace. Two different organizations facing similar technical problems and operating in the same industry are often found to take completely different approaches to what appear to be identical scaleup problems.

To the knowledge of the authors only a few books have been written and fewer courses have been taught with the broad topic of process scaleup as the primary focus. Indeed the rapid development of technological tools in recent years may have obscured the importance of scaleup as a distinct quantitative art. Our formal interest in scaleup began when one of us (AB) was asked by the Center for Professional Advancement to organize a short course on scaleup and found no suitable text. That was the origin of this book. However, we must acknowledge the pioneering text *Pilot Plants, Models, and Scale-up Methods in Chemical Engineering*, written by R. E. Johnstone and M. W. Thring in 1957. Those authors in turn, as do we, acknowledge their debt to the earliest book that discusses scaleup, *A Handbook in Chemical Engineering*, authored by G. E. Davis in 1901.

In our judgment it is not possible today for any individual to be an expert in all or even many aspects of scaleup. Except for very small companies or some simple problems, scaleup is achieved by a team of individuals operating with a common goal. This characteristic of collaboration is also evident in this book. One of the editors comes from industry and the other from a university, and each has substantial experience in the other's domain. Similarly, the other authors comprise a blend of academic and industrial backgrounds. Importantly, all have considerable experience in research, process development, manufacturing, and scaleup. The objective of this book is to turn their experience to the readers' advantage.

The book is designed to serve students, faculty, researchers, and practitioners alike. On the campus, courses are taught in stoichiometry, thermodynamics, kinetics, transport phenomena, unit operations, process control, and other special topics. Quite often these subjects are integrated in a course on process design. However, such courses are directed at design for process and economic evaluation and not design for construction and operation. Scaleup is directed at the design for construction and ultimately plant operation.

On the campuses, therefore, there is little preparation in any course today for the realities of scaleup. We hope that the thoughts in these pages can be brought effectively into the classroom. For researchers, on the other hand, portions of the book should provide perspective on some practical context of their work. Indeed the organization of the book is based on the assumption that few readers will read the entire book at one sitting, although those that do may enjoy it greatly. Rather, the reader is encouraged to read first Chapter 1 and then jump into the section which is relevant to the practical problem at hand.

Industrial users of this book may work for large companies with substantial resources and world-recognized experts or for small companies employing a few engineers. Since there are so many aspects to scaleup and because of the affiliations of the authors, the book may appear to represent only a large-company perspective. While it is true that large organizations are the sources of most scaleup expertise and that they benefit greatly from it, the potential contribution of such expertise to the success of smaller companies is far out of proportion to the size of these companies. This was evident to the editors from the affiliations of those attending the week-long courses on Scaleup in the Chemical Process Industries which we (and many of our contributing authors) taught at the Center for Professional Advancement, Somerville, New Jersey. Close attention to the expressed needs of those participants from smaller companies was made both in the conception and preparation of this book.

It is a difficult task to produce a coherent text with multiple authors. To solve this problem, the book has been organized in a parallel structure at three levels:

Each individual chapter.
Chapters 3–7 on the scaleup of chemical reactors.
The whole book.

Reader understanding of the parallel structure should greatly improve effective use of this book. Outlined in the first column below are the various elements of the structure:

Structural Elements

Element	Reactor Chapters	Whole Book
Major issues	3	1
Fundamentals	3	2–3
Prediction of performance	4–6	4–14
Idealization, assumptions	4–6	4–14
Rules of thumb, experience	7	15–17
Uncertainties		18

Within each chapter, the group of reactor chapters, and the whole book we follow the elements from problem definition, through the tools available, to the issues left unresolved. The absence of a chapter devoted to uncertainties in reactor scaleup (see the middle column) certainly does not imply the absence of uncertainty, just that uncertainty is dealt with in each of Chapters 3 through 7.

The overall structure will be most evident by comparing the last column of the above table to the Table of Contents for the book.

The progression through the elements of the parallel structure is most important within the individual chapters. Accordingly the authors have been guided by it. They have not been constrained to it, however, when a different organization was preferable. Familiar nomenclature has been used throughout the book; however, the diversity of subject matter made it desirable to unify the notation chapter by chapter rather than across the entire book.

For the most part, the philosophies and methods presented here are believed by the authors to be the best available today. However, two competing approaches to mass transfer are illustrated in the context of fluid–fluid–solid reactors (Chapter 6) and gas absorption (Chapters 12 and 13). In Chapter 6 we have an approach based upon transport phenomena which is more common in Europe than in the United States, and in Chapters 12 and 13 the approach is based upon unit operations. Together these three chapters show a field in transition.

Heat transfer and reactor design appear prominently throughout the book because of their profound influence on scaleup. However, topics such as heat exchanger design, which have become routine to some degree when design calculations are done, are not considered. Other topics omitted from the book, which are by no means routine, are fire and explosion hazards and toxicology. An approach to such peripheral but crucial subjects is given in Chapters 15 and 16 on environmental issues and corrosion.

Since scaleup is, by definition, applications oriented, there are parts in this book that can be understood by any technician and parts that challenge the most experienced practitioner. A practitioner would do well, as suggested earlier, to read the first and last chapters for perspective before going on to the individual chapters of greatest interest. Although some subject matter does carry over from chapter to chapter, individual chapters are constructed to be rather self-contained. Reading through the book would provide a comprehensive picture of the current status of scaleup. Unexpected "nuggets of wisdom" occur throughout the book and these should reward the thorough reader.

This book is more a sharing of experience than a classic teaching text. Nevertheless, we believe it could serve well for supplementary reading in conventional undergraduate plant design courses. It would also be appropriate as the primary text in an advanced course where design for construction would be the primary focus. The contents for such a course would obviously be determined in great part by the interest and background of the instructor. It is easy to envision a variety of projects for such a course that (as in actual practice) will require complete in-depth studies in some areas and little or no attention to others. The same is true of the chapters in this book. Many of the chapters provide detailed examples that should give ample opportunity for elaboration. Further, extensive reviews are provided through references to relevant literature.

One of us (RLK) wishes to thank the Pennsylvania State University for sabbatical leave during which most of his writing for this book was done and for invaluable facilities and secretarial assistance. Special appreciation is expressed to Mrs. Karin Miller and Miss Maryann MacRitchie for their typing, coordination, and gentle reminders to keep the work moving. In particular, we would like to acknowledge the participants in the Center for Professional Advancement short courses, who taught us far more than we could ever teach them about scaleup.

ATTILIO BISIO
ROBERT L. KABEL

Annandale, New Jersey
University Park, Pennsylvania
April 1985

ACKNOWLEDGMENTS

The following copyright holders have granted permission for their figures and tables to be reprinted: Academic Press (Figures 10-6, 10-8, 10-9, and 10-15), American Institute of Chemical Engineers (Figures 1-2, 5-1, 5-5, 5-6, 6-7, 7-3, 7-4, 10-2, 10-7, 10-11, 10-12, 10-27, 10-28, 12-3, 12-10, 12-14, 12-15, 13-7, 13-12, 13-17, and 13-19), American Chemical Society (Figure 12-7), Cambridge University Press (Figures 10-14 and 10-16), McGraw-Hill Book Co. (Figures 5-3, 6-2, 12-7, Tables 1-4 and 12-1), Ducon Co. (Figure 10-22), Elsevier Publishing Co. (Figures 10-2, 10-29, 10-31, and Table 10-2), Glitsch, Inc. (Table 13-2), Gordon and Breach (Figure 10-23), Koch Engineering Co. (Figure 13-3, Tables 13-1 and 13-2), Hydrocarbon Processing (Figures 12-8 and 13-18), Institute of Chemical Engineers (Figures 6-4, 10-5, and 12-9), Jacob Engineering (Appendix Chapter 1 and Figure 1-6), Munters Corp. (Tables 13-1 and 13-2), Norton Co. (Figures 13-8, 13-15, and Table 3-1), Pergamon Press (Figures 5-4, 6-3, 10-13, and 12-11), Petro Chem. Engineer (Figure 12-16), Van Nostrand Reinhold (Figures 10-20 and 10-21), Verlag Chemie GMBH (Figures 6-11, 6-12, 7-2, 10-32, and 13-20), John Wiley & Sons, Inc. (Figures 1-9, 5-1, 5-2, 5-4, 8-7, and 11-3(b)).

CONTENTS

1. Introduction to Scaleup   1
   A. Bisio
2. Mathematical Modeling   34
   D. M. Himmelblau
3. Reaction Kinetics   77
   R. L. Kabel
4. Homogeneous Reaction Systems   117
   R. L. Kabel
5. Reactors for Fluid-Phase Processes Catalyzed by Solids   167
   G. F. Froment
6. Fluid–Fluid Reactors   201
   Y. T. Shah and W.-D. Deckwer
7. Selection of Reactor Types   253
   R. L. Kabel
8. Flow Patterns and Residence Time Distributions   275
   E. B. Nauman
9. Mixing Processes   309
   J. Y. Oldshue
10. Fluidized Beds   347
    J. M. Matsen
11. Laminar Flow Processes   406
    E. B. Nauman
12. Stagewise Mass Transfer Processes   431
    J. R. Fair
13. Continuous Mass Transfer Processes   504
    J. R. Fair
14. Solid–Liquid Separation Processes   549
    L. Svarovsky

15.	The Environmental Challenges of Scaleup *P. B. Lederman*	594
16.	**Evaluating Materials of Construction in Pilot Plant Corrosion Tests** *P. E. Krystow*	620
17.	**Gaining Experience Through Pilot Plants and Demonstration Units** *F. G. Aerstin, L. A. Robbins, and A. J. Vogel*	655
18.	**Scaleup: Overview, Closing Remarks, and Cautions** *G. Astarita*	677

Index   691

1

INTRODUCTION TO SCALEUP

A. BISIO

I.	Major Issues in Scaleup	1
	A. Scaleup Ratio	6
	B. Heresies of Scaleup	10
	C. A Scaleup Experience	11
II.	Approaches to Scaleup	14
	A. Principle of Similarity	15
	B. Models	16
	C. What Method, When for Reactor Studies	17
III.	Describing a Process System	19
	A. Evolution of a Process System	23
	B. Process Documentation	26
IV.	Balanced Attack on Uncertainty	28
	References	30
	Appendix: Scope of Process Design	31

I. MAJOR ISSUES IN SCALEUP

When a new chemical process or a change in some part of a process moves from the laboratory to a commercial manufacturing operation, unexpected problems are often encountered. The problems may be of a physical nature, a chemical nature, or involve some aspects of both. Some examples of the difficulties that may be encountered in scaleup of a chemical process will illustrate the nature of the problems that can arise. Three examples will be considered: water as an impurity, determination of explosive limits, and the storage of unstable materials.

One of the most serious and frustrating problems that can be encountered in a commercial operation is the presence of impurities that were not considered or studied in the smaller scale laboratory or pilot plant studies. Some impurities can completely change the character of a catalytic process by deactivating the catalyst or by increasing the quantity of the by-products that are formed. Moreover, once a commercial installation has been built without giving adequate consideration to the removal of impurities from process streams, modifications can be made only with great difficulty and at significant expense.

Water is a common impurity in commercial hydrocarbon streams. In a large manufacturing unit there are many opportunities for water to "leak" into a process unit. While in principle water can be eliminated from process streams by rather straightforward manufacturing procedures and mechanical features, these can be costly and the mechanical features must be provided during the construction of a commercial plant. Otherwise, a small steam leak in a heat exchanger will result in sufficient water entering a process stream to hydrolyze chlorinated hydrocarbons, "kill" a catalyst, or seriously modify its performance. The critical features of water and similar impurities as well as their potential ramifications, if any, must be understood before the design of a commercial unit is undertaken.

Similarly, the explosive limits for hydrocarbon/oxygen/nitrogen mixtures as measured in small-size laboratory equipment are narrower (and, therefore, apparently safer and easier to handle) than when the same measurements are made in commercial size equipment. The apparent narrower explosive limits are the result of the higher heat transfer rates, especially through conduction and radiation, which are made to the walls and surfaces of laboratory equipment. Higher heat transfer rates slow down the catastrophic temperature-time buildup that leads to explosions. Moreover, small-scale equipment provides a larger heat sink relative to the combustion energy evolved. Soaking up energy also helps hold down the temperature rise.

Often, temperature-unstable materials can be stored and safely handled on a small scale with minimum difficulty by following well-established procedures. In a commercial unit the storage of such materials must be examined from a critical mass point of view to be certain that the heat removal capability of the equipment to be used is substantially greater than the potential spontaneous exothermic rate of heat release. Failure to recognize the limitations of small-scale data for unstable materials has led to a number of serious and expensive explosions and fires in plants utilizing air oxidation processes and those storing ammonium nitrate, wood chips, and powdered coal.

Implicit in all of these examples is the concept of *scaleup*. In this book scaleup is defined as:

The successful startup and operation of a commercial size unit whose design and operating procedures are *in part* based upon experimentation and demonstration at a smaller scale of operation.

The concept of *successful* must include production of the product at planned rates, at the projected manufacturing cost, and to the desired quality standards. Implicit in the term cost are not only the obvious factors such as the purchase prices for raw materials, the product yield, and the return on capital, but also the overall safety of the contemplated operation to plant personnel, the public, and the environment. The timing of project completion is also in most instances a critical factor. An experience in scaleup that results in the startup being completed later than planned is not a very successful experience.

To be successful at the scaleup of chemical processes requires the utilization of a broad spectrum of technical skills and a mature understanding of the total problem under study. Scaleup procedures do *not* involve only technical decisions and compromises. The selected compromise always has an economic aspect since it is never possible to establish exactly what an industrial process should be. There are always restrictions of time and money availability for the total development program of which scaleup is only a part. Therefore, *calculated risks* will have to be taken in the design, construction, and startup of a "first commercial unit." The scope of the risks (and, the resulting financial uncertainties) will have to be considered against the additional expenses required to improve still further one's knowledge of the process.

The required interplay of chemical engineering and the underlying sciences can be clearly seen in the decision process required to establish reactor geometry and mode of operation for economic performance. The specific combination of chemical kinetics and reactor type used to arrive at a level of reactor performance (an optimum one) can be determined entirely from either simple engineering methods or through a sophisticated analysis of the interacting physical and chemical rate phenomena. This interplay of scientific and engineering disciplines changes at each stage of the selection and development of a reactor, as shown in Figure 1-1. The preferred path of development is rarely the simple direct one based on either theory or empiricism but rather some hybrid.

Indeed, to follow a *direct path* from laboratory data to a commercial design requires either a fund of information that is often (or almost always) unavailable or scientific and engineering judgments beyond those normally considered possible or desirable. However, if only minor changes are contemplated in processes where there is considerable practical experience at several scales of operation, the direct path may be both feasible and desirable.

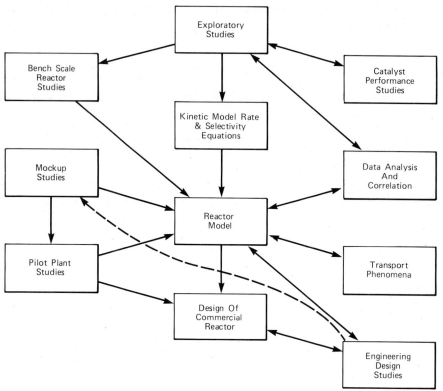

FIGURE 1-1 Structure of reactor design.

The first step in the development or modification of a reactor design must be the derivation of some rate and selectivity expressions from an analysis of laboratory data. Where the laboratory studies have only been exploratory in nature and directed primarily at determining the feasibility of reaction scheme, it is unlikely that even a simplified usable model can be extracted from the available data. At a minimum, reaction studies at the bench scale will be required to obtain even a marginally useful kinetic model.

However, proceeding from a kinetic model to a reactor design implies that the physical processes occurring in the reactor such as diffusion and velocity distribution are sufficiently well understood that their impact on conversion and selectivity can be allowed for. This is frequently *not* the case. When the reaction involves several phases, for example, not all of the phenomena of interest are influenced in the same manner by the dimensions of the physical equipment.

A qualitative indication of the impact of the major geometric dimensions on chemical reaction, mass transfer, and heat transfer in multiphase reactors is given in Table 1-1 (Levenspiel, 1979). Not surprisingly, a laboratory reactor

TABLE 1-1 Impact of Reactor Dimensions on Chemical Reaction, Mass Transfer, and Heat Transfer

Phenomenon	Reactor Volume	Length-to-Diameter Ratio	Surface-to-Volume Ratio
Chemical reaction	Significant and determining	Weak and indirectly	Only indirectly
Mass transfer	Only indirectly	Significant	Indirectly
Heat transfer	Weak and indirectly	Significant	Significant and directly

with a large surface/volume ratio can understate or even conceal a potential heat transfer problem. Pilot plant studies, therefore, are often necessary to ensure that the reactor model developed from laboratory studies can be related (and extrapolated) to the design and performance of a commercial reactor.

While small-scale experiments often permit the development and verification of fundamental hypotheses that will be used in developing a model, a note of caution is in order. As one proceeds with scaleup studies, it is critical to verify that the hypotheses developed from those early experiments are still valid. For example, for a tubular continuous reactor the hypothesis of plug flow must be verified both on small- and full-scale units. Failure to do this may result in laboratory data on conversion and selectivity apparently *not* being valid for the commercial unit. *Model studies utilizing mockups* (that is, flow without chemical reaction) are often used to study the hydrodynamic behavior of a system to help establish the commercial scale operating conditions required to ensure that a hypothesis derived from small-scale experiments is still valid.

Considerable research has been done in recent years on two-phase gas/liquid flow through catalyst beds. From an industrial point of view, however, much remains to be learned about this regime. Even pressure drop in many instances *cannot* be calculated with a reasonable degree of accuracy. Variations of ±100 percent are not unknown. Laboratory studies on pressure drop in small-diameter tubes must be supplemented by mockup studies on the distribution of the liquid and gas phases over the cross section of a catalyst bed. Only in this manner can one ensure that the distribution of liquid and gas phases will be reasonably uniform, a minimum requirement to achieve good catalyst efficiency and minimum reactor volume.

Mockup studies should help establish the variables that determine the flow conditions of each phase in a reactor. In a pilot reactor with a small inside diameter (often less than 5 cm), uniform gas/liquid distributions are obtained even where the linear velocities of both fluids are considerably lower than those in a commercial unit. Low linear velocities *can* have a marked impact on gas–liquid mass and heat transfer, particularly in processes such as hydrocracking where large quantities of hydrogen are consumed. An industrial unit may appear to have higher catalyst activity and stability as compared to pilot unit runs if the pilot unit runs have not been made at conditions where the fluid linear velocities are close to those in commercial units.

A. Scaleup Ratio

Implicit in our discussion of scaleup is the concept of a scaleup ratio; that is, the relationship between the size of the contemplated commercial unit and the largest small-scale unit in which data are obtained:

$$\text{scaleup ratio} = \frac{\text{commercial production rate}}{\text{pilot unit production rate}}$$

Obviously, one can also define analogous scaleup ratios between pilot plant units and laboratory scale equipment.

Scaleup ratios for a number of typical process systems as developed by Ohsol (1973) are given in Table 1-2. Scaleup ratios from laboratory equipment to pilot plant units of 500–1000 and from pilot plant to commercial units of 200–500 are not uncommon. However, the ratios tend to be lower for processes involving liquid and solid reactants as products rather than gases.

TABLE 1-2 Typical Scaleup Ratios

System	Scale of Operation, kg/hr		Scaleup Ratio	
	Laboratory	Pilot Plant	Laboratory to Pilot Plant	Pilot Plant to Commercial
Substantially gaseous (ammonia, methanol)	0.01–0.10	10–100	500–1000	200–1500
Gaseous reactants, liquid or solid products (sulfuric acid, urea, maleic anhydride)	0.01–0.2	10–100	200–500	100–500
Liquid and gaseous reactants, liquid products (benzene chlorination, oxidation)	0.01–0.2	1–30	100–500	100–500
Liquid reactants, solid or viscous liquid products (polymerizations, agricultural chemicals)	0.005–0.2	1–20	20–200	20–250
Solid reactants, solid products (phosphoric acid, cement, ore smelting)	0.10–1.0	10–200	10–100	10–200

Large scaleup ratios at a reasonable level of risk can be achieved only in those situations where there are "known" scaleup correlations (another way of saying considerable practical experience) or where a fundamental engineering science/chemistry approach is possible. However, the manufacture of uranium-235 by the gas diffusion process in the mid-1940s involved scaleup from a laboratory bench unit to a commercial unit, at a scaleup ratio in excess of 10^6. Obviously, the level of risk was significant but so were the perceived benefits. In addition, the technical and economic resources brought to bear on the problem were enormous.

Critical to the concept of a scaleup ratio is that the difference in the scale of the operations being compared is sufficiently great so meaningful differences in surface/volume and height/diameter ratios of packed towers, tower packing size/tower diameter ratios, agitator/vessel diameter ratios, and other factors are achieved. For example, in small-size packed columns, the gas may move in plug flow; in commercial size units there can be substantial backmixing. The variation in operating scale should be sufficient to uncover such a difference in behavior.

Where the observed results on moving from a smaller scale of operation to commercial size equipment are in good agreement with predictions the scaleup principles are well established. Then large scaleup ratios are possible. This is generally the case for processes that are substantially gaseous in nature. Gaseous flow processes can be scaled from laboratory to pilot plant and pilot plant to commercial units at scaleup ratios as high as 10,000. This is because accurate predictions of the flow regime, heat and mass transfer coefficients, and the like for any scale of operation can be calculated from physical data measured in the laboratory or from literature data for pure components.

However, in almost all cases of commercial interest, gaseous flow processes are utilized with a solid catalyst. The presence of the catalyst will have a significant impact on the flow and temperature distributions as well as on the heat and mass transfer. Now one must devise suitable small-scale measurements of catalyst performance which give information representative of commercial scale operation. One approach is to carry out catalyst evaluations in a single tube of commercial diameter, length, and materials of construction. Results from these studies can be extrapolated to commercial size reactors after making allowances for the perceived differences (at a minimum) in heat losses, flow rates, and feed compositions. However, it is critical that single tube experiments be related to commercial reactor experience for similar systems. Otherwise, one may find that the residence time and temperature distributions selected from a single tube experiment may be impractical in a commercial reactor or can result in unexpected side reactions such as coking.

In the study of new reforming catalysts, for example, the effectiveness of the catalyst is often studied in a small (less than one kilogram of catalyst) pilot unit. The catalyst performance obtained in a laboratory unit cannot be directly related to a commercial reactor. There is no relationship between the bed heights of the laboratory and commercial reactors if mass transfer *within* the catalyst particles controls the overall kinetics. Therefore, the transformation of the laboratory isothermal bed data to commercial bed operation involving temperature gradients is generally derived from empirical correlations.

There are many areas of importance where the available correlations are limited in their scope or our predictive capabilities are found wanting. For example, all the data (and resulting correlations) for liquid holdup and gas- or liquid-phase axial dispersion coefficients have been obtained with packings normally used in absorption or gas–liquid reaction processes. The packings are large and nonporous compared to those normally used in gas–liquid–solid catalytic reactions. Not surprisingly, one then finds a large degree of discrepancy in the literature. The usefulness of a given correlation to a specific catalytic reaction must be checked experimentally, if at all possible.

Similarly, theoretical predictions of contacting efficiency, flow regimes, and separation efficiency for solid–liquid, liquid–liquid, gas–liquid, and solid–gas systems are poor. If two-phase flow is involved, data at several scales of operation are nearly always going to be required. Also, not too surprisingly, some departure from the best of theoretical predictions is to be expected as one moves from small-scale to commercial size equipment.

Process systems involving solid reactants and particularly those phenomena that involve characteristics such as adhesion to surfaces, to other solids, or self-adhesion are quite variable and highly dependent on many factors. The contacting characteristics of solids need extensive evaluation particularly with respect to the impact of such variables as particle size and size distribution, traces of organic impurities, and even suppliers of the reactant. As a result, it is almost impossible to scale data from small-solids handling units to commercial units. That is, more than one pilot stage may be necessary for a new technology. Each of the intermediate-size units will have to be built, run in, the appropriate modifications made, and the unit tested.

Our experience suggests that there are recurring scaleup problems in the development of many process studies. Differences that may be of particular importance in moving from small-scale to commercial size equipment are:

Shape which can lead to differences in agitation, fluid short-circuiting, or stagnation zones.

Mode (and scale) of operation resulting in different residence time distributions

Surface-to-volume ratios, flow patterns, and geometry which result in significantly different gradients of concentrations and temperature.

Materials of construction resulting in different contaminant levels.

Flow stability.

Heat removal.

Wall, edge, and end effects.

Generalizations about the impact of the scale of operation on the importance and extent of these differences are difficult to make. Wall, edge, and end effects

MAJOR ISSUES IN SCALEUP

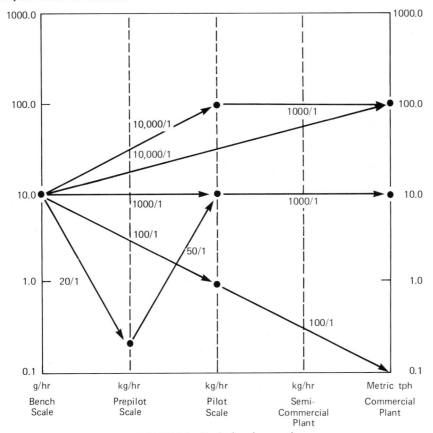

FIGURE 1-2 Typical scaleup paths.

are generally more important in small- than in commercial size equipment. As a result, there is some wall effect inherent in observations about concurrent flow in small equipment. Moreover, countercurrent operation involving two-phase flow in large-diameter vessels is generally less stable than that in small-size equipment.

Studying a process step on a pilot plant scale is *not* inexpensive. Indeed, the operating costs for a full-scale pilot unit or a semicommercial plant with a recycle system can be comparable to those for a small commercial size unit. There are large incentives for moving from bench scale to commercial scale operations with only one intermediate step and at a high scaleup ratio. A scaleup ratio of 1000, as shown in Figure 1-2, would permit a commercial size unit of 100 m t/hr to be designed from data obtained in a 100 kg/hr pilot unit. However, where the scaleup ratio is limited to less than 50–100 (such as could be the case with a unit handling solids), several intermediate stages of study may be needed to obtain data for a 1 t/hr unit.

The critical question is whether the benefits derived from the additional studies (inherent in low scaleup ratios) result in a significant improvement in commercial performance. That operation of a pilot unit permits experimental evaluation of the impact of a change in scale of operations on the relationships derived from bench-scale work is *not* a sufficient reason for building one. Many commercial units have been built without the benefit of data from a pilot plant.

Restated, the critical question is to what extent will commercial performance suffer if a process step is *not* pilot planted. Inadequate commercial performance is often first seen in an extension of time required to start up a commercial unit. The financial drain (and other consequences) of having a commercial unit continue in a startup mode for extended and unplanned periods of time is substantial. Experience suggests that a pilot program whose total cost is less than the costs incurred in three months of startup expenditures for a commercial unit can be a good investment. When the costs associated with the pilot plant operation are substantially higher, the decision to operate a pilot unit must arise from a specific technical/economic requirement involved in the design and operation of the commercial unit.

The Department of Energy (DOE) sponsored a significant number of coal gasification pilot units and demonstration prototype plants. The apparent justification was that ultimate commercial units would require significant state-of-art advances in a number of areas, particularly those involving solids handling. (Another way of stating this is that DOE, if we think in terms of Figure 1-2, felt that scaleup ratios of less than 100 should be considered in moving from bench-scale to full-scale commercial units.) In addition, DOE might have believed that there are some long-term environmental "potential" problem areas that could only be studied in demonstration prototype units. However, one may wonder whether the primary driving force behind the demonstration units was not a desire to convince Congress and the commercial entities who might have been the investors or operators of commercial plants that the processes indeed work. Here the size of the demonstration (or pilot unit) is being taken as a measure of technical success.

B. Heresies of Scaleup

The only defensible reason for scaleup studies is a reduction in the possibility of making expensive errors in the design or the operation of commercial size equipment. Scaleup studies are *not* the crowning achievement of a research and development program. They must be carried out in such a manner that the uncertainties that will face future design engineers and plant operators are reduced. Therefore, the starting point for scaleup studies, as discussed by Smith (1968), must be the ultimate commercial unit; the contemplated studies must be "scaled down" from the requirements and unknowns of that commercial unit. Scaleup from small-scale studies is a misleading concept.

Scaleup studies are *not* the place to economize on expenditures—even if operating a pilot unit is involved. The only permissible economy is *not* to do a scaleup study, a decision reached hopefully only after a thorough and sound technical and economic analysis. Economies at the scaleup stage are likely to be false. Moreover, an incomplete or misleading scaleup study is likely to lead to errors in either the design or the operation of the commercial unit. The costs associated with an error(s) in the commercial unit can be substantial and they will easily overwhelm any potential saving made in "short cutting" the scaleup study.

Where a pilot plant unit is involved there seems to be a belief in some quarters that the unit must be complete in every respect and indeed preferably that it have all of the systems that will be present in a commercial unit. This is *not* so. To begin with, it is often an impossibility. More importantly, scaleup studies should be directed primarily at attacking areas of doubt and uncertainty. Quite often this does not involve producing the product that will be sold from a commercial unit. The exception to this, of course, is where the properties of the product are of critical concern in the scaleup studies.

Scaleup studies involve modeling relevant phenomena, not the study of miniaturized commercial systems. In the scaleup of fixed-bed, small-tube catalyst processes, for example, a choice must be made between doing studies on a nest of tubes or a single tube. For a 4000-tube reactor a nest of at least 40–50 tubes is required to develop in-depth information about overall performance and control strategies. However, the vital information on hot spots, temperature gradients, and catalyst activity decay can be better obtained in a single tube. The single tube data can then be extrapolated to a commercial reactor design and operation.

The quest for optimization is one of the major sources of misconception about scaleup. Optimum performance to a significant degree depends on commercial, not technical, circumstances and events. These circumstances change during the development studies and certainly over the life of a commercial unit which can be in excess of 40 years. An optimum design is one that results from a careful consideration of the range of conditions over which the commercial plant will have to operate and the critical criteria to be met. A design that achieves the critical constraints at minimum *total cost* is optimum. This is not the same as the design whose objective is minimum capital investment. In thinking about the scope of the scaleup studies, it is important to think through carefully whether the data are being obtained to optimize the design or to optimize the performance. Often it is difficult in the same study to combine obtaining data directed at optimization of both design and performance.

C. A Scaleup Experience

The development of the HPO caprolactam process by Dutch State Mines as discussed by deRooij et al. (1977) and Simons (1978) provides excellent insights into the intricate nature of scaleup studies. In the caprolactam process two intermediates, cyclohexanone (anone) and hydroxylamine (hyam), are reacted to yield cyclohexanone oxime (oxime) as indicated symbolically by

Equation 1-1.

$$\text{(anone)} \quad C_6H_{10}=O + NH_2OH \rightarrow C_6H_{10}=NOH \quad \text{(oxime)} \tag{1-1}$$

(anone) (hyam) (oxime)

Cyclohexanone can be produced by the liquid-phase oxidation of cyclohexane or the vapor-phase hydrogenation of phenol.

Historically, the hyam intermediate was first used as an aqueous solution of hydroxylamine sulfuric acid and ammonium sulfate in equal amounts. The reaction between the anone and hyam mixtures then results in the manufacture of still another mole of ammonium sulfate. Moreover, the oxime is converted into caprolactam with the aid of oleum through an intramolecular Beckman rearrangement.

$$C_6H_{10}=NOH \xrightarrow[SO_3]{H_2SO_4} \underset{NH}{\underset{|}{(CH_2)_5}}\overset{CO}{\underset{|}{}} \tag{1-2}$$

The neutralization of the reaction mixture results in a crude lactam and additional ammonium sulfate. Several extensive purification steps, including extractions, are then needed to purify the lactam. For every metric ton of purified caprolactam, nearly 5 t of ammonium sulfate were (and still are in some circumstances) produced as a by-product.

The Dutch State Mines HPO process recognizes that if one is to avoid the sulfuric acid liberated during the formation of the hyam and the oxime, a closed acid loop must be introduced between the synthesis of the oxime and that of the caprolactam. The closed acid loop eliminates formation of the ammonium sulfate by-product.

In the HPO process the oxime is synthesized in a phosphate-buffered system. The key step in the HPO process is the production of hydroxylamine phosphate by the catalytic reduction of a nitrate solution in phosphoric acid with molecular hydrogen.

$$NO_3^- + 2H_3PO_4 + 3H_2 \xrightarrow{Pd/C} NH_3OH^+ + 2H_2PO_4^- + 2H_2O \tag{1-3}$$
$$\text{pH} \sim 0 \qquad\qquad\qquad \text{pH} \sim 2$$

Since the catalyst is a solid, palladium on active carbon, the following potential problem areas can be anticipated:

Mass transfer and dispersion will be important in this chemically complex, multiphase system.

Different catalyst activation techniques may influence performance.

Catalyst deactivation (poisoning) by impurities at low ppm levels may be critical; its nature and extent must be established.

Plastic mockups are particularly useful for studying the hydrodynamic behavior of a system, fluid flow in the commercial process being represented by fluid flow in the model.

Modeling with plastic mockups generally requires:

Shifting the range of operating conditions toward ambient temperatures and pressures.

Studying a single aspect of a complex phenomenon, each aspect being studied in mockups of different size.

Determination of scale factors for the phenomenon.

Simple visualization of a number of complex fluid phenomena, such as foaming, can often lead immediately to improved understanding. Moreover, when properly designed, mockups are quite inexpensive compared to pilot plant studies.

The ultimate outcome from a mockup study will be expressions involving dimensionless groups or some simple correlation. These can be put into use in mathematical models that are used to calculate:

Material balances and energy balances.

Chemical kinetics.

All heat transfer phenomena.

Mass transfer between two phases.

The equations in a mathematical model may result from theoretical (fundamental) studies. They may also be purely empirical, that is, derived from dimensional analysis expressions and correlations of experimental data.

Setting up a fundamental model for a complex chemical process is a difficult and often an impossible task. Whatever the difficulty, it is always worthwhile to proceed with at least the first steps in setting up such a model. The analysis, even if incomplete, will identify the most important phenomena to be considered in scaleup studies, the possible rate determining steps, and most importantly, the impact of changes in equipment size on each of the phenomena under study. A mathematical model study, however incomplete, should always precede the formulation of an empirical model.

C. What Method, When for Reactor Studies

The critical phenomena for study in a reactor scaleup program are:

Chemical kinetics.
Mass transfer.
Heat transfer.

TABLE 1-4 Reactor Scaleup Techniques

	Gas or Liquid		Gas and Liquid			Catalytic	
	Batch Tubular	Continuous Stirred	Liquid-Phase Controlling	Gas-Phase Controlling	Gas-Fixed Bed	Gas/Liquid- Fixed Bed	Fluidized or Moving Bed
Phenomena Under Study							
Chemical kinetics	2 3	1	3	1	1	1	1
Mass transfer	4 4	4	1	3	4	1	4
Heat transfer	1 1	2	4	3	3	2	3
Scaleup Methods							
Laboratory studies	1 4	1	3	2	3	2	1
Pilot plants	4 1	3	3	1	1	1	3
Mockups	4 4	3	1	1	4	1	1
Modeling	2 2	1	1	2	2	3	3

1. Critical/very important.
2. Necessary/important.
3. Desirable/some importance.
4. Little value/irrelevant.

However, the importance of these varies considerably from process to process as shown in Table 1-4, first developed by Corrigan and Mills (1956). Mass transfer, not surprisingly, is irrelevant to single phase noncatalytic processes. However, it is also of only slight importance for fluidized or moving bed catalytic reactors.

Similarly, specific scaleup methods tend to be used to study certain critical phenomena. Mockups, for example, are rarely used to study single phase processes. However, they are widely used in those situations where a number of phases are involved, such as fluidized and moving bed processes. Table 1-4 also shows the relative importance of the scaleup procedures we have been discussing.

III. DESCRIBING A PROCESS SYSTEM

Scaleup should also be approached from a knowledge of what it is believed the commercial unit will look like. Even at an exploratory research stage, this should include a flow sheet showing the dimensions of commercial equipment. The flow sheet would be based on laboratory results, literature data, practical experience with similar equipment, and reasonable judgments and extrapolations to the final results that are our goals. Then, by scaling down, judgments can be made as to the nature of the pilot plant and mockup studies that should be made. Only by identifying critical phenomena through scaledown analysis can the risk of failure of a development program be kept to a minimum.

Preparation of a process flow sheet is the major step in describing a process or change to a process. A process flow sheet, as shown in Figure 1-3, must address itself to the reactions, separations, and auxiliary operations that are required to transform raw materials into salable products. The auxiliary operations are concerned with transfers of materials and energy as well as conversions of energy. A properly formulated flow sheet showing the interrelationships is essential to the start of a scaleup program. Ultimately, with all the changes and revisions developed during the program, the "early" flow sheet serves as the basis for beginning a commercial plant design.

A process flow sheet is nothing more than a graphical representation of the equipment in a commercial plant and the path through that equipment that will be followed by the material being processed. The reaction, separation, and auxiliary operations shown schematically in Figure 1-3 along with the required utilities and by-product/waste streams must be shown in sufficient detail that the important interrelationships are obvious. Ultimately, a set of flow sheets will show the various processing steps, their flow relationship and sequence, the interconnection of all processes, mechanical and materials-handling equipment, as well as major instrument control circuits. In addition, information on stream flow rates or batch quantities, compositions of streams and their physical properties, as well as typical operating conditions will also be presented.

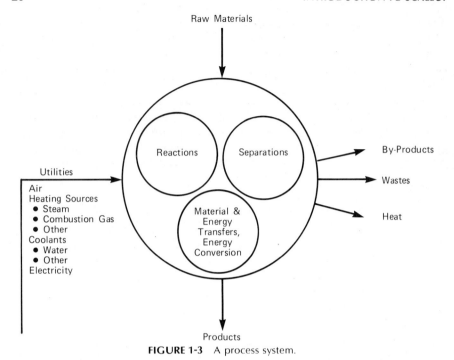

FIGURE 1-3 A process system.

No matter how detailed a process flow sheet appears, it is to some degree an abstraction, The flow sheet of the petroleum refinery shown in Figure 1-4 is an artistic abstraction. From this flow sheet one can obtain only a limited understanding about either the equipment involved or the precise nature of the processing steps. However, one does gain some appreciation of the process flow sequence. Considerable background and practical experience in each of the processes shown on the flow sheet would be required to fill in the details.

Another level of abstraction is the line/block flow diagram shown in Figure 1-5. Here each block can represent a processing step or an overall system. Flow sheets only evolve with time and repetition. A line/block flow diagram is a good starting point for a preliminary process as one develops a scaleup program.

Line/block flow diagrams are particularly valuable in developing a mathematical model. For example, a lumped parameter description of the two process elements shown in Figure 1-5 would consist of the material and energy balances for each element—one material balance for each component in the element and one energy balance. If values of F_0, F_1, F_2, F_3, C_{A0}, C_{B0}, K_1, and K_2 are known, then the set of four material balances and two energy balances can be "solved" for the six unknown concentrations. (what solution of a model means is described in Chapter 2).

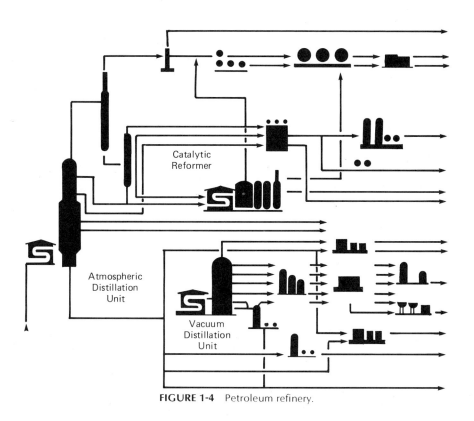

FIGURE 1-4 Petroleum refinery.

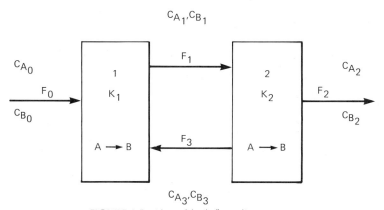

FIGURE 1-5 Line–block flow diagram.

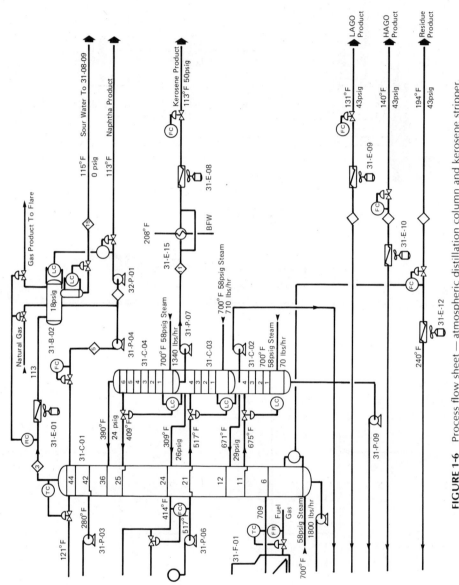

FIGURE 1-6 Process flow sheet — atmospheric distillation column and kerosene stripper.

DESCRIBING A PROCESS SYSTEM

A detailed process flow sheet for an atmospheric distillation column and kerosene product stripper typical of what would be found in a petroleum refinery is shown in Figure 1-6. From this flow sheet the following information can be obtained:

Precise process sequence.
Approximate nature of mechanical equipment used in each step.
Interrelationship of the mechanical equipment and the process flows.
Approximate operating conditions.
Magnitude of the major heat flows.
Relative position of the mechanical equipment.

Flow rate, stream composition, and other material balance information would have to be obtained from related documentation.

The detailed flow sheet shown in Figure 1-6 provides the basis for any number of scaleup studies. For example, one might be interested in establishing what impact foaming in the atmospheric distillation column could have on the kerosene stripper. Laboratory studies could be undertaken to define the conditions under which foaming occurs while a mockup could provide valuable information on possible interrelationships between the distillation column and the stripper.

A. Evolution of a Process System

For a commercial plant to be successful, a number of goals or objectives must be set: social, economic, technical, and temporal. The evolution of an acceptable process system, that is, the backbone of a commercial manufacturing unit, is a complex exercise in problem solving, as shown in Figure 1-7. Starting with an idea, there is a continual interaction between design/economic studies and experimental programs (laboratory, pilot plant, or mockup). Scaleup studies are involved in all of the activites that are crosshatched. In the unfolding of a project there is rarely a straightforward movement from the idea to the detailed plant design.

At the start of a development program and for some time thereafter, the complete process system "that will be" is only conceptual in nature in the collective minds of the individuals working in the area. Possibly the idea has been expressed as a process concept or even illustrated in a preliminary process design. However, the opportunities for misunderstandings or errors are substantial as discussed by Waters (1973).

All those associated with and working on a development can be considered to live in two worlds: a *conceptual* world where the ideas and concepts are synthesized from impressions of the real world, and the *perceptual* world, that is, the real world as viewed immediately through the senses. Pilot plant studies,

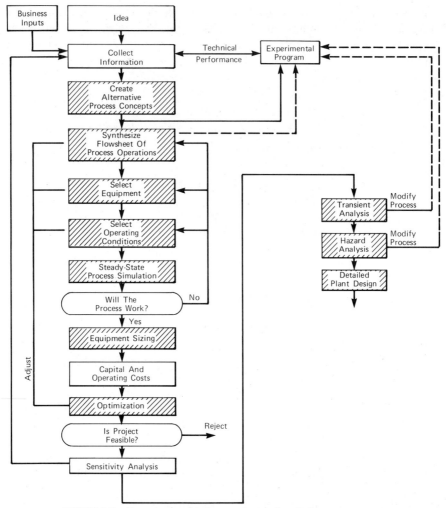

FIGURE 1-7 Steps in the development of a chemical process system.

market development sample production and marketing, and certain simulation studies can give us valuable information about the real world of our "idea." Throughout the development program we will find ourselves testing and adapting our conceptual world against the data bank we have been able to build up about the real world.

The development of a process system, then, can be regarded as the development of an idea at a conceptual level in a number of stages. Experimentation and model building are our windows on the real world. The final conceptual representation of our idea is the detailed plant and the associated operating information that can be straightforwardly translated into a commercial operating plant. The results obtained during the startup and early operation of the commercial plant are the real world. Then we will know whether our ideas were indeed "good ones."

Often for convenience and work planning the experimental part of an overall development program is divided into three successive phases: exploratory research, process research, and process development. The scope of these phases is neither completely clear nor distinct. However, the following definitions are often used.

Exploratory Research

The development program begins with an idea, perhaps a specific process scheme or concept that could be valuable and therefore worth developing. What are needed are data to illustrate or "prove out" the idea. In Figure 1-7, experimental studies are shown both during the *Collect Information* phase and immediately after the *Create Alternative Process Concepts* phase—this is exploratory research. Often the results from exploratory research can be used to make a preliminary evaluation of the idea.

Process Research

If we conclude that our idea has merit, extensive laboratory studies will be undertaken. Where a salable product is involved, limited quantities will be produced and subjected to laboratory evaluation. The influence of different process conditions on the reaction rate are determined, new catalysts are investigated, and the influence of impurities are explored. Attempts at modeling the phenomena of interest are undertaken and some preliminary models are developed. Limited mockups of certain critical process features may be utilized if only to set the basis for larger mockups. The data from these studies should be analyzed almost on a continual basis. A decision as to whether or not one believes the project is feasible is essential before proceeding further since the financial expenditures are beginning to escalate.

Process Development

Here the development and design of commercial equipment and the operation of mockups and pilot plants move together. The goal is to prove that the overall process is sound and meets the technical/economic objects established early in the program, possibly at the exploratory research phase. We must establish acceptable process ranges, prove out the relevant models that have been developed, and establish both the parameters determining product quality and the influence of impurities. Studies derived from the scaleup programs will set both the parameters for the design of commercial equipment and its ultimate operation. In some cases during the process development phase, substantial quantities of product are produced to help finalize market studies. Unfortunately, many have found to their sorrow that good quality scaleup studies and product manufacture are simply *not* compatible.

Throughout our program of moving from concept to commercial reality we have an interplay between experimentation and system/engineering studies as shown in Figure 1-7. Successful scaleup involves a series of steps (the size of the step being determined by our data base) with periodic checks and revisions. An acceptable scaleup (in both technical and economic terms) is *not* easy to achieve.

B. Process Documentation

To be certain of the nature of the dialogue that must go on between the personnel involved in a development program, let us consider a rather simple process such as the removal of H_2S and CO_2 with an aqueous monoethanolamine (MEA) solution. A simplified flow sheet typical of the many hundred MEA units in the natural gas industry is shown in Figure 1-8.

The gas to be purified is passed upward through the absorber countercurrent to an aqueous solution of 15–20 wt% MEA. As the gas passes upward, the H_2S and CO_2 in the gas are absorbed into the descending MEA solution. The rich amine solution containing the absorbed acid gases flows from the bottom of the absorber to a flash tank.

The rich amine solution from the flash tank recovers the sensible heat from the hot lean amine solution flowing from the bottom of the stripper and then flows downward through the stripping column. The contained acid gases are stripped from the rich amine solution as it descends the stripping column by heating it to 220–260°F with internally generated stripping steam. The acid gases and the stripping steam are passed through a condenser where the water vapor and any volatilized amine are condensed and returned to the top of the stripper as reflux. The acid gas stream, particularly when it contains H_2S, is processed further but let us stop here.

This process is quite simple in a mechanical sense—just a few pieces of equipment and pumps—but the data requirements to design and construct a

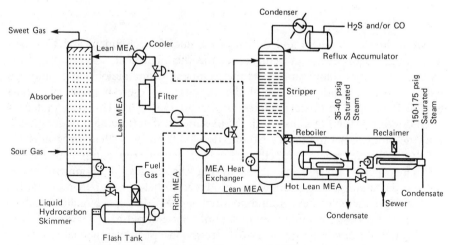

FIGURE 1-8 Typical process flow plan of MEA sweetening plant.

DESCRIBING A PROCESS SYSTEM

unit are significant. What are the work products of a design engineer as he works with laboratory and pilot unit data? An incomplete listing is:

Process alternatives.
Economic analysis.
Process flow sheets.
Process specifications.
Mechanical specifications.
Equipment catalog.
Startup instructions.
Operating and maintenance requirements.

Even if largely standardized in scope and content, as is true for MEA plants, this is still a formidable amount of information to assemble.

How does a design engineer go about developing the required information? Let us look at the absorber where the starting point is the desired acid gas cleanup. (One should keep in mind that once the MEA concentration, the lean solution temperature to the absorber, and the absorber pressure have been fixed, the lean solution loading will also be defined.) The minimum circulation rate will be limited by corrosion considerations. This means the design engineer must have access to pilot unit or commercial operating data. With the minimum circulation rate set by corrosion, the rich end (bottom of the absorber) composition must be calculated. This requires MEA–H_2S–CO_2 vapor–liquid equilibrium data over a range of temperatures, pressures, and compositions.

Since the equilibrium H_2S and CO_2 partial pressures at the bottom of the absorber must be at the rich solution loading and temperature, the heatup of the MEA solution caused by absorption of the acid gas must be known. This requires heat of absorption (reaction) data. Our data requirements are multiplying and as he continues his studies, the design engineer will have further need for data since he has to, at a minimum, establish the following information before *beginning* his process design:

Absorber pressure.
Lean MEA temperature into the absorber.
Rich MEA temperature leaving the absorber.
Product gas temperature leaving the absorber.
Feed gas temperature into the absorber.
Feed gas H_2S and CO_2 analysis.
Product gas H_2S and CO_2 analysis.
Lean MEA solution strength.
Lean MEA loadings—moles H_2S/mole MEA and moles CO_2/moles MEA.

Before the design of a commercial unit is completed, the data requirements will multiply manyfold.

Unfortunately, the concept of what is included in a process design is by no means standardized. Different firms and individuals have quite different concepts; however, a standardized scope of the information required to undertake the work (mechanical engineering and purchasing) needed to build a commercial unit is given in the Appendix.

An overall study of a process system, expressed as a flow sheet with relevant additional information, must be developed rather early in a development program since it provides the blueprint for the collective task to be undertaken. Certainly a flow sheet cannot replace verbal imagery and it should not; but a flow sheet does provide a common vocabulary and helps coordinate the overall activity. Without an early visualization of the overall process, one cannot complete in an effective manner the task of assembling all of the relevant information needed to build and operate a commercial plant.

IV. BALANCED ATTACK ON UNCERTAINTY

There is no foolproof system for forecasting future events from the limited information that is available at any given time. For example, let us consider the following future events:

The future demand for product.

The future variation that could be expected in environmental conditions in a given geographical area.

The rate of reaction in a proposed gas–liquid reactor.

The separation efficiency of a distillation process for a specific mixture.

Some cloud of uncertainty surrounds both the data and the techniques that would be used to make the needed prediction. While we *may* feel more comfortable making predictions of separation efficiency than reaction rate or market demand, that does *not* necessarily mean that our prediction of separation efficiency is more certain that that of rate.

In scaleup studies, the degree of uncertainty with the predictions we need to make will vary with the project development path we have chosen to follow in moving from laboratory studies to a commercial unit. Obviously, in choosing a path we have reflected upon our state of knowledge in the underlying technology of the process system and relevant financial parameters. The choice of path cannot be separated from that of business goals and financial resources available to reduce uncertainty.

In scaleup from the laboratory to a commercial operating unit, we can seek data from:

BALANCED ATTACK ON UNCERTAINTY

Commercial unit operation.

Semiworks or demonstration plants.

Vendor test programs on commercial but *not* full-scale equipment.

Pilot plant unit operations.

Mockup studies.

Simulation studies.

For different portions or sections of our overall process system, we may elect to emphasize the use of data from only one of these sources or we can choose to combine data from several.

Whatever approach we select is determined by:

Knowledge of any experience with related systems and technology.

Consequences of system failure in part or completely.

Possible alternatives and contingency plans in the event of system failure.

Cost of additional studies as compared to the value of potential reduction in uncertainty.

Impact of timing on financial success.

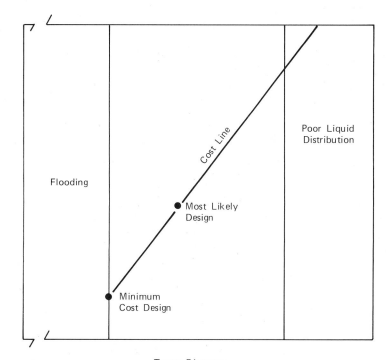

FIGURE 1-9 Optimization of MEA absorber tower diameter.

What we are doing is looking at the cost of developing information as contrasted to the possible benefit.

There is no easy answer to the problem of uncertainty, nor is there a formula or computer program to provide a quick answer. Much of the uncertainty that surrounds a process system can be swept away only after the system built and operated. Therefore, one finds the process and design engineer hedges against the uncertainty. Typically, one engineers on the safe side —processing equipment is purposely designed to be more durable, more flexible, and of greater capacity than is required on the basis of the best information available from smaller scale studies. One hopes that this will protect the system from unknown effects.

In many cases, engineering on the safe side means simply making equipment larger. This approach is "illustrated" in Figure 1-9 for an MEA absorber tower diameter where the areas of concern are limited to flooding at high gas velocities (small tower diameter) and poor liquid distribution at low liquid velocities. This minimum cost design would be a tower just on the borderline of flooding. However, since most cost optima are quite flat, making the tower diameter somewhat larger significantly reduces the uncertainty without a large increase in cost. The question of how much larger is a difficult one to answer—ideally it is decided on a case-by-case basis. However, the use of rules of thumb with all their drawbacks is not unknown.

In the end, we are left with a complex decision that will tax our practical experience and chemical engineering science knowledge. There is *no* simple answer.

REFERENCES

Corrigan, T. E., and Mills, W. C., "Reactor Design for Catalytic Reactions—I," *Chem. Eng.*, **63**, April, 197–202 (1956).

Corrigan, T. E., and Mills, W. C., "Reactor Design for Catalytic Reactions—II," *Chem. Eng.*, **63**, May, 203–206 (1956).

Kline, P. E., Vogel, A. J., Young, A. E., Towsend, D. I., Moyer, M. P., and Aerstin, F. G., "Guidelines for Process Scale-Up," *Chem. Eng. Progr.*, **70**(10), 67–70 (1974).

Levenspiel, O., *The Chemical Reactors Omnibook*, Oregon State University Book Stores, Corvallis, OR, 1979.

Ohsol, E. O., "What Does It Cost to Pilot a Process," *Chem. Eng. Progr.*, **69**(4), 17–20 (1973).

de Rooij, A. H., Dijkhuis, C., and Van Goolen, J. T. J., "A Scale-Up Experience," *Chemtech*, **7**(5), 309–315 (1977).

Simons, T. J. F., "Pulsed Packed Columns in the Production Routes to Caprolactam," *Chemistry and Industry*, **19**, 748–757 (1978).

Smith, J. M., "Scaledown to Research," *Chem. Eng. Progr.*, **64**(8), 78–82 (1968).

Waters, P. L., "Strategy for the Development of Chemical Processes," *Proc. Roy. Australian Chem. Inst.*, **40**(7), 183–191 (1973).

APPENDIX: SCOPE OF PROCESS DESIGN

Project Description

A summary of the design and the scope of the work.

Process Description

A detailed explanation of how the process operates. Generally used as the starting point for preparation of an operating manual for the plant.

Basis for the Commercial Plant

A statement of the design basis, including such items as production rate, battery limits requirements, fractionation specifications, relevant data, specification of feeds and products, and so on.

Heat and Material Balance

A heat, material, and pressure balance including all pertinent physical characteristics of the process fluids involved. All changes in temperature, pressure, heat content, vapor–liquid distribution, and so on, should be shown. Utility streams involved in heat transfer to the process system should also be shown.

Process Flow Sheet

Should show all major process lines including interconnection of major pieces of equipment. The flow sequence should be presented in its simplest form. However, principal control systems must be shown and all major equipment items clearly indicated. Spare equipment need not be shown. Operating conditions of temperature, pressure, and flow rates should be shown at principal points in the process. Exchange and furnace duties should be indicated.

Process Piping and Instrumentation Diagrams

a. All major equipment should be represented, with important process and mechanical detail shown.
b. All process lines and sizes (including bypasses, circulating lines, startup connections, inert gas, gas blanketing, pumpout lines, relief and safety valves, and their connections) should be shown. Vents and drains in process piping which are installed to suit the mechanical layout of the piping can be omitted. Drains to oily water sewers or similar items requiring special treatment for pollution abatement should be clearly indicated.

c. All instrumentation should be shown using standard ISA symbols to code the systems. Where control installation involves some definition of mechanical attachment to vessel, that is, level control, level glass, and so on, a typical detail should be developed.
d. Control valves and manifolds should be shown, including those valves in utility systems which have a process control function.
e. All valving in process systems and for tie-ins to utility systems should be shown.
f. General notes should be given which provide information or instruction pertinent to detailed engineering design, that is, notes regarding special sloping of lines for drainage, high-point bleeds and low-point drains, with reference to specifications covering any typical or special installation.

Utility Piping and Instrumentation

a. Utility systems should be shown on a separate flow diagram because of their dependence on plot layout. Utility control valves installed for process control should be shown on the process piping and instrumentation drawing.
b. The utility flow diagrams should show all utility lines and line sizes for cooling water, condensate, steam, fuel gas, and fuel oil systems. Electrical power and air piping need be shown only for control systems and where air piping serves a process function, such as air for drying.

Plot Plan

This should show recommended equipment layout. Ultimately it will be used as the basis for plot-oriented design work such as pressure balances and arrangement of utility and relief piping.

Major Equipment Outline Drawings and Specifications

Vessel sketches include specified pressure and temperature design ratings, a nozzle schedule specifying flange or coupling rating, elevations of all openings, and other pertinent dimensions. Metal thicknesses and nozzle orientation are shown. Any special internals required for vessels should be shown in a sketch.

Vapor–liquid loadings, stream gravities, and operating conditions must be developed for all trayed columns. Recommended tower diameters should be indicated.

Standard specification forms for major equipment items should be used wherever possible.

Process Safety

Since closed relief systems are dependent on plot layout, they should be shown on a separate drawing, superimposed on a plot plan drawing. This

APPENDIX: SCOPE OF PROCESS DESIGN

drawing should show all valves discharging to the closed system or subsystem, header piping, and any other safety equipment installed within the plant battery limits.

Utilities associated with the safety system, such as cooling water to an on-site blowdown drum, should be handled in the same manner as the process piping and instrumentation. Utility headers and all connections to the equipment should be shown on the utility flow diagram, with only tie-in piping, block valving, and instrumentation associated with process control shown on the process or on the closed relief system piping and instrumentation diagram.

A safety valve relief summary showing rates and stream properties for the various categories of relief should also be included here:

Fire.
Instrument air failure.
Cooling water failure.
Operator error or blocked outlet.
Reflux failure.
Others as required.

This summary should define the header flow rates for all important categories. Process specifications for relief valves should also be included in this section.

Utilities Requirements

Estimates are needed for consumption of steam, electrical power, cooling water, and fuel gas and should be presented on an itemized basis.

Environmental

Consideration of the potential emissions to air or water should be included. Where possible, a quantitative estimate of pollutants should be provided.

Instrument Specifications

Duty specification sheets should be provided for all instruments and controls.

Piping and Line Specifications

Line designation sheets (including operating and design conditions, flowing fluid, designation of pipe and flange ratings, and insulation requirements) should include all process and utility lines.

2

MATHEMATICAL MODELING

D. M. HIMMELBLAU

I.	Major Issues in Modeling	35
	A. How Does One Go About Modeling	36
	1. Problem Definition and Formulation Phase	36
	2. Preliminary Design Phase	38
	3. Detailed Design Phase	38
	4. Evaluation Phase	39
	B. General Problems in Modeling	39
	1. Model Simplification	39
	2. Model Verification	41
	3. Computer Code Verification	41
	4. Estimation of Model Parameters	41
	5. Random Nature of Process Variables	41
II.	Fundamental Principles for Mathematical Modeling	42
	A. Introduction	42
	1. Linear Versus Nonlinear	43
	2. Steady State Versus Nonsteady State	43
	3. Distributed Versus Lumped Parameters	44
	4. Continuous Versus Discrete Variables	44
	B. Models Based on Transport Phenomena Principles	45
	C. Classification of Models by Solution Procedure	54
	D. Models in the Frequency Domain	55
	1. The Transfer Function	56
	2. Empirical Transfer Functions	57
	E. Empirical Models	59
III.	Dimensional Analysis	62
	A. Making the Dimensional Equation Dimensionless	63

	B.	Short Cut Method of Obtaining Common Dimensionless Groups (by Inspection)	64
	C.	Correlation of Experimental Data	67
	D.	Scaleup	67
IV.	Practical Aspects of Modeling		69
	A.	Validation of Model	70
	B.	Precautions in Model Building	71
Nomenclature			72
References			75

I. MAJOR ISSUES IN MODELING

A fundamental premise underlying scaleup is that the basic physical principles of nature (such as mass and energy balances) apply to systems of different sizes. The major difficulty that exists in applying this basic concept is that our ability to describe processes by quantitative mathematical relations is of limited scope. Most process equipment is so complicated in operation that extension of simple principles leads to innumerable questions in modeling, many of which cannot be answered. Process models in the sense used here are mathematical representations of chemical processes that show the effects of those factors that are significant for the purpose of the analyst.

Because models contain less information than the equipment or processes they are presumed to represent, the purpose of a model determines whether or not a specific feature should be included in the model. Models are never completely faithful to the process they represent because if they were they would be the "real thing" itself or at least an exact replica. The process will contain information not present in the model and the model will contain information not present in the process. In preparing a model an engineer must decide which features of the process are of sufficient interest and at the same time are capable of being represented. One of the major advantages of a model is that it contains less information, and information in a different form, than the real process. An engineer must be able to eliminate those features of a real process not relevant to the purpose of the model, but must do so with great care if he is not to overlook some vital aspects of the process.

To select a model from the many that are available, one must weigh the complexity of the model with the degree of difficulty involved in its solution. It is necessary to consider (a) the "order" of the model, that is, the number of independent functions required to describe the process, (b) the number of parameters involved in the model, and (c) the number of independent variables to be included in the model. The simpler the model, the easier it is to solve analytically or numerically; the more complicated the model, the less likely it is that a simple solution can be found. But, the more complex the model, usually (but not always) the more representative it is of an actual process.

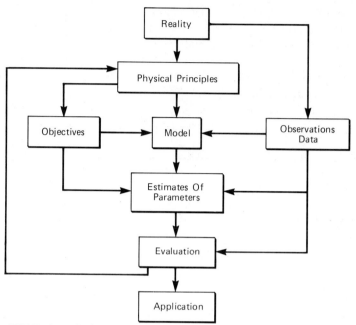

FIGURE 2-1 Cyclical nature of model development prior to application.

As a model necessarily departs from its real-life counterpart in many respects, its usefulness depends largely on the degree of simplification produced by this intentional or inadvertent departure. Although they are approximate, models can be reworked at will with the idea of improvement or of experimental verification, but they seldom if ever can be exactly verified because of the limits of experimental accuracy and because of the high cost (in time and money) of carrying out the required tasks. Figure 2-1 suggests how model building is an iterative process which encompasses continual refinement.

A. How Does One Go About Modeling

Modeling for scaleup can be divided for convenience of presentation into five phases: problem definition and formulation, preliminary analysis, detailed analysis, evaluation, and interpretation/application. But keep in mind the iterative nature of model building as presented in Figure 2-1.

1. *Problem Definition and Formulation Phase*

The following tasks need to be accomplished to develop an acceptable model:

Define the problem to be solved, including identifying various elements that pertain to the problem and its solution.

MAJOR ISSUES IN MODELING

Obtain a definite commitment for model development from the appropriate management involved.

Determine that a successfully developed model will indeed help solve the defined problem.

Obtain expert advice on the best approach to modeling the process; including whether modeling really would be appropriate.

Search the company and open literature for models already developed to solve the same or similar problems.

Estimate the frequency with which the model will be used (e.g., one time or repetitive use) and the possible need to update it in the future.

Determine the degree of accuracy needed from the model.

Estimate the benefits expected from using the model and the costs, if determinable at this stage, of developing and running the model.

Ascertain the qualifications and capabilities of the individuals or group that will be developing the model.

Determine whether the model will be developed within the organization or by an external organization, and identify the developer.

Determine the extent of training necessary to provide the user organization with the background to operate and maintain the model.

Determine the type and modeling techniques to be used.

Examine the requirements relative to ease of model use.

Ascertain from the available data what information exists that can be used in developing the model and verifying the model.

Define the extent of modular programming, that is, the extent the model will be segmented into self-contained units.

Prepare a model testing plan and specify evaluation criteria (e.g., specific test case and test data for the model) to determine if the model meets the user's needs.

Specify the estimated due dates for completing the preliminary design phase, the detailed design phase, and the evaluation phase.

Document requirements for the model.

Establish user-monitoring procedures and developer-reporting procedures during model development.

Develop a procedure for maintaining control over model code, test data, and documentation.

Specify the developer's training program for the user.

In the problem definition phase, you should acquire a clear definition of the problem and a firm concept of the character of the model to be developed. If the information acquired during this phase indicates that development should be stopped before the end of the phase, you should be prepared to terminate development at this point. To determine whether to continue into the next phase, you should consolidate and thoroughly review all of the work completed during this first phase.

2. Preliminary Design Phase

Activities in the preliminary design phase include specification of the information content, general programming logic, and algorithms necessary to develop a useful model, mathematical description, and simulation.

The following tasks need to be accomplished during this phase:

Define the input and output variables, and the desired formats to be used in the model.

Select a programming language for the model coding (if a computer code is to be used).

Verify the availability and adequacy of computer equipment and software.

Specify input and output media (e.g., punch cards, tape, CRT, plotter, or printer).

Describe the general program logic of the model, perhaps including basic flow charts with input, processing, and output described, or comment statements in the code.

Select the specific mathematical representation(s) to be used in the model.

Define the program modules and their structural relationships.

Specify the assumptions and limitations of the model, that is, any major differences that may result from translating the problem to the model or its coding.

Reevaluate the costs to be incurred and benefits to be realized from use of the model before proceeding to the next phase.

3. Detailed Design Phase

In this phase, you must continuously reevaluate the design being implemented, and make changes as necessary in the scope of the work. The following tasks need to be accomplished during this phase:

Complete all of the details of the model including initial and boundary conditions.

Prepare all of the data bank of parameters and inputs.

Complete all of the computer codes (if needed).

Systematically test the codes and modules.

Test the model for sensitivity of output to changes in inputs and parameters including the extreme values.

Prepare documentation for the user and other programming documentation.

MAJOR ISSUES IN MODELING

4. Evaluation Phase

This phase is intended as a final check of the model as a whole. Tests of individual model programs are conducted during earlier phases. Evaluation of the model is carried out according to the evaluation criteria and test plan established in the problem definition phase. Evaluation includes model validation and the determination of compliance with previously established requirements.

The following tasks need to be accomplished during this phase. A more complete discussion can be found in Section IV.

> Evaluate the model output with the evaluation criteria established in the problem definition phase.
>
> Evaluate the adequacy of the sensitivity testing and the results obtained.
>
> Determine the validity of the individual mathematical relationships in the model and whether all relationships are valid with respect to each other.
>
> Evaluate the model using actual data where possible instead of simply simulation.
>
> Apply common sense to the results of simulations.

The model is then ready for application and interpretation.

B. General Problems in Modeling

In this section we shall list and briefly describe some of the major issues with which you will be confronted as you carry out the activities described above.

1. Model Simplification

One of the most vexing issues faced by modelers concerns the extent to which simplifications are justified. Simplifications are forced on modelers because of the limitations in our tools of description, or because of cost or time limitations. What simplifications can be made? How are they justified? These questions are treated in Section II. Simplifications include omitting interactions, aggregating variables, eliminating random variables or replacing them by their expected values, or eliminating details in the mathematical description of a process. Typical simplifications are best understood by an example.

Example 2.1 Simplification of Reaction and Diffusion Model. Temperature and concentration profiles in a porous particle undergoing reaction can be described by the mass and energy balances appearing in rows 1 and 3 of Table 2-3. If no diffusional resistance exists nor temperature increase (decrease) because of the reaction, the reaction rate, R_1, would be a function of a uniform concentration, c_s, and temperature, T_s, at the exterior surface of the particle. Figures 2-2a and 2-2b show T_s and c_s. The actual reaction rate, R_2, can be

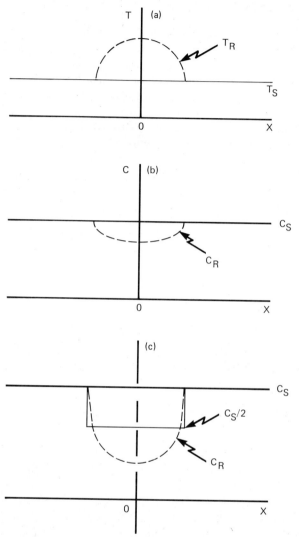

FIGURE 2-2 Temperature and concentration profiles for a porous particle in which a reaction takes place.

greater or less than R_1, and the ratio of the rates is the effectiveness factor η. A large number of articles have been published as to how to calculate η rigorously when quite complex reaction rate functions are substituted into the equations in Table 2-3. Typical temperature and concentration profiles within the pellet from rigorous calculations would appear as the dashed lines in Figures 2-2a and 2-2b.

Nevertheless, few plant engineers use the published results in the design of catalytic reactors. What they do is to simplify the rigorous physical model. For example, LeGoff and Zonalian (1976) suggested that the set of partial differential equations be replaced by a set of algebraic equations in which the curved concentration profile illustrated in Figure 2-2b is replaced by "equivalent" straight lines over the whole volume of the particle. If the concentration profile is quite steep as in Figure 2-2c, and the reaction takes place mainly in a thin layer under the interface, one can assume a reactive layer of constant concentration $c_s/2$ exists and zero concentration exists over the inner core. LeGoff and Zonalian (1976) compared the solutions of the rigorous and approximate models and found little difference between the two. □

2. Model Verification

Realistically, we rarely can compare the model performance with the process it is designed to represent. Accuracy of observation is always limited because of the instruments employed, and the measurements that are taken are always contaminated by uncontrollable extraneous influences. This problem is discussed in more detail in Section IV.

3. Computer Code Verification

If a computer code is to be used to make predictions about a process, one must always make sure that the code is correct. Verification is needed that what is intended to be coded is indeed coded, and that numerical errors, such as truncation and round-off errors, do not lead to blunders.

4. Estimation of Model Parameters

Even if the model form itself is exact, a model is no better (or worse) than the values of the parameters in the model. In the absence of adequate data taken directly from the process itself, can data from a small scale pilot be used? If no data at all are available either from laboratory or full-size experiments, what can be accomplished in model building? Some of these matters are discussed in each of the sections below, but no firm resolution of the problem is possible. Nevertheless, the more quantitative the analysis, the better a process will be understood.

5. Random Nature of Process Variables

A deterministic model is one in which each variable and parameter can be assigned a definite fixed number, or a series of fixed numbers, for any given set of conditions. However, most process variables are random variables, that is, repeated measurements of the same variable will not yield the identical number. A process model in which one or more of the variables, or the parameters, are random, is termed a *stochastic model*. Figure 2-3 shows how the output or the input of a model can be treated conceptually so as to include randomness (E). It would be quite possible for the equation(s) in the model to

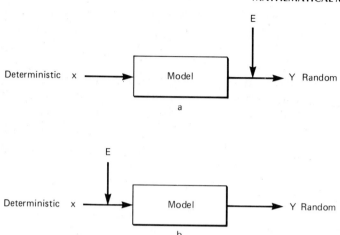

FIGURE 2-3 Block diagram representation of stochastic models.

yield random outputs because of random coefficient(s). While in theory, a stochastic model may be a better abstraction of the real process, in engineering practice all that is needed is to represent the process with reasonable faithfulness insofar as the mean (expected value) of the variable(s) of interest. As long as the deterministic model represents the real process sufficiently well so that the conclusions deduced from mathematical analysis of the model have the desired precision, the deterministic model is entirely adequate.

II. FUNDAMENTAL PRINCIPLES FOR MATHEMATICAL MODELING

A. Introduction

In this section, we examine various types of mathematical models as contrasting classes to reveal the interrelationships and various degrees of complexity among the models, and help in the selection of a suitable type of model for a given process. There are many ways to classify mathematical models. For our purposes, it is most satisfactory first to group the models into the opposite pairs:

Linear versus nonlinear.
Steady state versus nonsteady state.
Lumped parameter versus distributed parameter.
Continuous versus discrete variables.

FUNDAMENTAL PRINCIPLES FOR MATHEMATICAL MODELING

TABLE 2-1 Typical Linear and Nonlinear Functions

Operation	Example	Test Is	Classification
(a) Square	y^2	$(y_1 + y_2)^2 \stackrel{?}{=} y_1^2 + y_2^2$	NL
(b) Log	$\log(y)$	$\log(y_1 + y_2) \stackrel{?}{=} \log y_1 + \log y_2$	NL
(c) Partial derivative	$\dfrac{\partial u}{\partial x}$	$\dfrac{\partial(u_1 + u_2)}{\partial x} \stackrel{?}{=} \dfrac{\partial u_1}{\partial x} + \dfrac{\partial u_2}{\partial x}$	L
(d) Product of independent variable and derivative	$x\dfrac{\partial u}{\partial x}$	$x\dfrac{\partial(u_1 + u_2)}{\partial x} \stackrel{?}{=} x\dfrac{\partial u_1}{\partial x} + x\dfrac{\partial u_2}{\partial x}$	L
(e) Product of dependent variable (v_x) and derivative of dependent variable (c)	$v_x \dfrac{\partial c}{\partial x}$	$(v_{x_1} + v_{x_2})\dfrac{\partial(c_1 + c_2)}{\partial x}$ $\stackrel{?}{=} v_{x_1}\dfrac{\partial c_1}{\partial x} + v_{x_2}\dfrac{\partial c_2}{\partial x}$	NL

1. Linear Versus Nonlinear

Linear models exhibit the important property of superposition; nonlinear ones do not. Equations (and hence models) are linear if the dependent variables or their derivatives appear only to the first power; otherwise they are nonlinear. In practice, the ability to use a linear model for a process is of great significance, since the solution of linear models is an order of magnitude easier than the solution of nonlinear ones.

To test for the linearity versus nonlinearity of a model, you examine the equation(s) that represents the process. Check each term in the process model equations to see if it is linear or nonlinear, and if any one term is nonlinear, then the model itself is nonlinear. By implication, the process is nonlinear. Some examples of typical linear and nonlinear functions (L = linear, NL = nonlinear) that appear in models are given in Table 2-1.

How can you determine whether a real process is linear (or nonlinear)? One way is to introduce an input, such as a sinusoidal input, into the process, measure the output, and determine if the phase lag is independent of frequency. If so, the process is linear.

2. Steady State Versus Nonsteady State

Other synonyms for steady state are time invariant, static, or stationary. These terms refer to a process in which the point values of the dependent variables remain constant over time. Nonsteady-state processes are also called unsteady state, transient, or dynamic, and represent the situation in which the process dependent variables change with time. A typical example of an unsteady-state process might be the startup of a distillation column which would eventually reach a pseudosteady-state set of operating conditions. Examined in more detail, the column would prove always to be operating in the unsteady state with minor fluctuations in temperature, composition, and so on, taking place all the time, but ranging possibly about "average steady-state" values.

3. Distributed Versus Lumped Parameters

Briefly, a lumped-parameter representation means that spatial variations are ignored, and that the various properties and the state of the system can be considered homogeneous throughout the entire volume. A distributed-parameter representation, on the other hand, takes into account detailed variations in behavior from point to point throughout the system. All real systems are, of course, distributed in that there are some variations throughout them. As the variations often are relatively small, they may be ignored, and the system may then be "lumped."

The answer to the question as to whether lumping for a process model is valid is far from simple. A good rule of thumb is that if the response of the process is for all practical purposes instantaneous throughout the process, then the process model can be lumped. If the response shows instantaneous differences along the process (or vessel), then it should not be.

Because the mathematical procedures for the solution of lumped-parameter models are simpler than those for the solution of distributed-parameter models, one often approximates the latter by an equivalent lumped-parameter system. While lumping is often possible, one must be careful to avoid masking the salient features of the distributed element (hence building an inadequate model) by lumping.

4. Continuous Versus Discrete Variables

Continuous means that the variables can assume any values within an interval; discrete means the variable can take on only distinct values in the interval. For example, concentrations in a countercurrent packed bed are usually modeled in terms of continuous variables, whereas plate absorbers are modeled in terms of staged multicompartment models in which a concentration is uniform on each stage but differs from stage to stage in discrete jumps.

Figure 2-4 illustrates the two configurations. The left-hand figure shows the packed column modeled as a continuous system, whereas the right-hand figure represents the column as a sequence of discrete (staged) units. The concentrations in the left-hand column would be continuous variables; those in the right-hand column would involve discontinuous jumps. The tic marks in the left-hand column represent hypothetical stages for analysis. It is equally possible to model the packed column in terms of imaginary segregated stages and treat the plate column in terms of partial differential equations in which the concentrations are continuous variables.

FUNDAMENTAL PRINCIPLES FOR MATHEMATICAL MODELING

initial conditions, if applicable. Each equation usually includes one or more coefficients that are hopefully, but rarely, constant. The term *parameter* as used here will mean coefficient and possibly input or initial condition.

The principles underlying transport phenomena models are fairly widely known, but one should realize that these principles can be appled at various levels or strata of description, that is, you can portray the operation of a real process by models on a number of physical scales. Physicochemical models based on the degree of internal detail of the system encompassed by the model are classified in Table 2-2. The degree of detail about a process decreases as you proceed down the table. We will ignore the molecular description in this chapter. Keep in mind that the more detailed model is frequently, but not always, the more representative model. Also each *complete* model consists of the governing differential (or algebraic) equations plus appropriate boundary conditions.

All the models derived from physicochemical principles are based on a very general concept, which can be stated verbally as follows for either mass, momentum, or energy:

$$\begin{bmatrix} \text{accumulation} \\ \text{within} \\ \text{system} \end{bmatrix} = \begin{bmatrix} \text{net flow in through} \\ \text{system} \\ \text{boundaries} \end{bmatrix} + \begin{bmatrix} \text{net generation} \\ \text{within} \\ \text{system} \end{bmatrix} \quad (2\text{-}1)$$

The objective in model building is to transform the verbal concept into mathematical statements which are specific to the quantity of interest, namely, mass, momentum, or energy, and to an arbitrarily selected system.

The *microscopic description* assumes that the process acts as a continuum and that the mass, momentum, and energy balances can be written in the form of phenomenological equations. The microscopic description is useful for stagnant liquids, liquids in laminar flow, heat transfer by conduction and in laminar flow, and mass transfer in laminar flow and stagnant liquids. However susceptible these problems are to analysis by the microscopic equations, most practical processes in chemical engineering involve a degree of turbulent mixing which prohibits the use of the microscopic equations because the velocity distributions remain unknown. Equations for the microscopic description can be found in Bird, Stewart, and Lightfoot (1960).

The next level of description, the *dispersion* model, assumes that the process exhibits turbulent flow or that flow occurs in geometrically complex systems on a fine scale, such as packed beds. The values of the dependent variables are time-averaged values, and the basic balances include dispersion terms rather than molecular diffusion terms, but the terms still involve second partial derivatives. These dispersion terms also include coefficients that are functions of the flow regime as well as of the properties of the fluid.

The dispersion equations in rectangular coordinates are listed in Table 2-3. Development of these equations can be found in Chapter 10 of Slattery (1972).

TABLE 2-3 Dispersion Balances in Rectangular Coordinates

Mass Balance for αth Specie

$$\underbrace{\frac{\partial c_\alpha}{\partial t}}_{\substack{\text{Accumu-}\\\text{lation}}} + \underbrace{\frac{\partial}{\partial x}(v_x c_\alpha) + \frac{\partial}{\partial y}(v_y c_\alpha) + \frac{\partial}{\partial z}(v_z c_\alpha)}_{\substack{\text{Transport through surface}\\\text{by bulk flow}}}$$

$$= \underbrace{\frac{\partial}{\partial x}\left(\rho \tilde{D}_{\alpha x} \frac{\partial}{\partial x}\frac{c_\alpha}{\rho}\right) + \frac{\partial}{\partial y}\left(\rho \tilde{D}_{\alpha y}\frac{\partial}{\partial y}\frac{c_\alpha}{\rho}\right) + \frac{\partial}{\partial z}\left(\rho \tilde{D}_{\alpha z}\frac{\partial}{\partial z}\frac{c_\alpha}{\rho}\right)}_{\text{Transport through surface by dispersion}} + \underbrace{R_\alpha}_{\substack{\text{Genera-}\\\text{tion}}}$$

Momentum Balance (x direction)

$$\rho \left(\underbrace{\frac{\partial v_x}{\partial t}}_{\substack{\text{Accumu-}\\\text{lation}}} + \underbrace{v_x \frac{\partial v_x}{\partial x} + v_y \frac{\partial v_x}{\partial y} + v_z \frac{\partial v_x}{\partial z}}_{\substack{\text{Transport through surface}\\\text{by bulk flow}}}\right)$$

$$= -\frac{\partial p}{\partial x} + \underbrace{\left[\frac{\partial}{\partial x}\left(\tilde{\mu}_{xx}\frac{\partial v_x}{\partial x}\right) + \frac{\partial}{\partial y}\left(\tilde{\mu}_{yx}\frac{\partial v_x}{\partial y}\right) + \frac{\partial}{\partial z}\left(\tilde{\mu}_{zx}\frac{\partial v_x}{\partial z}\right)\right]}_{\text{Transport through surface by dispersion}} + \underbrace{\rho g_x}_{\substack{\text{Genera-}\\\text{tion}}}$$

Energy Balance

$$\rho C_p \left(\underbrace{\frac{\partial T}{\partial t}}_{\substack{\text{Accumu-}\\\text{lation}}} + \underbrace{v_x \frac{\partial T}{\partial x} + v_y \frac{\partial T}{\partial y} + v_z \frac{\partial T}{\partial z}}_{\substack{\text{Transport through surface}\\\text{by bulk flow}}}\right)$$

$$= \underbrace{\frac{\partial}{\partial x}\left(\tilde{k}_x \frac{\partial T}{\partial x}\right) + \frac{\partial}{\partial y}\left(\tilde{k}_y \frac{\partial T}{\partial y}\right) + \frac{\partial}{\partial z}\left(\tilde{k}_z \frac{\partial T}{\partial z}\right)}_{\substack{\text{Transport through surface}\\\text{by dispersion}}} + \underbrace{S_R}_{\substack{\text{Genera-}\\\text{tion}}}$$

The simplifications that are involved in using a dispersion model are best understood by an example.

Example 2.2 Dispersion Model. A fluid containing a solute penetrates into a permeable solid slab of wide radius compared to the slab thickness (so that end effects can be ignored and the process made one dimensional). This permeation process cannot be treated as a process of molecular diffusion because the transport is in a porous solid and is not molecular transport in a gas or liquid. However, it can be modeled by a dispersion model with effective dispersion coefficients.

FUNDAMENTAL PRINCIPLES FOR MATHEMATICAL MODELING

The statement of the physical problem leads to the following simplifications:

1. The process is one dimensional in the z direction.
2. No reaction occurs: $R_A = 0$.
3. No bulk flow occurs, only dispersion: $v_x = v_y = v_z = 0$.
4. The fluid density and dispersion coefficient are constant: ρ_A = constant, $\tilde{D}_{Az}$ = constant.

Introduction of these assumptions into the first equation in Table 2-3 and multiplying through by the molecular weight of species A gives as the mass balance (for one solute)

$$\frac{\partial \rho_A}{\partial t} = \tilde{D}_{Az} \frac{\partial^2 \rho_A}{\partial z^2} \qquad (2\text{-}2)$$

To complete the model, one initial and two boundary conditions must be written down. At the fluid–solid interface, the mass concentration of fluid is some saturated value (the maximum the solid can hold) while at the other boundary we will assume that the concentration is zero during the experiment; that is, the penetration of fluid is not too deep. If the solid is initially free of fluid, the initial and boundary conditions are:

$t = 0, \quad z \geq 0, \quad \rho_A = 0$ (initial condition)

$t \geq 0, \quad z = 0, \quad \rho_A = \rho_{\text{sat.}}$ (boundary condition 1)

$t \geq 0, \quad z = L, \quad \rho_A = 0$ (boundary condition 2)

Solutions of this problem can be found in Crank (1956) or Carslaw and Jaeger (1959). □

When dispersion models prove ineffective or too complicated for the analysis of a process, the simplest model that describes any of the internal features of a process may be used. It is called the *maximum gradient* or *plug flow* model. It includes no second derivatives, but solely first derivatives to express the spatial variation of concentration (or temperature) in one coordinate direction only. As an example, in the maximum gradient representation of a chemical reactor or gas absorber, only concentration gradients in the axial direction caused by the bulk flow are considered, and all radial gradients, diffusion, and so on, are ignored. This loss of detail greatly simplifies the mathematical description, but accompanying the simplification is a loss of information concerning the performance characteristics of the system that at times may make the model not very useful. Table 2-4 lists the maximum gradient equations.

MATHEMATICAL MODELING

TABLE 2-4 Maximum Gradient (Plug Flow) Balances

Mass Balance for ith Specie

$$\underbrace{\frac{\partial c_i}{\partial t}}_{\text{Accumulation}} + \underbrace{\frac{\partial (v_z c_i)}{\partial z}}_{\text{Bulk transport}} = \underbrace{R_i}_{\text{Generation}} + \underbrace{m_i^{(t)}}_{\substack{\text{Transport} \\ \text{through} \\ \text{surface}}}$$

Energy Balance

$$\rho C_p \left(\underbrace{\frac{\partial T}{\partial t}}_{\text{Accumulation}} + \underbrace{v_z \frac{\partial T}{\partial z}}_{\text{Bulk transport}} \right) = \underbrace{S_R}_{\text{Generation}} + \underbrace{E^{(t)}}_{\substack{\text{Transport} \\ \text{through} \\ \text{surface}}}$$

An analysis of a double pipe heat exchanger using maximum gradient balances is carried out in Example 2.3.

Example 2.3 Double Pipe Heat Exchanger. The classical problem of a double pipe heat exchanger (shown in Figure 2-5) is commonly modeled and solved by writing a steady-state maximum gradient balance for each stream with $E^{(t)}$ defined by an overall heat transfer coefficient U multiplied by the temperature difference between the streams. No heat is generated, that is, $S_R = 0$. The hot stream is designated by a superscript H and the cold stream by the superscript C.

$$\rho^C C_p^C v_z^C \frac{dT^C}{dz} = U \frac{P}{S^C} (T^H - T^C) \tag{2-3a}$$

$$\rho^H C_p^H v_z^H \frac{dT^H}{dz} = -U \frac{P}{S^H} (T^H - T^C) \tag{2-3b}$$

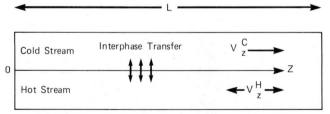

FIGURE 2-5 Double pipe heat exchanger in counterflow.

FUNDAMENTAL PRINCIPLES FOR MATHEMATICAL MODELING

where

$$\frac{P}{S} = \frac{\text{perimeter}}{\text{cross section}} = \frac{\text{wall area}}{\text{unit volume}}$$

Note that because of the definition of the direction for flow, v_z^H, will be positive for cocurrent flow and negative for countercurrent flow, and that $E^{(t)}$ is positive when entering a phase. These two equations when solved with appropriate boundary conditions for countercurrent or cocurrent flow give the familiar logarithmic temperature profiles (see Whitaker, 1977, for the derivation).

Cocurrent Flow

at $Z = 0$, $T^H = T_a$ (hot stream)

at $Z = 0$, $T^C = T_b$ (cold stream)

Countercurrent Flow

at $Z = L$, $T^H = T_a$ (hot stream)

at $Z = 0$, $T^C = T_b$ (cold stream)

where T_a and T_b are the inlet temperatures of the hot and cold streams, respectively. □

The final model, the *macroscopic model*, ignores all the detail within a system and merely makes a balance about the entire vessel. Only time remains as a differential independent variable in the general balances. The dependent variables, such as concentration and temperature, are not functions of position but represent overall averages through the volume of the system. The model is effective as long as detailed information internal to the system is not required in model building. Macroscopic and lumped mean the same thing. Macroscopic balances are listed in Table 2-5.

The macroscopic energy balance is used to model the process of heating a batch of liquid in a tank in Example 2.4.

Example 2.4 Heating of Liquid in a Tank. With no flow, transport through the surface is zero and $Q^{(m)}$ is zero. If no work is done, $W = 0$. If there is no heat generation, $S_R = 0$. The heat transfer term, Q, is usually written in the following form:

$$Q = UA(T_w - T) \qquad (2\text{-}4\text{a})$$

where

T_w = wall temperature

TABLE 2-5 Macroscopic Balances (Including Interphase Transport)

Mass Balance for αth Specie

$$\frac{d}{dt} m_{\alpha,\text{tot}} = -\Delta(\rho_\alpha \langle v \rangle S) + w_i^{(m)} + r_{\alpha,\text{av}} V_{\text{tot}}$$

Accumu- Transport through surface Genera-
lation   tion

$$\left(r_{\alpha,\text{av}} = \frac{1}{V} \int_V r_\alpha \, dV \right)$$

Momentum Balance

$$\frac{d}{dt} \mathbf{P}_{i,\text{tot}} = -\Delta\left(\rho \langle v^2 \rangle \mathbf{S}_i + \langle p \rangle \mathbf{S}_i \right) - \mathbf{F}_i^{(m)} + m_{\text{tot}} \mathbf{g}_i + \mathbf{F}_i$$

Accumu- Transport through surface Genera-
lation   tion

Energy Balance

$$\frac{d}{dt} E_{\text{tot}} = -\Delta\left[\left(\hat{H} + \frac{1}{2} \frac{\langle v^3 \rangle}{\langle v \rangle} + \hat{\phi} \right) (\rho \langle v \rangle S) \right] + Q - W + Q^{(m)} + S_R$$

Accumu- Transport through surface   Genera-
lation   tion

$$(E_{\text{tot}} = U_{\text{tot}} + K_{\text{tot}} + \phi_{\text{tot}})$$

Thus, the energy balance (with $dE = dU \simeq dH = C_p \, dT$, and ρ and C_p constant) reduces to

$$\rho C_p V_{\text{tot}} \frac{dT}{dt} = UA(T_w - T) \tag{2-4b}$$

With a starting temperature of T_0, the initial condition is

$$T(0) = T_0 \tag{2-4c}$$

It can be verified by differentiation that the solution of model is

$$T = T_w - (T_w - T_0) e^{-UAt/\rho C_p V_{\text{tot}}} \tag{2-4d}$$

In other words, the temperature rises in an exponential fashion, eventually reaching T_w. □

The macroscopic model and plug flow models each represent extremes of ideal mixing—from complete mixing in the former to no longitudinal mixing

TABLE 2-6 Common Boundary Conditions for Use with the General Transport Equations

Mass Balances

1. Concentration at a boundary is specified ($c = c_0$).
2a. Mass flux at a boundary is continuous

$$([n_i]_{x=0^-} = [n_i]_{x=0^+})$$

2b. Concentration on both sides of a boundary are related functionally

$$([c_i]_{x=0^-} = f[c_i]_{x=0^+})$$

3. Empirically determined mass (mole) flux at boundary is specified

$$([N_A]_{x=0} = k(c - c^*))$$

4. Rate of reaction at boundary surface is specified

$$-([N_A]_{x=0} = R_A)$$

Momentum Balances

1. Velocity at a boundary is specified (at a solid–fluid interface $\mathbf{v} = 0$).
2a. Momentum flux at a boundary is continuous. (At a liquid–liquid interface τ is continuous.)
2b. Velocity is the same on both sides of boundary

$$([v]_{x=0^-} = [v]_{x=0^+})$$

3. Momentum flux is specified (at a gas liquid interface momentum flux is approximately zero).

Energy Balances

1. Temperature at a boundary is specified ($T = T_0$).
2a. Heat flux at a boundary is continuous

$$([q]_{x=0^-} = [q]_{x=0^+})$$

2b. Temperature is the same on both sides of a boundary

$$([T]_{x=0^-} = [T]_{x=0^+})$$

3. Empirically determined heat flux at boundary is specified

$$([q]_{x=0} = h(T - T^*))$$

4. Heat flux at a boundary is specified ($\mathbf{q} = \mathbf{q}_0$).

(and complete radial mixing) at all in the latter. In a plug flow model each element of fluid (filling the cross section but differential in thickness) marches in "single file" without intermingling with other fluid elements. In a macroscopic, that is, lumped, model, on the other hand, perfect mixing makes the vessel contents and output homogeneous. Between those two extremes of flow patterns would fall the dispersion model.

Specification of the *boundary and/or initial conditions* is as important in model building as formulation of the differential equation(s). In order to calculate the values of the arbitrary constants that evolve in the solution of a differential equation, one generally needs a set of n boundary conditions for each nth order derivative with respect to a space variable, or initial conditions with respect to time. For example, the differential equation

$$\frac{\partial^2 c}{\partial x^2} = A(x)\frac{\partial c}{\partial t}$$

would require two boundary conditions on c concerned with x and one initial condition concerned with t.

Appropriate boundary conditions arise from the problem statement—they essentially are given or, more often, must be deduced from physical principles associated with the problem. These physical principles in general are nothing more than statements in mathematical form that the dependent variable at the boundary is at equilibrium, or, if some transport is taking place, that the flux of mass, momentum, or energy is conserved at the boundary. An additional type of boundary condition is often used that is based on a rate process taking place at the boundary in terms of an interphase transport coefficient and some type of driving force, such as Newton's law of cooling. Such descriptions involve an empirical parameter which must be evaluated before the boundary condition can be of value.

The common boundary conditions for use with mass, momentum, and energy balances are listed in Table 2-6; other conditions may also be used. Note the similarities among the three modes of transport. These boundary conditions apply to all the strata of description shown in Table 2-2 except the molecular and atomic.

C. Classification of Models by Solution Procedure

An alternate classification of models made from the viewpoint of the nature of the equations appearing in the model is given in Table 2-7. The classification is oriented toward the *solution of models*. As a rough guide, the complexity of solving the mathematical model roughly increases as we go down Table 2-7.

Since usually the more complex the mathematical description of a process is, the more difficult is its general solution, the art of model building is based on the ability to look at a problem in the "right" way. The engineer must be able to analyze the problem and represent it by a suitable compromise among the required detail, the available information on empirical parameters, the inherent

TABLE 2-7 Classification of Deterministic Transport Phenomena Models Based on Mathematical Structure

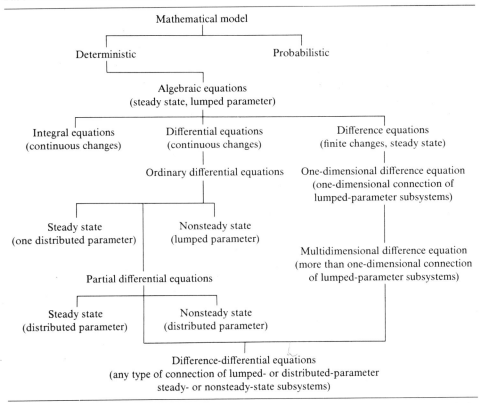

limits of the available mathematical tools, and his time. It becomes impossible to make statements about the suitability of models in general, and each process must be treated as a separate case. The danger is that simplifying assumptions that are not based on the physics of the process will be made purely to make the mathematics tractable. Also, there is little merit in a formal analytical solution to a problem, however elegant it may be, if it is so intractable that obtaining numbers from the answer is very difficult.

D. Models in the Frequency Domain

Frequency domain models are employed instead of time domain models for two primary reasons:

The analytical solution of the model may be simpler in the frequency domain than the related solution of the model in the time domain.

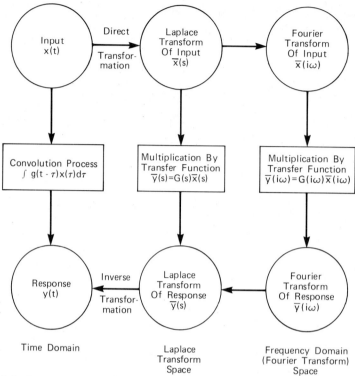

FIGURE 2-6 Representation of the relationships between the input and the output of a linear subsystem.

For some models, analytical particular solutions can be obtained for the frequency domain but not for the time domain. Also, the model response may be in the form of a complicated series which is hard to evaluate numerically.

Direct transformation between the time and frequency domains, as shown in Figure 2-6, is possible in general only for models linear in the time domain. However, satisfactory empirical models can be formulated directly in the frequency domain.

1. The Transfer Function

An alternate way of representing the relation between the inputs and outputs of constant-coefficient differential equations is by use of the transfer function. For example, consider a general nth-order ordinary differential equation (which includes the single first order equation as a special case)

$$\frac{d^n y}{dt^n} + a_{n-1}\frac{d^{n-1} y}{dt^{n-1}} + \cdots + a_1\frac{dy}{dt} + a_0 y = x(t) \qquad (2\text{-}5)$$

FUNDAMENTAL PRINCIPLES FOR MATHEMATICAL MODELING

The *transfer function*, which is defined as the ratio between the Laplace transform of the output divided by the Laplace transform of the input, can be found by taking Laplace transforms of each side of Equation 2-5 with $y(0) = y'(0) = \cdots y^{(n-1)}(0) = 0$,

$$s^n y(s) + a_{n-1} s^{n-1} y(s) + \cdots + a_1 s y(s) + a_0 y(s) = x(s) \qquad (2\text{-}6)$$

If we rearrange Equation 2-6 with $y(0)$ into the following form

$$\frac{y(s)}{x(s)} = \frac{1}{s^n + a_{n-1} s^{n-1} + \cdots + a_1 s + a_0} = g(s) \qquad (2\text{-}7)$$

we obtain the transfer function corresponding to the subsystem differential equation. If the input function $x(t)$ is generalized to include derivatives of $x(t)$, then a polynomial in s will appear in the numerator of the transfer function, Equation 2-7. Equation 2-7 can be written as

$$y(s) = g(s) x(s) \qquad (2\text{-}8)$$

Equation 2-8 states that the output can be found from the product of the input times the impulse response, all evaluated in Laplace transform space.

From another viewpoint, $g(t)$ is a weighting function as can be seen as a consequence of the so-called convolution theorem of Laplace transform theory

$$y(t) = \mathscr{L}^{-1}[g(s) x(s)] = \int_0^t g(t - \alpha) x(\alpha) \, d\alpha \qquad (2\text{-}9)$$

Figure 2-6 shows the relation between the convolution process and the transfer function. The weighting function and the transfer function are equivalent methods of representing linear system response, and each has its particular area of application. The transfer function is used mainly in control systems analysis where rather complicated sets of subsystems must be analyzed and combined.

To obtain the frequency response from the transfer function, you replace s by its imaginary part $i\omega$. Since $g(s)$ is an analytic function, its behavior along the imaginary axis in the s plane specifies its behavior over the entire complex plane. Conversely, experimentally evaluated frequency response values may be used to determine $g(s)$ for stable systems.

For transfer function representation of models involving partial differential equations, refer to Himmelblau and Bischoff (1965), Wen and Fan (1975), and Mecklenburgh and Hartland (1975).

2. Empirical Transfer Functions

Empirical transfer functions can be formulated directly from experimental data. From the response to a pulse type of input [Hougen (1964)], the transfer function, the amplitude ratio, and the phase angle can be evaluated. If we replace s by $i\omega$ in the Equation 2-8,

$$g(i\omega) = \frac{y(i\omega)}{x(i\omega)}$$

then $g(i\omega)$ can be interpreted as a ratio of Fourier transforms.

$$g(i\omega) = \frac{\int_{-\infty}^{\infty} y(t)e^{-i\omega t}\,dt}{\int_{-\infty}^{\infty} x(t)e^{-i\omega t}\,dt} \qquad (2\text{-}10)$$

Since the values of the input $x(t)$ and the output $y(t)$ will be zero for $t < 0$, the lower limit of both integrals in Equation 2-10 can be replaced by 0. Next, let t_x be the terminal time for the input and t_y (another time) be the terminal time for the observed experimental output. These values can be introduced as the respective upper limits in Equation 2-10 to yield

$$g(i\omega) = \frac{\int_{0}^{t_y} y(t)e^{-i\omega t}\,dt}{\int_{0}^{t_x} x(t)e^{-i\omega t}\,dt} \qquad (2\text{-}11)$$

Using the Euler identity,

$$e^{iz} = \cos z + i \sin z$$

Equation 2-11 can be written as

$$g(i\omega) = \frac{A_1 - iB_1}{A_2 - iB_2} = \frac{(A_1 A_2 + B_1 B_2) + i(A_1 B_2 - B_1 A_2)}{A_2^2 + B_2^2} \qquad (2\text{-}12)$$

where

$$A_1 = \int_0^{t_y} y(t)\cos(\omega t)\,dt \qquad A_2 = \int_0^{t_x} x(t)\cos(\omega t)\,dt$$

$$B_1 = \int_0^{t_y} y(t)\sin(\omega t)\,dt \qquad B_2 = \int_0^{t_x} x(t)\sin(\omega t)\,dt$$

The gain (amplitude ratio) and phase lag of $g(i\omega)$ are given, respectively, by

$$|g(i\omega)| = \frac{[A_1^2 + B_1^2]^{1/2}}{[A_2^2 + B_2^2]^{1/2}} \qquad \text{(gain)} \qquad (2\text{-}13)$$

$$\angle g(i\omega) = \tan^{-1}\frac{A_1 B_2 - A_2 B_1}{A_1 A_2 + B_1 B_2} \qquad \text{(phase lag)} \qquad (2\text{-}14)$$

For a particular frequency, A_1, B_1, A_2, and B_2 can be evaluated using a computer. The key to the success of the procedure is to accurately evaluate

FUNDAMENTAL PRINCIPLES FOR MATHEMATICAL MODELING

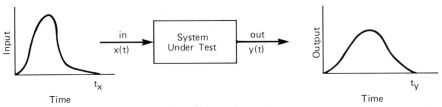

FIGURE 2-7 Pulse testing.

these integrals by using a sufficiently precise numerical integration routine. Suppose an input pulse $x(t)$ of fairly arbitrary shape is put into a process as illustrated in Figure 2-7. In essence, the pulse excites the system at all frequencies at once. The fast Fourier transform algorithm can be used very effectively to transform the time domain data into the frequency domain and vice versa. A range of frequencies can be selected, and the integrals calculated after which the calculated data can be used to prepare a Bode plot. Experimental data can be fit by statistical means to assumed forms of the transfer functions with time delays such as

$$g_1(s) = \frac{K}{as + 1} e^{-\tau s} \qquad \text{(first order system)}$$

$$g_2(s) = \frac{K}{(a_1 s + 1)(a_2 s + 1)} e^{-\tau s} \qquad \text{(factorable second order system)}$$

where K, a, a_1, a_2, and τ are constants. The dead time, τ, or time delay in the model, is evaluated by measuring the time to experimentally first detect a measurable output from the process for a pulse input.

E. Empirical Models

Empirical models are arbitrary functions or equations used by analysts to represent processes. Because they are not based on any of the basic principles of nature, their use in scaleup and extrapolation is hazardous at best. Typical forms of empirical models might be

$y = a_0 + a_1 x_1 + a_2 x_2 + \cdots$ (linear in variables and coefficients)

$y = a_0 + a_{11} x_1^2 + a_{12} x_1 x_2 + \cdots$ (linear in coefficients, nonlinear in variables)

$G(s) = \dfrac{1}{a_0 + a_1 s + a_2 s^2}$ (nonlinear)

$\text{Re} = a(\text{Pr})^b (\text{Sc})^c$ (nonlinear)

Models containing multiple equations with many dependent variables often are needed to represent a single process.

To build an empirical model when the form of the model is unknown is a lengthy process. Experiments must be carried out that might be quite time consuming and expensive. The following is a checklist of the factors that must be examined prior to and during the experimentation.

Statement of Objectives

Why is the work to be done? What questions will be answered by the experiment?

What are the consequences of a failure to find an effect or to claim one when it does not really exist?

What is the experimental space to be covered?

What is the time schedule?

What is the allowable cost?

What previous information is there about the model, experiment, or its results?

Is an optimum among the variables sought or only the effect of the variables?

Type of Model(s) To Be Used

Will linear or nonlinear models be used?

Is the form of the model correct or is the form to be determined?

What will be the independent and dependent variables?

Experimental Program

What are the variables to be measured? How will they be measured and in what sequence?

Which variables are initially considered most important? Which least important? Can the desired effect be detected?

What extraneous or disturbing factors must be controlled, balanced, or minimized?

What kind of control of the variables is desirable?

Are the variables independent or functions of other variables?

How much dispersion can be expected in the test results? Will the dispersion be different at different levels of the variables?

Replication and Analysis

What is the experimental unit and how are the experiments to be replicated —all at once, sequentially, or in a group?

What are the number and type of tests to be carried out?

How are the data to be analyzed and interpreted?

One way to proceed with the experimental program is to collect data and then determine which is the best model among all the possible models based on fitting the data. Such a procedure is not too efficient, and is more likely to be used with historical data than with an experimental program. It is better to carry out a sequential, iterative procedure to determine which model is best, and in particular the experiments should be planned to achieve two major objectives:

Estimate parameters in the model(s) within the desired precision.

Discriminate among rival models.

Each succeeding experiment should test possible suitable models as severely as possible.

What types of experimental designs should be used to reduce parameter uncertainty? Draper and Hunter (1967) show that maximizing the posterior probability density function for the coefficients in the model, given the experimental data, is equivalent to maximizing

$$\Delta \equiv \det[\mathbf{X}^T\mathbf{X} + \sigma_Y^2 \mathbf{C}^{-1}]$$

where $\mathbf{X}$ is the matrix of the observations of the independent variables for the sets of experiments, σ_Y^2 is the variance for the distribution of the dependent variable (assumed constant), that is, the measure of the experimental error, and $\mathbf{C}$ is the covariance matrix for the model coefficients (known, assumed, or estimated). After one series of experiments are completed, the setting of the independent variables for the next experiment would be those that minimize the confidence region associated with the model coefficients, a region that is inversely proportional to Δ.

If several competing models exist, the sequential strategy [Box and Hill (1967)] is recommended to discriminate among the models. A scalar discriminant function is formed summing over pairs (r, s) of models

$$D = \frac{1}{2} \sum_{r=1}^{v} \sum_{s=r+1}^{v} p_r^{(n)} p_s^{(n)} \left\{ \frac{(\sigma_r^2 - \sigma_s^2)^2}{(\sigma_y^2 + \sigma_r^2)(\sigma_y^2 + \sigma_s^2)} \right. $$
$$\left. + (\hat{Y}_r^{(n+1)} - \hat{Y}_s^{(n+1)})^2 \left[\frac{1}{(\sigma_Y^2 + \sigma_r^2)} + \frac{1}{(\sigma_Y^2 + \sigma_s^2)} \right] \right\}$$

where

$p_r^{(n)}, p_s^{(n)}$ = prior probability of model r or s being the correct model, respectively $[0 \leq p^{(n)} \leq 1]$

$\sigma_r^2, \sigma_s^2, \sigma_Y^2$ = variance of data using model r or s, and the experimental error variance, respectively

$\hat{y}_r^{(n+1)}, \hat{y}_s^{(n+1)}$ = predicted response for model r or s, after the $(n+1)$ experiment

The sequential procedure to discriminate among models can be summarized as follows:

Based on an experimental design selected in some arbitrary or suboptimal way, collect n data points.

Estimate the parameters in the v possible models by linear or nonlinear regression; estimate σ_Y^2 and calculate each σ_r^2 for each model.

Calculate the prior probabilities for the $(n + 1)$st run which are equal to the posterior probabilities for the nth run. The initial p's can all be equal to $1/v$ (equal probability) if no better choice is available. One way to obtain the posterior probability that model r is correct after taking n observations is to apply Bayes' theorem to each model successively.

Select the vector of experimental conditions (independent variables) for the $(n + 1)$st run, $\mathbf{X}^{n+1}$, by maximizing D using a numerical optimization routine.

Run an experiment at $\mathbf{X}^{n+1}$ and repeat.

The sequential procedure continues until one (or more) value of $p_r^{(n)}$ reaches a level that causes acceptance of the model by some criterion. Or the experimenter can just observe the trend of the changes in the $p_r^{(n)}$ as the number of experiments increases, dropping models with low values of $p_r^{(n)}$, and adding models, if he or she wishes, terminating the experiments when he feels satisfied with the discrimination actually achieved.

III. DIMENSIONAL ANALYSIS

Although the models described in Section II can be applied to a wide variety of practical problems, there still exist many types of processes to which the transport or empirical models cannot be applied. For example, because the velocity distribution is often impossible to obtain even experimentally in many vessels, a model that contains a velocity term can be written but not solved. In such cases engineers fall back on the use of dimensional analysis to obtain some type of relationship between the independent and dependent variables involved in the process even though the proper functional form of the relationship is not known.

DIMENSIONAL ANALYSIS

A. Making the Dimensional Equation Dimensionless

One way to obtain the dimensionless variables and groups (parameters) that describe a complex process is to transform the microscopic balances cited in section II into dimensionless form. The basic idea underlying the transformation is that physical laws should be independent of the units used for the variables [Becker (1976)].

We start with the premise that any variable expressed in one set of units can be converted to a dimensionless variable by multiplication by a conversion factor, that is, by multiplication by a scalar number and associated units. For example, you can convert length of a pipe in feet to a dimensionless length by dividing by the radius of the pipe in feet (multiplication by the inverse).

$$\frac{l \text{ (in feet)}}{R \text{ (in feet)}} = L \text{ (dimensionless)}$$

In any of the balances, such as, for example

$$\rho C_p \frac{\partial T}{\partial t} = k \frac{\partial^2 T}{\partial x^2}, \quad \text{that is,} \quad \frac{\rho C p}{k} \frac{\partial T}{\partial t} - \frac{\partial^2 T}{\partial x^2} = 0 \qquad (2\text{-}15)$$

we want to transform the dependent variable (T), and the independent variables (t, x) into dimensionless variables. Let us use the notation (*) to denote the dimensionless variables. We multiply the dimensional variables by appropriate scale factors (α_i) to get

$$T^* = \alpha_1 T$$

$$t^* = \alpha_2 t$$

$$x^* = \alpha_3 x$$

where α_1 might be the inverse of temperature, $1/T_0$, α_2 might be the inverse of some reference time, say $k/\rho C_p L^2$, and α_3 might be the inverse of the length of the vessel $1/L$. T_0 and L are known as the characteristic temperature and distance, respectively. Introduce T, t, and x into the differential Equation 2-15 so that the differential equation is transformed to

$$\frac{\partial T^*}{\partial t^*} - \frac{\partial^2 T^*}{\partial x^{*2}} = 0 \qquad (2\text{-}16)$$

Suppose the differential equation included a term containing the fluid velocity. Then if you wished to make time dimensionless, you might use for α a ratio of velocity and distance that will leave only inverse time as the net dimension, that is, v/L. The best procedure for choosing the α's, in case of doubt as to what to do, is to look in a textbook and see what more experienced engineers have used. Sometimes more than one choice is quite appropriate, such as the use of the length, radius, or diameter of a pipe to make the axial distance dimensionless.

Once you have put the equation into dimensionless form, it is always possible to substitute one dimensionless ratio or group for another by simply multiplying the group you would like to remove by another appropriate dimensionless group, and using the net product (still dimensionless!) in the equation.

Nothing has yet been said about the boundary and/or the initial conditions that must accompany the differential equation to make a complete model. In the above example, the boundary conditions might be

$$T(0,t) = T_0$$

$$T(L,t) = T_1$$

and the initial condition might be

$$T(x,0) = T_0$$

(Note how T_0 has been used in α_1 to make T dimensionless. We might equally well have used $\alpha_1 = 1/(T_1 - T_2)$.

To sum up, the final form of the differential equation contains three dimensionless variables and no (dimensionless) coefficients plus the dimensionless boundary conditions

$$T^*(0, t^*) = 1$$

$$T^*(1, t^*) = \frac{T_1}{T_0}$$

and the dimensionless initial condition

$$T^*(x^*, 0) = 1$$

Observe that the dimensional differential equation originally had one dependent variable, two independent variables, and one (group) coefficient. After making the equation dimensionless, only three variables remain, hence the number of degrees of freedom in the model has been reduced by one.

B. Short Cut Method of Obtaining Common Dimensionless Groups (by Inspection)

An easy way to ascertain the pertinent dimensionless groups for a process model is to look them up in Table 2.8a. In the Table, each of the three transport balances is shown (in vector/tensor notation) term by term under

TABLE 2-8a Development of and Relations Among Dimensionless Groups

		Rate of change of mass per unit volume	Rate of change of mass by convection per unit volume	Rate of change of mass by molecular transfer (diffusion) per unit volume	Generation per unit volume (chemical reaction)		Boundary Condition(s) (Interphase Transfer)		
	MASS BALANCE						Empirically determined flux specified (3)[a]	Concentration specified (1, 2b)[a]	Mass flux specified (2a, 4)[a]
	Symbols	$\frac{\partial}{\partial t} c_i$	$+ [\nabla \cdot c_i v] =$	$- [\nabla \cdot J_i]$	$+ R_i$		$N_A\|_{x=0} = K\Delta c$	$c = c_0$ or $c_1 = c_2$	$N_1 = N_2$ or $N_A = N_0$
	Dimensions[b]	$\frac{C}{\theta} = \frac{CV}{L}$	$\frac{VC}{L}$	$\frac{DC}{L^2}$	$R_i\ddagger$		$\frac{1}{L}(KC)$		

$\frac{LV}{D}$ = Mass Peclet

$\frac{KL}{D}$ = Sherwood

$\frac{R_i\ddagger L}{VC}$ = Damköhler I

$\frac{R_i\ddagger L^2}{DC}$ = Damköhler II

		Rate of change of momentum per unit volume	Rate of change of momentum by convection per unit volume	Rate of change of momentum by molecular transfer (viscous transfer) per volume	Generation per volume (External forces) (Ex: gravity)		Empirically determined flux specified (3)[a]	Velocity specified (1, 2b)[a]	Momentum flux specified (2a, 4)[a]
	MOMENTUM BALANCE					Pressure gradient			
	Symbols	"Inertial Forces" $\frac{\partial}{\partial t} \rho v$	"Viscous Forces" $+ [\nabla \cdot \rho vv] =$	$- [\nabla \cdot \tau]$	$+ \rho g$	$- \nabla p$	$\tau\|_{x=0} = \tau$ or $\tau = \frac{\sigma}{L}$	$v = 0$ or $v_1 = v_2$	$\tau_1 = \tau_2$ or $\tau = \tau_0$
	Dimensions[b]	$\frac{\rho V}{\theta} = \frac{\rho V^2}{L}$	$\frac{\rho V^2}{L}$	$\frac{\mu V}{L^2}$	ρg	$\frac{P}{L}$	$\frac{1}{L}(\tau\ddagger)$		

$\frac{LV\rho}{\mu}$ = Reynolds

$\frac{V^2}{gL}$ = Froude

$f = \left(\frac{\Delta P}{L}\right)\left(\frac{L}{\rho V^2}\right)$ = Friction Factor

$\frac{\tau\ddagger L}{\mu V}$ = Bingham

$\frac{\rho V^2 L}{\sigma}$ = Weber $(\sigma \rightarrow \tau L)$

		Rate of change of energy per unit volume	Rate of change of energy by convection per unit volume	Rate of change of energy by diffusion (conduction) per unit volume	Generation per volume (Ex: electrical, chem. rxn, etc.)	Other terms: rev. and irrev. transfer, viscous dissipation, etc.	Empirically determined flux specified (3)[a]	Temperature (1, 2b)[a]	Heat flux specified (2a, 4)[a]
	ENERGY BALANCE								
	Symbols	$\frac{\partial \rho \hat{C}_p T}{\partial t}$	$+ (\nabla \cdot \rho \hat{C}_p T v) =$	$- (\nabla \cdot q)$	S_R		$q\|_{x=0} = h\Delta T$	$T = T_0$ or $T_1 = T_2$	$q_1 = q_2$ or $q = q_0$
	Dimensions[b]	$\frac{T\ddagger \rho \hat{C}_p}{\theta} = \frac{T\ddagger V \rho \hat{C}_p}{L}$	$\frac{T\ddagger V \rho \hat{C}_p}{L}$	$\frac{kT\ddagger}{L^2}$	$S_R\ddagger$		$\frac{1}{L}(hT)$		

$\frac{hL}{k}$ = Nusselt

$\frac{h}{\rho \hat{C}_p v}$ = Stanton

$\frac{V\rho \hat{C}_p L}{k}$ = Heat Peclet

$\frac{S_R\ddagger L}{\rho V \hat{C}_p T\ddagger}$ = Damköhler III

Left side vertical relations:
- $Le \cdot Pr = \text{Schmidt}$
- $\text{Lewis} = \frac{\mu}{\rho D}$
- $\frac{\nu}{D} = \frac{\mu}{\rho D}$
- $\frac{k}{\rho \hat{C}_p D}$
- $\frac{\alpha}{D}$
- $\frac{Pr}{Re}$
- $\text{Prandtl} = \frac{\hat{C}_p \mu}{k}$
- $\frac{\nu}{\alpha}$

[a] From Himmelblau and Bischoff, *Process Analysis and Simulation*, Wiley, New York, 1965, p. 168.
[b] Notation is in Table 2-8b.

TABLE 2-8b Notation for Dimensionless Quantities in Table 2-8a

Variable	Equivalent Characteristic Quantity
c	C
t	θ
Length	L
R_i	$R_i^{\ddagger}$
v	V
p	P
τ	$\tau^{\ddagger}$
T	$T^{\ddagger}$
S_R	$S_R^{\ddagger}$

the description of the physical connotations of the respective terms. Underneath each term of the balances are the dimensions of the term as given by the symbols defined in Table 2-8b. The net dimensions of each term are dependent quantity (mass per volume, momentum per volume, energy per volume) per unit time.

Tables 2-8a and 2-8b assist in demonstrating how the common dimensionless groups used in the literature arise. In principle, one should take into account not only the differential equations but also the boundary conditions and the relative size of the system of interest. To obtain the dimensionless groups (exclusive of those for the independent and dependent variables) for any of the three equations, just divide one set of dimensions into all the others *including the boundary conditions* (remember they are part of the model). The numerator of a well-known group is identified by the arrowhead while the denominator is designated by the tail of the arrow. For example, the mass Peclet number represents the division of the dimensions representing the transport of mass by convection by the dimensions representing the transport of mass by molecular processes. When you divide, you reduce the groups of coefficients by at least one.

Table 2-8a has one more feature to examine. Look in the left-hand margin where the results of dividing one dimensionless group from one balance by a group from another balance appear. For example, the Peclet number (group) in the energy balance can be divided by the Reynolds number (group) in the momentum balance to give the Prandtl number.

Some examples will make use of the table clear:

What group evolves from dividing the dimensions of the mass boundary condition, the flux, by the molecular transport term dimensions?

$$\frac{kC/L}{DC/L^2} = \frac{Lk}{D} \quad \text{Sherwood number}$$

DIMENSIONAL ANALYSIS

What group evolves from dividing the dimensions of the mass boundary condition, the flux, by the reaction term dimensions?

$$\frac{kC/L}{R_i^{\ddagger}} = -\frac{kC}{LR_i^{\ddagger}}$$

What group evolves by dividing the Froude number by the mass Peclet number?

$$\frac{V^2/gL}{LV/D} = \frac{VD}{gL^2}$$

C. Correlation of Experimental Data

Once the dimensionless quantities for the differential equation(s) representing the process are determined, of what use are they? It still is assumed that the dimensionless groups are related to each other because of the physical law(s) they represent. Generally, one assumes some type of function exists for the relationship (completely arbitrary) and collects data to estimate the coefficients in the function. Chemical engineers usually propose that a given dimensionless group is equal to some constant times the product of other dimensionless groups each raised to a power, such as

$$\mathrm{Re} = a(\mathrm{Pr})^b (\mathrm{Sc})^c$$

This type of function in effect is a linear sum of the logarithms of the various groups. Chapter 10 in Churchill (1974) describes in detail ways of correlating the dimensionless groups.

D. Scaleup

The principle of similarity, first proposed by Newton, applies to chemical processes. Two systems are said to be geometrically similar when the ratios of corresponding dimensions in one system are equal to those in the other. Hence, geometrical similarity exists between two pieces of equipment of different sizes when both have the same shape. Kinematic similarity exists between two systems of different sizes when they are not only geometrically similar but when the ratios of velocities between corresponding points in each system are also the same. Dynamic similarity exists between two systems when, in addition to being geometrically and kinematically similar, the ratio of forces between corresponding points in each system are equal.

One can interpret similarity in terms of the dimensionless groups and variables in a process model: the numerical values of all of the dimensionless groups should remain constant during scaleup [Kline (1965)]. This is a fine principle but in practice frequently remains impossible to fulfill as the following example shows.

Example 2.5 Reactor Scaleup. Consider the scaleup of an endothermic reaction mixture in a 2-m-diameter pilot reactor to a 6-m-diameter process vessel, keeping the same geometrical shape and keeping the same heat transfer rate per unit mass in the large vessel as in the small vessel. The lengths are scaled up 3 times, the areas are scaled up 9 times, and the volumes 27 times. The heat required by the reaction mass depends on the amount of material present, which is 27 times greater than in the small unit, but the heating area is only 9 times greater on the big unit so that it is 3 times easier for the heat to get to the reaction mass in the small unit.

Suppose that the pilot and production tanks are cylindrical, baffled, geometrically similar, and equipped with turbine agitators. Take the height of liquid in each tank as equal to the tank diameter and the impeller diameter in each tank as one-third the tank diameter. Assume that the inside film heat transfer coefficient is low and controlling. The heating medium in the jacket is considered to be steam with a relatively high heat transfer coefficient.

Let the subscripts 1 and 2 refer to the pilot and production units, respectively. Then the ratio of the internal heat transfer film coefficients can be developed from the correlation

$$\mathrm{Nu} = 0.74 \mathrm{Re}^{0.67} \mathrm{Pr}^{0.33} \left(\frac{\mu_s}{\mu}\right)^{-0.14}$$

where

$\mathrm{Nu} = h_i D_T / k$
$\mathrm{Re} = \dfrac{\rho v d}{\mu} = \dfrac{\rho n D_i^2}{\mu}$
D_T = tank diameter
D_i = impeller diameter
n = number of revolutions per unit time of the impeller

The ratio is

$$\frac{(h_i D_T)_2}{(h_i D_T)_1} = \frac{(nD_i^2)_2^{2/3}}{(nD_i^2)_1^{2/3}}$$

It was specified that $D_T = 3D_i$. Therefore,

$$\frac{h_{i2}}{h_{i1}} = \frac{n_2^{2/3} D_{i2}^{1/3}}{n_1^{2/3} D_{i1}^{1/3}} = \left(\frac{n_2}{n_1}\right)^{2/3} \left(\frac{D_{i2}}{D_{i1}}\right)^{1/3}$$

PRACTICAL ASPECTS OF MODELING

For turbine agitators, the following approximate range of tip speeds indicates the degree of agitation produced in baffled tanks:

Tip Speed in m/s	Degree of Agitation
2.5–3.3	Low
3.3–4.0	Medium
4.0–5.6	High

The impeller tip speed $S = \pi D_i n$. Therefore, $n \propto S/D_i$, and

$$\frac{h_{i2}}{h_{i1}} = \left(\frac{S_2}{S_1}\right)^{2/3} \left(\frac{D_{i1}}{D_{i2}}\right)^{2/3} \left(\frac{D_{i2}}{D_{i1}}\right)^{1/3}$$

$$= \left(\frac{S_2}{S_1}\right)^{2/3} \left(\frac{D_{i1}}{D_{i2}}\right)^{1/3}$$

Now $D_{i1}/D_{i2} = \frac{1}{3}$, $h_{i2}/h_{i1} = 0.693$ for equal tip speeds on the two scales, and $h_{i2}/h_{i1} = 1.10$ for twice the tip speed in the production unit as in the pilot unit. For equal tip speeds, the internal heat transfer film coefficient for the 6-m-diameter tank is only about 70 percent of that in the 2-m one. In addition, the big tank has three times less heat transfer area per unit mass than the small one.

Increasing the degree of agitation offers little hope of obtaining the same degree of heat transfer per unit mass in the two cases, because doubling the tip speed is just about the maximum permissible increase, and in this case the heat transfer film coefficient on the big tank would be only 10 percent greater than that in the small tank.

The problem of obtaining the same degree of heat transfer per unit mass in the small and big tanks can, however, be solved by pumping the reaction mass through an externally located heat exchanger.

The difficulty in maintaining even geometrical similarity is no mean task. In fact, it is rarely possible to satisfy all or even a majority of the various similarity criteria in scaleup, particularly those related to the thermal and chemical dimensionless groups. The best possible compromise is to weight the effect of each group and carry out intermediate scale experiments to evaluate the effect of the critical groups. One can also scale down, that is, start with the final design of the full scale unit and base the pilot plant or laboratory unit on the similarity principle insofar as possible. □

IV. PRACTICAL ASPECTS OF MODELING

In this section we examine some of the factors that must be constantly kept in mind if the modeler is not to blunder in his work.

A. Validation of the Model

Finger and Naylor (1967) divide model validation into three parts:

Validation of the logic.
Validation of model behavior.
Validation of model assumptions.

Carrying out these tasks involves comparison with historical input–output data, or data in the literature, comparisons with pilot plant and/or future performance, and simulation studies. In general, data used in formulating the model should not be used to validate it. Model validation is an art in the sense that no formal technique exists for ascertaining the relative values of the criteria used for evaluation. There is nothing as good as an expert opinion in verification of models, that is, what is the impression of the model when reviewed by people who know the process being modeled?

No validation procedure is appropriate for all models. See Gass (1977) for a discussion of assessment criteria. Nevertheless, one can ask the question what would we really like the model to do? In the best of all possible worlds, we would like the model to predict the performance of the process with certainty, but this is an impossible goal. Consequently, the definition of validity which seems to be the most satisfactory is usefulness. A model is valid if it is useful for a clearly stated purpose. We can examine a model for its workability, clarity, representativeness (in some sense), and internal consistency as the major features of usefulness.

First, if the model is to serve any useful function other than the satisfaction of the intellectual curiosity of the modeler, then it must serve some practical function—identify problems, suggest a resolution of the problems, and produce results that compare well with the results of the process being modeled. We can call these factors criteria of workability. Second, it is implicit that a model cannot have good workability characteristics if it is not clear. Clarity must be such that the model is understandable to other modelers (who, for example, may provide advice and data to the users). For evaluation, this means that the model's behavior and its results must be translatable into terms commonly used in the plant environment.

Third, if a model is to have workability and clarity, it must be unambiguous in that the model behavior must correspond to the modeler's expectations as per its design. That is, the model must do whatever the modeler says it will do. (This is sometimes denoted as verification.) The model must behave in a manner consistent with the underlying logic or theory upon which it is built.

Fourth, there is a question of the relationship of the model to reality. The concept of validity which equates valid with true is not only popularly accepted but seems to be implicit in much of the literature. This notion of truth is fundamentally inaccurate if applied to models because all models are untrue;

PRACTICAL ASPECTS OF MODELING

it is impossible to ever prove a model to be true. Valid as true is also dangerously misleading. The process of model building is far more complex and the process of model testing far less conclusive than can ever be conveyed by a concept with such strong dichotomous overtones. Also, the search-for-truth approach to modeling ignores what may prove to be our strongest source of information in evaluating models—the consequences of using models to intervene in real world systems. It is reasonable to expect that the model results should agree reasonably well with accepted data and theories, and if they do not, then some satisfactory explanation should be forthcoming, for example, that the data are incorrect or incomplete or that the theories require modification. If predictions of the model compare favorably with empirical observations, one's faith in the appropriateness of the mathematical descriptions is enhanced. If the predictions are poor, then revision of the model or some other approach need be considered.

B. Precautions in Model Building

Model building has some significant limitations that must be recognized at the outset. The first is related to the availability of data and the accuracy of the data; that is, the success of model building depends heavily on the basic information available to the analyst. Process studies are only as accurate as the physical and chemical data that go into the model. In most cases the engineer finds a distinct shortage of data that can be employed in the model, and one of his major tasks after setting up the model is to evaluate the parameters in the model on the basis of experimental data. In various types of unit operations, the entire effort is directed toward more accurate estimation of parameters in models whose forms are now well established.

One of the areas of considerable importance to chemical engineers is that of process kinetics; this is also an area in which great uncertainty exists as to the "true kinetics" of the process. Kinetic coefficients are obtained by actually carrying out experiments in a small-scale reactor of some sort, where side effects may become quite important. Lack of information about the side effects may lead to inappropriate coefficients for a commercial reactor design. Another frequent problem is that impurities present in the plant do not always occur in the laboratory, which can lead to unpleasant surprises in the eventual plant operation.

For various separation processes, such as distillation, absorption, and evaporation, the efficiency of the apparatus when included as a design parameter is very uncertain and may degrade with time so that the macroscopic models that have been developed and are available in most common textbooks are of little use without reasonable estimates of the efficiency. For instance, in calculating the number of trays to be used in a distillation tower, the accuracy of methods of calculating the number of theoretical equilibrium stages far exceeds the accuracy of the methods for estimating the efficiency of an actual stage compared to a theoretical stage. Tray efficiencies depend on system properties such as liquid diffusivity, gas diffusivity, the viscosities of the two phases and their densities, the mass flow rates, the degree of mixing, and equilibrium relationships. This does not mean that the efficiency depends only on these values nor do these values always affect efficiency, since the actual mixing and mass transfer characteristics of the column are related to other variables impossible to measure on a macroscopic basis.

The accuracy with which parameters must be known depends to some extent on their influence in the overall process. In a general sense the ones that should be known with the greatest accuracy are the ones that have the greatest influence.

These remarks give some idea of one source of difficulty in the actual application of mathematical models. They apply both to scaling-up from laboratory or pilot plant data and to analysis of a commercial plant.

The second major limitation in modeling is the character of the tools available to manipulate the mathematical statements that compose the model. In general, these tools are taken from the field of mathematics, which in itself is somewhat limited. Quite complex models can be easily defined and described mathematically; however, our present-day tools cannot manipulate them because of limitations in theory or computational techniques. Under such conditions, although the model might be defined and might be completely appropriate, there would be no reasonable method of developing predictions.

In addition to the two limitations just described, if the building blocks for the model are not physically realizable, the danger exists that a concept intended merely as a technique of analysis can become endowed with a physical reality never intended by the inventor and for which no evidence exists. We must be wary of attaching to a model a general aura of validity that it does not merit. It may become acceptable as dogma with no real basis; this is a disease of mathematical modeling. Models that were originally models of something have been taken, dusted off, and applied as if they were known to be valid *a priori*, which emphatically is not the case.

A final danger in the use of models that should be mentioned, particularly with respect to empirical models, is to assume they represent the real system beyond the range of the variables that the model was intended to encompass. Such extrapolation may be a valuable aspect of the model, but it may also be very misleading.

The dangers cited above must be circumvented by constantly using common sense in the interpretation of the mathematical results.

NOMENCLATURE

a	Constant in general
A	Area
A_1, A_2	Constants

NOMENCLATURE

B_1, B_2	Constants
c	Concentration (subscript shows component)
c_s	Surface concentration
c_i, c_α	Concentration of species i or α
c^*	Equilibrium concentration with other phase
$\hat{C}_p, C_p$	Heat capacity
$\mathbf{C}$	Covariance matrix for model coefficients
D	Discriminant function
$\tilde{D}$	Dispersion coefficient in direction indicated by the subscript; subscripts L and R refer to axial and radial direction, respectively
D_T	Tank diameter
D_i	Impeller diameter
$E^{(t)}$	Interphase heat transfer
E_{tot}	Total energy
g	Acceleration of gravity (subscript indicates direction)
$g(t)$	Impulse response or weighting function
$g(s)$	Transfer function
$g(\omega), g(i\omega)$	Deterministic transfer function (frequency response) in frequency domain
$G(t)$	Empirical impulse response (weighting function)
$G(s)$	Empirical transfer function, Laplace transform of $G(t)$
h	Interphase heat transfer coefficient
h_i	Interphase heat transfer coefficient, inside
$\hat{H}$	Enthalpy per unit mass
i	Imaginary number
$\mathbf{J}_i$	Mass flux vector
k	Interphase mass transfer coefficient
$\tilde{k}_i$	Effective thermal conductivity in the i direction
K	Constant in general
l	Length
$\mathscr{L}$	Laplace transform
$\mathscr{L}^{-1}$	Inverse Laplace transform
L	Length
$m_i^{(t)}$	Interphase mass transfer for species i
$m_{\alpha,\text{tot}}$	Total mass of species α
n	Number of revolutions
n_i	Mass flux of species i at a boundary
Nu	Nusselt number

N_A	Mole flux of component A at a boundary
p	Pressure
$P_r^{(n)}$	Prior probability of model r being the correct model
P	Perimeter
Pr	Prandtl number
q	Heat flux (in a specified direction)
$\mathbf{q}$	Heat flux vector
Q	Interphase heat transfer
$Q^{(m)}$	Interphase energy transfer accompanying mass transfer
r_i, r_α	Reaction rate, deterministic variable, or i or α, mass/time
Re	Reynolds number
R	Radius
R_i	Reaction rate for i, or component i mol/time
s	Complex parameter in the Laplace transform
S	Cross-sectional area
Sc	Schmidt number
S_R	Rate of internal generation of energy
t	Time
T	Temperature, absolute temperature
T_s	Surface temperature
T^*	Temperature in equilibrium with the other phase
u	Dependent variable
U	Overall heat transfer coefficient
v	Velocity of flow (subscript shows direction)
V_{tot}	Total volume
w_i	Interphase mass transport term in macroscopic mass balance
W	Work
x	Coordinate direction in rectangular coordinates
$x(t)$	Process or model input
$\mathbf{X}$	Matrix of observations of independent variables $x_1, x_2, \ldots$
Y	Coordinate direction in rectangular coordinates
$\hat{Y}_r^{(n)}$	Predicted response for model r after the nth experiment
z	Independent variable
z	Axial coordinate

Greek

α	Delay factor in a time integral
α_i	Scale factors

Δ	Difference, and $\Delta X = X_{i+1} - X_i$
η	Effectiveness factor
μ	Viscosity in general; subscript s indicates surface
μ_{ij}	Effective viscosity (subscripts indicate associated velocity gradient)
ρ	Density
ρ_α	Mass density of species α
σ_r^2	Ensemble variance using model r
$\sigma_{Y_i}^2$	Ensemble variance response of Y_i (measure of experimental error)
τ	Momentum flux
τ	Residence time
$\boldsymbol{\tau}$	Momentum flux matrix
$\hat{\phi}$	Potential energy per unit mass
ω	Frequency in general

Overlays

$\char`\^$	Estimated or per unit mass

Superscript

(n)	Indicates order of derivative (1 = first derivative); also indicates the sequence or stage in an iterative calculation
*	Differs in some way from the usual definition of the parameter or variable, usually dimensionally

Other

$\|x(\omega)\|$	The modulus of a function of a complex variable
$\|g(i\omega)\|$	Gain (amplitude ratio) in the frequency domain
$\angle g(i\omega)$	Phase lag in the frequency domain
∇	Gradient of a function or matrix
∇^2	Hessian matrix of a function
$\langle \ \rangle$	Finite time average for continuous random variables

REFERENCES

Becker, H. A., *Dimensionless Parameters*, Applied Science Publishers, London, 1976.

Bird, R. B., Stewart, W. E., and Lightfoot, E. N, *Transport Phenomena*, Wiley, New York, 1960.

Box, G. E. P., and Hill, J. W., "Discrimination Among Mechanistic Models," *Technometrics*, **9**, 57 (1967).

Carslaw, H. S., and Jaeger, J. C., *Conduction of Heat in Solids*, Oxford Univ. Press, London, 1959.

Churchill, S. W., *The Interpretation and Use of Rate Data*, McGraw-Hill, New York, 1974.

Crank, J., *The Mathematics of Diffusion*, Oxford Univ. Press, London, 1956.

Draper, N. R., and Hunter, W. G., "The Use of Prior Distribution in the Design of Experiments for Parameter Estimation in Non-linear Situations," *Biometrica*, **54**, 147 (1967).

Finger, G. S., and Naylor, T. H., *Mang. Sci.*, **14**, 92 (1967).

Gass, S., "A Procedure for the Evaluation of Complex Models," *Comput. Oper. Res.*, **4**(1), 27–35 (1977).

Himmelbrau, D. M., and Bischoff, K. B., *Process Analysis and Simulation*, Wiley, New York, 1965.

Himmelblau, D. M., and Bischoff, K. B., *Process Analysis and Simulation*, Swift Publishing Co., Austin, TX, 1980.

Hougen, J. O., *Experiences and Experiments with Process Dynamics*, Chem. Engr. Progress Symp. Series, **60**, 1–89 (1964, No. 4).

Kline, S. J., *Similitude and Approximation Theory*, McGraw-Hill, New York, 1965.

LeGoff, P., and Zonalian, A., *The Chem. Engr. J.*, **12**, 33–46 (1976).

Mecklenburgh, J. C., and Hartland, S., *The Theory of Backmixing*, Wiley, New York, 1975.

Slattery, J. C., *Momentum, Energy, and Mass Transfer in Continua*, McGraw-Hill, New York, 1972.

Wen, C. Y., and Fan, L. T., *Models for Flow Systems and Chemical Reactors*, Decker, New York, 1975.

Whitaker, S., *Fundamental Principles of Heat Transfer*, Pergamon Press, New York, 1977.

3

REACTION KINETICS

R. L. KABEL

I.	Major Issues in Chemical Reactor Scaleup	78
II.	Fundamental Considerations	79
	A. The Rate Concept	79
	1. Rate of Change and Process Rate	79
	2. Deduction of Rate of Reaction	80
	3. Measurement of Rate of Change	81
	B. Thermodynamics	84
	1. Role and Limitations	84
	2. Heat Effects	85
	3. Maximum Conversion	87
	4. Role in Rate Equations	94
	C. Physical Data	94
	1. Phase Behavior	94
	2. Solubility	95
	3. Thermal Characteristics	96
	4. Hazards	96
	D. Analysis, Synthesis, and Data Handling	96
III.	Correlation of Rate Information	97
	A. Communication and Data Taking	97
	B. Rate Equations	98
	1. Definitions, Source, and Purpose	98
	2. Homogeneous Reactions	99
	3. Heterogeneous Reactions	102

		4. The Arrhenius Equation	109
		5. Thermodynamic Consistency	110
IV.	Uncertainties		112
	A.	Definition of the Reaction System	112
	B.	Data Base for Scaleup	112
		1. Extent of the Data Base	112
		2. Quality of the Data Base	113
	C.	Appearance of New Factors	113
Nomenclature			113
References			115

I. MAJOR ISSUES IN CHEMICAL REACTOR SCALEUP

Fifteen years ago when chemical reaction engineering was a much less developed discipline than it is today, a student of mine took a summer job with a prominent chemical company. There he was introduced to a pilot plant reactor which was producing a 30 percent conversion. The engineers associated with the pilot plant explained that their goal was to achieve a 90 percent conversion. Their scaleup technique was direct. They intended to make the reactor three times larger.

Several oversights or fallacies in their logic are immediately apparent. First of all, there could be thermodynamic limitations which would preclude any conversion larger than 30 percent. In fact, 30 percent conversion might be achievable by a reactor three times smaller. Even if there were no thermodynamic limitations, for example, an irreversible reaction, most reactions exhibit a decrease in rate as conversion increases. Therefore, a simple linear scaleup would not be expected to achieve the desired result. Further, a whole host of new factors can appear to complicate the scaleup. Among these are competing reactions, thermal effects, phase separation, and so on.

This example would be amusing if it were simply a historical relic. Unfortunately, the simplicity of the original solution is the magic lamp that many practitioners have in mind when they wish for a successful scaleup. Successful scaleups are common, but they are more likely to occur through careful attention to kinetics, thermodynamics, and other principles of chemical reaction engineering than by trial and error.

There are a number of major causes of scaleup problems which do not match the idealized assumptions upon which designs were based. Chapters 8, 9, and 11 give considerable attention to these aspects. Heat and mass transfer effects often complicate commercial reactor performance compared to the assumptions made in the design. These are particularly prominent in heterogeneous reactors and accordingly will be dealt with at length in Chapters 5 and 6.

FUNDAMENTAL CONSIDERATIONS

Fluidized beds, as discussed in Chapter 10, are used to facilitate heat and mass transfer and catalyst regeneration.

The presence and/or buildup of impurities influence the performance of commercial units to a far greater extent than they do the experimental studies upon which the original scaleups were based. Sometimes such matters can be anticipated and the problems avoided. In other cases, pilot planting such as is discussed in Chapter 17 becomes appropriate.

Apart from the many variations of Murphy's law, there is at least one more major cause of scaleup problems; the inadequate use of available engineering tools. One such tool is mathematical modeling, which is covered in Chapter 2. Also in this category are thermodynamics and kinetics. This chapter is the first of five on the scaleup of chemical reactors. In the last four, we deal with homogeneous reactors, fixed-bed catalytic reactors, fluid–fluid and fluid–fluid–solid reactors, and selection of reactor types. In this chapter we will deal with thermodynamics, kinetics, and other fundamental considerations which underlie all of the subsequent four chapters.

II. FUNDAMENTAL CONSIDERATIONS

A. The Rate Concept

The essence of effective chemical reactor scaleup resides in the recognition that chemical reactions are rate processes. The term "rate" is often used loosely or misunderstood. Therefore, a careful definition of the rate concept is essential at the outset.

1. Rate of Change and Process Rate

The dictionary defines rate as "a quantity, amount, or degree of something measured per unit of something else." The "something" in this definition may be anything, for example, temperature, position, or concentration. The "something else" is often time but can also be position or some other independent variable. This definition corresponds to the "rate of change," familiar from the calculus. The rate of change may or may not be important in chemical reactor scaleup. The "process rate," known as the rate of reaction, is always important. Rate of change and process rate are contrasted in Table 3-1. To illustrate the distinction between a rate of change and a process rate, consider the continuous flow stirred tank reactor system shown in Figure 3-1.

An unsteady-state mass balance on component A is given by Equation 3-1.

$$F_i - F_o - r_A V = \frac{dN_A}{dt} \tag{3-1}$$

TABLE 3-1 Comparison of Rate of Change and Process Rate

Rate of Change	Process Rate
Follows dictionary definition	Just a concept
Subject to measurement	Determined from measurement and theory
Examples:	Examples:
a. Change of position with time	a. Rate of reaction
b. Change of temperature with altitude or latitude	b. Rate of heat transfer
	c. Rate of inflation
Has derivative character	Just a concept usually denoted by r for reactions, q for heat transfer, etc.
$\dfrac{\Delta x}{\Delta t} \xrightarrow[\Delta t \to 0]{} \dfrac{dx}{dt}$	

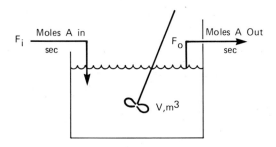

$$r_A = \text{Rate of reaction of A} = \frac{\text{Moles A Reacted}}{m^3 s}$$

$$r_A V = \frac{\text{Moles A Reacted}}{s}$$

FIGURE 3-1 Continuous stirred tank reactor.

where N_A is the number of moles of A in the reactor. The rate of reaction, r_A, is a process rate. The derivative, dN_A/dt, is the rate of change of moles of A with time.

2. Deduction of Rate of Reaction

Equation 3-1 can be solved for the process rate, r_A, as a function of the rate of change, dN_A/dt, and the other variables in the equation. A more familiar expression is obtained by restricting our attention to the case of steady state in which dN_A/dt is equal to zero. Then the mass balance can be written as shown in Equation 3-2.

$$r_A = \frac{F_i - F_o}{V} \tag{3-2}$$

FUNDAMENTAL CONSIDERATIONS

This equation shows how the rate of reaction may be determined experimentally in a CSTR operated at steady state. But note that a mass balance was required to arrive at this conclusion. One does not measure process rates directly, but only arrives at their values by some combination of measurement and theory.

A differential mass balance on an element of a plug flow reactor (see Chapter 4) is given by Equation 3-3:

$$r_A \, dV = F_A \, dX \qquad (3\text{-}3)$$

where F_A is the feed rate of A to the reactor and dX is the differential fractional conversion of A in the differential volume, dV. Equation 3-4

$$r_A = \frac{dX}{d(V/F_A)} \qquad (3\text{-}4)$$

gives the solution of Equation 3-3 for the rate of reaction, r_A. It can be seen that the process rate in this case is equal to the rate-of-change of conversion with V/F_A. The term V/F_A is often referred to as the reciprocal space velocity or when F has volumetric units as the space time. This relation between a process rate and a rate of change is frequently used to determine rates of reaction in tubular reactors.

As a final example we consider the case of a batch reactor. The mass balance given by Equation 3-1 applies here too. However, because there is no inflow or outflow in a batch reactor, the terms F_i and F_o are equal to zero. Thus the mass balance on a batch reactor tells us that the rate of reaction may be determined from the negative of the rate of change of N_A with time, divided by the reactor volume as implied by Equation 3-5.

$$r_A = -\frac{1}{V} \frac{dN_A}{dt} \qquad (3\text{-}5)$$

Here again theory, the mass balance, enables us to deduce a process rate from a measured rate of change. For the special case in which a batch reaction is conducted at constant volume, the rate of reaction can be seen from Equation 3-5 to equal the negative of the rate-of-change of concentration with time. This familiar result is by no means a definition of the reaction rate, but rather a direct result of a mass balance for a very special case. Excellent references on the ideas expressed above are Dixon (1970) and Churchill (1974).

3. Measurement of Rate of Change

We have seen how the rate of reaction may be calculated from measured variables. The simplest case is for a CSTR where an algebraic collection of easily measured parameters suffices (see Equation 3-2). This simple result is one reason why CSTRs are often used for experimental measurement of reaction rates. A much more critical problem arises when one wishes to determine the rate of reaction from a rate of change.

Example 3.1, adapted from Churchill (1974), shows the difficulties in obtaining good rate information in a batch reactor.

Example 3.1   Rate Data in a Batch Reactor. "The reaction between hydrogen bromide and diethyl ether in acetic acid solution in the presence of acetyl bromide at 25°C was studied by Mayo, Hardy, and Schultz (1941). The acetyl bromide maintained a constant concentration of HBr. The following data were obtained:

Original concentrations: Ether, 0.219 mol/L
 HBr,   0.604 mol/L

t, hr	0	21.0	28.5	51.2
x	0	0.065	0.080	0.124

The symbol x is the mol/L of ether which have disappeared at time t. Determine the reaction rate and tabulate as a function of composition."

Because the reaction is occurring in dilute aqueous solution, it will be at constant density (or constant volume in the batch reactor). A mass balance shows the reaction rate of ether to be equal to the rate of change of ether conversion with time, that is, dx/dt. The concentration of ether at any time will be $C_{E_0} - x = 0.219 - x$. Thus, the determination of the reaction rate in this case comes down to differentiation of the data.

Data differentiation is by no means a routine thing; in fact, it is somewhat of an art, involving substantial judgment and self-discipline. Graphical calculus suggests the most familiar method; that is, the slope of a graph of x versus t equals the derivative, dx/dt. Usually one plots the data, passes a smooth curve through them, and determines tangents to the curve at several points. Drawing the smooth curve involves judgment, if not prejudice. The opportunity to introduce bias into the results by this method is dramatically evident. However, the equivalent opportunity exists (perhaps more subtly) in all other methods as well. The advantage of data smoothing at this point in the calculation is that these are the data which have come directly from the experiment. Thus, more is known about their behavior and uncertainty than about quantities derived from them. □

Churchill (1974) discusses data differentiation and this example in great detail, emphasizing the calculation of average rates over an increment from adjacent data points. The incremental rate of change, for example, $\Delta x/\Delta t$, is the only "rate" that is really measured. These incremental rates are then

FUNDAMENTAL CONSIDERATIONS

plotted against the independent variable. At this point a curve is drawn through the steps such that the area, between the curve and the step, above the step is equal to that below the step. Accordingly, the method is referred to here as the equal area method. Points picked off this curve thus represent the instantaneous rates of change. With excellent data this method, like any other, works very well. Frequently, however, locating a smooth curve on the equal area graph presents a baffling challenge. Arbitrariness strikes again, only somewhat more remotely from the raw data. The reader is urged to explore this matter further in Churchill (1974). This author makes a practice of using both methods on any problem of real importance, for example, scaleup.

When a smooth curve has been drawn through the data to be differentiated, it remains to determine the slopes. This can result in the introduction of new error. Walas (1959) suggests some methods for dealing with this, including three- and five-point formulas for data differentiation. With the availability of computers, it has become common to fit the data with some appropriate analytical function (e.g., a polynomial equation) and to perform the differentiation analytically. Such sophisticated methods reduce the error in slope taking but the need for judgment and self-discipline in data fitting is as great as ever.

Returning to the example problem, the results of several calculations are shown in Table 3-2. The table compares values of dx/dt calculated quickly by the author using the slope (K slope) and equal area (K e.a.) methods. Also shown are the values given by Churchill (1974) in his solution manual. The variations found when the same person performs the calculation by different methods or when the same method is used by two different persons are typical. Probably greater care could improve the agreement somewhat. More data would improve the results by either method and might enable some judgment as to the quality of the data. The reader may wish to calculate his own values of dx/dt.

This example illustrates the difficulty of differentiation of data. Roughly an order of magnitude increase in error occurs when one differentiation is performed. Therefore, if first derivatives (i.e., rates of change and therefore process rates deduced therefrom) are to be satisfactorily accurate, very good integral data are needed. Imagine trying for a second derivative. Since the

TABLE 3-2   Hydrogen Bromide and Diethyl Ether In Aqueous Acid — Comparison of Calculated Rates of Change

t	x	C_E	$\left.\dfrac{dx}{dt}\right|_{K\text{ slope}}$	$\left.\dfrac{dx}{dt}\right|_{K\text{ e.a.}}$	$\left.\dfrac{dx}{dt}\right|_{C\text{ e.a.}}$
0	0	0.219	0.00445	0.00374	0.00402
21.0	0.065	0.154	0.00233	0.00249	0.00250
28.5	0.080	0.139	0.00210	0.00224	0.00216
51.2	0.124	0.095	0.00187	0.00165	0.00153

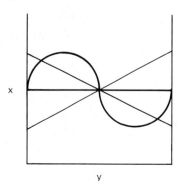

Figure 3-2 Integration of differential rate data.

underlying phenomenon, the chemical reaction, is a rate process, it is the rate of reaction that must be accurately known. This is a much more demanding requirement on the experimentor than the more obvious requirement of getting adequate integral data.

There is a corollary to the loss of accuracy inherent in differentiating data. It is that an order of magnitude decrease in error may be anticipated upon integration of any differential equation describing the reaction process. The impact of this statement is illustrated in Figure 3-2. The areas under all of these curves are identical and therefore the integrals represented by these areas are the same. Nevertheless, the underlying rate phenomena implied by the shapes of the curves are very different. If adequate data are obtained to describe properly the rate processes occurring, it can be expected that the overall (integral) result of the process can be accurately predicted. This is good news for scaleup, indeed. Unsatisfactory scaleup results most often from inadequate understanding of the underlying phenomena. Thus, poor rate data are the cause of many scaleup difficulties.

B. Thermodynamics

1. *Role and Limitations*

In a sense, thermodynamics is the study of situations in which nothing is happening, that is, equilibrium exists. The only possible purpose of commercial operation is to make something happen. Therefore, thermodynamics cannot tell you if you will succeed, but it can tell you if you have a chance or how to improve your chances. Thermodynamics is much more completely developed than kinetics since it is restricted to equilibrium conditions. Accordingly, most reactor scaleup problems should begin with a thorough analysis of the thermodynamics of the system. Powerful insights are available for the asking.

Thermodynamics impacts on chemical reaction engineering in three major ways. It may be used to quantify heat effects in reacting systems. It enables the calculation of the maximum conversion which may be obtained in a chemical reaction and the effects of feed composition, temperature, and pressure on that maximum conversion. Finally, it provides important guidance in the formulation of reaction rate equations. The thermodynamic analysis may be as simple as recognizing that the reaction is irreversible and neither produces nor consumes energy. In such a case, one can move directly to kinetic considerations. On the other hand, reversible and other simultaneous reactions complicate the picture greatly. The dominant feature of many reactor scaleups is the tremendous thermal effects that accompany reactions. It may be true that the more complicated the system becomes, the more important it is to conduct a thorough thermodynamic analysis. The methods and expected benefits of thermodynamic analysis are described in what follows. Ammonia synthesis will be taken as an example throughout much of this chapter because of its practical importance and since its analysis demonstrates the principles being discussed.

2. Heat Effects

In many reactor designs, the dominant issue is that the reactor be capable of transferring sufficient heat rather than being capable of achieving a certain degree of reaction. Many commercial reactors operate approximately adiabatically. That is, there is negligible heat exchanged with the surroundings compared to the huge flows of heat within the system. In such a case, the heat released or taken up by the reaction must be well defined so that the temperature history of the reacting fluid is known. Temperature runaways and reduced conversions relative to those possible at equilibrium are two problems that may arise in exothermic reactions. An endothermic reaction is capable of quenching itself. Even in a reactor that is to be operated isothermally, one needs to know how much heat must be exchanged with the surroundings in order to maintain isothermality.

The matter of calculation of heats of reaction is a part of all chemical engineering curricula and is covered in all texts on chemical engineering thermodynamics. An excellent example of such a text is by Smith and Van Ness (1975). Very few errors are made in the calculation of heats of reactions. Thus, only the results of such a familiar calculation will be provided here. Table 3-3 shows values for the heat of reaction for ammonia synthesis at two different temperatures. Several observations can be made with respect to Table

TABLE 3-3 Heat of Reaction in Ammonia Synthesis

$$N_2 + 3H_2 \rightleftharpoons 2NH_3$$

$\Delta H^\circ_{25°C} = -22{,}100$ cal/mol N_2 reacted ($-92{,}500$ J/mol)

$\Delta H^\circ_{538°C} = -25{,}800$ cal/mol N_2 reacted ($-108{,}000$ J/mol)

3-3. First, the negative enthalpy changes upon reaction imply that the products of the reaction contain less enthalpy than did the reactants. This means that heat must have been given off which is characteristic of an exothermic reaction. It is a common characteristic that reactions in which there is a decrease in the total number of moles are exothermic. Conversely, reactions in which the number of moles increases are usually endothermic, that is, energy must be added to the system in order to break the bonds. Cracking is a good example of an endothermic reaction. It will be seen that recognition of this relation between the change of the total number of moles and the exo- or endothermicity of the reaction can be a very valuable asset in anticipating reactor performance. Of course, a quantitative expression of the heat of reaction is

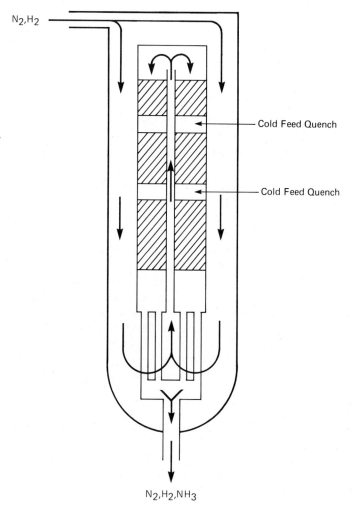

FIGURE 3-3 Reactor heat exchange.

FUNDAMENTAL CONSIDERATIONS

always important and can be achieved whether or not the reaction involves change in the total number of moles.

It should also be noted from Table 3-3 that even over a 500°C temperature range the heat of reaction varies only a little. This important simplification has significant ramifications as will be shown later. These very substantial heat effects mean that ammonia synthesis reactors must be designed with elaborate attention given to heat exchange. Figure 3-3 shows how this has been done in one example. Cold feed enters the shell of the reactor and is warmed in passing downflow outside of the catalyst beds. In the process the beds, where the exothermic reaction is occurring, are cooled. The feed then passes upflow, again in contact with the beds, accomplishing further heating of the feed and cooling of the beds. Finally, the feed passes downflow through the beds for the purpose of reaction. It will be noted that the earlier beds are smaller to preclude an excessive temperature buildup due to the higher rates of reaction occurring therein. Also, after each of the first two beds a cold feed quench is added to bring the temperature under control. Finally, the products depart the reactor.

3. Maximum Conversion

The area of chemical reaction thermodynamics in which errors are most likely to arise is in the calculation of chemical reaction equilibria. Regardless of the reactor type, size, or catalyst used, you can never do better than equilibrium and probably will not do as well. Thermodynamics gives a limit on what can be achieved. It also gives important clues toward optimal reaction conditions. A rather general development will be provided here highlighting the importance of careful attention to units and reaction stoichiometry. Then the principles will be illustrated for the case of ammonia synthesis.

Equation 3-6 gives a general reaction for our consideration.

$$aA + bB \rightleftharpoons cC + dD \tag{3-6}$$

For any reaction, Equation 3-7 gives the relationship between the thermodynamic equilibrium constant and the standard change of Gibbs free energy upon reaction.

$$\Delta F° = -RT \ln K_a \tag{3-7}$$

The origin of this equation is given in most physical chemistry and thermodynamics texts and is familiar to all chemical engineers. The equilibrium constant in this reaction is defined on the basis of thermodynamic activities as shown by the first equality in Equation 3-8.

$$K_a \equiv \frac{a_C^c a_D^d}{a_A^a a_B^b} = \frac{(\bar{f}_C/f_C°)^c (\bar{f}_D/f_D°)^d}{(\bar{f}_A/f_A°)^a (\bar{f}_B/f_B°)^b} \tag{3-8}$$

$\Delta F°$ is obtained from thermal data in a manner similar to the enthalpy change of reaction. It applies to the conversion of reactants in their standard states to products in their standard states. The standard states for reactants and products are the pure component at the temperature of interest and one atmosphere pressure. An important ramification of this standard state specification is that the standard free energy change may be a function of temperature but it will not be a function of total pressure. Therefore, Equation 3-7 shows that this equilibrium constant also will be independent of total pressure.

The activity of a component in solution is defined as the fugacity of that component in solution divided by the fugacity of that component in its standard state. This definition is used in expressing Equation 3-8 (the second equality) in terms of fugacities. The standard-state fugacities are conventionally set at one atmosphere, as shown by Equation 3-9,

$$f_i° \equiv 1 \text{ atm} \tag{3-9}$$

so that they can be dropped from Equation 3-8. Obviously, the remaining component fugacities also have units of atmospheres. Thus an equilibrium constant based on fugacities, K_f, results as indicated in Equation 3-10.

$$K_f = \frac{\bar{f}_C^c \bar{f}_D^d}{\bar{f}_A^a \bar{f}_B^b} [=] \text{ atm}^{(c+d)-(a+b)} \tag{3-10}$$

Clearly K_f will have the same numerical value as K_a. However, it should be realized that it will have units of pressure as indicated by Equation 3-10, whereas K_a will always be dimensionless.

Fugacity is still a difficult parameter to determine. For an ideal gas, the fugacity of a component is equal to the partial pressure of that component. If the gas is not ideal, then the component fugacity may be related to the partial pressure of the component by the fugacity coefficient as indicated in Equation 3-11.

$$\bar{f}_i = \phi_i P_i \tag{3-11}$$

The fugacity coefficients, ϕ_i, can be obtained from generalized fugacity coefficient charts available in many thermodynamics texts. Making the substitutions implied by Equation 3-11 into Equation 3-10 yields Equation 3-12

$$K_f = \frac{\phi_C^c \phi_D^d}{\phi_A^a \phi_B^b} \frac{P_C^c P_D^d}{P_A^a P_B^b} = K_\phi K_P \tag{3-12}$$

where K_ϕ and K_p are defined analogously to K_a and K_f. The partial pressure of a component is related to its mole fraction as indicated in Equation 3-13.

$$P_i = y_i P_T \tag{3-13}$$

FUNDAMENTAL CONSIDERATIONS

When this relation is inserted into the definition of K_p, Equation 3-14 results.

$$K_P = \frac{y_C^c y_D^d}{y_A^a y_B^b} P_T^{(c+d)-(a+b)} \tag{3-14}$$

Equation 3-14 finally arrives at a relationship among an equilibrium constant, the total pressure, and the composition of the mixture expressed in measurable terms of mole fractions. Combining Equation 3-12 and 3-14 gives the final result relating the variously defined equilibrium constants and the total pressure. At this point we recall that K_f has units of pressure but is numerically equal to K_a which was calculable from thermal data. Equation 3-15, like Equation 3-12, shows the relation between K_f and K_ϕ and K_p.

$$K_f = K_\phi K_P = K_\phi K_y P_T^{(c+d)-(a+b)} \tag{3-15}$$

It also gives an alternative expression encompassing K_ϕ, K_y, and the total pressure. Clearly, K_y is the term from which equilibrium compositions can be calculated as K_ϕ is the correction for nonideality of the gas.

A. Effect of Pressure

We observed that K_a, like $\Delta F°$, is independent of total pressure. Although K_f might have units of pressure (see Equation 3-10), it, too, is independent of total pressure by virtue of its numerical equality to K_a. The fugacity coefficients reflect the nonideality of the gas and accordingly may be a function of pressure as well as temperature. Because K_ϕ can be a function of pressure while K_f is independent of pressure, K_p can become a function of pressure due to the nonideality of the gases (see Equation 3-12).

If there is a change in the total number of moles upon reaction, then the exponent on P_T in Equation 3-15 is nonzero and this factor comes into play. It can be seen that for an ideal gas K_ϕ equals 1 and therefore K_f equals K_p. Also, for $K_\phi = 1$ we have a simple relationship between K_f, K_y, and the total pressure. If there is a change in the total number of moles upon reaction, the total pressure comes directly into play in Equation 3-15. However, because K_f is independent of pressure, K_y must display exactly the inverse characteristic to the total pressure term. Thus, if the total pressure increases, K_y must decrease correspondingly, and vice versa. Although we have seen that the equilibrium constant, K_a, is not a function of pressure, clearly the total pressure can have a tremendous influence on the equilibrium constant based on mole fractions, K_y. This is the source of the powerful influence of total pressure on the composition which can be achieved in chemical reaction systems.

From all of these equations and definitions it is clear that whenever an equilibrium constant is to be calculated, the exact stoichiometry of the reaction upon which it was based and the units of the concentration measures used in

TABLE 3-4 Free Energy Change in Ammonia Synthesis

$$N_2 + 3H_2 \rightleftharpoons 2NH_3$$

$$\Delta F°_{500°C} = 16800 \text{ cal/mol } N_2 \text{ reacted } (70{,}300 \text{ J/mol})$$

$$K_a = 1.74 \times 10^{-5}$$

---

its derivation must be supplied unambiguously. The failure of chemists and chemical engineers to recognize these inherent aspects in the definitions has been a major source of confusion and error in scaleup.

The effect of pressure on equilibrium composition will be illustrated for the case of ammonia synthesis. Table 3-4 shows the standard free energy change upon reaction at 500°C. This temperature is a common temperature for ammonia synthesis when catalyzed by conventional iron catalysts. The calculated value for K_a from Equation 3-7 is 1.74×10^{-5}. This extremely small equilibrium constant bodes poorly for the prospect of a commercial success in this reaction.

Table 3-5 shows the calculation of K_y from K_f at 1 atm. At this low pressure the ideal gas assumption is quite good. We see that for an ideal gas and a pressure of 1 atm, K_y is numerically equal to K_f and, of course, is dimensionless because the mole fraction is dimensionless. Once again, it is clear that operation at 1 atm looks very unattractive for ammonia synthesis.

The calculated value of K_y gives an indication of the extent of reaction at equilibrium. However, a more direct and easily visualized measure would be the percent conversion of one of the reacting species. In this case, nitrogen is chosen. A mole table is developed on the basis of a stoichiometric ratio of hydrogen to nitrogen in the feed. Table 3-6 develops the expressions for the mole fractions of the three components in the reaction. When the mole fractions from this table are substituted into the expression for K_y, the value of which was previously calculated at 1 atm, one obtains a 0.27 percent conversion of nitrogen. This confirms our expectation of an impractically low conversion.

TABLE 3-5 K_y at 1 Atmosphere

$$K_f = 1.74 \times 10^{-5} \text{ atm}^{-2} = \underbrace{K_\phi}_{\text{1 for ideal gas}} K_P$$

$$= \frac{P^2_{NH_3}}{P_{N_2} P^3_{H_2}} = \frac{y^2_{NH_3}}{y_{N_2} y^3_{H_2} P^2_T} = \frac{K_y}{P^2_T}$$

For $P_T = 1$ atm $K_y = 1.74 \times 10^{-5}$

FUNDAMENTAL CONSIDERATIONS

TABLE 3-6 Mole Table for Ammonia Synthesis

Basis: 4 mol feed ($H_2/N_2 = 3$)
   x = mol N_2 converted at equilibrium

Compound	N_{io}	N_i	y_i
N_2	1	$1 - x$	$(1 - x)/(4 - 2x)$
H_2	3	$3 - 3x$	$(3 - 3x)/(4 - 2x)$
NH_3	0	$2x$	$(2x)/(4 - 2x)$
Total	4	$4 - 2x$	1

When a pressure of 250 atm is inserted into Equation 3-14 one obtains a value of K_y equal to 1.09 as shown in Equation 3-16.

$$K_y = K_P P_T^2 = 1.74 \times 10^{-5} \text{ atm}^{-2} \times 250^2 \text{ atm}^2 = 1.09 \quad (3\text{-}16)$$

This value of K_y corresponds to a 35 percent conversion of N_2. This calculation was made assuming ideal gas behavior; however, at this pressure the gaseous components will not be ideal.

We can allow for nonideality by using the relationship $K_f = K_\phi K_p$. Table 3-7 shows the results obtained for K_ϕ from corresponding states theory. Allowing for nonideality, K_y can be calculated more accurately as indicated in Equation 3-17.

$$K_y = K_P P_T^2 = \frac{K_f P_T^2}{K_\phi}$$

$$= \frac{1.74 \times 10^{-5} \text{ atm}^{-2} \times 250^2 \text{ atm}^2}{0.59} = 1.84 \quad (3.17)$$

Substituting the mole fractions of the component species from the mole table into the expression for K_y as before indicates a 40 percent conversion of nitrogen. Most of this dramatic increase in conversion results from the application of high pressure. But it is interesting that the nonideality of the gas phase

TABLE 3-7 Fugacity Coefficient Correction for Nonideality[a]

$\phi_{N_2} = 1.01$	
$\phi_{H_2} = 1.14$	$K_\phi = \dfrac{0.94^2}{1.01 \times 1.14^3} = 0.59$
$\phi_{NH_3} = 0.94$	

[a] From Smith and Van Ness (1959).

TABLE 3-8 Effect of Pressure and Feed Composition on Conversion

T, °C	P_T, atm	H_2/N_2	% = 100x
500	1	3	0.27
500	250	3	40.0
500	250	4.5	52.5

yields a more favorable result from a practical point of view (40 percent) than would have been predicted assuming ideal gas behavior (35 percent).

The calculation can be repeated for a hydrogen to nitrogen mole ratio of 4.5. When this is done, the result is a 52.5 percent conversion of nitrogen. The results of these calculations are summarized in Table 3-8. All calculations are for a temperature of 500°C which is a typical operating temperature for an iron catalyst. The conversion at 1 atm is unthinkably low. The equilibrium conversion at 250 atm for a stoichiometric feed ratio is acceptable and explains why all ammonia synthesis processes operate at high pressure. There is a significant effect of increasing the hydrogen to nitrogen mole ratio. However, this effect is not exploited in practice because of the excessive cost of recompressing the recycled hydrogen.

B. Effect of Temperature

Thermodynamics has guided us to profound insights into the influence of pressure on chemical process performance. It is no less effective in indicating the potential influence of temperature. The starting point for analysis of the influence of temperature on equilibrium conversions is the van't Hoff equation, Equation 3-18.

$$\frac{d(\ln K_a)}{dT} = \frac{\Delta H°}{RT^2} \qquad (3-18)$$

From this equation one can deduce the effect of temperature on the equilibrium constant, K_a, for exothermic and endothermic reactions. We saw earlier that the enthalpy change for an exothermic reaction is negative. The resulting negative right-hand side of Equation 3-18 tells us that the value of the equilibrium constant must decrease as the temperature increases. This very direct result implies that increasing the temperature will reduce the maximum possible conversion in a reaction. The enthalpy change for an endothermic reaction is positive. The implication in this case is that as the temperature is increased the equilibrium constant also will be increased. Thus, increasing the temperature has a favorable influence on the maximum possible conversion for endothermic reactions.

One can also note that increasing the temperature almost invariably increases the rate of reaction. Therefore, for exothermic reactions, the selection

FUNDAMENTAL CONSIDERATIONS

of an operating temperature is a compromise between a fast approach to a low yield and a slow approach to a high yield. This compromise has tremendous influence on the selection of operating conditions. In a way, the situation for endothermic reactions is more favorable. Both rate and equilibrium are enhanced by high temperature. So the guideline is to operate as hot as possible. Of course, there are many caveats in the use of such a guideline. For example, materials of construction pose an ultimate limitation. Perhaps more significantly, as higher temperatures are used, undesired side reactions may occur. So once again, selection of an operating temperature requires consideration of many factors.

We saw in Table 3-3 that the enthalpy change on reaction was not a strong function of temperature. Thus, the assumption of constant $\Delta H°$ in Equation 3-18 enables the easy integration of this equation. The result is an expression for the effect of temperature on the equilibrium constant.

$$\ln \frac{K_{a_1}}{K_{a_2}} = \frac{\Delta H°}{R} \left(\frac{1}{T_2} - \frac{1}{T_1} \right) \tag{3-19}$$

One of the major advantages of thermodynamic analysis is the possibility of predicting an equilibrium constant entirely from thermodynamic data, that is, free energy data or enthalpy and entropy data. Equation 3-20 is an expression for the effect of temperature on the equilibrium constant

$$R \ln K = -\frac{\Delta H°}{T} + \Delta S° + \Delta a \left[\frac{T_0}{T} - 1 + \ln \frac{T}{T_0} \right]$$

$$+ \frac{\Delta b}{2} \left[\frac{(T - T_0)^2}{T} \right] + \frac{\Delta c}{6} \left[T^2 - 3T_0^2 + \frac{2T_0^3}{T} \right] \tag{3-20}$$

which is more rigorous than Equation 3-19 in that heat capacity effects have been allowed for (Kabel and Johanson, 1961) rather than assuming that the enthalpy change on reaction is temperature independent. Another advantage of Equation 3-20 is that it shows how the equilibrium constant can be predicted purely from thermal data. It often happens that no measurements of equilibrium compositions or constants are available and yet information on the magnitude of the equilibrium constant would be very desirable. Even in situations where thermal data are unavailable, they can be predicted by methods such as are described by Reid, Prausnitz, and Sherwood (1977).

Measured equilibrium constants are to be preferred whenever they are available. If good and complete thermodynamic data are available, one can expect to predict equilibrium constants which are accurate within an order of magnitude. If one must use estimation methods to obtain the thermal data, errors in the equilibrium constants of several orders of magnitude may occur. The situation is not as futile as it may seem. For example, if an equilibrium constant is predicted to be 10^{-14} and an error of six orders of magnitude is possible, one still knows with a certainty that the equilibrium conversion will be impracticably low. Likewise, if the predicted equilibrium constant is 10^{10} and the error is expected to be of several orders of magnitude, the approximation that the reaction is irreversible is still a good one. In any case, such *a priori* calculations represent an excellent guide to experimentation and correlation.

We have seen that ammonia synthesis is a strongly exothermic reaction. This means that much improved equilibrium constants could be achieved at lower temperatures. Indeed, the equilibrium constant at 25°C is 6.80×10^5. Clearly this equilibrium constant is higher than any of those calculated even at high pressures. The problem is that at such a low temperature the rate of reaction is impracticably slow. So what is needed is an improved catalyst such that the ammonia synthesis can be conducted at lower temperatures. The tremendous benefit to be gained by this would be the achievement of comparable conversions to those presently obtained, but at a much lower pressure. The lower pressures would result in vastly improved cost savings. In fact, commercial ammonia synthesis operations have been through this cycle of improved catalysts resulting in improved economics through reducing the cost of compression several times.

4. Role in Rate Equations

The final important role of thermodynamics is the guidance it provides in structuring the reaction rate equations. This issue will be taken up in detail when rate equations are discussed later in the chapter.

C. Physical Data

Reactor scaleup is much more than simply a matter of thermodynamics and kinetics. It is just as much a matter of fluid flow, heat transfer, and mass transfer. A whole host of such peripheral, but crucial, concerns are manifested in the reactor designer's need for a wide variety of physical data. For example, the material presented in Chapter 16 on materials of construction enters into almost every reactor scaleup. Some particularly prominent areas are highlighted below.

1. Phase Behavior

Even in the simplest case of single-phase, or homogeneous, reactors, phase behavior is an important consideration. Properties such as density, viscosity, thermal conductivity, diffusivity, heat capacity, and other less frequently used ones are often required. Chapter 11 on laminar flow processes is very much concerned with the property of viscosity. Non-Newtonian behavior can be important in reactor scaleup, for example, in polymerization reactors, but not if the viscosity is low.

FUNDAMENTAL CONSIDERATIONS

Property tables are often available in such reference sources as the *Chemical Engineers Handbook* (1973) and the *Handbook of Chemistry and Physics* (see Weast, ed., 1978). For familiar compounds such references may suffice, but the data are often included more on the basis of tradition than rigorous review and evaluation. Usually it is not enough to simply look up the needed physical properties in a handbook. Chemical reactors are inherently concerned with chemical mixtures. There are very few good predictive correlations for physical properties of mixtures. Similarly, chemical reactors often operate under changing and extreme conditions. Thus, data at 1 atm and 25°C will hardly suffice for use in a reaction being conducted at 250 atm and 600°C. For example, even a simple parameter as gas density requires sophisticated information on nonideal behavior.

More specialized tabulations exist as well. An excellent source of physical property information for hydrocarbons and hydrocarbon mixtures is the American Petroleum Institute's *Technical Data Book–Petroleum Refining*. A similar data book is being prepared for all chemical compounds under the auspices of the AIChE DIPPR Committee. These two references contain not only evaluated data, but also thoroughly tested prediction techniques. The selection of recommended data and prediction methods is described in supplementary documentation reports. An especially versatile source of estimation techniques for physical properties is the book by Reid, Prausnitz, and Sherwood (1977). They also provide a rather extensive tabulation of data and correlation constants. Reid et al. review the recent literature sufficiently that the reader achieves a strong sense of the potential uncertainty which exists in the prediction of any property.

If the problem is complex for the case of homogeneous reactors, it is ever so much more so for heterogeneous reactors. Not only do multiphase reactors involve more phases for which properties must be known, but they also involve interfaces with their own peculiar properties such as surface tension and vapor pressure. The above comments relate to those properties as well.

2. Solubility

Somewhat related to the matter of phase behavior is the more specific issue of solubility. This turns out to have major influence in some reactor scaleup situations. The issue will be illustrated by considering the removal of carbon dioxide from an acid gas stream by scrubbing with a hot potassium carbonate solution. The process is often thought of as gas absorption, but the absorber and stripper units are clearly reactors as well. The predominant chemical reaction is given by Equation 3-21,

$$CO_2 + H_2O + K_2CO_3 \rightleftharpoons 2KHCO_3 \qquad (3\text{-}21)$$

wherein carbon dioxide reacts with an aqueous solution of potassium carbonate to produce potassium bicarbonate. Not only is water consumed in this process, but more to the point, potassium bicarbonate is much less soluble than potassium carbonate. Thus, a precipitate tends to form as the reaction progresses toward completion. Although a solution as concentrated as possible in potassium carbonate would be capable of scrubbing the maximum amount of CO_2, the process is limited to initial carbonate solution concentrations below 40 wt.% to preclude potassium bicarbonate precipitation at high conversion [Riesenfeld and Kohl (1974)]. The details vary from one reaction system to the next, but such complications are a common fact of life.

3. Thermal Characteristics

It has been mentioned already that reactors often must function as heat exchangers as well. The heat exchange may simply be through the wall of a reactor, or it may be to or from a heat transfer device inserted in the reactor. Further, heat transfer between the vapor and liquid phases in an acid gas absorber or between solid and fluid phases in a fixed-bed reactor may also be of considerable importance. Such matters will be dealt with in Chapters 5 and 6 where it will be seen that various heat transfer coefficients will play major roles in addition to the roles already mentioned for heat capacity and thermal conductivity.

4. Hazards

Chemical reactors provide extreme hazard potential. Every reaction system presents a different situation and the hazards posed by each require thorough consideration. Unfortunately, it has not been possible in this book to deal with either fire and explosion hazards or the handling of toxic substances.

D. Analysis, Synthesis, and Data Handling

It sometimes helps to have a systematic way of looking at the nature of a project. Table 3-9 shows a tabular scheme of project characterization. The issue is the interrelation between the input and output of a process when only two of the three factors are known. The problem treated most commonly in textbooks and academic studies is analysis, that is, answering the question "what is the output when a given input is sent to a known process?." The synthesis, or design, problem is "what process can produce a specified output from a given input?." This problem is most often the one facing the scaleup engineer.

TABLE 3-9   Project Characterization

Input + process = ?	Analysis
Input + = output	Synthesis
? + process = output	Data handling

Synthesis receives some attention in academic programs, but much less than analysis. If you can see the output and are able to identify the process, then the question remains, "what was the input?." For want of a better name, this problem is called the data handling problem; data are available and you wonder what they mean. An example of such a problem is given by Graboski et al. (1982). Little is taught on this subject, yet it along with synthesis is the major concern of the scaleup engineer. Perhaps experience is the best teacher.

III. CORRELATION OF RATE INFORMATION

A. Communication and Data Taking

Almost all scaleups are based on previous experience. Some organizations make it a practice to assign a single person or group to develop a process all the way from the lab bench to commercial operation. In this way, the benefits of continuity are realized. Other organizations choose to exploit the unique talents of personnel with special training, by moving a project via individuals or teams working at different scales (e.g., bench, pilot plant, commercial). Such a scheme as this takes advantage of functional specialization of personnel but it produces a problem in communications.

Communication is a two-way street, and the importance of it cannot be overemphasized. Perhaps the most important thing is for a person involved in a later stage of scaleup to inform those at earlier stages what is needed. How often we hear statements like "I wish they had measured the temperature profile in the laboratory reactor" or "it would have helped if we had the weight of the catalyst at the end of the reaction as well as at the beginning." Even more basic than this is the need to specify the reaction conditions for which rate information is required.

In a recent experience, the author requested pH measurements to accompany an otherwise routine experiment. Accuracy to at least two decimal places was requested. This appeared to be no problem with the new digital pH meters. However, we failed to communicate that it was accuracy, more than precision, that was required. The necessary accuracy would require more elaborate calibration of the pH meters than is normally done. This was achieved easily enough once the prerequisite communication had taken place. In short, the scaleup engineer should at least have a look at the unit from which the data were obtained.

Communication often breaks down in the transmission of information from one investigator to a subsequent one. For example, chemists often work in units conventional to their discipline and, accordingly, do not always report them explicitly. The engineer collects information from many disciplines, each with its own units conventions. His deduction of the units may or may not be correct. An incorrect deduction can lead to a flawed scaleup. A case in which this lack of communication commonly occurs is in the reporting of chemical reaction equilibrium constants. As explained earlier, an equilibrium constant must have associated with it an equation giving the reaction stoichiometry. Further, depending on the measured variable, reaction stoichiometry, the phase in which the measurements were made, and other factors, the equilibrium constant reported may well have units. If these units are not clearly specified, rational use of the equilibrium constant becomes impossible.

B. Rate Equations

1. Definitions, Source, and Purpose

We saw earlier that the rate of reaction is a process rate and appears in conservation equations as a generation or consumption term. In our earlier examples, the reaction rate had units of moles of A reacted per second per cubic meter. Units of moles per unit time are sufficient to characterize the rate of generation or consumption of a species, but a size parameter is introduced into the definition to keep the rate from being dependent upon the size of the system. The size parameter is usually volume, in this case, cubic meters. When the reaction occurs at a surface, for example on the vessel wall, the area of the surface may be used as the size parameter. In solid-catalyzed fluid-phase reactions the mass of the catalyst is often chosen as an easily reproducible size parameter. It is becoming increasingly common to try to express catalytic reaction rates on a "per site" basis. The choice of the size parameter is arbitrary but crucial to effective scaleup. With the proper definition of reaction rate, as discussed earlier and with an appropriate size parameter specification, the rate of reaction is a function only of the reaction environment and not of the size or type of reactor. The superiority of a reaction rate so defined to the commonly used contact time (volume/volume/hr) will be demonstrated in Chapter 4.

Equation 3-22 represents the desired rate equation,

$$r = f(T, P_T, X, \text{catalyst}) \qquad (3\text{-}22)$$

showing the reaction rate to be a function of temperature, pressure, composition, and catalyst. The goal of much of kinetics research is to work out the functional form of such a rate equation. When this is done, the combination of the rate equation and conservation equations enables the scaleup of chemical reactors. Much of what follows is concerned with the determination of the appropriate functional form for the rate equation.

Research to determine such functional forms has continued for over a century. No attempt at exhaustive analysis of this subject will be made here. Physical chemistry texts and a number of books in chemical kinetics often provide elaborate treatments of the subject. Much can be learned from these sources; however, they have one major drawback. Almost without exception, their treatment is restricted to consideration of batch reactions. This is a severe restriction indeed when one considers that the goal of many scaleup projects is to design flow systems and that flow reactors are frequently used in the laboratory as well. Kabel (1981) discusses this restriction and its carryover into chemical reaction engineering texts. Nevertheless, the many fine chemical reactions engineering texts do treat successfully the analysis and design of flow reactors.

2. Homogeneous Reactions

The central idea of chemical kinetics is that reactions occur when the species to be reacted come together in the presence of the appropriate energetics. In an elementary reaction, the law of mass action asserts that the rate of reaction will be proportional to the concentrations of the reacting species, each raised to an exponent which is the stoichiometric coefficient for that species in the reaction equation. Thus, for the bimolecular reaction indicated by Equation 3-23, the rate equation might be written as indicated in Equation 3-24.

$$A + B \to C \tag{3-23}$$

$$r_A = k_C C_A C_B = k_P P_A P_B = k_y y_A y_B \tag{3-24}$$

Of course, observable reactions usually comprise many elementary steps. These elementary steps are seldom identifiable and therefore one can hardly hope to discern and exploit the true mechanism of a chemical reaction.

Nevertheless, we take guidance from the fundamental ideas of chemical kinetics in seeking rate equations. For example, even if Equation 3-23 represents an overall reaction comprised of many elementary steps, an equation of the form of Equation 3-24 might still correlate the rate data successfully. If not, other exponents on the concentration terms or other functional forms may prove to be suitable. It should be noted that three different concentration measures have been used in Equation 3-24. The choice is quite arbitrary; however, it should be recognized that the proportionality coefficients take their units from the defining equations. Recognizing that such equations can become almost totally empirical, one can become quite cynical about their meaning. Healthy skepticism is certainly warranted. However, successful scaleup is much more likely if one is continually seeking the most thorough chemical understanding possible.

Hill (1977) does a nice job in showing how to combine the rate equations for a collection of elementary steps into a single rate equation for an overall reaction. He makes use of the equilibrium and steady-state approximations. The same principles can be utilized to describe systems in which more reactions than one occur simultaneously. In this case, one simply writes the rate equation for each of the simultaneous reactions. The resulting collection of equations is solved simultaneously.

A special case of simultaneous equations is the reversible reaction. The synthesis of phosgene from carbon monoxide and chlorine is given as an illustration by Equation 3-25.

$$CO + Cl_2 \rightleftharpoons COCl_2 \qquad (3\text{-}25)$$

The rate of reaction of carbon monoxide is given by the first term on the right-hand side of Equation 3-26.

$$r_{CO} = k_f C_{CO} C_{Cl_2}^{3/2} - k_r C_{COCl_2} C_{Cl_2}^{1/2} \qquad (3\text{-}26)$$

The rate of formation of carbon monoxide by the reverse reaction is given by the last term of Equation 3-26. Therefore, the net rate of reaction of carbon monoxide is equal to its rate of reaction minus its rate of formation. The exponents in Equation 3-26 do not always correspond to the stoichiometric coefficients of Equation 3-25. Therefore, Equation 3-25 must not be an elementary reaction itself, but must consist of other elementary reactions.

Our analysis of the phosgene reaction can be carried further to illustrate the arduousness of a search for the reaction mechanism and a method of treatment of simultaneous reactions. Hill (1977) suggests the following alternative mechanisms for the gas phase reaction given by Equation 3-25.

Mechanism I

$$Cl_2 \underset{}{\overset{K_1}{\rightleftharpoons}} 2Cl \quad (\text{rapid}) \qquad (3\text{-}27)$$

$$Cl + CO \underset{}{\overset{K_2}{\rightleftharpoons}} COCl \quad (\text{rapid}) \qquad (3\text{-}28)$$

$$COCl + Cl_2 \xrightarrow{k_3} COCl_2 + Cl \quad (\text{slow}) \qquad (3\text{-}29)$$

$$Cl + COCl_2 \xrightarrow{k_4} COCl + Cl_2 \quad (\text{slow}) \qquad (3\text{-}30)$$

Mechanism II

$$Cl_2 \underset{}{\overset{K_1}{\rightleftharpoons}} 2Cl \quad (\text{rapid}) \qquad (3\text{-}31)$$

$$Cl + Cl_2 \underset{}{\overset{K_3}{\rightleftharpoons}} Cl_3 \quad (\text{rapid}) \qquad (3\text{-}32)$$

$$Cl_3 + CO \xrightarrow{k_5} COCl_2 + Cl \quad (\text{slow}) \qquad (3\text{-}33)$$

$$Cl + COCl_2 \xrightarrow{k_6} Cl_3 + CO \quad (\text{slow}) \qquad (3\text{-}34)$$

CORRELATION OF RATE INFORMATION

In each mechanism the first two steps are considered to be rapid. Thus, it is postulated that these steps are virtually at equilibrium and can be characterized by the equilibrium constants, K_1, K_2, and K_3. The last two steps in each mechanism are assumed to be slow and can be seen to be simply the reverse of each other. Obtaining an overall rate equation for the phosgene reaction is simply a matter of simultaneous solution of the equations quantifying each of the steps in the postulated mechanism.

For mechanism I the net rate of formation of phosgene is

$$r_{COCl_2} = k_3 C_{COCl} C_{Cl_2} - k_4 C_{Cl} C_{COCl_2} \tag{3-35}$$

The concentrations of the unmeasured intermediate species, COCl and Cl, can be obtained from equilibrium relationships as follows.

$$K_1 = \frac{C_{Cl}^2}{C_{Cl_2}}, \quad C_{Cl} = \sqrt{K_1 C_{Cl_2}} \tag{3-36}$$

$$K_2 = \frac{C_{COCl}}{C_{Cl} C_{CO}}, \quad C_{COCl} = K_2 C_{CO} C_{Cl} = K_2 C_{CO} \sqrt{K_1 C_{Cl_2}} \tag{3-37}$$

Combining Equations 3-35 to 3-37 gives

$$r_{COCl_2} = k_3 K_2 \sqrt{K_1}\, C_{CO} C_{Cl_2}^{3/2} - k_4 \sqrt{K_1}\, C_{COCl_2} C_{Cl_2}^{1/2} \tag{3-38}$$

A similar treatment for mechanism II gives

$$r_{COCl_2} = k_5 K_3 \sqrt{K_1}\, C_{CO} C_{Cl_2}^{3/2} - k_6 \sqrt{K_1}\, C_{COCl_2} C_{Cl_2}^{1/2} \tag{3-39}$$

Although some of the constants are different, the functional forms of Equations 3-38 and 3-39 are identical with each other and Equation 3-26. Thus, the alternative mechanisms are indistinguishable on the basis of kinetics studies involving only reactants and products. While a reaction mechanism is helpful in discerning the appropriate functional form of a rate equation, only the rate equation itself is required for scaleup.

Returning to our objective of ascertaining the functional form appropriate for Equation 3-22, we can now consider how the rate equations presented in Equations 3-24 and 3-26 relate to the process variables indicated in Equation 3-22. Through the law of mass action, the rate of reaction has been related to the concentrations of the reacting species. The major effect of the total pressure is to alter the concentration terms in the rate equation. This is especially significant for gas-phase reactions, and much less so for liquid-phase reactions. The effect of temperature is accounted for in the rate coefficient and will be discussed later in this chapter. The discussion in this section has been limited to homogeneous reactions and no mention has been made of catalysts. However, homogeneous catalysis is possible. If a fundamental enough mechanism can be written to describe the catalytic process, the concentration of the catalyst may appear in the rate equation or equations describing the process. More commonly, the effect of a catalyst is allowed for in the reaction rate coefficient.

3. Heterogeneous Reactions

Heterogeneous systems contain more than one phase and hence at least one interface. As in homogeneous reactions, it is still necessary for reacting molecules to come together. The difference is that the point of contact in a fluid–solid system is at the fluid–solid interface. In a fluid–fluid system, the reaction may take place at the interface, but it may also occur in one or the other of the phases after diffusional transport. These two cases will be taken up individually.

A. Fluid–Solid Systems

Fluid–solid reaction systems exist in staggering variety. Sometimes the solid is inert. Its purpose in such cases is usually the facilitation of heat and/or mass transfer, for example, in pebble heaters and packed absorption columns. Such matters will be dealt with elsewhere in the book. Often the solid is a reactant or product, for example, in the burning of coal and the calcination of limestone. Such processes may produce a second solid phase surrounding the reacting solid. Usually such processes are on such a tremendous scale that a whole technology has evolved around a single fluid–solid noncatalytic reaction. At the moment rate equations are almost wholly empirical and extensive pilot planting with modest scaleup ratios predominates in the commercialization of such reactions between fluids and solids. Gradually, and in some areas, empiricism is being supplanted with fundamental understanding and methods. A good reference is Carberry (1976).

Solid catalyzed heterogeneous reactions, for example, catalytic cracking and reforming of hydrocarbons, play a massive role in our technological society. The importance and diversity of application of such reactions is increasing at an astounding rate. At the same time, rapid advances are occurring in the study and understanding of catalytic kinetics. These advances are reflected in the successful development of rate equations for heterogeneous catalytic processes. A particularly comprehensive treatment is given by Carberry (1976). The key concepts will be first presented and then illustrated with an example.

The extensively developed and classical theory of homogeneous reaction rates has proved less than satisfactory in providing for heterogeneous systems the adequacy of rate data correlation and the degree of physical representation that is desirable. In particular, it seems clear that the concept of the law of mass action should be applied to the catalyst surface (where the reaction occurs) rather than to the fluid phase.

CORRELATION OF RATE INFORMATION

It is generally believed that catalysts function by providing a reaction path having a lower activation energy than the homogeneous reaction. For solid catalysts the energetic alterations are usually attributed to the adsorption of reacting species. To be more specific, it is chemisorption or activated adsorption, as compared to weaker physical adsorption, which can provide the requisite energy effects. Seven steps, which must occur in series, can be postulated for a catalytic reaction to be consummated. They are:

1. Transport of the reactant molecule through the bulk fluid phase to the catalyst particle.
2. Diffusion of the molecule within the particle to a catalyst "site."
3. Adsorption of the reactant on the catalyst surface.
4. Reaction of the adsorbed molecule alone, with other adsorbed molecules, or with molecules from the fluid phase on the catalyst surface to form an adsorbed product molecule.
5. Desorption of the product.
6. Diffusion of the product molecule out through the catalyst.
7. Transport of the product from the catalyst particle to the bulk fluid phase.

All of these steps occur at finite rates. At steady state the rates of all of the processes in series must be the same. In principle then, a quantitative expression for the rate of the overall process might be obtained by the resistances in series approach.

In practice, however, one or more of the steps may be negligible for physical reasons. For example, steps 2 and 6 would not be involved if the catalyst were nonporous. Also, some steps may be much faster than others. An example would be bulk transport in a vigorously agitated system. In fact, it is sometimes found that one individual step is much slower than all of the others. Then the rate of this slow step determines uniquely the rate of the overall process and this step is called the "rate determining step." Since all other steps are presumed to be rapid compared to the rate determining step, they are assumed to achieve equilibrium in approximating their effects on the overall process. This is obviously not rigorous but it serves its purpose well. From experience with heat transfer, one might expect to have to account for the resistances of more than one step to adequately describe the process. But the rates of the chemical processes of interest here may vary over a much wider range than the rates of those physical processes extant in the heat exchanger. More important, perhaps, is the reality that reaction rate data are seldom of adequate accuracy to justify the complexity resulting from consideration of multiple rate determining steps. In any case, the concept of a single rate determining step has proved fruitful indeed.

Mass transfer in fluid phases and pore diffusion are addressed in Chapter 5. Well-designed reactors and catalysts minimize the importance of these factors whenever possible. If they must be taken into consideration, techniques are available for doing so. For example, see Froment and Bischoff (1979). The treatment here will be limited to those processes taking place on the catalyst surface, steps 3, 4, and 5. Extensive use of the Langmuir theory of adsorption will be made to characterize the rates of adsorption and desorption processes and to relate surface concentrations to fluid phase concentrations. This approach will be recognized as the familiar "Langmuir–Hinshelwood model" or "Hougen and Watson rate equations." Such characterizations as these of catalytic reaction rates should be no better than the extremely limiting assumptions inherent in the Langmuir isotherm. They have proved to be surprisingly durable and useful, however [see Carberry (1976)].

For purposes of illustration, ammonia synthesis is chosen. It is known from a variety of studies on many catalysts that the adsorption of atomic nitrogen is the slow step for this reaction. The hydrogen and ammonia do not appear to be adsorbed. From the information given, the implied mechanism can be written

$$N_2 + 2s \rightleftharpoons 2N \cdot s \tag{3-40}$$

$$2N \cdot s + 3H_2 \rightleftharpoons 2NH_3 + 2s \tag{3-41}$$

Note that Equation 3-41 suggests fifth order behavior, which is mechanistically ridiculous. However, this step occurs after the rate determining step and may reflect the combined effect of many rapid elementary steps. Since the rate determining step is the adsorption of nitrogen,

$$r_{N_2} = k_{N_2} P_{N_2} \Theta_V^2 - k_{-N_2} \Theta_N^2 \tag{3-42}$$

Because it is inherently rapid, the surface reaction for ammonia formation, Equation 3-41, may be taken to be at equilibrium

$$r_s = 0 = k_s \Theta_N^2 P_{H_2}^3 - k_{-s} \Theta_V^2 P_{NH_3}^2$$

$$= k_s \left[\Theta_N^2 P_{H_2}^3 - \left(\frac{\Theta_V^2 P_{NH_3}^2}{K_s} \right) \right] \tag{3-43}$$

Hence,

$$\Theta_N^2 = \frac{\Theta_V^2 P_{NH_3}^2}{K_s P_{H_2}^3} \tag{3-44}$$

Combining Equations 3-42 and 3-44,

$$r_{N_2} = k_{N_2} \Theta_V^2 \left[P_{N_2} - \left(\frac{P_{NH_3}^2}{K_{N_2} K_s P_{H_2}^3} \right) \right] \tag{3-45}$$

CORRELATION OF RATE INFORMATION

At equilibrium

$$\left(\frac{P_{NH_3}^2}{P_{N_2}P_{H_2}^3}\right)_{eq} = K_{eq} = K_{N_2}K_s \quad (3\text{-}46)$$

Thus Equations 3-45 and 3-46 may be combined to give

$$r_{N_2} = k_{N_2}\Theta_V^2 \left[P_{N_2} - \left(\frac{P_{NH_3}^2}{K_{eq}P_{H_2}^3}\right) \right] \quad (3\text{-}47)$$

Since nitrogen is the only component adsorbed

$$1 = \Theta_V + \Theta_N = \Theta_V + \sqrt{\frac{\Theta_V^2 P_{NH_3}^2}{K_s P_{H_2}^3}} \quad (3\text{-}48)$$

But $K_s = K_{eq}/K_{N_2}$ from Equation 3-46, so

$$1 = \Theta_V + \left(\Theta_V \sqrt{\frac{K_{N_2}}{K_{eq}}} \frac{P_{NH_3}}{P_{H_2}^{3/2}}\right) \quad (3\text{-}49)$$

and

$$\Theta_V = \frac{1}{\left[1 + \left(P_{NH_3}/P_{H_2}^{3/2}\right)\sqrt{K_{N_2}/K_{eq}}\right]} \quad (3\text{-}50)$$

From Equations 3-47 and 3-50 we obtain the desired rate equation

$$r_{N_2} = \frac{k_{N_2}\left[P_{N_2} - \left(P_{NH_3}^2/K_{eq}P_{H_2}^3\right)\right]}{\left[1 + \left(P_{NH_3}/P_{H_2}^{3/2}\right)\sqrt{K_{N_2}/K_{eq}}\right]^2} \quad (3\text{-}51)$$

If Equation 3-51 is multiplied and divided by $P_{H_2}^3$, the result is

$$r_{N_2} = \frac{k_{N_2}\left[P_{N_2}P_{H_2}^3 - \left(P_{NH_3}^2/K_{eq}\right)\right]}{P_{H_2}^3\left[1 + \left(P_{NH_3}/P_{H_2}^{3/2}\right)\sqrt{K_{N_2}/K_{eq}}\right]^2} \quad (3\text{-}52)$$

The numerator of this equation is of the familiar form often found in homogeneous kinetics. However, the concentration dependence exhibited by the denominator results from the use of the Langmuir theory to relate the measurable fluid-phase concentrations to the adsorbed-phase concentrations, which determine the rate of the catalytic reaction.

This Langmuir–Hinshelwood or Hougen–Watson (LHHW) rate equation shown in Equations 3-51 and 3-52 satisfactorily correlates data on ammonia synthesis. It has often been observed (Weller, 1956) that power function (homogeneous) type rate equations correlate catalytic reaction rate data as well, and more simply than, the LHHW type derived here. Boudart (1956) advanced arguments supporting the rational use of the LHHW approach. His demonstration that, in fact, the two approaches may often be mathematically equivalent is indicated below.

To obtain the power function form of rate equation, the forward and reverse terms of Equation 3-51 are separated.

$$r_{N_2} = \frac{k_{N_2} P_{N_2}}{\left[1 + \sqrt{K_{N_2}/K_{eq}}\left(P_{NH_3}/P_{H_2}^{3/2}\right)\right]^2} - \frac{(k_{N_2}/K_{eq})(P_{NH_3}^2/P_{H_2}^3)}{\left[1 + \sqrt{K_{N_2}/K_{eq}}\left(P_{NH_3}/P_{H_2}^{3/2}\right)\right]^2}$$

$$= \frac{k_{N_2} P_{N_2}}{[1 + bx]^2} - \frac{b^2 x^2}{[1 + bx]^2} \qquad (3\text{-}53)$$

where

$$b = \sqrt{\frac{k_{N_2}}{K_{eq}}} \quad \text{and} \quad x = \frac{P_{NH_3}}{P_{H_2}^{3/2}}$$

From the mathematical equivalency, Cx^n ($0 < n < 1$) $\approx bx/(1 + bx)$, one can write $1/(1 + bx)^2 = C^2(x^{n_1}/bx)^2$ and $b^2 x^2/(1 + bx)^2 = C^2(x^{n_2})^2$ where $0 < n_1 < 1$ and $0 < n_2 < 1$. Thus

$$r_{N_2} = k_{N_2} P_{N_2} C^2 \left(\frac{x^{n_1}}{bx}\right)^2 - C^2(x^{n_2})^2$$

$$= k_{N_2} P_{N_2} C^2 \left(\frac{x^{n_1 - 1}}{b}\right)^2 - C^2(x^{n_2})^2$$

$$= \frac{k_{N_2} C^2 P_{N_2} x^{2(n_1 - 1)}}{b^2} - C^2 x^{2n_2} \qquad (3\text{-}54)$$

Substituting in for x and b gives

$$r_{N_2} = \frac{k_{N_2} C^2 P_{N_2} \left(P_{NH_3}/P_{H_2}^{3/2}\right)^{2(n_1 - 1)}}{(K_{N_2}/K_{eq})} - C^2 \left(\frac{P_{NH_3}}{P_{H_2}^{3/2}}\right)^{2n_2} \qquad (3\text{-}55)$$

CORRELATION OF RATE INFORMATION

Squaring the appropriate terms and recalling that n_1 is a positive fraction,

$$r_{N_2} = \frac{C^2 k_{N_2} K_{eq} P_{N_2}}{K_{N_2}} \left(\frac{P_{H_2}^3}{P_{NH_3}^2} \right)^{1-n_1} - C^2 \left(\frac{P_{NH_3}^2}{P_{H_2}^3} \right)^{n_2} \quad (3\text{-}56)$$

This equation is identical in form to the Temkin–Pyzhev (1940) equation

$$r_s = k_s P_{N_2} \left(\frac{P_{N_2}^3}{P_{NH_3}^2} \right)^m - k_d \left(\frac{P_{NH_3}^2}{P_{H_2}^3} \right)^n \quad (3\text{-}57)$$

with

$$r_{N_2} = r_s, \quad \frac{C^2 k_{N_2} K_{eq}}{K_{N_2}} = k_s, \quad 1 - n_1 = m, \quad C^2 = k_d, \quad \text{and} \quad n_2 = n$$

The determination of a power function form of rate equation is often done because of the familiarity and sheer convenience of doing so. As shown above, the result is often satisfactory. This may be all that is required for a successful scaleup. However, such equations provide very little insight into the reaction process. It often occurs that greater insight is invaluable and the development of a LHHW model can provide that.

The deduction of the rate controlling step is often a problem with the usual steady-state data. Hsu and Kabel (1974) have shown how dynamic (unsteady-state) experiments in a batch heterogeneous catalytic reactor can lead to direct indications of the rate controlling step. Perti and Kabel (1982) have carried this idea much farther in using transient measurements in a fixed-bed reactor to establish a plausible mechanistic interpretation for all important reaction steps as well as to ascertain the degree of control that the various steps exert under different reaction conditions. Although unsteady-state experiments are more difficult to conduct than their steady-state counterparts, they need not be so well done in order to yield far more information. It is strongly urged that careful attention is paid to transient phenomena rather than concentrating solely on conventional steady-state measurements. In addition to the benefits indicated above, rewards will be found during startup, changes in operating conditions, and shutdown of the intended commercial installations.

Some cautionary notes should be included. LHHW equations have many constants and therefore may be expected to be successful in correlating data regardless of their theoretical validity. Thus disciplined use of them is vital. A successful correlation should not be considered to be proof of the assumed reaction mechanism. As with homogeneous reactions, different mechanisms may lead to identical functional forms. Here again the dynamic experiments suggested above can advance understanding and protect against inadvertent overinterpretation.

It can be seen that the power function and LHHW equations express the rate of the solid catalyzed reaction in terms of the fluid-phase concentrations. The importance of this will be shown in Chapter 5 where it will be seen that reactor models of the pseudohomogeneous type (see Chapter 4) can be used for heterogeneous reactions.

B. Fluid–Fluid Systems

Many examples of fluid–fluid reactions exist, but the vessels in which they occur are often not thought of as reactors. Both gas absorption and liquid extraction processes are often accompanied by a reaction in one of the phases. An example is the carbon dioxide scrubbing by potassium carbonate solutions referred to earlier. The reactive nature of such a system is even more profound than implied by the single reaction of Equation 3-21. While even that reaction is actually a composite of several rate and equilibrium processes, most acid gas removal processes also involve an amine as a promoter or catalyst. The amine is not consumed in the process, but it has a very definite chemical role to play. Similarly, liquid extraction systems may involve only simple molecular phenomena, but they often involve reactions as well.

Such processes have historically been designed as absorption with chemical reaction or liquid extraction with chemical reaction. As such, the designs are usually based upon the equilibrium stage concept discussed in Chapter 12 or upon the mass transfer coefficient approach to continuous separation processes described in Chapter 13. These approaches have to be modified to allow for the effects of chemical reaction.

Other fluid–fluid reactions are viewed as clear-cut chemical reactor problems. An example would be hydrogenation of unsaturated oils. It would be possible for such a reaction to occur in the vapor phase between a gas and a volatile liquid. In this case, the rate of evaporation of the liquid might be as important a factor as the rate of reaction in the gas phase. Alternatively, the reaction might occur in the liquid phase between absorbed gas and a nonvolatile liquid whose presence is restricted to the liquid phase. Even if the reaction were inherently very fast, the rate at which it could occur might be limited by the rate of diffusion of the dissolved gas into the liquid medium. Often the solubility of the gas in the liquid plays a role as well.

From the foregoing discussion, it should be clear that the overall rate of a fluid–fluid reaction will be influenced not only by the rate of the chemical reaction, but also by the rates and equilibria involved in accompanying mass transfer processes. The rate of reaction occurring in either fluid phase can be correlated by conventional homogeneous rate equations. Interaction between the diffusion and reaction phenomena determines where in a given phase the reaction occurs. Astarita (1967) and Danckwerts (1970) published early books on this subject. Probably the most thorough treatment in conventional chemical reaction engineering texts today is by Carberry (1976). A major revision of Astarita's book with a strong industrial input is now available [Astarita, Savage and Bisio (1983)]. This subject will be considered in greater detail in Chapter 6. The area is very much a research topic and the practitioner interested in scaleup of a fluid–fluid reactor should not expect to find a cookbook.

For example, some interesting things occur when simultaneous reactions occur in a two-fluid-phase system. Return to the acid gas absorption case where CO_2 reacted with K_2CO_3 and water to form $KHCO_3$, Equation 3-21. If the acid gas also contains H_2S (a common situation), the H_2S is taken up in the absorber through the reaction given in Equation 3-58.

$$H_2S + K_2CO_3 \rightleftarrows KHCO_3 + KHS \qquad (3\text{-}58)$$

In normal operation the acid gas is absorbed in a solvent (the potassium carbonate solution), which is then regenerated by steam stripping. For a single acid gas, the reaction given by Equation 3-21 or 3-58 is simply driven in the forward or reverse direction in the absorber or regenerator, respectively. Increasing the steam rate in the regenerator decreases the amount of the acid gas in the solvent.

When both acid gases are present, increasing the steam rate results in a decrease in the CO_2 held by the stripped solvent but an increase in the amount of H_2S. This counterintuitive result can be understood by qualitative reasoning. The reactions involving H_2S are proton transfer reactions and are thus much faster than the reactions involving CO_2. Thus Equation 3-58 will always be near equilibrium while Equation 3-21 will be rate controlling. The stripping of H_2S depends on the presence of $KHCO_3$ to react with KHS. The stripping of CO_2 depletes the solution of $KHCO_3$ and enriches it in K_2CO_3. Thus, as the CO_2 is stripped out in the regenerator, the equilibrium in Equation 3-58 shifts from H_2S toward KHS. The result of this is that the absorber is less effective than might be expected in removing H_2S. This can be a real problem because the cleanup specifications for H_2S are usually much more severe than for CO_2.

For ease of understanding, the example above was kept qualitative. For further (and more quantitative) insight into this situation, the paper by Astarita and Gioia (1965) is recommended for study. The paper deals with the simultaneous absorption of H_2S and CO_2 into aqueous hydroxide solutions and is remarkable for its clarity.

4. *The Arrhenius Equation*

It was mentioned earlier that the effect of temperature on chemical reactions was accounted for by the reaction rate coefficients in the rate equations. Other influences of temperature come into play with equilibrium constants which may also appear in rate equations. Equations 3-18 and 3-19 showed how equilibrium constants were related to temperature. An interpretation of the van't Hoff equation for elementary reactions leads to the conclusion that the reaction rate coefficients should be correlated by an exponential function similar to that describing the equilibrium constants. It should be noted that the interpretation of this function is quite different in the case of a rate constant than in the case of an equilibrium constant (e.g., contrast the activation energy and the heat of reaction). Equation 3-59 shows the familiar Arrhenius equation.

$$k = Ae^{-E/RT} \qquad (3\text{-}59)$$

This equation has proved, over many decades, to be amazingly successful in correlating the effect of temperature on rates of reaction. As with equilibrium constants, a plot of the log of the rate coefficient against reciprocal of the absolute temperature can be expected to produce a linear or nearly linear graph over most temperature ranges of practical interest.

The success of Equation 3-59 in correlating reaction rate coefficients with temperature has led to many attempts to interpret its parameters, the frequency factor and the activation energy. Unfortunately, it is still not possible to predict these parameters with any certainty. Further, complex reactions or influences of mass transfer can falsify the usual interpretations [e.g., see Froment and Bischoff (1979)].

5. Thermodynamic Consistency

Thermodynamics has one more very important role to play in the scaleup of chemical reactors. It suggest to us the form that rate equations must take in the limiting case of a reaction which has gone to equilibrium. That is, if a rate equation is to be successful over the entire possible conversion range, then it must be thermodynamically consistent. Thermodynamic consistency can be reliably expected only for elementary reactions. As mentioned earlier, we will seldom be dealing with elementary reactions. Accordingly, it may be difficult, or even impossible, to obtain a rate equation which is thermodynamically consistent for the reaction of interest to us. However, in a great many instances, thermodynamic consistency is achievable and therefore it is a goal which should be sought and against which every rate equation should be tested.

An example of thermodynamic consistency is given for the case of the phosgene reaction, Equation 3-25, where the rate equation, Equation 3-26, showed that the reaction was not an elementary one. The thermodynamic consistency of Equation 3-26 can be tested. At equilibrium, the reaction rate, r_{CO}, will be zero and the two terms on the right-hand side can be equated. When this equation is solved for the ratio of the forward to the reverse reaction rate coefficients, it can be seen that the resulting ratio of concentrations is by definition equal to the equilibrium constant according to Equation 3-25. This result is shown in Equation 3-60

$$\frac{k_f}{k_r} = K_C = \frac{C_{COCl_2}}{C_{CO}C_{Cl_2}} \qquad (3\text{-}60)$$

CORRELATION OF RATE INFORMATION

and is a demonstration of the thermodynamic consistency of the rate equation, Equation 3-26. Knowing that $k_f/k_r = K_C$, one can eliminate k_r from Equation 3-26. The result is shown in familiar form as Equation 3-61.

$$r_{CO} = k_f \left(C_{CO} C_{Cl_2}^{3/2} - \frac{C_{COCl_2} C_{Cl_2}^{1/2}}{K_C} \right) \tag{3-61}$$

For such a thermodynamically consistent rate equation, it is only necessary to obtain one rate coefficient from kinetic data. The equilibrium constant can be obtained from thermodynamic considerations. Sometimes K_a or K_C can be predicted without recourse to experiment. Even when experiments are necessary, equilibrium experiments are often much easier to conduct than rate experiments. Hawes and Kabel (1968) describe an especially convenient method. Finally, a proof of thermodynamic consistency increases one's confidence in the validity and utility of a rate equation.

As another example of thermodynamic consistency, consider the Temkhin–Pyzhev rate equation for iron-catalyzed ammonia synthesis. Their derived equation is given as Equation 3-57 wherein k_s, k_d, m, and n remain to be determined experimentally.

Equation 3-57 can be tested for thermodynamic consistency as follows. At equilibrium $r_s = 0$. Equating the two terms on the right-hand side gives Equation 3-62.

$$k_s P_{N_2} \left(\frac{P_{H_2}^3}{P_{NH_3}^2} \right)^m = k_d \left(\frac{P_{NH_3}^2}{P_{H_2}^3} \right)^n \tag{3-62}$$

Solving for the ratio of the forward and reverse reaction rate coefficients gives Equation 3-63.

$$\frac{k_s}{k_d} = \frac{P_{NH_3}^{2n} P_{NH_3}^{2m}}{P_{N_2} P_{H_2}^{3m} P_{H_2}^{3n}} = \frac{P_{NH_3}^{2(n+m)}}{P_{N_2} P_{H_2}^{3(n+m)}} \tag{3-63}$$

For the ammonia synthesis stoichiometry as shown in Table 3-4, the equilibrium constant in terms of partial pressures is given by Equation 3-64.

$$K_P = \frac{P_{NH_3}^2}{P_{N_2} P_{H_2}^3} \tag{3-64}$$

Comparing Equations 3-63 and 3-64, it can be seen that if $(m+n) = 1$, then $k_s/k_d = K_p$. Experimentally, it was found that $m = n = \frac{1}{2}$, which proves the thermodynamic consistency of the Temkhin–Pyzhev rate equation. As a sidelight, some have called ammonia synthesis the most successful scaleup in history.

IV. UNCERTAINTIES

The chapters on actual reactor scaleup are yet to come. Nevertheless, there are many uncertainties involved in simply getting the basic information necessary for reactor scaleup. Many of these have been touched on in the preceding paragraphs. Others follow.

A. Definition of the Reaction System

We have seen in earlier sections that a rigorous mass balance is crucial in the determination of accurate rate data. Thus, the reaction system in which an investigation is being carried out must be well defined. This brings us back to the matter of communication and data taking. Not only are very accurate data needed, but the circumstances of their attainment must be well known, hence, the need for good communication. Very often rate data are obtained in ill-defined systems. Commercial installations almost always fall into this category, but they need not and some would not if the value of the data potentially available were fully appreciated. This is especially true in the case of smaller scale chemical manufacturers, who often work in batch and semibatch equipment. The data from such apparatus are often used in scaleup. And yet how well known are the mixing and thermal characteristics of such systems? Often it would be possible, at least for a limited time, to perform very controlled experiments on such apparatus without losing production. These controlled experiments could form the basis for much more confident scaleup. Further, knowledge gained might enable operators to improve the performance of existing equipment without significant cost.

Proceeding to a still smaller scale of reactor, there are questions that can be asked of laboratory reactors as well. For example, what is the state of mixedness in a paddle-stirred, three-neck flask? Does the temperature measured by the thermometer inserted into the side arm represent the temperature of the reaction environment? Did plug flow exist in the flow tube packed with a small amount of catalyst? Was there bypassing? Without satisfactory answers to questions like these, scaleup is certainly an uncertain business.

B. Data Base for Scaleup

1. Extent of the Data Base

There is always a question of when sufficient data have been obtained upon which to base a scaleup. Unfortunately, there is no universal answer to this question. For some conventional operations (homogeneous vapor-phase reaction in a tube, dilute aqueous-phase reaction in a stirred tank), enough information exists in the literature to proceed almost immediately to full-scale design. For example, Liederman et al. (1980) report a fixed-bed process for converting methanol to gasoline which is ready to grow from a 4 barrel/day demonstration unit directly to commercial scale. More commonly, rate and equilibrium studies at several levels must be carried out before one can confidently proceed on to the next level. Neither endless trial-and-error nor interminable prescaleup investigation can be tolerated. A decision based on economics will be the ultimate guide as to when the data base is extensive enough.

2. Quality of the Data Base

The extent of the data base can be reduced considerably if the data are of high enough quality. Very often the role of quality is not fully appreciated. An engineer has a particular goal, and he wants to do the quickest experiment he can to answer what he perceives to be the essential question. Unfortunately, if the question is not well posed, or the experiment is only roughly controlled, the engineer will get only a partial answer. New experiments may be necessary and the original experiment may contribute little to the ultimate data base. Therefore, quality of experimentation, and especially quality of thought, are crucial to reducing the necessary extent of a scaleup data base and in reducing the uncertainty in scaleup.

C. Appearance of New Factors

The whole point of all of the foregoing material in this chapter is to obtain a predictable result; that is, a scaled up system that works the way you want it to. This demands alertness at all phases of the project to indications of potential problems. The earlier problems are faced, the less wasted commitment there will be to hopeless approaches. Nevertheless, as a project proceeds to larger scales, longer durations of operation, less pure raw materials, economical materials of construction, and other factors too numerous to mention, new problems will appear. This is a fact of life in scaleup. Nevertheless, if you have done your job well, fewer unforeseen problems will arise and those that do will be more easily solved because of your more thorough understanding of the system.

NOMENCLATURE

a, b, c	Coefficients in heat capacity correlation $C_p = a + bT + cT^2$
a_i	Activity of component i, dimensionless
A	Frequency factor, same units as k
C_i	Concentration of component i, mol/m^3
C_p	Molar heat capacity at constant pressure, J/mol K
E	Activation energy, J/mol

f_i°	Standard state fugacity of component $i \equiv 1$ atm
$\bar{f}_i$	Fugacity of component i in solution, atm
ΔF°	Standard free energy of reaction, J/mol
F_A	Feed rate of A, mol/s
F_i	Feed rate, mol A in/s
F_o	Outflow rate, mol A out/s
ΔH°	Standard heat of reaction, J/mol
k	General reaction rate coefficient, units from defining rate equation
k_C	Reaction rate coefficient based on concentrations
k_f, k_i, k_s	Reaction rate coefficient for forward reaction
k_P	Reaction rate coefficient based on partial pressures
k_r, k_{-i}, k_d	Reaction rate coefficient for reverse reaction
k_y	Reaction rate coefficient based on mole fractions
K_a	Equilibrium constant based on activities
K_C	Equilibrium constant based on concentrations
K_{eq}	Equilibrium constant (Equation 3-46)
K_f	Equilibrium constant based on fugacities
K_{N_2}	Adsorption equilibrium constant $= k_{N_2}/k_{-N_2}$ (Equation 3-42)
K_P	Equilibrium constant based on partial pressures
K_s	Equilibrium constant for surface reaction (see Equation 3-43)
K_y	Equilibrium constant based on mole fractions
K_ϕ	Equilibrium constant based on fugacity coefficients
m, n	Exponents in Temkin–Pyzhev rate equation (Equation 3-57)
N_A	mol A present, mol
P_i	Partial pressure of component i, atm
P_T	Total pressure, atm
r_A	Rate of reaction of A, mol A reacted/m³ s
R	Gas constant $= 1.987$ cal/mol K $= 8.314$ J/mol K
s	A catalyst site
ΔS°	Standard entropy of reaction, J/mol K
t	Time, s
T	Temperature, K
T_0	Standard-state temperature, 298.2 K
V	Reactor volume, m³
X	Fractional conversion, mol A converted/mol A fed
y_i	Mole fraction of component i, dimensionless

Greek

Θ_i Fraction of sites occupied by component i, dimensionless
Θ_v Fraction of sites that are vacant, dimensionless
ϕ_i Fugacity coefficient of component i, dimensionless

REFERENCES

Astarita, G., *Mass Transfer with Chemical Reaction*, Elsevier, New York, 1967.

Astarita, G., and Gioia, F., "Simultaneous Absorption of Hydrogen Sulfide and Carbon Dioxide in Aqueous Hydroxide Solutions," *Ind. Eng. Chem. Fundam.*, **4**, 317–320 (1965).

Astarita, G., Savage, D. W., and Bisio, A., *Gas Treating with Chemical Solvents*, Wiley, New York, 1983.

Boudart, M., "Kinetics on Ideal and Real Surfaces," *AIChE J.*, **2**, 62–64 (1956).

Carberry, J. J., *Chemical and Catalytic Reaction Engineering*, McGraw-Hill, New York, 1976.

Churchill, S. W., *The Interpretation and Use of Rate Data: The Rate Concept*, McGraw-Hill, New York, 1974.

Danckwerts, P. V., *Gas Liquid Reactions*, McGraw-Hill, New York, 1970.

Dixon, D. C., "The Definition of Reaction Rate," *Chem. Eng. Sci.*, **25**, 337–338 (1970).

Froment, G. F., and Bischoff, K. B., *Chemical Reactor Analysis and Design*, Wiley, New York, 1979.

Graboski, M. S., Kabel, R. L., Danner, R. P., and Al-Amelri, R. S., "Process Input Analysis," *Chem. Eng. Commun.*, **17**, 137–149, 1982.

Hawes, R. W., and Kabel, R. L., "Thermodynamic Equilibrium in the Vapor Phase Esterification of Acetic Acid with Ethanol," *AIChE J.*, **14**, 606–611 (1968).

Hill, C. G., Jr., *An Introduction to Chemical Engineering Kinetics and Reactor Design*, Wiley, New York, 1977.

Hsu, S. M., and Kabel, R. L., "Adsorption and Kinetics in a Batch Heterogeneous Catalytic Reactor," *AIChE J.*, **20**, 713–720 (1974).

Kabel, R. L., "Rates," *Chem. Eng. Commun.*, **9**, 15–17 (1981).

Kabel, R. L., and Johanson, L. N., "Thermodynamic Equilibrium in the Ethyl Alcohol-Ethyl Ether-Water System," *J. Chem. Eng. Data*, **6**, 496–498 (1961).

Liederman, D., Yurchak, S., Kuo, J. C. W., and Lee, W., "Mobil Methanol-to-Gasoline Process," 15th Intersociety Energy Conversion Engineering Conference, Seattle, WA (August 18–22, 1980).

Mayo, F. R., Hardy, W. B., and Shultz, C. G., "The Cleavage of Diethyl Ether by Hydrogen Bromide," *J. Am. Chem. Soc.*, **63**, 426–436 (1941).

Perry, R. H., and Chilton, C. H., eds., *Chemical Engineers' Handbook*, 5th ed., McGraw-Hill, New York, 1973.

Perti, D., and Kabel, R. L., "Dynamic Discernment of Catalytic Kinetics," ACS Symposium Series, No. 196, Chemical Reaction Engineering—Boston, American Chemical Society, Washington, D.C., 271–282 (1982).

Reid, R. C., Prausnitz, J. M., and Sherwood, T. K., *The Properties of Gases and Liquids*, 3rd ed., McGraw-Hill, New York, 1977

Riesenfeld, F. C., and Kohl, A. L., *Gas Purification*, 2nd ed., Gulf Publishing Company, Houston, 1974.

Smith, J. M., and Van Ness, H. C., *Introduction to Chemical Engineering Thermodynamics*, 2nd ed., McGraw-Hill, New York, 1959.

Smith, J. M., and Van Ness, H. C., *Introduction to Chemical Engineering Thermodynamics*, 3rd ed., McGraw-Hill, New York, 1975.

Technical Data Book—Petroleum Refining, American Petroleum Institute, New York.

Temkin, M. I., and Pyzhev, V., *Acta Physicochim. U.R.S.S.*, **12**, 327 (1940).

Walas, S. M., *Reaction Kinetics for Chemical Engineers*, McGraw-Hill, New York, 1959.

Weast, R. C., ed., *CRC Handbook of Chemistry and Physics*, 59th ed., CRC Press, West Palm Beach, FL, 1978.

Weller, S., "Analysis of Kinetic Data for Heterogeneous Reactions," *AIChE J.*, **2**, 59–62 (1956).

4

HOMOGENEOUS REACTION SYSTEMS

R. L. KABEL

I.	Major Issues in Homogeneous Reactions			118
II.	Mass Balances			120
	A.	General Equations of Change		120
	B.	By Reactor Type		120
		1.	Batch	120
		2.	Tubular Plug Flow	124
		3.	Tubular Laminar Flow	128
		4.	Continuous Stirred Tank	130
		5.	Semibatch	133
III.	Energy Balances			134
	A.	Nature of the Balances		134
		1.	Isothermal	134
		2.	Adiabatic	135
		3.	Other	135
	B.	By Reactor Type		136
		1.	Batch	136
		2.	Tubular	138
		3.	Continuous Stirred Tank	138
		4.	Semibatch	142
IV.	An Example of Homogeneous Reactor Scaleup			142
	A.	Kinetics		143
	B.	Mass Balances and Isothermal Performance		149
	C.	Energy Balance and Thermal Effects		156
	D.	Size, Shape, and Performance		161

V. Uncertainties 162
   A. Nonideal Flow 162
   B. Heat and Mass Transfer Effects 163
   C. Competing Reactions and Impurities 163
   D. Reactor Instability 164
Nomenclature 164
References 166

I. MAJOR ISSUES IN HOMOGENEOUS REACTORS

Scaleup of reactors in a chemical process involves two, somewhat interdependent, aspects. One, the selection of the type of reactor to be used, is treated at length in Chapter 7. The other is the actual design of the selected type of reactor. This chapter deals with the design of homogeneous reactors and provides also the conceptual bases for the heterogeneous reactors, whose scaleup is described in subsequent chapters. For example, pseudohomogeneous reactor models are used extensively in Chapter 5 which is almost wholly concerned with fixed-bed catalytic reactors.

Figure 4-1 illustrates some of the reactor possibilities which may be a part of scaleup decisions. The processes shown in this figure progress from batch to continuous as one moves from left to right. To some extent this progression corresponds to frequent practice in scaleup; that is, there is often a substantial economic advantage in continuous processing for large-scale production. Such a generalization may be misleading, however, as the choice of reactor type may be much more dependent upon the peculiarities of a particular process. For example, consider the giant batch reactors involved both in brewing and penicillin production.

Batch reactors are inherently unsteady-state devices; that is, conditions within the reactor change with time. Semibatch or semiflow reactors are a hybrid of batch and continuous operation. Typically one of the reacting species is charged to the tank all at once and a second reactant is fed in gradually, or

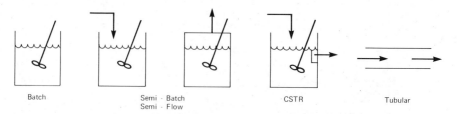

FIGURE 4-1 Some reactor possibilities.

MAJOR ISSUES IN HOMOGENEOUS REACTORS

perhaps a product of the reaction is continuously withdrawn. In either case, these processes too are inherently unsteady state. A great many commercial processes begin with such modes of operation.

Continuous processing is often conducted in continuous flow stirred tank reactors (CSTR) or in tubular reactors. Almost always the operation of these continuous reactors is intended to be in the steady state; that is, conditions at a given point do not change with time. However, these reactors too will be operated under transient conditions when started up or shut down as well as when conditions are changed for whatever purpose. Reactor dynamics is becoming an increasingly important aspect of reactor scaleup as intentional transient operation of continuous reactors is beginning to be used.

There are three main elements involved in the scaleup of any particular kind of homogeneous reactor. The first, reaction kinetics, was treated in detail in Chapter 3. The second and third elements are mass and energy balances. Mass balances for various reactor types are treated in virtually every text on chemical reaction engineering. Several familiar and excellent examples are Froment and Bischoff (1979), Hill (1977), Carberry (1976), Levenspiel (1972), Smith (1970, 1981), Aris (1969), and Walas (1959).

These books also provide substantial treatments of energy balances in nonisothermal reactors. That is, thermal effects are usually covered quite comprehensively for batch reactors, for steady-state plug-flow reactors, and for transient and steady-state CSTRs. Less attention is given to mass balances for semibatch reactors and only Aris (1969) and Smith (1981) provide an adequate energy balance for such systems. Of course, any desired degree of rigor is available from the methods described in Chapter 2. For analysis and design of semibatch reactors, the author's preference is for mass and energy balances derived individually for the specific case of interest rather than the use of generalized equations. In this way a stronger feeling for the physical situation is often obtained.

In this chapter, generalized treatment of homogeneous reactors will be left to the text books. Emphasis will be on illustrative examples and points of special interest in scaleup. Several examples comprise only partial results of problems adapted from other sources. The reader is encouraged to work out the omitted details. The result should be greater understanding and confidence in scaleup.

The major issues in homogeneous reactor scaleup are the determination of the size, shape, and performance of the reactor. In performance, such matters as conversion, selectivity, and stability are of primary concern. In this chapter they are addressed by presenting first the appropriate mass and energy balances for a variety of ideal reactor types. Brief examples are included in the discussion and a detailed example of a semibatch operation follows. The principles involved in the example are characteristic of both small- and large-scale commercial operations. Finally, the complexities of nonideal flow, heat and mass transfer, simultaneous reactions, and reactor stability and dynamics are considered.

II. MASS BALANCES

A. General Equations of Change

The general equations of change, as described in Chapter 2, are capable of comprehensive description of reactor performance. The equations apply both to steady-state and transient processes. Many problems do not demand such rigor, sophistication, and generality. In fact, the general equations may not be solvable and simpler models, such as are also described in Chapter 2, can be preferable.

Frequently both mass and energy balances are required to satisfactorily describe the reactor behavior. If the reaction of interest involves no significant heat effects, the thermal inertia of the reaction system may be sufficiently large that whatever heat effects exist result in a negligible change in temperature or sufficient heat can be transferred to or from the reactor to maintain a constant temperature. In such cases, the energy balance is unnecessary for the determination of size and performance of the reactor. That is, only one or more mass balances and a reaction rate equation are required for design.

B. By Reactor Type

1. Batch

The mass balance for a batch reactor was derived in Chapter 3 as Equation 3-5. An example will be presented to show the use of this mass balance in the design of an isothermal batch reactor.

The reaction is given by Equation 4-1:

$$\underset{(H)}{CH_3\overset{O}{\overset{\|}{C}}-OH} + \underset{(OH)}{C_2H_5OH} \xrightarrow[100^\circ C]{HCl} \underset{(E)}{CH_3\overset{O}{\overset{\|}{C}}-OC_2H_5} + \underset{(W)}{H_2O} \quad (4\text{-}1)$$

The charge to the batch reactor consists of equal masses of a 90 wt% aqueous solution of acetic acid and a 95 wt% aqueous solution of ethyl alcohol. The objective is to find the size of a batch reactor to produce 1000 kg of ethyl acetate per day in a reactor which is allowed to reach 80 percent of the equilibrium conversion before being dumped. A downtime of 30 min between batches is assumed. This illustration is an adaptation of Problem 2-10 originally given by Smith (1970).

The rate equation, presumably obtained by methods outlined in Chapter 3, is given by Equation 4-2,

$$r = 4.76 \times 10^{-4} [H][OH] - 1.63 \times 10^{-4} [E][W] \quad (4\text{-}2)$$

MASS BALANCES

TABLE 4-1 Mole Table for Esterification Reaction[a]

Compound	m_{i_0}, kg	M_i	N_{i_0}, mol	N_i, mol
Acid (H)	0.90	60.05	14.98	14.98 − x
Alcohol (OH)	0.95	46.07	20.62	20.62 − x
Water (W)	0.15	18.02	8.32	8.32 + x
Ester (E)	0.00	88.10	0.00	x
Total		2.00		43.92	43.92

[a] Basis: 1 kg each of alcohol and acid solutions, $x =$ mol H converted.

where the units are liters, gram moles, and minutes. The reactor design equation, Equation 4-3, is obtained by combining the mass balance and the rate equation:

$$-\frac{1}{V}\frac{dN_H}{dt} = r_H = k_f\left(\frac{N_H}{V}\right)\left(\frac{N_{OH}}{V}\right) - k_r\left(\frac{N_E}{V}\right)\left(\frac{N_W}{V}\right) \quad (4\text{-}3)$$

Often the assumption is made at this point that the volume of the reacting mass in the batch reactor is constant. This assumption enables the volume term to be taken inside of the derivative and all of the N/V terms in Equation 4-3 to be expressed as concentrations. Since the mass contained by the batch reactor does not vary, a constant volume implies a mass density that does not change with progress of the reaction.

In many commercial operations significant mass density changes do occur as the result of concentration and/or temperature variations as the reaction proceeds. For this example, the constant density approximation is probably acceptable and will be made to keep the example from becoming cumbersome. Nevertheless, Equation 4-3 will be retained as shown so that one can see how to allow for the effects of varying volume.

To proceed further we need to prepare the mole table, Table 4-1. Since there are fewer moles of acetic acid than ethyl alcohol in the charge, the acetic acid is the limiting reagent, and the conversion, x, is specified as the gram moles of acid converted. Thus, the moles of any reacting species can be expressed, as shown, as a function of the conversion. There is no change in the total number of moles with progress of the reaction.

If Equation 4-3 is multiplied though by V and the N_i's from Table 4-1 are substituted, the result is Equation 4-4:

$$\frac{dx}{dt} = \frac{k_f}{V}(14.98 - x)(20.62 - x) - \frac{k_r}{V}(8.32 + x)(x) \quad (4\text{-}4)$$

Since the reactor volume, V, is a function of the conversion, x, Equation 4-4 may be solved by separating the variables as shown in Equation 4-5:

$$\int_0^x \frac{V(x)\,dx}{k_f(14.98-x)(20.62-x) - k_r(8.32+x)(x)} = \int_0^t dt = t \quad (4\text{-}5)$$

At any value of conversion the composition of the liquid mixture is known, and hence the density of a solution of that composition, $\rho(x)$, can be obtained by experiment or by some predictive method. Once $\rho(x)$ is available, the volume occupied by the 2 kg of solution is known. With this information, the left-hand side of Equation 4-5 can be integrated graphically or numerically to obtain the conversion as a function of time.

As anyone experienced in scaleup will attest, large amounts of effort can be expended obtaining physical property information. In this instance, the simple illustration of obtaining the function, $V(x)$, will be demonstrated. The assumption is made that the volumes of all of the components of the reaction mixture are additive and that there are no volume changes on mixing. The volume of any component in the pure state is then given by Equation 4-6.

$$V_i = \frac{N_i M_i}{\rho_i} \quad (4\text{-}6)$$

In accordance with our assumption, the total volume is given by Equation 4-7.

$$V = \Sigma V_i = \Sigma \frac{N_i M_i}{\rho_i} \quad (4\text{-}7)$$

Combining Equations 4-6 and 4-7 with the expressions for N_i from Table 4-1 results in Equation 4-8.

$$V(x) = \left(\frac{14.98 M_H}{\rho_H} + \frac{20.62 M_{OH}}{\rho_{OH}} + \frac{8.32 M_W}{\rho_W} \right)$$

$$+ \left(\frac{M_W}{\rho_W} + \frac{M_E}{\rho_E} - \frac{M_H}{\rho_H} - \frac{M_{OH}}{\rho_{OH}} \right) x \quad (4\text{-}8)$$

Equation 4-8 shows $V(x)$ to be a linear function of the conversion, x. Such a simple function makes analytical integration of Equation 4-5 routine. Additional insight is available by carrying this development further. The densities of acetic acid, ethanol, water and ethyl acetate at 20°C are readily found to be 1049, 789, 1000, and 901 g/L, respectively. Substituting all available numerical values into Equation 4-8 yields Equation 4-9.

$$V(x) = 2.211 + 0.0002x \quad (4\text{-}9)$$

From Equation 4-9 one sees immediately that the volume of the reacting mass

MASS BALANCES

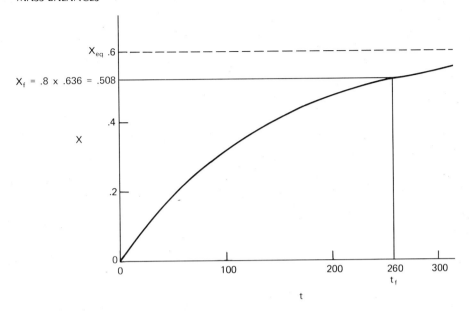

FIGURE 4-2 Conversion in a batch reactor producing ethyl acetate.

is only a negligibly slight function of the conversion, x, as anticipated. Therefore, the constant volume approximation is a good one for this case.

Greater sophistication could be brought to bear on the prediction of $V(x)$. Nevertheless, calculation of a constant value of V for the 2 kg of reacting mass, taking into account approximately the effect of temperature and volume changes on mixing, gave a value of $V = 2.23$ L. This value is used in all subsequent calculations. All numerical values are now available for the analytical integration of Equation 4-5. The time required to achieve any desired conversion is given by the integrated result shown in Equation 4-10.

$$t = 182 \ln \frac{(3.68 - 75.3x)}{(3.68 - 383x)} \qquad (4\text{-}10)$$

This $x(t)$ function is plotted on Figure 4-2.

The desired conversion was specified to be 80 percent of the equilibrium conversion. At equilibrium there will be no further conversion with increased reactor time. Therefore, at equilibrium, the right-hand side of Equation 4-4 must go to 0. (Reactor volume cancels out for this calculation.) Solving for the ratio of the forward to the reverse reaction rate coefficient gives Equation 4-11.

$$\frac{k_f}{k_r} = \frac{4.76 \times 10^{-4}}{1.63 \times 10^{-4}} = 2.92 = \frac{(8.32 + x_{eq})(x_{eq})}{(14.98 - x_{eq})(20.62 - x_{eq})} \qquad (4\text{-}11)$$

The result is a quadratic equation in x_{eq} with two roots, 49 and 9.54. Clearly 49 is impossible because it would mean that more moles were converted than existed initially (see Table 4-1). Therefore the equilibrium conversion, $x_{eq} = 9.54$, and the fractional conversion of the limiting reagent, acetic acid, is $9.54/14.98 = 0.636$.

Eighty percent of this equilibrium conversion gives a final conversion of $x_f = 0.508$. From Figure 4-2 it is seen that the batch reaction must be allowed to run for 260 min in order to achieve this conversion. Adding in the 30-min downtime gives a total cycle time of 290 min.

Finally, these calculations may be used to obtain the size of the batch reactor required to achieve the desired production rate of ethyl acetate. Equation 4-12 shows the calculation of the amount of ester produced per liter of reaction mixture.

$$\frac{0.508 \text{ mol H reacted}}{\text{mol H charged}} \times \frac{14.98 \text{ mol H charged}}{2.23 \text{ L}} \times \frac{1 \text{ mol E produced}}{\text{mol H reacted}}$$

$$\times 88.10 \frac{\text{g E}}{\text{mol E}} = 301 \frac{\text{g E produced}}{\text{L}} \qquad (4\text{-}12)$$

Equation 4-13 shows how this value is used to calculate the volume of the batch reactor.

$$\text{batch volume} = \frac{1000 \text{ kg E/day}}{301 \text{ g E/L}} \times \frac{1000 \text{ g}}{\text{kg}} \times \frac{\text{day}}{24 \times 60 \text{ min}} \times \frac{290 \text{ min}}{\text{batch}}$$

$$= 670 \text{ L/batch} = 0.67 \text{ m}^3/\text{batch} \qquad (4\text{-}13)$$

2. Tubular Plug Flow

A common goal in scaleup is to convert a batch process to a continuous operation. Often the continuous reactor selected is a tubular reactor. In this instance a familiar scaleup concept is to assume that the residence time in a tubular reactor, operating in plug flow, would be identical to the batch reaction time without the downtime. The volume of the tubular reactor would then be equal to the product of this residence time and the volumetric feed rate necessary to give the desired production rate.

The tubular reactor volume required to achieve the same production rate as specified in the previous batch reactor example can be calculated by multiplying the volume of the batch reactor by the ratio of the batch reaction time to the batch cycle time. Therefore,

$$\text{tubular reactor volume} = 670 \times \frac{260}{290} = 600 \text{ L} \qquad (4\text{-}14)$$

The advantages of continuous operation are the evident reduction in reactor

MASS BALANCES

volume resulting from the elimination of the downtime, a substantial saving in labor costs, and possibly improved quality control.

This familiar scaleup concept is usually convenient in a qualitative sense, sometimes correct in a quantitative sense, and potentially dangerous if used promiscuously. The plug flow approximation can be badly flawed as will be shown later in this chapter. It is retained for now. If the reaction occurs at constant mass density, as was demonstrated in the above example, the residence time in a plug flow reactor is indeed the same as the batch reaction time and this scaleup method gives accurate results. Unfortunately, many tubular reactor calculations are not so straightforward.

Many batch reactors (tanks for liquids, autoclaves for gases) do operate at constant volume and therefore constant mass density. On the other hand, tubular reactors usually operate at an externally fixed pressure with a constant mass flow rate. If an increase in total moles occurs in a parcel of reacting gas at constant pressure, the parcel must expand. That is, there must be a decrease in the mass density of the parcel. To maintain a constant mass flow rate, an increase in fluid velocity must accompany this expansion or mass density decrease. A similar effect occurs when liquid reactants are heated, especially if a phase change accompanies the heating and/or the reaction.

Figure 4-3 shows a schematic diagram of a tubular reactor. This diagram will help make some definitions clearer. The ratio of the reactor volume, V, to the volumetric feed rate, Q_0, is called the space time; its reciprocal is the space velocity. Similarly, the space time can be calculated as the length, L, of a reactor of constant cross-sectional area divided by the fluid velocity at the inlet of the reactor, u_0. If the reaction occurs at constant mass density, the volumetric flow rate at any point in the reactor is equal to the volumetric feed rate and therefore the residence time would equal the space time, V/Q_0. As we have said, for our example, the batch reaction time is equal to the space time, and the plug flow reactor volume, V, can be calculated for a specified volumetric feed rate, Q_0. However, if the mass density changes with progress of the reaction then the space time, V/Q_0, will not be equal to the ratio V/Q and therefore will not be equal to the batch time, t_f.

Where an increase in the total number of moles of gas occurs at constant pressure, it is clear that the increased fluid velocity will result in an actual residence time lower than expected on the basis of the inflow rate. This reduced residence time will lead to a conversion lower than that achieved in the batch process or expected on the basis of the inflow rate to the tubular reactor. Thus a familiar scaleup concept can lead to a serious scaleup error, in this case a reactor performance worse than expected. Of course, a scaleup engineer might be luckier with a reaction that is accompanied by a decrease in the total

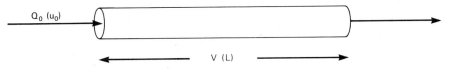

FIGURE 4-3 Schematic diagram of a tubular reactor.

number of moles such as in the dimerization of butadiene. In this instance, the reactor performance would be better than anticipated, unless subsequent polymerization occurred, producing an inferior product. Optimal scaleup is to know what you want and get exactly that.

The arguments illustrating this scaleup error have been qualitative so far, but they can be made quantitative as follows. For the reaction of an ideal gas in an isothermal, isobaric, fixed-bed reactor of uniform cross section; Equation 4-15 gives an exact expression for the true residence time.

$$t_r = \frac{\varepsilon P_T F_{t_0}}{RT} \int_0^x \frac{dx}{F_t r} \tag{4-15}$$

In this equation r is the reaction rate, mol i converted/volume time and x is the conversion, mol i converted/total mol fed. Although Equation 4-15 is derived for a fixed-bed reactor, it applies as well to a homogeneous reactor by setting the void fraction, ε, equal to unity. If in Equation 4-15 the total molar flow rate at every point, F_t, is equal to the feed rate, F_{t_0}, then the true residence time will equal the batch time. As has been stated before, F_t will equal F_{t_0} only when the mass density is constant. For a specific example, consider the ammonia synthesis reaction given by Equation 4-16.

$$N_2 + 3H_2 \rightleftharpoons 2NH_3 \tag{4-16}$$

Defining the conversion, x, as the moles of nitrogen reacted per total moles fed, the stoichiometry of the reaction leads to the following relation for the molar flow rate at any point in the reactor.

$$F_t = F_{t_0}(1 - 2x) \tag{4-17}$$

Substitution of Equation 4-17 into Equation 4-15 yields

$$t_r = \frac{\varepsilon P_T}{RT} \int_0^x \frac{dx}{(1 - 2x)r} \tag{4-18}$$

The usual situation of an H_2 to N_2 ratio of 3 in the feed, corresponding to the reaction stoichiometry, will be assumed. For complete conversion, $x = 0.25$. For zero conversion, of course, $x = 0$. Thus the factor $(1 - 2x)$ varies as indicated by Equation 4-19.

$$0.5 \leq (1 - 2x) \leq 1.0 \tag{4-19}$$

Thus if the reactor is operated at very low conversion, little error results from the constant density assumption. On the other hand, the error would be about 100 percent for nearly complete conversion. For conversion levels typical of commercial ammonia synthesis processes, the scaleup error in assuming constant density would be on the order of 25 percent.

MASS BALANCES

There must be a better basis for the design of tubular reactors than to scaleup directly from the batch reaction time. There is, and it does not involve residence time at all. This basis is the mass balance on a plug flow reactor, which is given in Equation 4-20 for the unsteady state:

$$-\frac{\partial N_A}{\partial (W/F_{t_0})} - r_A = \frac{1}{N_t} \frac{P_T}{RT} \frac{\varepsilon}{\rho_c} \frac{\partial N_A}{\partial t} \qquad (4\text{-}20)$$

where

N_A = mol A present/total mol feed
N_t = total mol present/total mol feed
r_A = mol A reacted/time mass of catalyst

Once again the equation is written for a fixed-bed reactor, but it is easily converted to a form appropriate for a homogeneous reactor. To make this conversion the void fraction, ε, is set equal to 1 and the equation is multiplied through by ρ_c, which converts the reaction rate to a volume basis and results in the replacement of the mass of catalyst, W, by the reactor volume, V.

Time, t, appears in Equation 4-20. However, this is the time elapsed since the initiation of the unsteady state, and is not related to the residence time. Later, the use of Equation 4-20 in describing the reactor dynamics will be discussed but most design studies emphasis is on reactor operation at steady state. For steady state the number of moles per mole of feed at any point in the reactor does not change with time; therefore

$$\frac{\partial N_A}{\partial t} = 0 \qquad (4\text{-}21)$$

Equation 4-20 reduces to Equation 4-22.

$$r_A = -\frac{dN_A}{d(W/F_{t_0})} \qquad (4\text{-}22)$$

Since $dN_A = -dx$, where x = mol A reacted/total mol fed,

$$r_A = \frac{dx}{d(W/F_{t_0})} \qquad (4\text{-}23)$$

Equation 4-23 is the fixed-bed reactor equivalent of Equation 3-4.

When Equation 4-23 is integrated to a specified conversion, x, the result is the familiar plug flow reactor design equation.

$$\frac{W}{F_{t_0}} = \int_0^x \frac{dx}{r_A(x)} \qquad (4\text{-}24)$$

Sometimes the indicated integration in Equation 4-24 can be done analytically, but it is always possible to do it graphically or numerically. It is clear from Equation 4-24 that the amount of catalyst, W, or the size of a reactor, V, can be obtained without reference to any residence time.

Equation 4-20 describes the operation of a tubular reactor in the unsteady state. Rather than be intimidated by such a partial differential equation, it is better to think of reactor dynamics as a friend. Reactors must be started up and shut down. Designers and operators will want to know how the system responds to some change in conditions. Knowledge of reactor dynamics is crucial to effective control. Further, it is often possible to learn more about the phenomena underlying a steady-state behavior by observing the dynamics of the reactor. For example, mechanistic insights and rate determining steps that could only have been guessed from extensive steady-state data have been obtained from a few transient experiments [Perti and Kabel (1982)].

It may even be possible to improve a reactor's performance by exploiting its dynamics. In one case tight control on flow rate and loose control on temperature was shown to increase production rate (Denis and Kabel, 1970). As surely as the rates of individual reactions can be influenced by transient conditions, the selectivity in simultaneous reactions can also be altered by operating intentionally in the unsteady state. Improvement of selectivity is of much greater consequence than merely increasing production rate, and has been demonstrated by Renken (1974).

3. Tubular Laminar Flow

Most tubular reactors of commercial size and flow rate operate in the turbulent regime. This is important because turbulent flow promotes radial heat and mass transfer, and a corresponding semblance of plug flow. In contrast, when the reaction mixture is a viscous polymer, laminar flow will certainly exist near the wall and may be obtained throughout the tube. If molecular diffusion is neglected, laminar flow is a state of complete segregation of fluid elements both radially and longitudinally. The result is that different fluid elements remain in the reactor for different lengths of time and therefore react to different extents.

The resulting residence time distribution almost always gives rise to a lower overall conversion than would be obtained in plug flow at the same mean flow rate. The reason is that, for a reaction whose rate decreases with increasing conversion, the additional conversion experienced by elements present beyond the mean residence time does not fully compensate for potential conversion lost by elements leaving in less than the mean residence time. The argument can be made more quantitative as follows.

The mass balance for a first order irreversible reaction occurring in a differential length of an annular element is

$$u(r)\frac{dC_A(r)}{dl} + kC_A(r) = 0 \tag{4-25}$$

For an isothermal, Newtonian fluid in laminar flow the familiar parabolic velocity profile is

$$u(r) = 2\bar{u}\left[1 - \left(\frac{r}{R}\right)^2\right] \tag{4-26}$$

Substituting Equation 4-26 into Equation 4-25, separating the variables, and

MASS BALANCES

integrating over the length of the reactor with the boundary condition that $C_A = C_{A_0}$ at $l = 0$ and all r gives

$$\ln \frac{C_{A_0}}{C_A(r)} = \frac{kL}{2\bar{u}\left[1 - (r/R)^2\right]} \tag{4-27}$$

or

$$\frac{C_A(r)}{C_{A_0}} = \exp\left[-\frac{kL/2\bar{u}}{1 - (r/R)^2}\right] \tag{4-28}$$

Inherent in this simple integration is the idealization that the velocity at a given radius, r, does not vary with distance (and hence with conversion level) down the reactor. A more serious limitation is the invalidation of Equation 4-26 from changes in viscosity resulting from changing temperature or degree of polymerization.

Equation 4-28 gives the normalized radial concentration profile at L, $C_{A_L}(r)/C_{A_0}$. The average normalized concentration over the reactor cross section can be obtained by numerically or graphically performing the following indicated integration.

$$\overline{\left(\frac{C_{A_L}}{C_{A_0}}\right)} = \frac{\int_0^R \left(\frac{C_{A_L}(r)}{C_{A_0}}\right) u(r) 2\pi r \, dr}{\int_0^R u(r) 2\pi r \, dr}$$

$$= 4 \int_0^1 \exp\left[-\frac{kL/2\bar{u}}{1 - (r/R)^2}\right]\left[1 - \left(\frac{r}{R}\right)^2\right]\left(\frac{r}{R}\right) d\left(\frac{r}{R}\right) \tag{4-29}$$

The integration has been performed by Cleland and Wilhelm (1956). They also calculated the normalized concentration (or fraction of reactant remaining) for a plug flow reactor operating at the same overall flow rate. A few of their results for different values of $kL/2\bar{u}$ are shown in Table 4-2. From the dimensionless group, $kL/2\bar{u}$, it can be seen that an increase in reaction rate coefficient or reactor length or a decrease in average flow rate results in an increase in conversion (decrease in fraction of reactant remaining), as would be

TABLE 4-2 Comparison of Plug and Laminar Flow Reactor Performance

	(C_{A_L}/C_{A_0})	
$kL/2\bar{u}$	Plug	Laminar
0.01	0.9802	0.9810
0.2	0.6703	0.7037
1.0	0.1353	0.2194
2.0	0.0183	0.0603

expected. Further the discrepancy between the calculated results for plug and laminar flow increases with increasing conversion. As reasoned earlier, the conversion in laminar flow is always less than that for plug flow.

Cleland and Wilhelm went further, considering the effects of diffusion, heat transfer, and free convection and comparing their theory to experimental data for the pseudo-first order hydrolysis of acetic anhydride. Carberry (1976) gives a nice treatment of laminar flow reactors, including a worthwhile summary of the work of Johnson (1970).

4. Continuous Stirred Tank

The mass balance for a well-mixed continuous flow tank reactor (CSTR) was derived in Chapter 3; both unsteady-state and steady-state versions were given. The fact that the steady state mass balance is an algebraic equation and the unsteady state mass balance is an ordinary differential equation results in easy mathematical analysis and description of CSTRs. Accordingly extensive literature exists to assist the scaleup engineer in the design of CSTR systems to achieve his objectives. Denbigh (1965), Levenspiel (1972), Hill (1977), and Froment and Bischoff (1979) are particularly good references to begin with.

Continuous stirred tank reactors can be operated individually or as tanks in series, according to the purpose. The equations derived in Chapter 3 apply to an individual CSTR, whether it is operated alone or as one of a group. Analysis of a battery of stirred tank reactors simply becomes a matter of solving simultaneous equations. Analytical, graphical, and numerical methods for solutions are commonplace. Because the equations are linear, an elaborate theoretical basis exists for the solution of such systems of equations or for the consideration of advanced topics, as will be touched on later. So powerful are such treatments that residence time distributions are often modeled by tanks in series (see Chapters 2 and 8).

Some important conceptual ideas in CSTR scaleup are illustrated here with quantitative results from an adaptation of Problem 4-4 of Walas (1959). Reaction rate data were available. The volumes of a single tank and of two, three, and four tanks in series were to be calculated for a given flow rate, feed concentration, and conversion level. The results of the calculations are given in Table 4-3. The volume units in this table are cubic feet, but the units are unimportant to the purpose of this example.

Naturally the volume of an individual tank in a series of tanks decreases as the number of tanks in the series increases. More noteworthy is the fact that the total volume of all of the tanks in a battery also decreases significantly as the number of tanks in the battery increases. Also the volume required to achieve the same specified performance is exceptionally high for a single tank. This is the direct result of the complete mixing which exists in the tanks. If all of the reaction must occur in a single well-mixed tank, then the tank must be very large because the reaction will occur at the low rate corresponding to the low outlet reactant concentration.

For two tanks in series, this logic applies to the second tank. In the first tank, however, the reaction may proceed at the somewhat higher rate corre-

MASS BALANCES

TABLE 4-3 Typical CSTR Calculation Results

Number of Stages	Volume per Stage	Total Volume
1	281	281
2	67.6	135.2
3	35.7	107.2
4	23.4	93.4
∞^a	0	62.8

$^a\infty$ corresponds to batch or plug flow.

sponding to the outlet concentration from that tank. As the number of tanks is increased to three and then four ore more, a smaller and smaller proportion of the reaction occurs at the ultimate outlet concentration of the battery and more and more occurs at concentrations closer to the feed concentration. If this idea is extended to an infinite number of stages, each having infinitesimal volume, the limit of the tubular reactor is reached. For this constant density reaction system, the volume of a plug flow reactor (or a batch reactor with negligible downtime) for the desired performance can be calculated. It is fascinating to note that the product of infinity times zero is 62.8 as indicated in Table 4-3.

The results of Table 4-3 are plotted as a continuum in Figure 4-4. Such a graph gives a dramatic indication of the benefit of adding a second tank, and perhaps a third and fourth. It also makes clear the diminishing return which accompanies the addition of an endless succession of tanks as the limit of plug flow is approached. If costs are attached to the decreasing volume and increasing complexity which accompany additional stages, it is clear that an economic optimum will exist at some finite number of stages. This matter is treated in some detail by Froment and Bischoff (1979). The attractiveness of the simplicity of tubular reactors as one scales up to increasingly large throughputs is also evident from these ideas.

The size advantage of a plug flow reactor, as compared to a single CSTR, is demonstrated in Figure 4-5. These particular results correspond to a first order reaction as calculated for Problem 5-5 of Walas (1959). At 0 percent conversion the volume ratio is unity because the reaction occurs in both reactors at the feed concentration. As the conversion increases, however, the volume of a tubular reactor becomes increasingly smaller than that of a CSTR because all of the reaction in the CSTR must occur at the final desired conversion, whereas only the last increment of reaction in the tubular reactor occurs under such unfavorable circumstances. Finally, at complete conversion, the advantage of the tubular reactor becomes infinite as a finite CSTR cannot reach this conversion.

This illustrative graph is for a first order reaction only. Levenspiel (1972) presents generalized graphs of this nature for a wide variety of reaction orders. As might be expected the impact increases for the more concentration-dependent (high-order) reactions. Further, Figure 4-5 compared the plug flow reactor volume to the volume of a single CSTR only for a first order reaction. It should

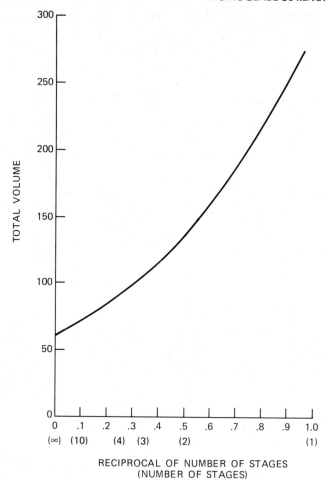

FIGURE 4-4 Total volume of CSTRs in series.

be realized that qualitatively similar, but quantitatively different, behavior can be expected for tanks in series. Levenspiel (1972) also gives generalized charts of this nature for varying numbers of tanks in series. Such charts might occasionally be used in design but they are of more value for the trends they display.

In all of the cases discussed in this section, the reaction rate decreased as the reactant concentration decreased. This is by far the most common circumstance, but it should be realized that in some situations (exothermic reactions at some conditions, autocatalytic reactions, special selectivity requirements in competitive reactions, and some kinds of heterogeneous catalytic behavior), a CSTR could yield a smaller volume than a plug flow reactor. Thus one should be careful about following generalized ideas too far without a thorough understanding of the underlying phenomena. Such surprises would result in poor scaleup for sure.

MASS BALANCES

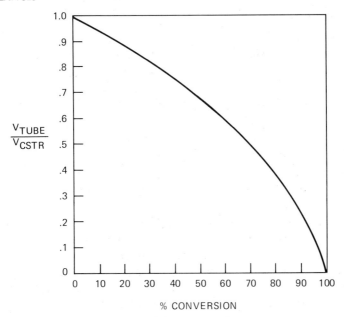

FIGURE 4-5 Comparison of reactor volumes for first order reaction.

5. Semibatch

Semiflow or semibatch reactors come in great variety. Accordingly, the mass balances will take different forms. For example, if a liquid-phase reaction gives off a gaseous product that is continuously removed, a reaction that would otherwise be a batch process would become a semiflow process with no continuous feed stream but with a continuous withdrawal stream. The rate of withdrawal would, of course, depend upon the rate of reaction. A second common situation is where one liquid reactant is charged to a reactor tank and a second reactant is fed in slowly. Such a procedure is often used to maintain thermal control of an exothermic reaction. If the thermal control thus achieved results in an isothermal condition, then a mass balance is sufficient for design of the semibatch reactor. Carberry (1976) gives a rather complete analysis of this type of semibatch reactor.

Semibatch reactors may take forms as varied as the process engineer's imagination will allow. The appropriate mass balance may differ from one case to the next. Smith (1970, 1981) provides useful guidance in the establishment of mass balance equations for such reactors. Froment and Bischoff (1979) discuss at some length the general species continuity equations described here in Chapter 2. Many scaleup engineers will find it quite worthwhile to derive mass balances for the processes of interest to them. The difficulties of working with completely rigorous formalisms and the dangers of working with cookbook methods, for which the underlying assumptions are unclear, are legion. In Section IV of this chapter a scaleup example involving a semibatch reactor will be considered in detail.

III. ENERGY BALANCES

A. Nature of the Balances

The rates of most chemical reactions depend exponentially on temperature (see the discussion of the Arrhenius equation in Chapter 3). This powerful influence of temperature, and other more subtle effects, means that thermal aspects of reactive processes must be quantitatively described. Thus energy balances come into play. It was noted in Chapter 3 that heat exchange in reactors often receives as much attention as kinetics.

1. Isothermal

The size of an isothermal reactor can be determined from a rate equation and a mass balance. However, an energy balance will be necessary if the rate of heat addition or removal required to maintain isothermality must be determined.

Laboratory reactors are usually isothermal, and for good reason. Their purpose is usually to gather kinetic data from which reaction rate correlations may be ascertained. Because the influence of temperature on reaction rate is so powerful, it is especially important that a constant known temperature be maintained in kinetic studies. In this way the more subtle but very important influences of species concentrations can be discerned. Later the effect of temperature can be studied in isothermal runs at other temperatures. Further, laboratory reactors are usually small and therefore more easily maintained at a constant temperature without excessive costs.

If some nonisothermality is noted in a laboratory study, it is usually possible to correct for this by any of a number of techniques. For example, the stirring rate in a thermostatic bath may be increased, a diluent may be added to the reacting mixture to provide or carry away heat, or the shape of the reactor may be altered to provide a higher surface area to volume ratio. The last method may provide two benefits:

> It may enhance the area-dependent heat transfer process relative to the volumetric effect of heat generation or consumption.
>
> The velocity of the fluid flowing through a tube may be increased with resulting improved heat transfer.

In tanks special cooling or heating coils can be inserted.

When it is desirable to maintain a commercial scale reactor at constant temperature, any of the techniques used in the laboratory studies as well as others may be employed. For example, boiling liquid reactors are especially effective in maintaining isothermality. But effective scaleup can hardly tolerate the trial-and-error approach so common in the laboratory. Therefore, effective use of the energy balance to quantify heat effects is essential. For example, one can calculate whether the surface area of a tentatively specified reactor is sufficient to accomplish the heat exchange necessary to maintain isothermality.

ENERGY BALANCES

If not, the energy balance can be used to determine what size and shape of reactor would suffice.

2. Adiabatic

In contrast to the frequent occurrence of isothermal reactors on the laboratory scale, commercial reactors very often approach adiabatic operating character. Adiabatic implies no heat exchange between a system and its surroundings. Although commercial sized exothermic reactors transfer large amounts of heat to their environment, this amount of heat is minuscule compared to the tremendous heat flows through the system. Thus, the temperature profiles in commercial reactors are very nearly what they would be if the reactors were truly adiabatic. For nonisothermal reactors the simultaneous solution of a rate equation, one or more mass balances, and an energy balance is necessary for satisfactory reactor design. If the reactor approximates adiabatic operation the energy balance is considerably simpler and the calculations less tedious than if heat exchange with the surroundings must be taken into account.

3. Other

Scaleup of reactors involving significant thermal effects provides ample opportunity for imaginative engineering. Paradoxically it may be necessary to heat up a reaction mixture to initiate the reaction and then to provide vigorous cooling to keep the reaction from running away. Such situations are described in Chapter 5. It is possible to compute an optimal temperature profile for a reactor in order to achieve a minimum reactor size. Commercial operations can seldom afford the cost or complexity of providing such optimal profiles, but by careful consideration of their heat exchange media and thermal control devices they can at least operate in the realm suggested by such optimization concepts. This process is often called temperature programming.

An illustration of temperature programming is given here based upon an adaptation of Problem 7-8 of Walas (1959). The calculations were made for a 40 percent conversion of ethylene and hydrogen chloride to form ethyl chloride in a fixed-bed reactor. As noted before, and as discussed in Chapter 5, a pseudohomogeneous model is often satisfactory for fixed-bed reactor design. Therefore all of the principles and results of this illustration apply equally well to homogeneous reactors. The reaction is exothermic, characterized by favorable rate at high temperatures and favorable equilibrium at low temperatures. These two competing characteristics require some kind of compromise in the design of exothermic reactors.

Figure 4-6 shows the results of the reciprocal space velocity, W/F, for isothermal operation at various temperatures. At low temperature the rate of reaction is small and therefore the size of the reactor must be large or the feed rate must be kept small. A large reactor size or a small feed rate is also the result at high temperature, but in this case the effect is due to the unfavorable equilibrium which exists. The minimum reactor size or maximum feed rate can be seen from Figure 4-6 to be achievable by isothermal operation at 195°C.

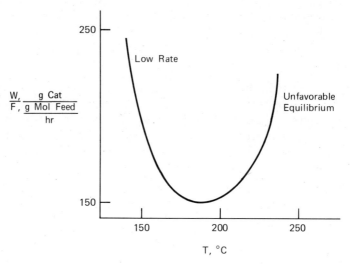

FIGURE 4-6 Competing effects in design of an exothermic reactor.

However, isothermal operation is not optimal either. If three catalyst beds (or reactor sections) were maintained at 250, 200, and 150°C the same conversion could be achieved with a still smaller reactor or for a still larger flow rate. Extending this idea to a tube and shell homogeneous reactor with countercurrent flow of reacting gases and coolant fluid, the reactants and coolant could be brought in at 200 and 150°C, respectively. Near the reactant inlet (coolant outlet), the high rate of the exothermic reaction might be expected to produce a temperature rise even with the cooling available. The temperature might even approach 250°C but would then decline because of the reduced reaction rate corresponding to the increasing conversion. (For greater depth in such considerations, see about temperature runaway in Chapter 5.) Although the resulting temperature profile may be far from optimal, it represents an effective engineering compromise among the competing constraints of reaction rate, equilibrium, heat transfer, and economics. Such compromises are part of effective scaleup.

B. By Reactor Type

As was the case with mass balances, the familiar textbooks of chemical reaction engineering also deal in considerable detail with energy balances, both in general and with respect to various reactor types. In this chapter, some highlights will be presented in the form of illustrations.

1. Batch

To illustrate the use of the energy balance in a batch reactor design, Problem 5-1 of Smith (1970) is adapted here. The reaction of acetic anhydride in water

ENERGY BALANCES

to produce acetic acid, Equation 4-30, is selected.

$$(CH_3CO)_2O + H_2O \rightarrow 2CH_3\overset{=O}{C}-OH \quad (4\text{-}30)$$
$$\text{(AA)} \qquad \text{(W)} \qquad\qquad \text{(A)}$$

Because of the large excess of water in the reacting mixture, the reaction rate may be thought of as pseudo-first order in acetic anhydride. Thus the rate equation can be written as Equation 4-31 where the acetic anhydride concentration has units of gram moles per cubic centimeter.

$$r_{AA} = k[AA] \quad (4\text{-}31)$$

The reaction rate coefficient is given as a function of absolute temperature, K, by

$$k = 2.24 \times 10^7 \exp\left[-\frac{5.6 \times 10^3}{T}\right] \text{min}^{-1} \quad (4\text{-}32)$$

For a constant density batch reaction the mass balance is

$$r_{AA} = -\frac{d[AA]}{dt} \quad (4\text{-}33)$$

Of course, if the temperature rise in the reactor is too great, the constant density assumption may introduce significant error, in which case a more rigorous form of the mass balance should be used. For this dilute aqueous solution, the temperature dependence of the solution density can be taken as that of water and found in any handbook. For adiabatic operation the energy balance of a batch reactor is

$$\rho V C_V \frac{dT}{dt} = -\Delta H_R r_{AA} V \quad (4\text{-}34)$$

If Equations 4-31 and 4-32 are combined to eliminate the rate coefficient k, and the resulting equation for the rate of reaction, r_{AA}, is substituted into Equations 4-33 and 4-34, it is seen that two equations remain with three unknowns (T, [AA], and t). In general, one of these three unknowns is specified and the two differential equations are solved simultaneously for the other two unknowns. Table 4-4 gives the parameters and initial conditions for this example calculation.

An analytical solution of these two differential equations is achieved by eliminating the time differential, dt, and solving the resulting differential equation with the initial conditions specified in Table 4-4. A 70 percent final conversion is specified and the final temperature resulting from the adiabatic operation comes out to be 23°C. The profile of temperature as a function of

TABLE 4-4 Parameters for Adiabatic Batch Reaction of Acetic Anhydride to Acetic Acid

$V = 200$ L	$[AA]_0 = 2.16 \times 10^{-4}$ mol/cm^3
$C_V = 0.9$ cal/g °C	$\rho = 1.09$ g/cm^3
$\Delta H_R = -50{,}000$ cal/mol	$T_0 = 15°C = 288$ K

acetic anhydride concentration, available from the analytical solution, can be substituted into either the mass balance or the energy balance. Solution of the resulting differential equation gives 11.4 min as the time required to achieve a 70 percent conversion in adiabatic operation. If the reactor is operated isothermally at the same initial temperature of 15°C, solution of the mass balance shows the time required to achieve a 70 percent conversion to be 14.9 min.

Less time is required to achieve the specified conversion for the same starting conditions under adiabatic operation than for isothermal operation. The reason for this is that the exothermic reaction produces a rise in temperature in the adiabatic reactor as the reaction proceeds. This rise in temperature produces an increased rate of reaction and therefore a decreased time to achieve a specified conversion than would be the case if the temperature were held constant at the lower initial temperature. Of course, if the reaction had been endothermic, adiabatic operation would have required a longer processing time to reach the specified result.

2. Tubular

Some qualitative discussion of thermal effects, and hence the need for energy balances, in tubular reactors has already been presented. Because of the distributed nature (see Chapter 2) of tubular reactors, the fact that large quantities of wall heat exchange are usually involved, and the likelihood that a significant change in the mass density of the reacting fluid will occur, a quantitative example of the analysis and design of a nonisothermal tubular reactor would be too lengthy to present here. This problem is addressed in considerable detail, however, in Chapter 5. The reader is also referred to Smith (1970, 1981) where the calculations involved in nonisothermal tubular reactors are performed in great detail.

3. Continuous Stirred Tank

We have seen already that in a well-mixed CSTR the concentrations anywhere in the tank are the same as the concentrations in the stream leaving the tank. These concentrations were the result of a balance between the in- and outflow rates of reacting components and their rates of consumption or production by reaction. Similarly, the temperature anywhere in a well-mixed CSTR will be

ENERGY BALANCES

the same as the temperature of the exiting stream. This temperature is the result of a balance between the heat flowing with the in- and outflow streams and the heat released or consumed by the reaction. Because the contents of the tank exist at the outflow levels of concentration and temperature, the rate of reaction will be dependent upon these outflow conditions.

When a significant thermal effect exists (that is, if the outlet temperature differs significantly from the inlet temperature), then an energy balance must be added to the mass balance and rate equation for the complete description of a CSTR. Just as the mass balance for a steady-state CSTR is an algebraic expression, so is the energy balance. The fact that all equations involved in the description of a CSTR at steady state are algebraic means that analyses of this reactor type are mathematically quite simple. Even in the unsteady state, the mass and energy balances lead only to ordinary differential equations. The result of this simplicity of analysis is not only exceptionally thorough coverage in the literature of the characteristics of continuous stirred tank reactors but also some highly esoteric studies. Those matters of greatest importance in scaleup are well covered in the familiar chemical reaction engineering texts.

An example is chosen here to illustrate the potential for the existence of multiple steady states in a CSTR. It will be seen that this example suffices also to describe the situation in which a unique steady state exists. For simplicity the reaction is taken to be first order and irreversible. The reaction rate coefficient is assumed to follow the Arrhenius equation. When the rate equation is combined with the mass balance, the result in conversion form is

$$X = \frac{\bar{\theta} A e^{-E/RT}}{1 + \bar{\theta} A e^{-E/RT}} \tag{4-35}$$

The energy balance for an adiabatic reactor can be written as

$$X = \frac{\rho C_p}{C_{A_0}(-\Delta H_R)} (T - T_0) \tag{4-36}$$

For a given reactor and reaction mixture these two equations can be solved simultaneously for the two unknowns, conversion and temperature.

In most cases only one pair of values of X and T will satisfy these equations simultaneously. In such cases a unique steady state exists and the reactor performance has been completely defined. If the problem was one of design, the desired conversion might be specified and Equation 4-36 used to calculate the corresponding reaction temperature. This temperature and conversion could then be substituted back into Equation 4-35 to determine the volume of reactor required for a particular flow rate. It is also a straightforward matter to apply these principles to each tank in a series of tanks, hence designing a stirred tank reactor battery.

Under certain circumstances, it is possible for the simultaneous solution of these two equations to yield more than one root and therefore more than one

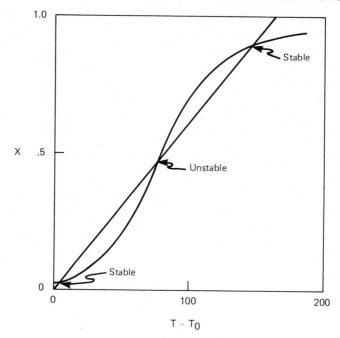

FIGURE 4-7 Multiple steady states in a CSTR.

steady state. This possibility is most easily visualized graphically. From Equation 4-35 it can be seen that the mass balance gives rise to a sigmoidal curve on a graph of conversion, X, versus temperature rise, $T - T_0$. On the other hand, Equation 4-36 shows that the energy balance yields a straight line on the same graph. The graph resulting from this type of analysis as applied to Problem 5-7 in Smith (1970) is shown as Figure 4-7. There it is seen that the mass and energy balance curves have three intersections, or three simultaneous solutions. The implication is that any of three steady states can satisfy the describing equations and therefore could occur in the CSTR.

Dynamic analysis of the CSTR behavior shows that only the lower and upper steady states are stable and that the intermediate steady state cannot be maintained without some kind of control. At the lower steady state it is seen that a very small temperature rise exists but also that the conversion is impracticably low. At the upper steady state an acceptable conversion is achieved; however, a very large temperature rise must be tolerated. The unstable steady state gives an indication of what conditions would be required to establish the reactor operation at either the upper or lower steady state. If the reactor is started up with a temperature rise less than about 90°, the process will naturally move to the lower steady state. If in the startup the temperature rise is artificially raised above 90° (for example, by a heating coil) the reactor operation will move to the upper steady state. If, during stable operation at the upper steady state, a disturbance is introduced giving $T - T_0$ less than 90°, the reaction would move spontaneously to the lower steady state.

ENERGY BALANCES

Such a movement from a high to a low steady state is often called quenching. The reverse movement is called ignition. From what has been said it is seen that the steady state of operation attained depends upon the history of the system. This is the reason why dynamic analysis is required to assess stability.

Lest it appear that multiple steady states are something that occurs only in theory, consider the case of the familiar Bunsen burner. With the gas flow and all environmental conditions the same, there can be a flame or no flame depending upon whether a spark has been imposed. Either state is stable and the spark provides the mechanism of ignition. If one now disturbs the ignited steady state by blowing the flame off the top of the burner, the reaction will be quenched to the lower steady state. After the quenching, the gas flow and environmental circumstances will be the same as they were before. Multiple steady states are most common in strongly exothermic reaction systems. However, Carberry (1976) notes that they can arise in isothermal, and even endothermic, cases as well.

Although it is clear that multiple steady states can exist, they are not frequently observed. The reason for this can be seen from Figure 4-7. The mass and energy balance curves are readily shifted by variation of such parameters as the reactor volume, heat of reaction, and others evident in Equations 4-35 and 4-36. If the heat of reaction is significantly less the energy balance line will be steeper. Then its only intersection with the mass balance curve will be in the region of the low steady state. Alternatively a much higher heat of reaction could result in a less steep energy balance line intersecting the mass balance curve only at a point of high conversion.

Effective design will very often arrive at a desirable unique steady state without consideration of multiple steady states. But the prudent scaleup engineer should check for multiple steady states because of the consequences of discovering them by accident. It would be quite a nuisance to have a reactor that is designed to operate at a high conversion but keeps getting quenched by perturbations that move it beyond the unstable steady state to a condition of little reaction. Much more dramatic is the potential ignition of a reaction intended to operate at a low steady state which for some reason moves toward an upper steady state with disastrous consequences. Concern with such matters is not simply a recent thing. See, for example Liljenroth (1918) who was concerned with starting and stability phenomena of ammonia oxidation.

If multiple reactions occur, the appropriate rate equations are established for each reaction according to the methods in Chapter 3. Further, mass balances are written for each of the reactive species in the CSTR. It was shown in Chapter 3 how the mass balances could be derived for the unsteady state, as well. What is seldom seen in conventional textbooks is an energy balance for simultaneous reactions occurring in a CSTR in the transient state. Such an equation is given by Aris (1969) and is repeated here, with illustrative units, as Equation 4-37.

$$\theta \frac{dT}{dt} = T_f - T + \theta \sum_{i=1}^{R} J_i r_i - Q \qquad (4\text{-}37)$$

where

$$J_i = \frac{-\Delta H_i}{C_p}$$

i = identifier of the ith simultaneous reaction
R = number of simultaneous reactions
ΔH_i = enthalpy change with reaction, cal/g mol reacted
C_p = total heat capacity per unit volume of reaction mixture, cal/cm^3 K = C_p (mass basis) (cal/g K)ρ(g/cm^3)
T = reaction mixture temperature, K
T_f = feed temperature, K
$\theta = V/q$ = residence time, min
V = reactor volume, cm^3
q = volumetric feed rate, cm^3/min
r_i = rate of reaction i, g mol reacted/cm^3 min
$Q = Q^*/qC_p$
Q^* = rate of heat removal by deliberate cooling, cal/min

In his derivation Aris assumes V, q, ρ, C_p, and ΔH_i to be constant. These assumptions are not very restrictive. Completion of the analysis or design requires the simultaneous solution of the several ordinary differential equations indicated above. Presently evolving computer software makes this a much less formidable problem than it once was.

4. Semibatch

Because of the great variety in the modes of operation of semibatch reactors, an energy balance general enough to satisfy all cases becomes unwieldy. As with the mass balance for semibatch reactors, it is recommended that the energy balance be derived for the specific case of interest. It should be recognized that one of the main reasons for using semibatch reactors is temperature control. Accordingly, the energy balance should be expected to play a paramount role. No more will be said at this point because the scaleup example to follow will deal with a semibatch reactor.

IV. AN EXAMPLE OF HOMOGENEOUS REACTOR SCALEUP

This example will focus on a semibatch reactor of a type not considered elsewhere in this book or in familiar textbooks. The process is the sulfonation of benzene to benzene sulfonic acid by sulfuric acid. Countless modes of operation exist for such sulfonations. The mode chosen for this illustration is similar to that originally put forth by Guyot [Gilbert and Groggins (1958)].

AN EXAMPLE OF HOMOGENEOUS REACTOR SCALEUP

Sulfuric acid is charged to a cast iron reaction vessel and heated to the desired reaction temperature. Benzene vapors are passed through the acidic liquid and some of the benzene is converted to benzene sulfonic acid (BSA).

$$C_6H_6 + H_2SO_4 \rightleftharpoons C_6H_5SO_3H + H_2O \quad (4\text{-}38)$$
$$\text{(B)} \quad \text{(A)} \quad \text{(S)} \quad \text{(W)}$$

The water formed would soon dilute the acid, quenching the reaction. However, the passage of excess benzene vapor through the reaction mass strips the water formed from the mixture. The result is that the reaction goes to completion, converting virtually all of the sulfuric acid to the sulfonic acid with little side reaction and waste products.

There are practical problems with this mode of operation. Benzene and sulfuric acid are nearly insoluble in one another; however, the product BSA is a solubilizing agent and when present will contribute to miscibility. Further, vigorous agitation by the bubble flow of large amounts of benzene and by mechanical stirring allow the reaction to be considered homogeneous. The results of Crooks and White (1950) suggest that agitation sufficient to eliminate mass transfer resistances is feasible. In Chapter 6 similar situations where mass transfer is not negligible are discussed.

It has been shown earlier that the essential ingredients of chemical reactor analysis and design are a rate equation, species mass balances, and an energy balance. These will be taken up in turn as this example is developed.

A. KINETICS

The kinetics of sulfonation reactions have been reviewed at length by Gilbert (1965) and much instructive insight is to be gained therefrom. But it would be pretty much a matter of luck if the scaleup engineer were to find just what he needed in the literature. Typically, neither Gilbert nor the subsequent literature gives a satisfactory rate equation for all conditions that might be considered in a design for commercial operation.

The closest one can find is a rate equation given by Crooks and White (1950) based upon data from an experimental reactor operated rather like the one selected for this example.

$$r_B = 118 C_A [x_A - \tfrac{1}{2}x_W + \tfrac{1}{4}x_S]^{-9.239 + (5349/T)} \quad (4\text{-}39)$$

r_B = rate of benzene sulfonation, mol BSA/hr L
C_A = concentration of H_2SO_4, mol/L
x_A = mol fraction H_2SO_4
x_W = mol fraction water
x_S = mol fraction BSA
T = temperature, K

They studied temperatures from 90 to 140°C. No concentration dependence on C_6H_6 was included because in all of their work they maintained a C_6H_6 fugacity of 1 atm by bubbling benzene through the reacting mixture. The reaction was found to be first order in sulfuric acid concentration although another concentration dependent term

$$[x_A - \tfrac{1}{2}x_W + \tfrac{1}{4}x_S]$$

based on the now-outmoded "pi-value" concept, was required to complete the correlation of their and other's data.

Correlation graphs of Crooks and White suggest that the observed temperature dependence be handled by Equation 4-39. However, the temperature dependence is accounted for by an exponent on a concentration term, a scheme that contradicts the concepts presented in Chapter 3. Equation 4-39 is acceptable only if our region of interest were within the realm of the correlated data. In these studies, x_W was on the order of x_A giving values of the bracketed term from 0.1 to 0.4. With a fractional value in the brackets, an increase in temperature corresponds to an increase in reaction rate which is what they observed. If the initial condition ($x_W = x_S = 0$, $x_A = 1$) is considered, Equation 4-39 shows no dependence whatever of the rate on temperature. Further, as the reaction proceeds, with $x_W \cong 0$, Equation 4-39 predicts that the rate should become increasingly dependent on temperature. Such behavior is quite unlikely and is contrary to experience in sulfonation. The dangers of using such a semiempirical equation beyond the range of its validation are clear.

For successful scaleup obviously kinetic studies over the complete range of anticipated operating conditions should be performed. For the purposes of this illustration the Crooks and White rate equation will be adopted and the process conditions constrained to fall within its range of validation.

Before Equation 4-39 can be used it is necessary to know what values of C_A, x_A, x_W, and x_S correspond to the conditions of operation. A rather general development will be presented at this point for use later in the example. The concentration of any species in the liquid phase is given by

$$C_i = \frac{N_i}{V} \tag{4-40}$$

where N_i is the mol i present in the liquid volume, V. The mass density of any species is

$$\rho_i = \frac{N_i M_i}{V_i} \tag{4-41}$$

For simplicity it is assumed that the species volumes are additive, hence

$$V = \sum_i V_i = \sum_i \frac{N_i M_i}{\rho_i} \tag{4-42}$$

AN EXAMPLE OF HOMOGENEOUS REACTOR SCALEUP

and

$$C_i = \frac{N_i}{\sum_i \frac{N_i M_i}{\rho_i}} \tag{4-43}$$

All the concentrations can be expressed in terms of the fractional conversion, X, defined here for sulfuric acid (A) as

$$X = \frac{N_{A_0} - N_A}{N_{A_0}} \tag{4-44}$$

Therefore

$$N_A = N_{A_0}(1 - X) \tag{4-45}$$

From the reaction stoichiometry and the physical situation it may be reasoned that 1 mole of benzene sulfonic acid (S) appears in the liquid for each mole of sulfuric acid which is reacted. Hence for no S in the initial charge

$$N_S = N_{A_0} - N_A = N_{A_0} X \tag{4-46}$$

It is proposed to control the rate of benzene vapor flow so that the concentration of aqueous sulfuric acid remains at 90 wt% as charged. Thus, the mass of water (W) equals one-ninth the mass of H_2SO_4 and

$$N_W = \frac{N_A M_A}{9 M_W} = N_{A_0}(1 - X)\frac{M_A}{9 M_W} \tag{4-47}$$

The amount of benzene (B) in the liquid depends upon the equilibrium with the benzene vapor being passed through the reaction mass. The operating pressure will be specified as 1 atm, which is the condition in the Crooks and White studies as well as the Guyot process. We will assume now, and verify later, that the flow rate of benzene vapor is much greater than the rate of formation and removal of water. Again, this is consistent with successful operating practice. From these two assertions, it follows that the partial pressure of benzene equals nearly 1 atm. Also, Raoults law will be taken to characterize the benzene vapor–liquid equilibrium. (In addition to the usual inadequacies of Raoults law, remember that there may be miscibility problems at low conversions.)

$$P_B = y_B P_T = x_B V P_B \quad \text{or} \quad x_B = \frac{1}{V P_B} \tag{4-48}$$

By definition,

$$x_B = N_B/(N_A + N_S + N_W + N_B) \tag{4-49}$$

Combining Equations 4-45 through 4-49 results in

$$N_B = \frac{N_{A_0}}{VP_B - 1}\left[1 + \frac{(1-X)M_A}{9M_W}\right] \quad (4\text{-}50)$$

Inserting Equations 4-45, 4-46, 4-47, and 4-50 into Equation 4-43 enables the determination of the liquid phase concentrations as given below.

$$C_A = \frac{(1-X)}{\dfrac{(1-X)M_A}{\rho_A} + \dfrac{XM_S}{\rho_S} + \dfrac{(1-X)M_A M_W}{9M_W \rho_W} + \dfrac{M_B}{(VP_B - 1)\rho_B}\left[1 + \dfrac{(1-X)M_A}{9M_W}\right]}$$

$$= \frac{(1-X)}{D + EX} \quad (4\text{-}51)$$

$$C_S = \frac{X}{\dfrac{(1-X)M_A}{\rho_A} + \dfrac{XM_S}{\rho_S} + \dfrac{(1-X)M_A M_W}{9M_W \rho_W} + \dfrac{M_B}{(VP_B - 1)\rho_B}\left[1 + \dfrac{(1-X)M_A}{9M_W}\right]}$$

$$= \frac{X}{D + EX} \quad (4\text{-}52)$$

$$C_B = \frac{\left[1 + \dfrac{(1-X)M_A}{9M_W}\right]/[VP_B - 1]}{\dfrac{(1-X)M_A}{\rho_A} + \dfrac{XM_S}{\rho_S} + \dfrac{(1-X)M_A M_W}{9M_W \rho_W} + \dfrac{M_B}{(VP_B - 1)\rho_B}\left[1 + \dfrac{(1-X)M_A}{9M_W}\right]}$$

$$= \frac{\dfrac{9M_W + M_A}{9M_W(VP_B - 1)} - \dfrac{M_A}{9M_W(VP_B - 1)}X}{D + EX} \quad (4\text{-}53)$$

$$C_W = \frac{\dfrac{(1-X)M_A}{9M_W}}{\dfrac{(1-X)M_A}{\rho_A} + \dfrac{XM_S}{\rho_S} + \dfrac{(1-X)M_A M_W}{9M_W \rho_W} + \dfrac{M_B}{(VP_B - 1)\rho_B}\left[1 + \dfrac{(1-X)M_A}{9M_W}\right]}$$

$$= \frac{\dfrac{M_A}{9M_W} - \dfrac{M_A}{9M_W}X}{D + EX} \quad (4\text{-}54)$$

AN EXAMPLE OF HOMOGENEOUS REACTOR SCALEUP

where

$$D = \frac{M_A}{\rho_A} + \frac{M_A}{9\rho_W} + \frac{M_B}{(VP_B - 1)\rho_B}\left[1 + \frac{M_A}{9M_W}\right] \quad (4\text{-}55)$$

$$E = \frac{M_S}{\rho_S} - \frac{M_A}{\rho_A} - \frac{M_A}{9\rho_W} - \frac{M_B M_A}{(VP_B - 1)\rho_B 9M_W} \quad (4\text{-}56)$$

From the definition of the liquid-phase mole fraction

$$x_i = \frac{N_i}{\sum_i N_i} \quad (4\text{-}57)$$

and Equations 4-45 to 4-47 and 4-50 the liquid-phase mole fractions may be shown (similarly to the concentrations) to be

$$x_A = \frac{(1 - X)}{\left(1 + \dfrac{M_A}{9M_W} + \dfrac{9M_W + M_A}{9M_W(VP_B - 1)}\right) - \left(\dfrac{M_A}{9M_W} + \dfrac{M_A}{(VP_B - 1)9M_W}\right)X} \quad (4\text{-}58)$$

$$x_S = \frac{X}{\left(1 + \dfrac{M_A}{9M_W} + \dfrac{9M_W + M_A}{9M_W(VP_B - 1)}\right) - \left(\dfrac{M_A}{9M_W} + \dfrac{M_A}{(VP_B - 1)9M_W}\right)X} \quad (4\text{-}59)$$

$$x_B = \frac{\dfrac{9M_W + M_A}{9M_W(VP_B - 1)} - \dfrac{M_A}{9M_W(VP_B - 1)}X}{\left(1 + \dfrac{M_A}{9M_W} + \dfrac{9M_W + M_A}{9M_W(VP_B - 1)}\right) - \left(\dfrac{M_A}{9M_W} + \dfrac{M_A}{(VP_B - 1)9M_W}\right)X} \quad (4\text{-}60)$$

$$x_W = \frac{\dfrac{M_A}{9M_W} - \dfrac{M_A}{9M_W}X}{\left(1 + \dfrac{M_A}{9M_W} + \dfrac{9M_W + M_A}{9M_W(VP_B - 1)}\right) - \left(\dfrac{M_A}{9M_W} + \dfrac{M_A}{(VP_B - 1)9M_W}\right)X} \quad (4\text{-}61)$$

It will be convenient to know the mass density of the liquid as a function of conversion.

$$\rho = C_A M_A + C_S M_S + C_B M_B + C_W M_W \qquad (4\text{-}62)$$

Combining Equations 4-51 to 4-56 and Equation 4-62 and simplifying

$$\rho = \frac{\left(M_A + \dfrac{(9M_W + M_A)M_B}{9M_W(VP_B - 1)} + \dfrac{M_A}{9}\right) + \left(M_S - M_A - \dfrac{M_A M_B}{9M_W(VP_B - 1)} - \dfrac{M_A}{9}\right)X}{D + EX}$$

$$(4\text{-}63)$$

In Table 4-5 are given the properties at a temperature of 140°C that are needed for Equations 4-51 to 4-54, 4-58 to 4-61, and Equation 4-63. Making the appropriate substitutions, we obtain

$$C_A = \frac{1 - X}{0.114 + 0.0392 X} \qquad (4\text{-}64)$$

$$C_S = \frac{X}{0.114 + 0.0392 X} \qquad (4\text{-}65)$$

$$C_B = \frac{0.442 - 0.167 X}{0.114 + 0.0392 X} \qquad (4\text{-}66)$$

$$C_W = \frac{0.605 - 0.605 X}{0.114 + 0.0392 X} \qquad (4\text{-}67)$$

$$x_A = \frac{1 - X}{2.047 - 0.772 X} \qquad (4\text{-}68)$$

$$x_S = \frac{X}{2.047 - 0.772 X} \qquad (4\text{-}69)$$

$$x_B = \frac{0.442 - 0.167 X}{2.047 - 0.772 X} \qquad (4\text{-}70)$$

$$x_W = \frac{0.605 - 0.605 X}{2.047 - 0.772 X} \qquad (4\text{-}71)$$

$$\rho = \frac{143 + 36.1 X}{0.114 + 0.0392 X} \qquad (4\text{-}72)$$

The initial and final liquid compositions and densities calculated from these equations are given in Table 4-6.

AN EXAMPLE OF HOMOGENEOUS REACTOR SCALEUP

**TABLE 4-5 Properties of Reacting Species
Sulfonation of Benzene**

Compound	M_i	VP_i, atm	ρ_i, g/L
H_2SO_4 (A)	98	Nonvolatile	1750
$C_6H_5SO_3H$ (S)	158	Nonvolatile	1270
C_6H_6 (B)	78	4.63	744
H_2O (W)	18	0.015^a	926

^{a} Partial pressure of water over 90 wt% H_2SO_4.

**TABLE 4-6 Liquid Density and Composition at 140°C
Sulfonation of Benzene**

Parameter	Initial Value ($X = 0$)	Final Value ($X = 1$)
C_A, mol/L	8.77	0
C_S, mol/L	0	6.53
C_B, mol/L	3.88	1.80
C_W, mol/L	5.31	0
x_A	0.488	0
x_S	0	0.784
x_B	0.216	0.216
x_W	0.296	0
ρ, g/L	1250	1170

The density changes by about 6 percent from start to finish. By comparison to other potential errors in the analysis, it appears that the constant density approximation would be tolerable if made. Actually it is not made in this illustration. Comparison of the values in Table 4-6 to those in Crooks and White's work shows the conditions of this illustration to be consistent with those for which the rate equation was developed.

B. Mass Balances and Isothermal Performance

A semibatch reaction is inherently a dynamic process. For the moment isothermal operation at 140°C will be considered. Accordingly, unsteady-state mass balances are the key to the determination of the reactor performance, specifically the time dependence of the fractional conversion of sulfuric acid. The reaction occurs in the period between the charging of the sulfuric acid and the dumping of the benzene BSA product mixture. The condensation of benzene vapor to saturate the liquid mixture is taken to be rapid. To some extent at least it is a means of heating the acid charge. During the reaction period there is no in- or outflow of sulfuric acid or BSA and no inflow of

water. Thus the mass balances for each of the reacting species are
for benzene (B)

$$F_{t,\text{in}} y_{B,\text{in}} - F_{t,\text{out}} y_{B,\text{out}} - r_B V = \frac{dN_B}{dt} \qquad (4\text{-}73)$$

for water (W)

$$-F_{t,\text{out}} y_{W,\text{out}} + r_B V = \frac{dN_W}{dt} \qquad (4\text{-}74)$$

for H_2SO_4 (A)

$$-r_B V = \frac{dN_A}{dt} \qquad (4\text{-}75)$$

for BSA (S)

$$r_B V = \frac{dN_S}{dt} \qquad (4\text{-}76)$$

All four of these equations are used to find the concentrations and flow rates of all species in all phases but only one of them is required to find $X(t)$. Combining Equations 4-39, 4-40, 4-45, 4-64, 4-68, 4-69, and 4-71 readily yields

$$r_B V = 118(1 - X) \left[\frac{0.698 - 0.448 X}{2.047 - 0.772 X} \right]^{-9.239 + (5349/T)} N_{A_0} \qquad (4\text{-}77)$$

Differentiation of Equation 4-45 gives

$$\frac{dN_A}{dt} = -N_{A_0} \frac{dX}{dt} \qquad (4\text{-}78)$$

Combining the sulfuric acid mass balance, Equation 4-75, with Equations 4-77 and 4-78 gives

$$118(1 - X) \left[\frac{0.698 - 0.448 X}{2.047 - 0.772 X} \right]^{-9.239 + (5349/T)} = \frac{dX}{dt} \qquad (4\text{-}79)$$

For the isothermal case, the temperature (413K) is not a function of conversion and Equation 4-79 can be integrated by separating the variables.

$$t = \int_0^t dt = \int_0^X \frac{1}{118(1 - X) \left[\frac{0.698 - 0.448 X}{2.047 - 0.772 X} \right]^{3.713}} dX \qquad (4\text{-}80)$$

The integration of the right-hand side can be done graphically or numerically. The calculated results are presented in Table 4-7 and in Figure 4-8.

AN EXAMPLE OF HOMOGENEOUS REACTOR SCALEUP 151

**TABLE 4-7 Conversion in an Isothermal Semibatch Reactor
Sulfonation of Benzene**

X	t, hr	X	t, hr
0	0	0.8	1.7
0.1	0.05	0.9	3.2
0.2	0.12	0.95	5.0
0.3	0.20	0.99	10
0.4	0.31	0.995	13
0.5	0.45	0.999	19
0.6	0.67		
0.7	1.0		

From Table 4-7 it can be seen that 70 percent of the acid can be converted in 1 hr. Two more hours would be required to reach a 90 percent conversion. Ten hours are required to convert all but 1 percent of the original acid. The decision on where to stop the reaction depends on downstream processing, product specifications, desired production rate, waste disposal, and a host of other economic factors.

Obviously the availability of $X(t)$ information enables the calculation of liquid-phase density and concentrations of all species as functions of time

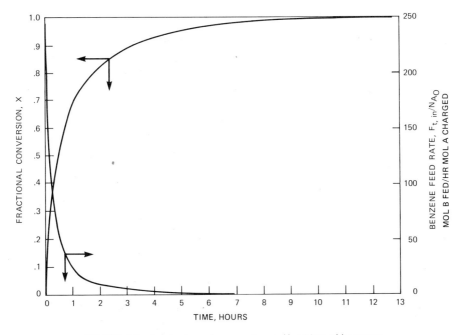

FIGURE 4-8 Feed rate and conversion, sulfonation of benzene.

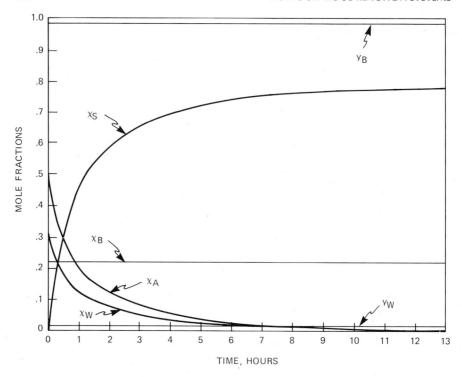

FIGURE 4-9 Composition of vapor and liquid phases, sulfonation of benzene.

through Equations 4-64 to 4-72. The component mole fractions in the liquid and vapor phases are plotted as functions of time in Figure 4-9. In obtaining the vapor-phase mole fractions, water in the reacting solution was assumed to exert the same partial pressure that it would over 90 wt% aqueous sulfuric acid, namely 0.015 atm. Thus,

$$y_W = \frac{P_W}{P_T} = \frac{0.015}{1} = 0.015 \tag{4-81}$$

and

$$y_B = \frac{(P_T - P_W)}{P_T} = \frac{(1 - 0.015)}{1} = 0.985 \tag{4-82}$$

The demonstrated preponderance of benzene in the outflow stream validates an approximation made earlier in developing Equation 4-48.

It may be noted that Equations 4-75 and 4-76 confirm the intuitive fact that whatever H_2SO_4 reacts is replaced in the liquid by BSA (i.e., $dN_A/dt = -dN_S/dt$). Indeed, the problem could have been solved just as well by using Equation 4-76 instead of 4-75.

AN EXAMPLE OF HOMOGENEOUS REACTOR SCALEUP

Equation 4-73 may be recognized as the unsteady-state CSTR mass balance, which is appropriate for benzene. Equation 4-50 showed N_B to be a function of temperature-dependent parameters VP_B and X as well as N_{A_0}, M_A, and M_W. For later use in the analysis of a nonisothermal case, dN_B/dt will be derived allowing for variation of VP_B with time as well as X. The chain rule of differentiation gives

$$\frac{dN_B}{dt} = \frac{\partial N_B}{\partial VP_B} \frac{dVP_B}{dt} + \frac{\partial N_B}{\partial X} \frac{dX}{dt} \qquad (4\text{-}83)$$

Performing the indicated operations on Equation 4-50 leads to

$$\frac{dN_B}{dt} = -\frac{N_{A_0}}{(VP_B - 1)^2}\left[1 + \frac{(1-X)M_A}{9M_W}\right]\frac{dVP_B}{dt} - \frac{N_{A_0}M_A}{(VP_B - 1)9M_W}\frac{dX}{dt}$$

$$(4\text{-}84)$$

In the isothermal case presently under consideration, temperature does not change with time and hence $dVP_B/dt = 0$. Thus,

$$\frac{dN_B}{dt} = -\frac{N_{A_0}M_A}{(VP_B - 1)9M_W}\frac{dX}{dt} \qquad (4\text{-}85)$$

The derivative, dX/dt, is already known from Equation 4-79.

Combining Equations 4-73, 4-77, 4-79, and 4-85 gives for pure benzene vapor inflow ($y_{B,in} = 1$)

$$\frac{F_{t,in} - F_{t,out}y_{B,out}}{N_{A_0}} = \left[1 - \frac{M_A}{(VP_B - 1)9M_W}\right]118(1 - X)$$

$$\times \left[\frac{0.698 - 0.448X}{2.047 - 0.772X}\right]^{-9.239 + (5349/T)} \qquad (4\text{-}86)$$

For the isothermal case, the conversion is known as a function of time from Table 4-7. The unknowns are $F_{t,in}$, $F_{t,out}$, and $y_{B,out}$. However, $y_{B,out} = 0.985$ from Equation 4-82.

For more information we can turn to the water mass balance, Equation 4-74. The accumulation term, dN_W/dt, can be found by differentiating Equation 4-47 to give

$$\frac{dN_W}{dt} = -\frac{N_{A_0}M_A}{9M_W}\frac{dX}{dt} \qquad (4\text{-}87)$$

This equation can be combined with Equations 4-74, 4-77, and 4-79 to give

$$\frac{F_{t,\text{out}} y_{W,\text{out}}}{N_{A_0}} = \left[1 + \frac{M_A}{9M_W}\right] 118(1 - X) \left[\frac{0.698 - 0.448X}{2.047 - 0.772X}\right]^{-9.239 + (5349/T)}$$

(4-88)

Unlike Equation 4-86, this equation is not restricted to a constant temperature. Equation 4-88 represents an additional equation (making two). Its additional unknown, $y_{W,\text{out}} = 0.015$ can be obtained from Equation 4-81.

This completes the pair of equations necessary to find the flow rates, $F_{t,\text{in}}$ and $F_{t,\text{out}}$. From Equations 4-86 and 4-88 it will be noted that the vapor flow rates are proportional to the initial sulfuric acid charged as is the product, $r_B V$. Thus the larger the initial charge of sulfuric acid, the larger the reactor volume and benzene flow rates.

From Equation 4-88 the moles of water leaving the reactor per hour per mole of acid charged ($F_{W,\text{out}}/N_{A_0}$) can be calculated for any conversion, X. Multiplying this result by $y_B/y_W = 65.67$, from the values given in Equations 4-81 and 4-82, gives the moles of benzene leaving the reactor per hour per mole of acid charged ($F_{B,\text{out}}/N_{A_0}$) at any X. At this point $F_{t,\text{out}}/N_{A_0} = (F_{W,\text{out}} + F_{B,\text{out}})/N_{A_0}$ is known. What remains to be found is $F_{t,\text{in}}/N_{A_0}$. From Equation 4-86 this is

$$\frac{F_{t,\text{in}}}{N_{A_0}} = \frac{F_{B,\text{out}}}{N_{A_0}} + \left[1 - \frac{M_A}{(VP_B - 1)9M_W}\right] 118(1 - X)$$

$$\times \left[\frac{0.698 - 0.448X}{2.047 - 0.772X}\right]^{-9.239 + (5349/T)}$$

(4-89)

Calculated results for 140°C, using parameter values from Table 4-5, are shown in Table 4-8.

As time goes on the rate of water removal, $F_{W,\text{out}}/N_{A_0}$, decreases drastically. Thus the rate at which benzene must be passed through the reactor decreases by orders of magnitude as the conversion goes from 0 to 1.

Compared to commercial operations the benzene feed rate per unit of sulfuric acid charged is unreasonably large. For example, if the reaction were run only to 70 percent conversion,

$$\int_0^1 \frac{F_{t,\text{in}}}{N_{A_0}} dt = 70 \text{ mol } C_6H_6/\text{mol } H_2SO_4 \text{ (56 kg } C_6H_6/\text{kg } H_2SO_4\text{)}$$

would be required compared to typical ratios (Gilbert and Groggins, 1958) of 10 mol/mol or less. For a 99 percent conversion 100 mol C_6H_6/mol H_2SO_4 would be needed. The reason for this is that the temperature of the process in

AN EXAMPLE OF HOMOGENEOUS REACTOR SCALEUP

TABLE 4-8 Flow Rates in Semibatch Process[a]

X	t, hr	$F_{W,\text{out}}/N_{A_0}$	y_B/y_W	$F_{B,\text{out}}/N_{A_0}$	$F_{t,\text{out}}/N_{A_0}$	$(F_{t,\text{in}} - F_{B,\text{out}})/N_{A_0}$	$F_{t,\text{in}}/N_{A_0}$
0	0	3.49	65.67	229	232	1.81	231
0.1	0.05	2.83	65.67	186	189	1.47	187
0.2	0.12	2.24	65.67	147	149	1.16	148
0.3	0.20	1.72	65.67	113	115	0.89	114
0.4	0.31	1.28	65.67	84.1	85.4	0.66	84.8
0.5	0.45	0.900	65.67	59.1	60.0	0.47	59.6
0.6	0.67	0.594	65.67	39.0	39.6	0.31	39.3
0.7	1.0	0.356	65.67	23.4	23.8	0.19	23.6
0.8	1.7	0.182	65.67	12.0	12.2	0.09	12.1
0.9	3.2	0.0662	65.67	4.35	4.42	0.03	4.38
0.95	5.0	0.0275	65.67	1.81	1.84	0.01	1.82
0.99	10	0.00467	65.67	0.307	0.312	0.00	0.307
0.995	13	0.00228	65.67	0.150	0.152	0.00	0.150
0.999	19	0.000449	65.67	0.0295	0.0299	0.00	0.0295
1	∞	0	65.67	0	0	0	0

[a] Flow rate units are mol flowing/hr mol acid charged.

this illustration was limited to 140°C or less by the range of validity of the rate equation. At a higher temperature the vapor pressure of water over the acid solution would increase greatly, thereby significantly reducing the amount of stripping benzene required. Of course, even in the present case the benzene would be recycled and thus would not be wasted. Nevertheless the cost of vaporizing the recycle stream would be excessive unless the water could be removed from the vapor without condensation of the vapor benzene.

Another way to reduce the benzene flow rate would be to use a more dilute acid. The reaction stops for sulfuric acid less than 78 wt% because of chemical complexing with the water (Groggins and Gilbert, 1958). At 80 wt%, however, the partial pressure of water over the acid would be considerably higher, thus reducing the quantity of stripping benzene required. At this acid concentration the reaction rate would be much lower and a longer reaction time would result. Again, a higher temperature would relieve this somewhat. For 80 wt% acid at 180°C the water vapor pressure is about 0.5 atm compared to the value of 0.015 atm used for 90 percent acid at 140°C. Clearly reaction rate data at temperatures up to 180°C should be obtained. The alternative is trial-and-error process development.

In this illustration the vapor–liquid equilibrium has been rather cavalierly treated as an ideal miscible system, when in fact immiscibility is anticipated early in the reaction and few systems are ideal. The results and trends obtained are probably qualitatively correct but their quantitative accuracy should be viewed skeptically until a more definitive analysis is performed. Even if theoretical work were halted at this point, however, far greater insight would have been gained than if it never had been done at all.

C. Energy Balance and Thermal Effects

To analyze the potential thermal effects in this semibatch reaction an energy balance is required. For greatest rigor a general enthalpy balance might be preferred, but the rigor is lost if the available data are inadequate (as they are in this instance). A more phenomenological energy balance is presented here.

$$F_{t,\text{in}}Cp_{\text{in}}M_{\text{in}}(T_{\text{in}} - T_0) - F_{t,\text{out}}Cp_{\text{out}}M_{\text{out}}(T_{\text{out}} - T_0) + r_BV(-\Delta H_R) + Q$$
$$= \frac{d}{dt}N_tCp_tM_t(T_{\text{out}} - T_0) \quad (4\text{-}90)$$

The five terms represent the rate of heat (a) flowing in with the benzene feed, (b) flowing out with the vapor, (c) generation by the reaction, (d) transfer from the surroundings, and (e) accumulation in the reaction liquid. For coherence with a rigorous enthalpy balance the heat of reaction should be calculated at the datum temperature, T_0.

On the basis of what is already known quite a few simplifications are possible in Equation 4-90. The in- and outflow streams have nearly identical flow rates and are pure and nearly pure benzene, respectively. Therefore,

$$F_{t,\text{in}} = F_{t,\text{out}}$$
$$Cp_{\text{in}} = Cp_{\text{out}} = Cp_B$$
$$M_{\text{in}} = M_{\text{out}} = M_B$$

The first two terms can be combined to give $F_{t,\text{in}}Cp_BM_B(T_{\text{in}} - T_{\text{out}})$ and the product, $Cp_BM_B = 27.6$ cal/mol K, can be recognized as the molar heat capacity of vapor benzene. From Table 4-8, $F_{t,\text{in}}/N_{A_0}$ is available as a function of conversion and time.

In the accumulation term of Equation 4-90, the product Cp_tM_t may be recognized as the molar heat capacity of the liquid mixture. Reid et al. (1977) suggest a mole fraction average of pure component molar hear capacities for liquid mixtures. Therefore

$$Cp_tM_t = \sum_i Cp_iM_ix_i \quad (4\text{-}91)$$

Table 4-9 gives the data necessary for the use of Equation 4-91.

The mole fractions, x_i, can be obtained for any temperature from Equations 4-58 to 4-61 or from Equations 4-68 to 4-71 for isothermal operation at 140°C. Experience shows that the variation of pure component, liquid heat capacities

AN EXAMPLE OF HOMOGENEOUS REACTOR SCALEUP

TABLE 4-9 Liquid Heat Capacity Data at 140°C Sulfonation of Benzene

Compound	Cp_i, cal/g K	M_i	$Cp_i M_i$, cal/mol K
C_6H_6	0.528	78	41.2
H_2O	1.10	18	19.8
H_2SO_4	0.398	98	39.0
BSA	0.5	158	79.0

with temperature has a very minor impact on the characterization of thermal effects in process systems as long as values representative of the regime of interest are used. Combining Equations 4-68 to 4-71, the data of Table 4-9, and Equation 4-91 gives

$$Cp_t M_t = \frac{69.2 - 21.1X}{2.047 - 0.772X} \tag{4-92}$$

Equation 4-92 is limited to 140°C because the numerical coefficients are functions of the vapor pressure of benzene. The total moles of liquid can be calculated from Equations 4-45 to 4-47 and 4-50 by $N_t = \Sigma_i N_i$ to give in general

$$\frac{N_t}{N_{A_0}} = \left(1 + \frac{M_A}{9M_W} + \frac{9M_W + M_A}{9M_W(VP_B - 1)}\right) - \left(\frac{M_A}{9M_W} + \frac{M_A}{(VP_B - 1)9M_W}\right)X \tag{4-93}$$

and for 140°C

$$\frac{N_t}{N_{A_0}} = 2.047 - 0.772X \tag{4-94}$$

Combining Equations 4-92 and 4-94 gives the $N_t Cp_t M_t$ product in the accumulation term at 140°C to be

$$N_t Cp_t M_t = (69.2 - 21.1X) N_{A_0} \tag{4-95}$$

For any temperature a similar, but more general, development leads to

$$\frac{N_t Cp_t M_t}{N_{A_0}} = Cp_A M_A + \frac{Cp_W M_A}{9} + \frac{Cp_B M_B}{(VP_B - 1)}\left(1 + \frac{M_A}{9M_W}\right)$$

$$+ \left(Cp_S M_S - Cp_A M_A - \frac{Cp_W M_A}{9} - \frac{Cp_B M_B M_A}{(VP_B - 1)9M_W}\right)X$$

$$\tag{4-96}$$

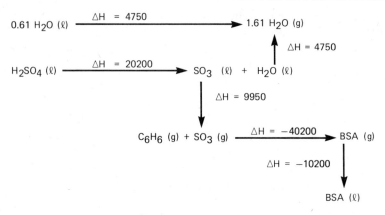

FIGURE 4-10 Thermal effects accompanying the sulfonation of benzene.

Turning to the reaction term, we recall that $r_B V$ is given as a function of X by Equation 4-77. That equation was derived for 140°C but it can also be used to account for some nonisothermal character. The real difficulty comes in estimating the heat of reaction. The net reaction process in the reaction of 1 mole of H_2SO_4 is shown in Figure 4-10. The net process consists of reacting 1 mole of liquid sulfuric acid with 1 mole of vapor benzene to give 1 mole of liquid BSA and 1 mole of water vapor. Further the constraint that the aqueous sulfuric acid concentration is to be maintained at 90 wt% H_2SO_4 means that 0.61 mole of water of dilution must be removed for each mole of acid reacted. The pathways leading to this net process, as shown in Figure 4-10, are for calculational convenience.

The data available in the open literature are inadequate to characterize accurately all of the thermal effects. All enthalpy changes indicated were taken at room temperature where a choice existed, had units of cal/mol, and came from standard tables unless discussed below. The effect of temperature on heats of reaction was discussed in Chapter 3. The heat of the benzene sulfonation reaction was taken to be the same as the measured value for sulfonation of dodecylbenzene reported by Gilbert et al. (1953). The heat of condensation of benzene sulfonic acid was taken to be the same as for benzoic acid. The heat required to separate the water and sulfur trioxide in sulfuric acid came from Gilbert and Groggins (1958). The net heat of reaction is given by Equation 4-97.

$$\Delta H_R = \underset{\substack{\text{(splitting)} \\ H_2SO_4}}{20{,}200} + \underset{\substack{\text{(vaporizing)} \\ SO_3}}{9950} + \underset{\substack{\text{(vaporizing)} \\ H_2O}}{7650} - \underset{\substack{\text{(sulfonation)} \\ \text{reaction}}}{40{,}200} - \underset{\substack{\text{(condensing)} \\ BSA}}{10{,}200}$$

$$= -12{,}600 \text{ cal/mol} \qquad (4\text{-}97)$$

Although the sulfonation reaction is strongly exothermic, the net reaction is

AN EXAMPLE OF HOMOGENEOUS REACTOR SCALEUP

only mildly so because of the endothermic processes that accompany it in the present mode of operation.

The energy balance, Equation 4-90, can now be rewritten with these various quantitative values and simplifications included.

$$N_{A_0}\left(\frac{F_{t,\text{in}}(X)}{N_{A_0}}\right)(27.6)(T_{\text{in}} - T_{\text{out}})$$

$$+ 118(1-X)\left[\frac{0.698 - 0.448X}{2.047 - 0.772X}\right]^{-9.239+(5349/T)} N_{A_0}(12,600) + Q$$

$$= \frac{d}{dt} N_{A_0}(69.2 - 21.2X)(T_{\text{out}} - T_0) \qquad (4\text{-}98)$$

Equation 4-98 can be used at temperatures other than 140°C, but in doing so one should remember that the effects of the dependence of the benzene vapor pressure, VP_B, on temperature at several locations within the model would not be reflected. No doubt some, perhaps nearly all, useful insight can be obtained without incorporating this rigor. But the vapor pressure is a very strong function of temperature and the issue (unexplored so far) is how this function will propagate through the simultaneous solution of the more rigorously derived rate equation, mass balances, and energy balance. For anyone interested in investigating any nonisothermal case more rigorously, all of the information necessary to do so has been presented here except for the details of the solution.

The study of thermal effects in isothermal operation at 140°C is of interest in itself. The datum temperature, T_0, can be set to any desired value; 140°C ($= T_{\text{out}}$) is particularly convenient. Referring to Equations 4-90 or 4-98, it can be seen that for this specification the flow and accumulation terms drop out. What remains is

$$\left(-\frac{Q}{N_{A_0}}\right) = (118)(12,600)(1-X)\left[\frac{0.698 - 0.448X}{2.047 - 0.772X}\right]^{3.713} \qquad (4\text{-}99)$$

Equation 4-99 tells how many calories of heat must be removed per hour per mole of H_2SO_4 charged in order to maintain isothermality. Calculated values of $(-Q/N_{A_0})$ as a function of conversion level, X, are shown in Table 4-10.

Inasmuch as 140°C was chosen as the datum temperature for this calculation, the tacit assumption has been made that the $\Delta H_R = -12,600$ cal/mol is for the reaction occurring at 140°C. The effect of temperature on heat of reaction is not large and the estimation of ΔH_R was so crude that allowing for variation with temperature is unwarranted. Nevertheless, ΔH_R was estimated at room temperature, that is, about 25°C. Thus the omission of the accumulation term in Equation 4-98 implies some error. This error is subject to

TABLE 4-10 Heat Removal Rate in Isothermal Sulfonation of Benzene

X	$(-Q/N_{A_0})$, cal/hr mol A	X	$(-Q/N_{A_0})$, cal/hr mol A
0	27400	0.8	1430
0.1	22200	0.9	520
0.2	17600	0.95	216
0.3	13500	0.99	36.6
0.4	10000	0.995	17.9
0.5	7070	0.999	3.52
0.6	4670	1	0
0.7	2800		

evaluation as follows. Dropping the flow term as before and dividing by N_{A_0} gives

$$\frac{Q}{N_{A_0}} = -12{,}600 \times 118(1-X)\left[\frac{0.698 - 0.448X}{2.047 - 0.772X}\right]^{-9.239+(5349/T)}$$

$$+ \frac{d}{dt}(69.2 - 21.2X)(T_{\text{out}} - T_0) \qquad (4\text{-}100)$$

The heat of reaction (second) term in Equation 4-100 has the magnitude given for Q/N_{A_0} in Table 4-10. For isothermal operation, $(T_{\text{out}} - T_0)$ is a constant and the accumulation term can be simplified to

$$\frac{d}{dt}(69.2 - 21.2X)(T_{\text{out}} - T_0) = -21.2(T_{\text{out}} - T_0)\frac{dX}{dt} \qquad (4\text{-}101)$$

To allow for the accumulation term where the impact is the greatest and to illustrate how rigorous calculations could be performed if desired, we consider the initial condition where conversion is zero. From Figure 4-8, the slope of the $X(t)$ curve at $t = 0$ is found to be

$$\frac{dX}{dt} = 2 \qquad (4\text{-}102)$$

Thus the accumulation term is

$$(-21.2)(140 - 25) \times 2 = -4880 \text{ cal/hr mol acid charged} \qquad (4\text{-}103)$$

which is 18 percent of the value of 27,400 given for $(-Q/N_{A_0})_{X=0}$ in Table 4-10. This discrepancy is considerably smaller than the percentage uncertainty in ΔH_R. Calculating Q/N_{A_0} from Equation 4-100, however, gives

$$Q/N_{A_0} = -27{,}400 - 4880 = -32{,}300 \text{ cal/hr mol} \qquad (4\text{-}104)$$

D. Size, Shape, and Performance

Eventually in any scaleup the reactor dimensions, flow rates, and so on must be specified. Let $N_{A_0} = 38{,}700$ mol H_2SO_4 be the initial charge to the reactor. In practice the desired production rate would be specified. Assume that the reaction is to be run until 99 percent of the H_2SO_4 has reacted to form benzene sulfonic acid. This will take about 10 hr and will allow one batch to be processed and worked up per day. Thus a production rate of BSA of just over 6 metric tons per day ($38{,}700 \times 0.99 \times 158 \times 10^{-6} = 6.05$) can be achieved. From Table 4-8 the flow rates of benzene ($38{,}700 F_{t,\text{in}}/N_{A_0}$) to the reactor initially and at 99% conversion would be 8.94×10^6 and 1.19×10^4 mol/hr, respectively. The origin of the impractically high initial rate was discussed earlier. The maximum rate of heat release and hence the maximum required heat exchange occurs at $X = 0$ (see Equation 4-99). From Table 4-10, $(-Q_{\max}) = 27{,}400 \times 38{,}700 = 1.06 \times 10^9$ cal/hr. With such a high flow rate of benzene this amount of heat could be carried away with very little rise in T_{out}. With more realistic flow rates, however, the goal would be to see if adequate heat exchange surface exists to maintain isothermality by rejecting all of the heat of reaction. For this purpose a jacketed reactor, such as is commonly used, would be specified.

From relationships presented earlier the volume occupied by the liquid reacting mass can be shown to be

$$V = \sum_i V_i = \sum_i \frac{N_i M_i}{\rho_i} = N_{A_0}(D + EX) \tag{4-105}$$

For 140°C this becomes

$$\frac{V}{N_{A_0}} = 0.114 + 0.0392 X \tag{4-106}$$

Obviously the maximum volume requirement occurs at the end of the reaction when $V = [0.114 + 0.0392 (0.99)]38{,}700 = 5910$ L $= 5.91$ m³. The actual reactor volume will need to be somewhat larger than this, but this value will be used for the volume enclosed by the cooling jacket. The reactor is taken to be cylindrical with hemispherical ends with the height of the cylindrical section of the jacketed portion equal to the diameter. Thus the volume enclosed by the jacket is $V = \pi D^3/4 + \pi D^3/12 = \pi D^3/3$. For $V = 5.91$ m³, the diameter, D, comes out to be 1.78 m. The surface area of the reactor available for heat exchange is $A = \pi D^2 + \pi D^2/2 = 3\pi D^2/2 = 14.9$ m².

The question is, is this heat transfer surface sufficient to accommodate the maximum rate of heat release? Cooling water will be assumed to be available at 15°C and the overall heat transfer coefficient will be taken as 4.89×10^5 cal/hr m² °C (100 Btu/hr ft² °F). The amount of heat that can be rejected is $Q = UA \, \Delta T = 4.89 \times 10^5 \times 14.9 \times (140 - 15) = 9.11 \times 10^8$ cal/hr. This value is almost identical with the maximum rate of heat release of 1.06×10^9 cal/hr, so the heat exchange capability is satisfactory. Some caution about this conclusion is warranted since if the heat transfer requirement is taken from Equation 4-104, $(-Q_{max}) = 32{,}300 \times 38{,}700 = 1.25 \times 10^9$ cal/hr. This result suggests that cooling coils or some other heat exchange augmentation may be required.

This illustration has considered just one operating mode of a single semibatch reaction. The analysis was drastically simplified and the directions to greater rigor were pointed out. Places where the theory and supporting data were inadequate were identified. Nevertheless, the analysis provides considerable insight into the nature of the process. If the indicated design were commercialized some of the pitfalls suggested in the analysis would be manifested as startup or other operating problems. Even so, the analysis would be invaluable in understanding and solving these problems. Further the analysis can be extended to a variety of other modes of operation, eventually leading to the optimal process. From the specific case considered here and succeeding variations of it, there arise specific areas for further study (characterizing the vapor–liquid equilibrium, extending the temperature range of the rate equation, better definition of the heat of reaction, etc.). Laboratory investigations of these areas would be much more cost effective than wrestling with the global problem by trial and error in an integrated pilot plant.

V. UNCERTAINTIES

This chapter has dealt with many aspects of analysis, design, and scaleup of homogeneous reaction systems. Still, the cases considered were necessarily idealized to a considerable extent. A number of uncertainties remain to plague the scaleup engineer.

A. Nonideal Flow

The stirred tanks considered earlier in this chapter were always assumed to be perfectly mixed. For economic reasons or error this is often not achieved. The mean residence time in a poorly mixed tank may differ considerably from that in a well-mixed tank. The whole matter of residence time distribution is considered in Chapter 8. At this point it may be noted that a single stirred tank can sometimes be successfully modeled as some combination of perfectly mixed and plug flow reactors with dead space. Effective presentations of such combined models are made by Levenspiel (1972), Himmelblau and Bischoff (1968), and Wen and Fan (1975).

Another form of ideal flow was the assumption of plug flow in tubular reactors. The plug flow assumption is often excellent in fixed beds and quite satisfactory for fluids flowing in a thoroughly turbulent regime. When the deviations from plug flow are not too great, dispersion models are often effective in characterizing the back mixing. Viscous fluids, however, may experience extreme residence time distributions. Even vigorous turbulence in the bulk fluid can be accompanied by very slow laminar flow near the walls. The references given above for stirred tank nonidealities are also pertinent for tubular reactors. But such models can hardly predict or account for the "gunk" formation at the wall of a polymerization reactor. Scaleup of mixers for particularly difficult situations is considered in Chapter 9.

Departures from ideal flow behavior have in the past been a major impediment to effective startup. However the effective use of mockups, residence time distribution theory, and more sophisticated system models have enabled a rational and reasonably successful solution to these problems. For an indication of how residence time distribution theory is used to treat the laminar flow reactor problem discussed earlier, see Hill (1977) or Smith (1970). As has been shown, residence time distributions can arise in well-mixed (micromixed) and segregated flow systems. After the mean residence time, the residence time distribution is the major factor governing the conversion in a nonideal reactor. The degree of segregation or micromixing can have a third level effect. When mixing is poor, however, temperature gradients as well as micro- and macroscale concentration gradients arise. The resulting thermal effects may dwarf in importance the increasingly subtle concentration effects.

B. Heat and Mass Transfer Effects

In homogeneous reaction systems, complexities due to interfacial mass transfer are absent. In poorly mixed systems with partially segregated flow, such as in polymerization reactors, diffusional resistances can play a role. The amelioration of such a problem is less likely to be found in theoretical analysis than in the provision of better mixing, for example, with static mixers. Much more troublesome is the matter of heat transfer. In some cases this will be simply a matter of heat exchange and the providing of sufficient surface area to accomplish it. However, temperature gradients within the reactor can become quite a problem for viscous materials with high heats of reaction or for temperature sensitive materials. Removal of heat in such a situation can be difficult.

Addition of heat to an endothermic system is somewhat easier. When turbulent flow exists radial mixing of both mass and heat in tubes is usually excellent. Also, although substantial longitudinal temperature gradients may exist, the heat transfer due to these gradients is usually negligible compared to the heat carried along by the bulk fluid flow; hence axial profiles are seldom a major stumbling block. Chapter 5 deals extensively with heat and mass transfer effects in fixed-bed reactors. Many of the ideas presented there are adaptable to homogeneous reactors as well.

C. Competing Reactions and Impurities

One could get the impression from reading textbooks that the major problem in chemical reaction engineering is to determine the size of a single reactor to carry out a single reaction. In developing commercial processes, however, great attention is given to the suppression of undesired side reactions and by-products. The analysis of simultaneous reactions may be prohibitively complex for a fundamental approach to be successful. Therefore, a more empirical model of scaleup is frequently adopted. Nevertheless clever engineering is often successful in ameliorating the problems caused by competing reactions and impurities. By far the best treatment in the area of optimizing selectivity is given by Carberry (1976). The matter of selectivity has great impact on the selection of reactor type, the subject of Chapter 7.

D. Reactor Instability

Two kinds of reactor instability are treated in some detail in this book. One is the earlier treatment in this chapter on multiple steady states in CSTRs. A second is the problem of runaway in exothermic reactions, treated in Chapter 5. In more complicated cases theoretical analysis of reactor dynamics and instability is quite difficult and therefore such cases pose problems in scaleup. Even if definitive quantification is out of the question, the value of qualitative insights and scaleup decisions based upon them should not be underestimated. In any case, careful attention to the details of process control is essential.

NOMENCLATURE

A	Frequency factor
C_i	Concentration of component i
C_{i_0}	Concentration of component i at inlet
Cp	Heat capacity at constant pressure
Cp_i	Heat capacity of component i
Cp_t	Heat capacity of solution
C_V	Heat capacity at constant volume
D	A constant defined by Equation 4-55
E	A constant defined by Equation 4-56
E	Activation energy
F_i	Molar flow rate of component i at any point
F_t	Total molar flow rate at any point
F_{t_0}	Total molar flow rate at inlet
ΔH_R	Heat of reaction
$[i]$	Concentration of component i
$[i]_0$	Initial concentration of component i
k, k_f, k_r	Reaction rate coefficient
l	Axial position coordinate

NOMENCLATURE

L	Length of reactor
m_{i_0}	Initial mass of i present
M_i	Molecular weight of component i
M_t	Molecular weight of mixture
N_i	mol i present/total mol feed
N_i	Moles of i present
N_{i_0}	Initial moles of i present
N_t	Total moles present
P_i	Partial pressure of component i
P_T	Total pressure
Q	Rate of heat transfer from surroundings
$(-Q_{max})$	Maximum rate of heat removal
Q_0	Volumetric feed rate
r	Radial position coordinate
r, r_i	Rate of reaction of component i
R	Radius of reactor
R	Gas constant
t	Time
t_f	Final reaction time
t_r	True residence time
T	Absolute temperature
T_0	Initial, feed, or datum temperature
u	Fluid velocity at any point
u_0	Fluid velocity at reactor inlet
$\bar{u}$	Mean fluid velocity over cross section
V	Reactor volume
V_i	Volume of component i
VP_i	Vapor pressure of component i
W	Mass of catalyst
x	mol i converted/total mol feed
x	Moles converted
x_{eq}	Moles converted at equilibrium
x_f	Final moles converted
x_i	Mole fraction of component i in liquid phase
X	Fractional conversion
y_i	Mole fraction of component i in vapor phase
ε	Void fraction

Greek

$\bar{\theta}$ Mean residence time in CSTR $= V/Q_0$
ρ Density of mixture
ρ_c Bulk density of catalyst
ρ_i Density of pure component i

REFERENCES

Aris R., *Elementary Chemical Reactor Analysis*, Prentice-Hall, Englewood Cliffs, NJ, 1969.

Carberry, J. J., *Chemical and Catalytic Reaction Engineering*, McGraw-Hill, New York, 1976.

Cleland, F. A., and Wilhelm, R. H., "Diffusion and Reaction in Viscous-Flow Tubular Reactor," *AIChE J*, **2**, 489–497 (1956).

Crooks, R. C., and White, R. R., "Rate of Sulfonation of Benzene with Sulfuric Acid," *Chem. Eng. Progr.*, **46**, 249–257 (1950).

Denbigh, K., *Chemical Reactor Theory*, Cambridge University Press, London, 1965.

Denis, G. H., and Kabel, R. L., "The Effect of Temperature Changes on a Tubular Heterogeneous Catalytic Reactor," *Chem. Eng. Sci.*, **25**, 1057–1071 (1970).

Froment, G. F., and Bischoff, K. B., *Chemical Reactor Analysis and Design*, Wiley, New York, 1979.

Gilbert, E. E., and Groggins, P. H., "Sulfonation and Sulfation" in *Unit Processes in Organic Synthesis*, Groggins, ed., 5th ed., McGraw-Hill, New York, 1958.

Gilbert, E. E., *Sulfonation and Related Reactions*, Interscience Publishers, New York, 1965.

Gilbert, E. E., Veldhuis, B., Carlson, E. J., and Giolito, S. L., "Sulfonation and Sulfation with Sulfur Trioxide," *Ind. Eng. Chem.*, **45**, 2065–2072 (1953).

Hill, C. G., Jr., *An Introduction to Chemical Engineering Kinetics and Reactor Design*, Wiley, New York, 1977.

Himmelblau, D. M., and Bischoff, K. B., *Process Analysis and Simulation*, Wiley, New York, 1968.

Johnson, M. M., "Correlation of Kinetic Data from Laminar Flow-Tubular Reactors," *Ind. Eng. Chem. Fundam.*, **9**, 681–684 (1970).

Levenspiel, O., *Chemical Reaction Engineering*, Wiley, New York, 1972.

Liljenroth, F. G., "Starting and Stability Phenomena of Ammonia Oxidation and Similar Reactions," *Chem. & Metal Eng.*, **19**, 287–293 (1918).

Perti, D., and Kabel, R. L., "Dynamic Discernment of Catalytic Kinetics," ACS Symposium Series, No. 196, Chemical Reaction Engineering-Boston, American Chemical Society, Washington, D.C., 271–282 (1982).

Reid, R. C., Prausnitz, J. M., and Sherwood, T. K., *The Properties of Gases and Liquids*, 3rd ed., McGraw-Hill, New York, 1977.

Renken, A., "Verbesserung von Selektivität und Ausbeute durch periodische Prozessführung," *Chem. Ing. Technik.*, **46**, 113 (1974).

Smith, J. M., *Chemical Engineering Kinetics*, 2nd ed., McGraw-Hill, New York, 1970.

Smith, J. M., *Chemical Engineering Kinetics*, 3rd ed., McGraw-Hill, New York, 1981.

Walas, S. M., *Reaction Kinetics for Chemical Engineers*, McGraw-Hill, New York, 1959.

Wen, C. Y., and Fan, L. T., *Models for Flow Systems and Chemical Reactors*, Dekker, New York, 1975.

5

REACTORS FOR FLUID-PHASE PROCESSES CATALYZED BY SOLIDS

G. F. FROMENT

I.	Major Issues in Design of Fixed-Bed Reactors	168
	A. Single Fluid Phase	168
	1. Control of Temperature Change	168
	2. Limits on Pressure Drop	171
	3. Deactivation	171
	B. Multiple Fluid Phases	172
	1. Countercurrent Versus Cocurrent Flow	172
II.	Direct Experimental Simulation	173
	A. Similarity	173
	B. Mockups	176
III.	Mathematical Modeling	178
	A. Single Fluid Phase	179
	1. One-Dimensional Models	180
	2. Two-Dimensional Models	186
	3. A Note on Cell Models	187
	B. Multiple Fluid Phases	188
	1. Principles	188
	2. Industrial Models	190
IV.	Problem Areas	191
	A. Modeling of Kinetics for Complex Feedstocks	191
	B. Catalyst Deactivation	192
	C. Maldistribution of Fluid	194
Nomenclature		195
References		197

This discussion of the scaling up of reactors packed with solid and through which one or more fluid phases are flowing will deal exclusively with reactions catalyzed by the solid. Reactors in which the fluid reacts with the solid itself or in which the solid is inert, as encountered in absorbers, will be considered in Chapter 6. Further, the attention will be focused almost exclusively on fixed beds. Reactors in which the solid is moving or fluidized will only occasionally be dealt with in this chapter; they are treated in Chapter 10.

I. MAJOR ISSUES IN DESIGN OF FIXED-BED REACTORS

The design of a commercial size reactor based upon information gathered on a smaller scale involves many facets and several stages of evolution. It has to be kept in mind that the reactor is to achieve a certain conversion with optimal selectivity at a given temperature and pressure level with minimum downtime and hazard and maximum profit. Before the definitive sizing can be undertaken, several major issues have to be settled. Decisions concerning the type of reactor and its configuration, or about the desirability of recycling, have to be made at an early stage to avoid costly detailed studies of unfruitful side tracks. At such a preliminary stage only qualitative or at best semiquantitative information is available for the selection between possible alternatives. Still, in many cases, this should permit the narrowing down of alternatives to one or two and the identification of the main items to be investigated in more detail.

A. Single Fluid Phase

The first reactor type to be considered is an adiabatic bed. This is generally a simple vessel, easy to construct and to operate. Simple adiabatic operation is not always possible, however, when the heat effect of the reaction is large. With endothermic reactions the resulting temperature drop may be so severe that the reactor would have to be excessively long to compensate for the low reaction rates. This problem is encountered with steam reforming of hydrocarbons on nickel catalysts for hydrogen production, catalytic reforming on platinum catalysts for gasoline upgrading, or ethylbenzene dehydrogenation to styrene. With strongly exothermic reactions the adiabatic temperature rise may be unacceptable because of detrimental effects on the selectivity or because of catalyst deactivation resulting from coke formation or sintering.

1. Control of Temperature Change

To limit the ΔT, be it positive or negative, the adiabatic reactor can be staged. There are many examples of application of multibed adiabatic reactors, for example, in the processes mentioned above, in ammonia synthesis, in SO_2 oxidation, in CO conversion, and in hydrodesulfurization. The degree of staging depends not only on the desired conversion and the allowable ΔT per bed, but also on the temperature at the inlet of each bed. Intermediate cooling can be achieved by means of internal or external heat exchangers, by injection of cold reacting fluid, or by a combination of both means. With large-capacity reactors operating at atmospheric pressure, as in SO$_2$ oxidation, the intermediate heat exchangers are located outside the reactor. With high-pressure operation, as in ammonia synthesis, they are internal, to avoid external connecting pipes at high pressure and temperature. One of the problems associated with external heat exchangers is the even distribution of the fluid over the downstream catalyst bed. Special devices have been developed for this. Yet, additional hydrodynamic studies may be required. Cooling by cold-shot injection is applied in SO$_2$ oxidation, modern large-capacity ammonia synthesis reactors, and hydrotreaters. Evidently, this mode of operation shifts a more important fraction of the conversion to lower stages of the reactor. To what extent this occurs can only be predicted by quantitative modeling.

With fast exothermic reactions the first bed may cause problems. The amount of catalyst in this bed has to be accurately determined, since there is no external action to correct for too large a temperature rise and possible runaway.

With exothermic equilibrium reactions (ammonia and methanol synthesis, CO conversion, SO$_2$ oxidation) high temperatures do not necessarily lead to high conversions, because of equilibrium limitations. This is clearly seen from the location of the equilibrium curve Γ_e in the (T, x) plane of Figure 5-1.

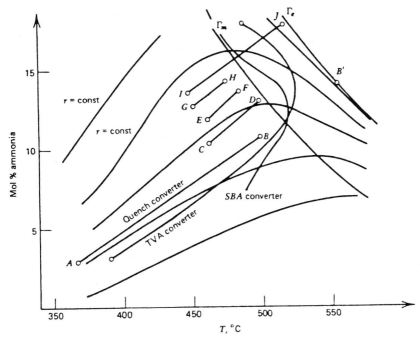

FIGURE 5-1 Temperature-conversion plane for exothermic equilibrium reaction showing reaction paths in multibed adiabatic and multitubular reaction.

To achieve high conversions with reasonable amounts of catalyst the temperature would have to decrease progressively with increasing conversion, ideally according to the Γ_m curve, also shown in Figure 5-1. This is the locus of points in the reactor in which the rate is maximum by the appropriate adaptation of the temperature (i.e., $\partial r/\partial T = 0$) and is roughly parallel to the Γ_e curve. Also shown is a T–x trajectory, ABCDEFGHIJ, typical for a multibed adiabatic NH_3 synthesis reactor. Clearly, such a mode of operation cannot closely fit the Γ_m trajectory. This curve could be better approximated, in principle at least, by a reactor with continuous heat exchange.

Two types of construction are encountered for exothermic reactions. In the first the coolant flows through pipes inserted into the bed, in the second the catalyst is packed into tubes, while the coolant flows around these. Both types are commonly called multitubular reactors.

With strongly exothermic reactions the possibility of runaway, which would lead to near adiabatic operation, requires the catalyst to be packed in narrow tubes. In phthalic anhydride synthesis these have an internal diameter of only 2.5 cm. To achieve the required production, modern reactors may contain 10,000 or more parallel tubes. This synthesis is carried out at temperatures between 350 and 420°C and the only convenient coolant at that temperature level is a molten salt mixture. To achieve a uniform flow of coolant around each of the 10,000 tubes in a reactor with a diameter of say 4.5 m is a problem in itself.

Another way of attempting to reduce the problems associated with highly exothermic reactions is to resort to fluidized bed operation, which enables high heat transfer rates. Examples are acrylonitrile synthesis, naphthalene oxidation to phthalic anhydride, and oxychlorination of ethylene to vinyl chloride. Scaling up fluidized bed reactors is more complicated, however, and the technology is more involved as discussed in Chapter 10.

In some cases the reacting gases are cooled by means of the feed flowing around the catalyst tubes. If the heat of reaction is sufficient for preheating the feed to the desired reaction temperature, the operation is called autothermal. This is the case for ammonia and methanol synthesis reactors. Co- or countercurrent flow is possible for the feed and the reacting gases. In ammonia synthesis cocurrent flow has been favored, since the trajectory fits the Γ_m curve more closely, in particular in the inlet region of the catalyst bed. In this region considerable overshoot is observed with countercurrent flow, as is also shown in the TVA curve of Figure 5-1.

In the case of endothermic reactions, like steam reforming, the reaction mixture has to be heated as rapidly as possible to the highest allowable temperature level, to achieve maximum production with a given amount of catalyst. Steam reforming tubes are mounted in a furnace in which the heat flux may be as high as 78.5 kJ/m^2 s. To heat the flow up to a sufficiently high temperature over the entire cross section of the tube, in other words, to avoid important radial gradients, the internal diameter is limited to 10 cm. A typical tube length is 10 m. A furnace for a daily hydrogen production of 300,000 m^3 at standard conditions would have 60 such tubes in parallel and its dimensions would be of the order of 20 m × 20 m × 20 m. One might have been inclined to think the evolution in steam reforming had come to a standstill. Yet, the introduction of pressure combustion and a complete rethinking of the catalyst tube recently enabled the development of a much more compact arrangement [Minet and Olesen (1979)]. In such a steam reformer, heat is partly transferred by convection, so that a much more closely spaced configuration is possible. The catalyst is packed in an annular space. The reacting gases flow upward through the catalyst bed and then downward through an internal annulus, thus permitting the feed to be preheated.

2. Limits on Pressure Drop

Pressure drop over the bed must be limited, particularly when a stream is recycled and the compressor costs have to be minimized, as in ammonia synthesis or platinum reforming. For given flow rate, catalyst volume, and particle dimensions, lowering the pressure drop implies reducing the bed height. With axial flow this would lead to excessive bed diameters. To avoid this, reactors with radial flow were introduced. The bed thickness has to be sufficient, however, to avoid bypassing of the catalyst, which would lower conversion. For a given flow rate the pressure drop can also be reduced by increasing the particle size, of course, but there are drawbacks to this. The first one is related to the flow pattern. If the ratio of tube diameter to particle diameter is too small the void fraction will be high, particularly close to the wall, and an important fraction of the reaction mixture will bypass the catalyst. The second drawback relates to chemical kinetics. With relatively fast reactions, important concentration gradients develop inside the catalyst particle. In other words: the catalyst effectiveness factor or utilization factor drops below one. The larger the particle size, the greater the effect. Clearly, a compromise has to be found between pressure drop and catalyst utilization, but again this optimization problem can only be solved by a more quantitative approach.

3. Deactivation

An important issue in the choice of reactor type or mode of operation is the deactivation of the catalyst by coke deposition. To maintain the production capacity at the desired level the catalyst has to be periodically regenerated by burning off the coke. Such an intermittent operation of the reactor definitely presents drawbacks. Rather than installing a number of parallel reactors, moving or fluidized bed operation may be envisaged. There are two well-known examples of this: catalytic cracking of gas oil in petroleum refining and reforming on platinum catalysts. Catalytic cracking was started with fixed-bed reactors, switched after a few years to moving bed reactors, which were in turn abandoned in favor of fluidized beds. Although deactivation is normally much slower in catalytic reforming, Universal Oil Products and Institut Francais du Petrole recently introduced moving bed technology in this process, to cope with faster coking resulting from more severe operating conditions.

B. Multiple Fluid Phases

There are various ways in which a gas–liquid reaction catalyzed by a solid can be carried out. Provided that the catalyst size permits it, fixed-bed operation is preferred to slurry type operation, since it leads to gas and liquid flow patterns which are closer to plug flow. Slurry type reactors would be preferred when the catalyst particle size has to be kept small and when temperature control is critical.

1. Countercurrent Versus Cocurrent Flow

Specific aspects that have to be considered with multiphase fixed-bed reactors are cocurrent or countercurrent flow of gas and liquid. One of the elements in the choice is the available driving force for mass and heat transfer. Countercurrent flow, very common in absorption, is normally not used for catalytic reactions. High flow rates of gas and liquid lead to flooding. Cocurrent downflow is the most common type of operation used today in fixed-bed catalytic reactors. Depending on whether the main transfer resistance is located in the gas or in the liquid, these reactors are operated either with a distributed liquid phase and continuous gas phase or vice versa. The first alternative is encountered in the trickle bed regime, the second in the bubble flow regime. For a given gas flow rate both regimes may lead to pulsed flow when the liquid flow rate is high, as can be seen from Figure 5-2 [Froment and Bischoff (1979)]. No definitive correlation is available for foaming situations.

The most important area in which fixed beds with cocurrent downflow are used today is hydrotreating and hydrocracking of petroleum fractions. They generally operate in the trickle flow regime and under pressure, to ensure a high concentration of the gas in the liquid phase. Commercial reactors have large diameters, so that they are essentially adiabatic. Again, intermediate

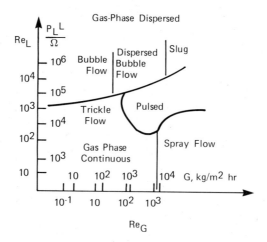

FIGURE 5-2 Flow regimes in multiphase fixed-bed reactors with cocurrent downflow and nonfoaming liquids.

cooling by cold hydrogen injection is applied, to prevent excessive temperature rise. The length to diameter ratio is limited to avoid maldistribution or channeling and, therefore, uneven wetting of the catalyst. Often redistribution of the liquid has to be resorted to.

There are evidently more variables involved in multiphase than in single phase reactors, so that scaleup is more complicated. Weekman (1976) compared hydrocracking in bench scale equipment having a volume of 100 cm^3 with that in a 7-m-high pilot unit of 3200 cm^3. The latter reactor led to the same conversion at much lower temperatures than the former, because of the higher mass loading. Ross (1965), on the other hand, observed a higher efficiency in a pilot hydrotreater than in a commercial reactor and attributed this to a better liquid distribution. When the liquid superficial velocity was increased the efficiency of the commercial reactor was improved, presumably because of a higher hold up.

With cocurrent upflow, the contacting between gas, liquid, and catalyst is improved, but the pressure drop is higher, too. Montagna and Shah (1975a) have observed the removal of vanadium and sulfur in resid hydrodesulfurization to be superior with upflow operation than with downflow. Ebulliated beds have been introduced to limit the pressure drop. The flow regimes encountered in cocurrent upflow operation are less thoroughly investigated than in downflow. A tentative diagram for air–water mixtures has been developed by Shah (1979).

II. DIRECT EXPERIMENTAL SIMULATION

In the preceding discussion the main issues arising in the selection of a reactor type and in its mode of operation have been outlined, both for single phase and multiphase fluid flow. It was also pointed out where a more quantitative approach was necessary to answer important questions in the ultimate design stage. The information required for this can be obtained from investigation on a smaller scale by:

Experimental simulation applying the rules of similarity.
Mathematical simulation based on a more detailed mathematical model.

The distinction between those two routes is not always clear-cut in practice. Let us say that the first approach is more global and would probably involve more than one intermediate scale of operation, whereas the second would decompose the process into several aspects which would be studied separately in the most appropriate equipment, for example, mockups for hydrodynamic aspects, before a reliable model would be developed.

A. Similarity

Similarity theory has been applied successfully in hydrodynamics and quite early in the development of fundamental chemical engineering has attracted the attention of Damköhler (1936, 1937), who came to essentially definitive conclusions as to its use in scaling up chemical reactors. Similarity between small- and large-size equipment requires a number of dimensionless groups to be identical. These groups can be derived by means of the Buckingham Pi theorem from a list of variables presumed to influence the processes in the equipment of interest. Another approach starts from the conservation equations, provided sufficient insight is available to develop these. The equations are made dimensionless by substitution of variables and/or by dividing each term by other terms of the equations or the boundary conditions as discussed in Chapter 2.

Himmelblau and Bischoff (1968) applied dimensionless conservation equations to a fixed-bed catalytic reactor with one fluid phase in which they assumed the pressure to be constant. To account for important radial gradients of temperature and concentration, effective conduction and diffusion mechanisms were superposed upon plug flow heat and mass transfer. Differences in temperature and concentration between the bulk of the gas and the catalyst surface were also accounted for. This model differs only in details from the two-dimensional heterogeneous model presented by DeWasch and Froment (1971), and discussed later in this chapter (p. 188).

Himmelblau and Bischoff derived the following controlling groups from their analysis:

$$\frac{2D_{er}}{u_i d_t}$$

$$\frac{2\lambda_{er}}{\rho_g u_s c_p d_t}$$

$$\frac{k_g a_v d_t}{2u_s}$$

$$\frac{h_f a_v d_t}{2u_s \rho_g c_p}$$

$$\frac{\Delta H}{RT_0}$$

$$\frac{(1-\varepsilon)k}{k_g a_v}$$

$$\frac{(1-\varepsilon)k c_0 \Delta H}{h_f a_v T_0}$$

$$\frac{a_w d_t}{2\lambda_{er}}$$

For Reynolds numbers exceeding 400, $N_{Pe_m} = d_p u_i / D_{er}$ and $N_{Pe_h} = \rho_G u_s c_p d_p / \lambda_{er}$ are constant. Applications of the similarity condition to the group $2 D_{er} / u_i d_t$ and $2 \lambda_{er} / \rho_g u_s c_p d_t$ leads

$$\left(\frac{d_p}{d_t}\right)_1 = \left(\frac{d_p}{d_t}\right)_2 \tag{5-1}$$

The index 1 refers to the smaller experimental scale and 2 to the larger industrial scale. Since the industrial tube diameter is usually larger than the pilot scale diameter this means that the particle diameter for the industrial scale is larger than the pilot particle diameter. This can lead in some cases to important internal gradients in the industrial catalyst. With Equation 5-1 and a correlation of the type presented by Gamson (1951) for k_g and h_f in beds of spherical particles, the groups $k_g a_v d_t / 2 u_s$ and $h_f a_v d_t / 2 u_s \rho_g c_p$ lead to the similarity condition

$$\frac{(u_i)_1}{(u_i)_2} = \left(\frac{\varepsilon}{1-\varepsilon}\right)_2^5 \left(\frac{1-\varepsilon}{\varepsilon}\right)_1^5 \frac{(d_p)_2}{(d_p)_1} \frac{\varepsilon_1}{\varepsilon_2} \tag{5-2}$$

where $u_s = \varepsilon u_i$.

Since ε is nearly identical for both sizes, this would mean that $u_1 > u_2$, which is unrealistic. The fifth group requires the temperature profiles to be identical. When the Gamson correlation for k_g and h_f is used with $(1-\varepsilon) k / k_g a_v$ and $(1-\varepsilon) k c_0 \Delta H / h_f a_v T_0$, the following relation is obtained:

$$\frac{k_1}{k_2} = \frac{(d_p^2)_2}{(d_p^2)_1} \left(\frac{1-\varepsilon}{\varepsilon}\right)_1^4 \left(\frac{\varepsilon}{1-\varepsilon}\right)_2^4 \frac{\varepsilon_1}{\varepsilon_2} \tag{5-3}$$

Since the scaling factor for the catalyst diameter has already been set, Equation 5-3 leads to the unrealistic recommendation that the catalyst chosen for the commercial size operation should be less active than that for the smaller scale. Use of the group $a_w d_t / 2 \lambda_{er}$ will lead to conclusions similar to those derived from Equation 5-2.

Clearly, complete similarity is not possible between two nonisothermal and nonisobaric reactors and the scaling up will involve some compromise and some risks. For multiphase reactors Charpentier (1978) advocates an approach developed for gas–liquid absorbers by Danckwerts and Gillham (1966) and Danckwerts and Alper (1975). It consists of an experimental simulation of the commercial equipment by means of a laboratory model that does not necessarily bear any resemblance with it, but has similar contact times. A packed column, for example, would be experimentally simulated by a stirred vessel, calibrated against the large-scale packed column. For multitubular reactors, safe extrapolation requires experimentation on a tube of the size used in the commercial unit. This is not a challenge. However, a single tube does not address the problem of equalizing heat transfer over a multiplicity of tubes.

Industrial reactors with large diameters operate adiabatically. This is practically impossible to achieve on bench or even pilot scale. The awareness that complete similarity is practically impossible has frequently led to development procedures involving several steps, with progressive increases in the size of the equipment. It should be clear by now that, no matter what the intermediate size, the results will never exactly or completely reproduce those obtained in the commercial unit. The pilot plant is no guarantee for the successful scaleup if it is not used for producing better insight and fundamental knowledge about the process.

The pilot unit may not even be the most appropriate type of equipment for generating such information. It is not likely that a single apparatus will be adequate for studying such different phenomena as chemical kinetics, heat transfer, and degree of mixing. Chemical kinetics are preferably determined in an isothermal reactor, even if the commercial reactor operates adiabatically. Hydrodynamic aspects are preferably studied in the absence of reaction, for example, in so-called "mockups." These pieces of information can then be tied together in a mathematical model that can be used to simulate the process in the commercial reactor. Prior to proceeding to the mathematical simulation as a tool for scaling up chemical reactors, a couple of examples of the use of mockups will be given.

B. Mockups

Mockups or homologous models are quite helpful for investigating hydrodynamic aspects of processes. These are preferably studied in the absence of reaction and under more favorable conditions for convenient observation, for example, at low temperature or pressure. When viscosity or surface tension are relatively unimportant the fluids used in the process may be replaced by others with similar properties, but typically less corrosive or inexpensive. Mockups are often made of transparent plastic to enable visualization of the phenomena. Since they are not too costly they may be built in different sizes, to investigate the effect of scaling up on the phenomena. The information gained in this way may be processed through dimensional analysis or through mathematical modeling.

The use of mockups in studies of hydrodynamic aspects of the fluidized state is well known. The behavior of bubbles and the importance of the wake were observed in this way [Davidson and Harrison (1971); DeGroot (1967)]. Recently Trambouze (1979) reported on the use of mockups in both fixed- and moving bed processes developed at the French Petroleum Institute.

The first mockup was built to study the distribution of the fluid phases over a cross section in the fixed-bed hydrotreatment of a petroleum fraction. However nitrogen was used instead of hydrogen. The size of the catalyst was identical to that used in the commercial unit, but the diameter of the mockup

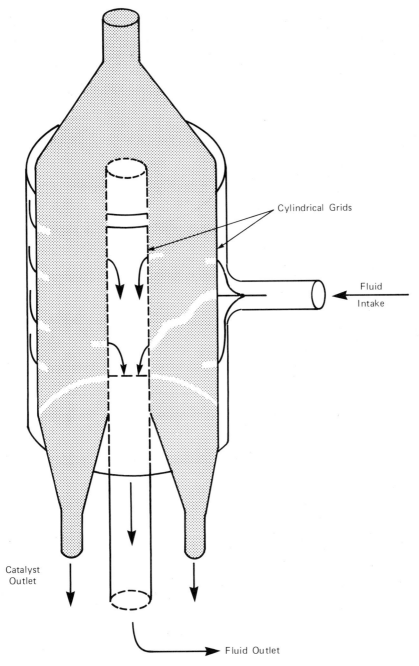

FIGURE 5-3 Radial flow reactor for moving bed catalytic reforming process of IFP.

was only 60 cm, instead of the 3.6 m of the commercial unit. Several types of distributor plates for the gas and liquid feed were investigated. The bed height was also varied. The gas and the liquid were collected by 22 exit tubes at the bottom of the bed and the flow rate in each of them was measured.

To increase the production of unleaded gasoline with a high octane number, the severity of catalytic reforming has to be increased. As a result, coking is accelerated and more frequent regenerations are necessary. To avoid the difficulties associated with fixed-bed regeneration, the French Petroleum Institute developed a staged moving bed process with continuous regeneration in a separate vessel.

The regenerated catalyst is fed to the first reactor stage, slowly moves downward by gravity, and is then carried to the next stage by a pneumatic lift. The pneumatic transfer of the catalyst, the downflow of it through the reactor stages, and different types of valves were studied in mockups. The reactor is schematically represented in Figure 5-3. The catalyst moves down between two concentric cylindrical screens and the gas flows radially through and cocurrently with the catalyst, to limit the pressure drop and avoid fluidization at the high gas feed rates.

Three mockups were constructed to guide the development of the moving bed. The possibility of having catalyst particles stick to the gas exit screen was examined in a laboratory size unit. A rectangular mockup containing 350 L of catalyst provided information on the relation between the width of the fixed-bed and the gas-phase pressure drop. The large mockup, made of plexiglass and containing 2500 L of catalyst, was only a sector of the cylinder, to avoid too large a consumption of gas.

The flow of solid was observed by means of a layer of colored tracer beads. The catalyst size was the same as that used in the commercial unit. The geometry of the commercial reactor, the maximum values for the flow rates, and the pressure drop were determined at ambient temperature and pressure. Results concerning optimal operating conditions obtained in pilot units and the conclusions on solid transport drawn from the mockups were combined into the successful design of an industrial unit.

III. MATHEMATICAL MODELING

Intensive research on the fundamentals of transport processes in fixed beds and the increasing usage of computers have led to considerable progress in fixed-bed reactor analysis and design (Froment, 1971, 1974; Ray, 1972; Weekman, 1976). Mathematical models for fixed-bed reactors with one fluid phase now range from the very simple ones developed before 1960 to some very sophisticated ones presented in the seventies. Models for fixed-bed reactors with two fluid phases are less well developed. In fact, this is an area which has received intensive attention in recent years only, parallel with the growing importance of hydrotreating and hydrocracking processes. In what

MATHEMATICAL MODELING

follows, reactors with one and two fluid phases will again be dealt with separately.

A. Single Fluid Phase

Froment's (1972, 1974) classification of fixed-bed reactor models is shown in Table 5-1 and has been widely accepted. It distinguishes between two broad categories:

Pseudohomogeneous.
Heterogeneous models.

Pseudohomogeneous models make no explicit distinction between the solid and the fluid. Separate conservation equations are written for each phase in heterogeneous models.

The basic ideal model, AI, assumes that concentration and temperature gradients only occur in axial direction. Transport of mass and heat only occurs by an ideal type of convection called plug flow. If a certain degree of axial mixing is superposed on the plug flow, model AII is obtained. With the deep beds and high flow rates normally encountered in commercial reactors, the effect of axial mixing is negligible, so that model AII will not be discussed any further here. The two-dimensional model AIII is derived from AI when radial gradients are accounted for. This has to be done for reactions having a pronounced heat effect which are carried out in tubular reactors and exchange heat through the walls.

With fast reactions and low flow velocities, interfacial gradients of concentration and temperature may develop. Extending model AI to account for this leads to the first heterogeneous model, BI. If, in addition, gradients occur inside the catalyst, model BII should be used. Model BIII is the heterogeneous equivalent of the two-dimensional model AIII. The most advanced version of this model was developed by De Wasch and Froment (1971), but more detailed experimental work on heat transfer in packed beds is required to provide accurate values for some of the model parameters. For this reason it will not be considered any further in this chapter.

TABLE 5-1 Model Classification, Fixed-Bed Reactors with One Fluid Phase

	Pseudohomogeneous Models $T = T_s; C = C_s$	Heterogeneous Models $T \neq T_s; C \neq C_s$
One-Dimensional	AI basic, ideal AII + axial mixing	BI + interfacial gradients BII + intraparticle gradients
Two-Dimensional	AIII + radial mixing	BIII + radial mixing

1. One-Dimensional Models

a. BASIC EQUATIONS

Consider first the basic one-dimensional pseudohomogeneous model. For a single reaction at steady state, the continuity equation for the reactant A and the energy and momentum equations may be written:

$$-u_s \frac{dC}{dz} - \rho_B r_A = 0 \tag{5-4}$$

$$u_s \rho_g c_p \frac{dT}{dz} - (-\Delta H)\rho_B r_A + 4\frac{U}{d_t}(T - T_w) = 0 \tag{5-5}$$

$$\frac{dp_t}{dz} = f(u_s, d_p, \rho_g, \varepsilon, u) \tag{5-6}$$

with $C = C_0$, $T = T_0$, and $p_t = p_{t_0}$ at $z = 0$. For adiabatic operation the term $4(U/d_t)(T - T_w)$ of Equation 5-5 is zero.

Earlier correlations for heat transfer through the wall are generally empirical and consider the bed as a black box [Leva (1948)]. In recent years several detailed models for heat transfer in packed beds have been developed [Yagi and Kunii (1957, 1960); Kunii and Smith (1960); Zehner and Schlünder (1970, 1972, 1973); Hennecke and Schlünder (1973); Yagi and Wakao (1959); Balakrishnan and Pei (1979)]. They permit the pseudohomogeneous internal-wall side-heat transfer coefficient to be calculated in terms of a number of mechanisms operating in the solid and in the gas phase. A static and a dynamic contribution must then be considered with the latter varying linearly with the Reynolds number (Yagi and Wakao, 1959). Such a relation was experimentally confirmed by De Wasch and Froment (1971).

Pressure drop equations were recently reviewed by Macdonald et al. (1979), who propose the following modified version of the Ergun equation:

$$\frac{\Delta p_t}{L} \frac{d_p'}{\rho_g u_s^2} \frac{\varepsilon^3}{1-\varepsilon} = 180 \frac{(1-\varepsilon)}{N_{Re'}} + a \tag{5-7}$$

where $a = 1.8$ for smooth and 4.0 for rough particles. Equations 5-4 and 5-6 can then be used to predict concentration, temperature, and pressure profiles for various operating conditions and reactor configurations.

In Section I of this chapter, the problem of hot spots in tubular reactors was dealt with in a qualitative way. Equations 5-4, 5-5, and 5-6 allow a quantitative study of this phenomenon, through studies of parametric sensitivity. Convenient *a priori* rules have been derived for selecting operating conditions which avoid excessive reactor sensitivity and runaway with first order reactions [Barkelew (1959); Van Welsenaere and Froment (1970); Dente and Collina

(1964)]. For other reaction orders a number of simulations based on the system of differential Equations 5-4 to 5-6 have to be performed. First order kinetics could also be forced upon the reaction rate correlations, of course, but this distorts the activation energy. This can be a very important parameter in determining whether or not a reactor is operating in the sensitive region.

b. APPLICATION TO AUTOTHERMAL REACTORS

Several reactors used in commercial processes operate in a so-called autothermal manner. This implies that heat of reaction is sufficient to preheat the feed to the required inlet temperature. The modeling of autothermal reactors necessitates the addition of another equation to the set Equations 5-4 to 5-6 which expresses the internal or external preheating of the feed. Autothermal reactors can be subject to runaway caused by parametric sensitivity, just like the simple tubular reactor previously discussed. However, in this case runaway can also be induced by true instabilities associated with multiple steady states which result from the feedback of heat.

Computer simulation of autothermal reactors can easily locate areas of operation in which such undesirable situations could develop [Baddour et al. (1965); Shah (1967)]. Recently, Inoue (1978) presented analytical criteria, applicable for nth order irreversible and reversible reactions. The following dimensionless groups are used:

$$\beta = \frac{(-\Delta H)C_{A_0}}{\rho_g c_p T_0} \tag{5-8}$$

$$\gamma = \frac{E}{RT_0} \tag{5-9}$$

$$N = \frac{4U}{d_t} \frac{C_{A_0}}{c_p \rho_g r_A(T_0) \rho_B} \tag{5-10}$$

$r_A(T_0)$ is the rate of reaction evaluated at T_0.

The criteria for stability may then be written as:

$N > \beta\gamma - n$ (nth order irreversible reactions)

$N > \beta\gamma - n\kappa$ with $\kappa = \dfrac{1+K}{K}$ (nth order reversible reactions)

Inoue also derived a somewhat simpler criterion stating that when $\beta\gamma$ is smaller than 8, there is no risk of instability with nth order irreversible reactions. However, a hot spot may develop at small values of N. The corresponding rule for reversible reactions is $\beta\gamma < 8\kappa$.

Optimal operating conditions of autothermal reactors are close to the so-called blow-off conditions, where the supply of external heat is necessary

[Baddour et al. (1965)]. Ampaya and Rinker (1977a, b) relate both the critical feed temperature leading to blow-off, T_F, and the critical catalyst bed inlet temperature, T_0, to the operating and design parameters for irreversible and reversible first order reactions. At blow-off, for irreversible first order reaction,

$$(\gamma - 1)\tau \exp \gamma = 1 \quad \text{with} \quad \tau = \frac{2U}{d_t c_p} \frac{\Delta T_{ad} R}{k \rho_B P_t} \tag{5-11}$$

The required degree of preheating then is

$$\frac{T_0 - T_F}{\Delta T_{ad}} = \frac{US}{F_0 c_p'} - \frac{1}{\gamma - 1} \frac{T_0}{\Delta T_{ad}} \tag{5-12}$$

For reversible reactions the critical bed inlet temperature T_0 also depends on the equilibrium conversion at T_0 as well as on the temperature derivative of the equilibrium constant. Ampaya and Rinker have also shown how these results can be plotted in convenient diagrams.

Baddour et al. (1965) simulated the TVA–ammonia synthesis reactor with internal countercurrent heat exchange using the Temkin–Pyzhev kinetic equations. Figure 5-4 compares the experimental and simulated temperature profiles.

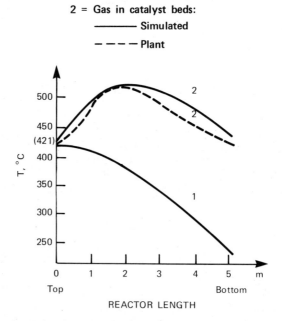

FIGURE 5-4 Experimental and simulated temperature profiles in a TVA–ammonia synthesis reactor.

MATHEMATICAL MODELING

Examples of simulation of multibed adiabatic reactors for NH_3 synthesis, SO_3 synthesis, and the water gas shift reaction and illustrations of some of their characteristics can be found in the literature [Froment and Bischoff (1979); Shipman and Hickman (1968)]. Optimization calculations can be performed along with the simulations. Murase et al. (1970) have calculated optimal temperature profiles in a multitubular ammonia synthesis reactor, applying Pontryagin's maximum principle. Unfortunately, the proposed profiles imply heat transfer rates which are unlikely to be attained.

The optimum size of catalyst beds and intermediate heat exchangers in multibed adiabatic reactors can be determined by coupling Bellman's dynamic programming algorithm with the reactor simulation model. Aris (1960), Roberts (1964), and Froment and Bischoff (1979) have given examples of such applications of the modeling approach. An accurate experimental optimization would evidently require a much larger effort.

C. Accounting For Temperature and Concentration Differences Between Gas and Solid

In the system of Equations 5-4 to 5-6, r_A represents the true rate of the reaction. The pseudohomogeneous model AI, therefore, ignores temperature and concentration differences between the gas and the solid. When these occur, separate continuity and energy equations for each phase have to be written, leading to the one-dimensional heterogeneous model BI:

For the Fluid

$$-u_s \frac{dC}{dz} = k_g a_v (C - C_s^s) \tag{5-13}$$

$$u_s \rho_g c_p \frac{dT}{dz} = h_f a_v (T_s^s - T) - 4\frac{U}{d_t}(T - T_w) \tag{5-14}$$

For the Solid

$$\rho_B r_A = k_g a_v (C - C_s^s) \tag{5-15}$$

$$(-\Delta H)\rho_B r_A = h_f a_v (T_s^s - T) \tag{5-16}$$

The inlet boundary conditions are the same as those for Equations 5-4 to 5-6.

Correlations for the mass and heat transfer coefficients were recently reviewed [Froment and Bischoff (1979); Schlünder (1978)]. The fluid–solid heat transfer coefficients reported in older studies were determined from experiments in packed beds heated through the walls [Gamson et al. (1939); Baumeister and Bennett (1958); De Acetis and Thodos (1962); Gupta and Thodos (1962); Handley and Heggs (1968)]. These data are a conglomerate resulting from several mechanisms. There is now a trend in the literature to decompose the global heat transfer coefficient, h_f, into three contributions:

The true fluid to solid heat transfer coefficient.
The particle to particle heat transfer coefficient.
The coefficient for heat transfer through the solid.

By heating the solid with microwaves Bhattacharya and Pei (1975) managed to measure the true fluid to solid heat transfer coefficient. Equations 5-15 and 5-16 do not provide for any coupling between particles. Therefore, for fixed beds exchanging heat through the wall, use of the global h_f is consistent with the model. However, it is not necessarily the most accurate way of dealing with the phenomenon.

In industrial reactors the flow velocity is generally so high that the ΔT and ΔC over the fluid film around the catalyst particles are very small and even negligible [Baddour et al. (1965); Cappelli et al. (1972)]. However, there are exceptions. These are encountered when either a component of the catalyst itself or coke deposited on the catalyst is involved in the process. Hatcher et al. (1978) have simulated the reoxidation of a nickel catalyst in a commercial secondary reformer and have shown how the temperature difference between gas and solid may exceed 100°C. Olson et al. (1968) have simulated temperature transients in the regeneration of a coked catalyst in a fixed bed.

d. GRADIENTS INSIDE THE CATALYST PARTICLE

Concentration gradients are much more likely inside than outside the catalyst particle. Their magnitude depends upon the rate of reaction and the rate of transport inside the porous structure. Internal concentration gradients are encountered in many industrial processes like ethylbenzene dehydrogenation into styrene, catalytic cracking of gasoil, and methanol synthesis.

For steady state and a single reaction, a set of equations accounting for internal as well as external gradients may be written as follows:

For the Fluid

$$-u_s \frac{dC}{dz} = k_g a_v (C - C_s^s) \qquad (5\text{-}17)$$

$$u_s \rho_g c_p \frac{dT}{dz} = h_f a_v (T_s^s - T) - 4\frac{U}{d_t}(T - T_w) \qquad (5\text{-}18)$$

For the Solid

$$\frac{D_e}{\xi^2} \frac{d}{d\xi}\left(\xi^2 \frac{dC_s}{d\xi}\right) - \rho_s r_A(C_s, T_s) = 0 \qquad (5\text{-}19)$$

$$\frac{\lambda_e}{\xi^2} \frac{d}{d\xi}\left(\xi^2 \frac{dT_s}{d\xi}\right) + \rho_s(-\Delta H) r_A(C_s, T_s) = 0 \qquad (5\text{-}20)$$

With Boundary Conditions

$$z = 0 \quad C = C_0 \quad T = T_0$$

$$\xi = 0 \quad \frac{dC_s}{d\xi} = \frac{dT_s}{d\xi} = 0$$

$$\xi = \frac{d_p}{2} \quad k_g(C_s^s - C) = -D_e \frac{dC_s}{d\xi}$$

$$h_f(T_s^s - T) = -\lambda_e \frac{dT_s}{d\xi}$$

Equation 5-20 can often be dropped from the model, since the particle is practically always isothermal [Kehoe and Butt (1972); Froment and Bischoff (1979); Carberry (1975)]. D_e is an effective diffusivity for transport by bulk flow and by molecular or Knudsen diffusion. D_e depends on the internal void fraction of the particle and on the tortuosity factor which characterize the structure of the network.

When there are concentration gradients inside the particle an effectiveness factor, η, is frequently used to characterize the efficiency at which a catalyst particle is operating in a given environment. The effectiveness factor is usually defined as the ratio of the actual rate of reaction to that which would be observed if the concentration inside the particle were uniform and equal to that at the surface [Froment and Bischoff (1979); Petersen (1965)]. Some authors have defined a different effectiveness factor, sometimes represented by η^*, which accounts also for external gradients. This means that η^* is referred to bulk fluid conditions (C, T), and η to surface conditions (C_s^s, T_s^s).

The effectiveness factor by definition permits replacing the continuity equation for the key component in the solid phase, Equation 5-19, by a simple algebraic equation

$$k_g a_v (C - C_s^s) = \eta \rho_B r_A (C_s^s, T_s^s) \tag{5-21}$$

The effectiveness factor η is related to a modulus ϕ containing the particle size and the ratio of rate coefficient and effective diffusivity. If this relation is analytical then there is a definitive advantage in the use of η as far as the design calculations are concerned. If not, Equation 5-21 is just another way of writing Equation 5-19. The appropriate boundary conditions and η must be obtained by numerical integration of Equation 5-19 for each node used for the integration of the fluid field Equations 5-17 and 5-18. Some authors have preferred rewriting Equation 5-19 in an integral form which can be more convenient for calculation [Cappelli et al. (1972)]. Modern collocation techniques have also been successfully used to integrate Equation 5-19 [Villadsen and Michelsen (1978)].

The parametric sensitivity of the model has been studied by McGreavy and Adderley (1973, 1974) and by Rajadhyaksha and Vasudeva (1975). Froment (1980) recently reviewed this problem. A lot of attention has been devoted to multiplicity of steady states and associated instabilities, but so far there is no industrial evidence for such "pathological phenomena."

A successful application of Equations 5-17 to 5-19 to a multibed adiabatic methanol synthesis reactor has been published by Cappelli et al. (1972). Dumez and Froment (1976) used an even more complex model to simulate 1-butene dehydrogenation into butadiene in an adiabatic reactor. This reaction is accompanied by fast coking which deactivates the catalyst and requires very frequent regeneration. An optimization revealed that the production part of the cycle, which usually does not exceed 20 min, could be prolonged to 1 hr.

2. Two-Dimensional Models

Exothermic reactions conducted in narrow tubular reactors cannot be accurately simulated by the one-dimensional models AI, BI, or BII. Indeed, pronounced radial gradients, mainly of temperature, may develop, so that the temperature at the axis can significantly exceed that of the wall. The reverse would be true for endothermic reactions, of course.

The continuity equation for a reacting component and the energy equation would be

$$D_{er}\left(\frac{\partial^2 C}{\partial r^2} + \frac{1}{r}\frac{\partial C}{\partial r}\right) - u_s \frac{\partial C}{\partial z} - r_A \rho_B = 0 \quad (5\text{-}22)$$

$$\lambda_{er}\left(\frac{\partial^2 T}{\partial r^2} + \frac{1}{r}\frac{\partial T}{\partial r}\right) - u_s \rho_g c_p \frac{\partial T}{\partial z} + (-\Delta H) r_A \rho_B = 0 \quad (5\text{-}23)$$

With Boundary Conditions

$$C = C_0 \text{ and } T = T_0 \text{ at } z = 0 \quad 0 < r < R_t$$

$$\frac{\partial C}{\partial r} = 0 \text{ at } r = 0 \text{ and } R = R_t$$

$$\frac{\partial T}{\partial r} = 0 \text{ at } r = 0$$

$$\frac{\partial T}{\partial r} = -\frac{\alpha_w}{\lambda_{er}}(T - T_w) \text{ at } r = R_t$$

The first terms in the left-hand side of Equations 5-22 and 5-23 account for limited rates of radial mass and heat transfer. Mass and heat fluxes in the radial direction are considered to result, respectively, from diffusion- and conduction-like mechanisms. Mass transfer is expressed in terms of a Fick type law, with the effective diffusivity, D_{er}, as a proportionality factor between the flux and gradient. The axial component of this effective diffusion may be neglected in commercial reactors.

The heat flux in the radial direction results from several mechanisms operating in both the fluid and the solid phases. These are lumped into a conduction-like mechanism, described by a Fourier type equation with a proportionality factor called the effective conductivity, λ_{er}. Heat transfer in the immediate vicinity of the wall is expressed in terms of a wall heat transfer coefficient, α_w, which is not to be confused with the internal heat transfer coefficient of the one-dimensional models. The axial component of the effective conduction flux is negligible with respect to the heat flux by the overall convection.

Experimental correlations for λ_{er} and D_{er} are available in the literature and have been reviewed [Froment (1967); De Wasch and Froment (1971); Froment (1974)]. Fundamental models have also been developed [Kunii and Smith (1960); Zehner and Schlünder (1970, 1972, 1973); Yagi and Wakao (1959); Hennecke and Schlünder (1973); Balakrishnan and Pei (1979)]. Recently, Dietz (1979) proposed an equation for heat transfer in packed beds in the absence of flow and radiation, while Vortmeyer (1979) reviewed radiation in fixed beds. The model has been extensively used for the simulation of very exothermic reactions like the oxidation of o-xylene and naphthalene to phthalic anhydride [Froment (1967, 1971); Smith and Carberry (1974)].

Since radial temperature gradients can be important, while the radial concentration gradients are generally negligible, and continuity Equation 5-22 can be simplified to Equation 5-4 [Froment (1967); Smith and Carberry (1974)]. Froment has compared average cross-sectional values of temperature, calculated from the two-dimensional model AIII, with the radially uniform values predicted by the one-dimensional model, AI. To do so the heat transfer coefficient of model AI has to be related to λ_{er} and α_w of model AIII [Froment (1961, 1971)]. One result was that the one-dimensional model predicted safe operation under some conditions where the two-dimensional model had excessive parametric sensitivity and indicated a runaway.

Radial variation in flow velocity is closely related to variations in void fraction. This aspect is important for reactions having a large heat effect, carried out under severe conditions [Lerou and Froment (1977)].

3. A Note on Cell Models

The above models were of the continuum type, but a description of a reactor in terms of discrete cells is also possible. Cell models consider complete mixing of the fluid in the voids between the particles. The effluent of each cell is split into two streams which are fed into the next row of cells. In the two-dimensional array of cells, alternate rows are offset half a stage to allow for radial mixing [Deans and Lapidus (1960); McGuire and Lapidus (1965)].

Early models considered heat to be transferred by convection through the fluid only and this can be shown to lead to important errors. More elaborate models include conduction through the solid and radiation [Kunii and Furusawa (1972)]. However, the computations get much more involved and there is no advantage over the effective transport model that lumps these phenomena into λ_{er}. Further discussions of this topic can be found in a paper by Sundaresan et al. (1980).

B. Multiple Fluid Phases

1. Principles

A fairly general mathematical model for the simulation of multiphase reactors can be developed in a manner analogous to that for heterogeneous models with one fluid phase. The model must distinguish between gas, liquid, and solid as well as account for deviations from plug flow. Deviations from plug flow while not excessive are nevertheless more pronounced than in fixed beds with a single fluid phase. The mass flux in the axial direction is described by a plug flow contribution and an effective diffusion superposed on it.

The steady-state continuity equations for the reacting components, A from the gas phase and B from the liquid phase, are as follows:

For A in the Gas Phase

$$(\varepsilon - \varepsilon_L)D_{eA_G}\frac{d^2C_{A_G}}{dz^2} - (\varepsilon - \varepsilon_L)u_{iG}\frac{dC_{A_G}}{dz} - K_L a'_v\left(\frac{p_A}{H} - C_{A_L}\right) = 0$$

(5-24)

where the overall mass transfer coefficient in terms of the liquid concentration gradient is given by $1/K_L = 1/k_L + 1/Hk_G$ and ε_L is the liquid holdup.

For A in the Liquid Phase

$$\varepsilon_L D_{eA_L}\frac{d^2C_{A_L}}{dz^2} - \varepsilon_L u_{iL}\frac{dC_{A_L}}{dz} + K_L a'_v\left(\frac{p_A}{H} - C_{A_L}\right) - k_{l_A}a''_v\left(C_{A_L} - C^s_{A_S}\right) = 0$$

(5-25)

For Transfer of A from the Bulk Liquid to the Catalyst

$$k_{l_A}a''_v\left(C_{A_L} - C^S_{A_S}\right) = \eta r_A \rho_B \qquad (5\text{-}26)$$

For B in the Liquid Phase

$$\varepsilon_L D_{e_L}\frac{d^2C_{B_L}}{dz^2} - \varepsilon_L u_{iL}\frac{dC_{B_L}}{dz} - k_{l_B}a''_v\left(C_{B_L} - C^s_{B_S}\right) = 0 \qquad (5\text{-}27)$$

For Transfer of B from the Bulk Liquid to the Catalyst

$$k_{l_B}a''_v\left(C_{B_L} - C^s_{B_S}\right) = \frac{b}{a}\eta r_A \rho_B \qquad (5\text{-}28)$$

Normally it is assumed that $D_{eA_L} = D_{eB_L}$ and $k_{l_A} = k_{l_B}$.

The relevant boundary conditions are easily derived. An energy equation and a pressure drop equation may also have to be considered.

Effective diffusivities D_{eG} and D_{eL} in the trickle flow regime have been determined by Hochman and Effron (1969) and by Elenkov and Kolev (1972). The mass transfer coefficients k_G, k_L, and k_l were measured and correlated by Reiss (1967), by Charpentier (1978), and by Van Krevelen and Krekels (1948). Puranik and Vogelpohl's correlation (1974) can be used to estimate a'_v. Holdup correlations were derived by Otake and Okada (1963) and also by Midoux et al. (1976). Pressure drop equations were proposed by Larkins et al. (1961), by Sweeney (1967), by Midoux et al. (1976), and by Turpin and Huntington (1967).

The deviations from plug flow, revealed by residence time distribution measurements, are probably mainly caused by preferential paths and stagnant zones. Deviations of this type are not adequately described by an effective diffusion model. A two-zone model would be more appropriate. Models of this type were developed for fluidized bed operation in the early sixties. The underlying picture is that one fraction of the liquid flows in a more or less orderly way through the bed, exchanging mass at each height with another fraction which is well mixed in a stagnant zone.

This continuity equation for A in the gas phase is either that given by Equation 5-24 or a simplified version that neglects the effective diffusion term. The liquid-phase continuity equations for A may be written as follows [Froment and Bischoff (1979)]:

For the Orderly Flowing Fluid

$$-\varepsilon^f_L u_{iL}\frac{dC^f_{A_L}}{dz} - k_T\left(C^f_{A_L} - C^d_{A_L}\right) + k_L a'_v\left(\frac{p_A}{H} - C^f_{A_L}\right) - k_l a''_v\left(C^f_{A_L} - C^s_{A_S}\right) = 0$$

$$(5\text{-}29)$$

For the Well-Mixed Liquid in the Corresponding Slice of the Stagnant Zone

$$k_T\left(C^d_{A_L} - C^f_{A_L}\right) = k'_l\left(C^d_{A_L} - C^s_{A_S}\right) \qquad (5\text{-}30)$$

For Transport to and Reaction in the Catalyst

$$k_l a''_v\left(C^f_{A_L} - C^s_{A_S}\right) + k'_l\left(C^d_{A_L} - C^s_{A_S}\right) = \eta r_A \rho_B \qquad (5\text{-}31)$$

Analogous equations may be written for the reacting component of the liquid B.

The transfer coefficients k_T and k'_l between the flowing and stagnant fractions and between the stagnant fluid and the solid, respectively, contain interfacial areas which have not been measured so far. Notice that the models given above assume that the catalyst is completely wetted. This is not always true, particularly at low flow rates.

Michell and Furzer (1972) modeled trickle flow conditions by considering laminar film flow over the solid and mixing as well as bypassing at solid junctions. A more refined hydrodynamic model was recently published by Crine et al. (1979). Schmalzer and Hoelscher (1971) described the flow of fluid elements as a Markov process.

2. Industrial Models

The models used in industrial practice are far less elaborate. In hydrotreating petroleum fractions the liquid is generally saturated with hydrogen. When the mass transfer effects between liquid and solid and the effective diffusion in the liquid are not considered in the data treatment the model described by Equations 5-24 to 5-28 reduces to the following pseudohomogeneous equation:

$$-\varepsilon_L u_{iL} \frac{dC_{B_L}}{dz} = \frac{b}{a} \eta r_A \rho_B \qquad (5\text{-}32)$$

Even this equation has been simplified by assuming that there is no resistance to mass transfer inside the catalyst particle and that the reaction is first order. This is a bold assumption as catalytic reactions seldom follow first order kinetics if only because of adsorption. Even if each reaction between hydrogen and the various sulfur compounds was first order in nature, lumping them into a single reaction will lead to a higher order with respect to the lumped equation.

The kinetic coefficient of the reaction, calculated according to Equation 5-32 with $\eta = 1$, increased when both L and W were doubled. This could be ascribed to mass transfer effects in the liquid and inside the catalyst. Henry and Gilbert (1973) associated the observed effect with an increase in liquid holdup. Satterfield et al. (1969), however, indicate that liquid holdup is proportional to $L^{1/3}$. After integration then, Equation 5-32, for a first order reaction, becomes

$$\ln \frac{C}{C_0} \sim k L^{-2/3} \qquad (5\text{-}33)$$

This equation indeed correlates Henry and Gilbert's data. Mears (1974), however, criticized this concept and proposed instead to correlate the rate of reaction with the wetted area of the catalyst. To do this the equation of Puranik and Vogelpohl (1974) or, at higher flow rates, that of Onda (1968) was used. Mears thus arrived the following expression,

$$-\ln\frac{C}{C_0} = \frac{k\eta}{\text{LHSV}}\left\{1 - \exp\left[-KL^{0.4}(\text{LHSV})^{0.4}\right]\right\} \quad (5\text{-}34)$$

where K is a factor incorporating the effects of viscosity, surface tension, density, and particle diameter, and where LHSV is the liquid hourly space velocity.

In their study of hydrodesulfurization of atmospheric and vacuum resids, Montagna et al. (1977) and Montagna and Shah (1975b) showed the Mears equation to be superior to that of Henry and Gilbert.

What degree of sophistication is required for accurate scaleup is still an area of judgment. However, it would seem that for organic reactions carried out under trickle bed conditions some of the simplifying assumptions introduced above for hydrotreating may not be adequate. More refined models must be used (Hofmann, 1978).

Selection of the degree of sophistication necessary and justified to arrive at a successful model is a problem frequently encountered in design. There is no generally valid answer. The approach depends on both the reaction scheme and on the process sensitivity to perturbations in the operating conditions. Of equal importance is the degree of accuracy with which the kinetic and transport parameters are known. And finally, as must be clear by now, the reactor type is of paramount importance in determining the degree of sophistication of the modeling.

IV. PROBLEM AREAS

A. Modeling of Kinetics for Complex Feedstocks

So far, no attention has been paid to the kinetics of the reaction. Clearly, this topic is not specific for the scaleup of fixed-bed reactors only. The importance of accurate kinetics data as a basis for a successful scaleup cannot be sufficiently stressed, however. A couple of cases discussed below will illustrate this.

A kinetic study has to be performed in the most appropriate type of equipment, not necessarily identical in geometry or mode of operation to the industrial reactor. Too often also there is a lack of interaction between the group performing the kinetic study and that responsible for the design of the reactor.

Determining reliable kinetics is by no means a simple matter, certainly not with complex reactions and complex feedstocks. The growing research capabilities of large companies, the automated operation of their experimental facilities, and the possibilities of modern analytical techniques and computers have enabled the derivation of fairly complicated kinetic models for industrial

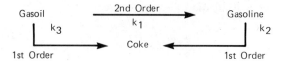

FIGURE 5-5 Three lump model for catalytic cracking of gasoil.

processes like catalytic reforming [Kmak (1971); Smith (1959)] and catalytic cracking [Jacob et al. (1976); Weekman and Nace (1970)].

The kinetic modeling of catalytic cracking has been discussed in detail by Weekman (1979). His case history is of interest in the present context because it deals with two levels of sophistication comparing possibilities and shortcomings of a simple three-lump model as compared with a more complex ten-lump model.

The three-lump model, shown in Figure 5-5, was developed early. The rate coefficient for each of the three lumped reactions contained a deactivation function to account for the effect of coke on the catalyst activity. Simulations based on this model revealed that, because of the deactivation, a moving bed reactor would lead to a higher conversion than the time-averaged conversion from a fixed-bed reactor (Weekman and Nace, 1970).

The model was successfully used in design with substantial improvements in the operation of existing plants. However, its parameters had to be re-

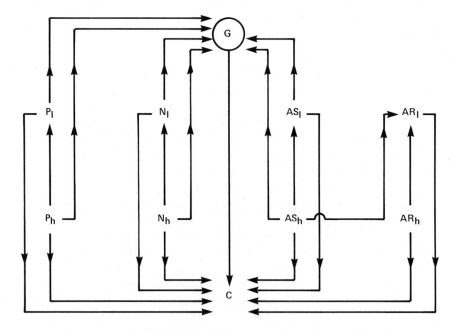

FIGURE 5-6 Mobil's ten lump model for catalytic cracking of gasoil.

PROBLEM AREAS

determined for each new feedstock. The rate coefficient k_1 and the deactivation parameters were successfully correlated against the aromatics/naphthene ratio. However, the correlations could not be used for nonvirgin feedstocks.

To cope with this problem, a ten-lump model involving 22 reactions was developed by Jacob et al. (1976). A crucial feature of the model, represented in Figure 5-6, is the introduction of separate lumps for aromatic rings and for side chains. Side chains are so readily split off that they essentially react as separate species. Derivation of this model was based upon experimentation in a fluid bed reactor. As a result, coke lay down was uniform over all the catalyst. The experiments covered not only a wide variety of feedstocks, but also synthetic charge stocks, rich in each of the chemical species.

A similar development occurred in catalytic reforming. Smith (1959) used a four-lump model based on the PONA analysis to simulate the process. Kmak (1971) presented a far more complex model distinguishing between hydrocarbons according to their chemical nature and carbon number. Graziani and Ramage (1978) published a 13-lump model also accounting for deactivation. It is now used routinely in the monitoring of Mobil's commercial units [Ramage et al. (1980)].

B. Catalyst Deactivation

In the two examples dealt with above, the modeling had to account for catalyst deactivation. This is a frequently encountered phenomenon which can be caused by poisoning, by structural modifications to the catalyst, and by coking. The latter are particularly difficult to avoid. Unfortunately any description of sintering is still rather empirical [Flynn and Wanke (1974, 1975); Ruckenstein and Pulvermacher (1973a, b)].

Coking has received increasing attention in recent years [Butt (1972, 1978); Froment (1976)]. Froment and Bischoff (1961, 1962) stressed the importance of relating the rate of coke formation to both the fluid composition and the operating variables. They showed that in tubular reactors with plug flow and isothermal operation the coke is deposited according to a profile.

Parallel coking—decreasing with downstream distance when the coke is formed from a feed component or from a component in equilibrium with it.

Consecutive coking—increasing from zero onward when the coke originates from a reaction product or an intermediate component in equilibrium with it.

With complex reactions the deactivation may cause the selectivities to vary because the reactions can be affected in different ways. Quantitatively, deactivation is allowed for through deactivation functions that multiply the rate coefficients of the reactions. Froment and Bischoff used empirical exponential or hyperbolic functions of the coke content of the catalyst. DePauw and Froment (1975) and Dumez and Froment (1976) applied the approach to the isomerization of the pentane on a platinum-reforming catalyst and to the dehydrogenation of 1-butene into butadiene. Weekman and Nace (1970) used a time-dependent deactivation function, although time is not the true deactivating agent.

Recently, Beeckman and Froment (1979) attempted to explain experimentally observed deactivation in terms of mechanisms. They developed a model relating the deactivation by coking to surface coverage and pore blockage. An approach of this type is expected to lead ultimately to fundamental coking parameters which will enable safer scaling up and extrapolation to different operating conditions.

C. Maldistribution of Fluid

A very important problem in scaling up is a maldistribution of the fluid that can develop with time. Ponzi and Kaye (1979) investigated the effect of flow maldistribution on conversion and selectivity in a radial flow fixed-bed reactor with a single fluid phase. For a second order irreversible reaction and isothermal operation, the conversion debit depends upon the conversion level and the nature of maldistribution. However, it can be over 10 percent.

Local hot spots may develop in fixed-bed reactors because of flow maldistribution caused by physical obstructions, agglomerated catalyst particles, or other means. An analysis of such a situation, encountered in a commercial hydrocracker, has been published by Jaffe (1976). A local hot spot was detected in one of the stages of a multibed adiabatic reactor, wherein normally only a monotonic temperature increase is possible. The fixed position of the thermocouple sheath did not allow measuring the extension of the hot spot in the bed. Jaffe's simulation led to an estimated hot spot of 120°C with a radius for the affected region of 8.5 cm.

Whether or not such a hot spot will propagate through the entire reactor and eventually cause runaway depends on the intensity of mixing with colder fluid coming from unaffected regions of the bed. Jaffe's simulation also revealed the importance of reactor internals which can cause regions of low flow velocities where hot spots are able to develop.

Each of the topics discussed here, complex feedstocks and kinetics, catalyst deactivation, and complex hydrodynamics, can cause problems during development and scaleup of a process. Even worse, they are usually encountered simultaneously in one and the same process. A multidisciplinary task force has to be set up to cope with such a challenge, preferably at an early stage of the development. In doing so, the different aspects of the process which may cause problems can be addressed in the appropriate way and specialists can be consulted to ensure an optimal concept of the pilot plant and a well-defined scope for the expensive pilot experimentation.

NOMENCLATURE

a	Stoichiometric coefficient
a_v	External particle surface area per unit reactor volume, m_p^2/m_r^3
a_v'	Gas–liquid interfacial area per unit packed volume, m_i^2/m_r^3
a_v''	Liquid–solid interfacial area per unit packed volume, m_i^2/m_r^3
b	Stoichiometric coefficient
C_A, C_B	Molar concentration of species A, B, $kmol/m^3$
C_s	Molar concentration of fluid reactant inside the solid, $kmol/m_f^3$
C_s^s	Molar concentration of fluid reactant in front of the solid surface, $kmol/m_f^3$
c_p	Specific heat of fluid, kJ/kg K
c_p'	Specific heat of fluid, kJ/kmol K
D_e	Effective diffusivity for transport in continuum, m_f^3/m s
D_{eG}	Gas-phase effective diffusivity in axial direction in a multiphase packed bed, m_G^3/m_r s
D_{eL}	Liquid-phase effective diffusivity in axial direction in a multiphase packed bed, m_L^3/m_r s
D_{er}	Effective diffusivity in radial direction, m_f^3/m_r s
d_p	Particle diameter, m
d_t	Tube diameter, m
F	Total molar flow rate, kmol/s
H	Henry's law constant, Nm/kmol
$-\Delta H$	Heat of reaction, kJ/kmol
h_f	Heat transfer coefficient for film surrounding a particle, kJ/m_p^2 s K
K	Factor in Equation 5-34; equilibrium constant
K_L	Overall mass transfer coefficient, m_L^3/m_i^2 s
k	Reaction rate coefficient
k_g	Gas-phase mass transfer coefficient, m_f^3/m_p^2 s, when based on concentrations, $kmol/m_p^2$ s, when based on mole fractions, $kmol/m^2$ s (N/m^2)
k_l	Mass transfer coefficient between liquid and catalyst surface, m_L^3/m_i^2 s, referred to unit interfacial area
k_l'	Transfer coefficient between the stagnant fluid and the solid, m_L^3/m_i^2 s
k_T	Transfer coefficient between the flowing and stagnant fraction, m_L^3/m_i^2 s
L	Volumetric liquid flow rate, m^3/hr
LHSV	Liquid hourly space velocity, m_f^3/m_r^3 hr

N	Dimensionless group, $\dfrac{4UC_{A_0}}{d_t c_p \rho_g \rho_B}$
N_{Pe_h}, N_{Pe_m}	Peclet number based on particle diameter for heat and mass transfer, respectively, $\dfrac{\rho_g u_s c_p d_p}{\lambda_{er}}$, $\dfrac{d_p u_i}{D_{er}}$
N_{Re}	Reynolds number, $d_p u_s \rho_g / \mu$
n	Reaction order
p_A, p_B	Partial pressures of components A, B, N/m²
p_t	Total pressure, N/m²
R	Gas constant, kJ/kmol K
r	Radial coordinate in reactor, m$_r$
r_A	Rate of reaction of component A per unit volume, kmol/m³ s
T	Temperature, K
T_s, T_s^s	Temperature inside solid, at solid interface, respectively, K
T_w	Bed temperature at radius $d_t/2$, K
S	Heat transfer surface, m²
x	Conversion
U	Overall heat transfer coefficient, kJ/m² s K
u_{iG}, u_{iL}	Interstitial velocity of gas, liquid, respectively, m$_r$/s
u_s	Superficial velocity, m$_f^3$/m$_r^2$ s
W	Total catalyst mass, kg cat
z	Axial coordinate in reactor, m$_r^2$
Z	Length of packed bed, m

Greek

α_w	Convective heat transfer coefficient in the vicinity of the wall, kJ/m² s K
β	Dimensionless group, $\dfrac{(-\Delta H)C_{A_0}}{\rho_g c_p T_0}$
Γ_e	Locus of equilibrium conditions in x–T diagram
Γ_m	Locus of the points in x–T diagram where the rate is maximum
γ	Activation energy, E/RT
ε	Void fraction of packing, m$_f^3$/m$_f^3$
ε_L	Liquid holdup, m$_f^3$/m$_r^3$
η	Effectiveness factor for solid particle
κ	Dimensionless group
λ_e	Effective thermal conductivity in a solid or solid particle, kJ/m s K

$\lambda_{ea}, \lambda_{er}$	Effective thermal conductivity in axial, radial direction, respectively, kJ/m s K
ν	Dynamic viscosity, kg/m s
ξ	Radial coordinate inside particle, m_p
ρ_B	Catalyst bulk density, kg cat/m_r^3
ρ_g	Gas density, kg/m_f^3
ρ_s	Solid density, kg/m_s^3
τ	Dimensionless group $\dfrac{2U}{d_t c_p} \cdot \dfrac{\Delta T_{ad} R_g}{P}$
ψ	Sphericity of a particle
Ω	Cross section of reactor or column, m²

Subscripts

ad	Adiabatic
G	Gas-phase conditions
L	Liquid-phase conditions
S	Solid-phase conditions
O	Inlet conditions

Superscripts

d	Stagnant fraction of fluid
f	Flowing fraction of fluid
s	Conditions at external surface

REFERENCES

Ampaya, J. P., and Rinker, R. G., *Chem. Eng. Sci.*, **32**, 1327 (1977a).

Ampaya, J. P., and Rinker, R. G., *Ind. Eng. Chem. Process Des. Dev.*, **16**, 63 (1977b).

Aris, R., *The Optimal Design of Chemical Reactors*, Academic Press, New York, 1960.

Baddour, R. F., Brian, P. L. T., Logeais, B. A., and Eymery, J. P., *Chem. Eng. Sci.*, **20**, 281 (1965).

Balakrishnan, A. R., and Pei, D. C. T., *Ind. Eng. Chem. Process Des. Dev.*, **18**, 30 (1979). Ibid, **18**, 40 (1979). Ibid, **18**, 47 (1979).

Barkelew, C. R., *Chem. Eng. Progr., Symp. Series*, **55**(25), 38 (1959).

Baumeister, E. B., and Bennett, C. O., *AIChE J.*, **4**, 69 (1958).

Bhattacharyya, D., and Pei, D. C. T., *Chem. Eng. Sci.*, **30**, 293 (1975).

Beeckman, J. W., and Froment, G. F., *Ind. Eng. Chem. Fundam.*, **18**, 245 (1979).

Beek, J., *Adv. Chem. Eng.*, **3**, 303 (1962).

Bellman, R., *Dynamic Programming*, Princeton University Press, Princeton, N.J., 1957.

Butt, J. B., *Adv. Chem. Series*, **109**, 259 (1972).

Butt, J. B., *Chemical Reaction Engineering Reviews Houston*, A.C.S. *Symp. Ser.*, **72** (1978).

Cappelli, A., Collina, A., and Dente, M., *Ind. Eng. Chem. Process Des. Dev.*, **11**, 184 (1972).

Carberry, J. J., *Ind. Eng. Chem. Fundam.*, **14**, 129 (1975).

Charpentier, J. C., in *Chem. React. Eng. Reviews, ACS Symp. Series*, **72**, D. Luss and V. W. Weekman, eds. (1978).

Crine, M., Marchot, P. and L'Homme, G. A., *Comput. Chem. Engng.*, **3** (1979).

Damköhler, G., *Zeitschrift für Elektrochemie*, **42**, 846 (1936).

Damköhler, G., in *der Chemie Ingenieur*, Band III, Teil I, A. Eucken and M. Jakob, eds., Akad. Verlag, Leipzig, 1937.

Danckwerts, P. V., and Gillham, A. J., *Trans. Inst. Chem. Engrs.*, **44**, T42 (1966).

Danckwerts, P. V., and Alper, E., *Trans. Inst. Chem. Engrs.*, **52**, T34 (1975).

Davidson, J. F., and Harrison, D., eds., *Fluidization*, Academic Press, New York, 1971.

De Acetis, J., and Thodos, G., *Ind. Eng. Chem.*, **52**, 1003 (1960).

Deans, H. A., and Lapidus, L., *AIChE J.*, **6**, 656 (1960).

De Groot, J. H., *Proc. Int. Symp. on Fluidization*, Netherlands University Press, Eindhoven, 1967.

Dente, M., and Collina, A., *Chim. et Industria*, **46**, 752 (1964).

De Pauw, R. P., and Froment, G. F., *Chem. Eng. Sci.*, **30**, 789 (1975).

De Wasch, A. P., and Froment, G. F., *Chem. Eng. Sci.*, **26**, 629 (1971).

Dietz, P. W., *Ind. Eng. Chem. Fundam.*, **18**, 283 (1979).

Dumez, F. J., and Froment, G. F., *Ind. Eng. Chem. Process Des. Dev.*, **15**, 291 (1976).

Elenkov, D., and Kolev, N., *Chem. Eng. Technol.*, **44**, 845 (1972).

Flynn, P. C., and Wanke, S. F., *J. Catal.*, **34**, 390 (1974). Ibid, **34**, 400 (1974).

Flynn, P. C., and Wanke, S. F., *Cat. Rev. Sci.*, **12**, 93 (1975).

Froment, G. F., and Bischoff, K. B., *Chem. Eng. Sci.*, **16**, 189 (1961). Ibid, **17**, 105 (1962).

Froment, G. F., *Chem. Eng. Sci.*, **7**, 29 (1961).

Froment, G. F., *Ind. Eng. Chem.*, **59**, No. 2, 18 (1967).

Froment, G. F., *Periodica Polytechnica (Budapest)*, **15**, 219 (1971).

Froment, G. F., *Proc. 5th Eur. Symp. Chem. React. Engng.*, Amsterdam, 1972.

Froment, G. F., *Chem. -Ing. -Techn.*, **46**, 374 (1974).

Froment, G. F., *Proc. 6th Int. Congress on Catalysis*, London (1976).

Froment, G. F., and Bischoff, K. B., *Chemical Reactor Analysis & Design*, Wiley, New York, 1979.

Froment, G. F., in *Chemistry and Chemical Engineering of Catalytic Processes*, R. Prins and G. C. A. Schuit, eds., Sijthof & Noordhof, Alphen aan de Rijn, Nederland, 1980.

Gamson, B. W., Thodos, G., and Hougen, O. A., *Trans. AIChE*, **39**, 1 (1939).

Gamson, B. W., *Chem. Eng. Progr.*, **47**, 19 (1951).

Graziani, K. R., and Ramage, M. P., *ACS Symp. Series*, **65**, 282 (1978).

Gupta, A. S., and Thodos, G., *AIChE J.*, **8**, 608 (1962).

Handley, D., and Heggs, P. J., *Trans. Inst. Chem. Engrs.*, **46**, T251 (1968).

Hatcher, W. J., Viville, L., and Froment, G. F., *Ind. Eng. Chem. Process Des. Dev.*, **17**, 491 (1978).

Hennecke, F. W., and Schlünder, E. U., *Chem. -Ing. -Techn.*, **45**, 277 (1973).

Henry, H. C., and Gilbert, J. B., *Ind. Eng. Chem. Process Des. Dev.*, **12**, 328 (1973).

Himmelblau, D. M., and Bischoff, K. B., *Process Analysis and Simulation*, Wiley, New York, 1968.

Hochman, J. M., and Effron, E., *Ind. Eng. Chem. Fundam.*, **8**, 63 (1969).

Hofmann, H., *Cat. Rev. -Sci. Eng.*, **17**, 71 (1978).

REFERENCES

Inoue, H., *Chem. Eng. J. Japan*, **11**, 40 (1978).

Jaffe, S. B., *Ind. Eng. Chem. Process Des. Dev.*, **15**, 410 (1976).

Jacob, S. M., Gross, B., Voltz, S. E., and Weekman, V. W., *AIChE J.*, **22**, 701 (1976).

Kehoe, J. P. G., and Butt, J. B., *AIChE J.*, **18**, 347 (1972).

Kmak, W. S., paper presented at AIChE National Meeting Houston (1971).

Kunii, D., and Furusawa, T., *Chem. Eng. J.*, **4**, 268 (1972).

Kunii, D., and Smith, J. M., *AIChE J.*, **6**, 71 (1960).

Larkins, R. P., White, R. R., and Jeffrey, D. W., *AIChE J.*, **7**, 231 (1961).

Lerou, J. J., and Froment, G. F., *Chem. Eng. Sci.*, **32**, 853 (1977).

Leva, M., *Ind. Eng. Chem.*, **40**, 747 (1948).

Macdonald, I. F., El-Sayed, M. S., Mow, K., and Dullien, F. A. L., *Ind. Eng. Chem. Fundam.*, **18**, 199 (1979).

Mc Greavy, C., and Adderley, C. I., *Chem. Eng. Sci.*, **28**, 577 (1973).

Mc Greavy, C., and Adderley, C. I., *Adv. Chem. Series*, **133**, 519 (1974).

Mc Guire, M., and Lapidus, L., *AIChE J.*, **11**, 85 (1965).

Mears, D. E., *Adv. Chem. Series*, **133**, 218 (1974).

Michell, R. W., and Furzer, I. A., *Trans. Inst. Chem. Engrs.*, **50**, T334 (1972).

Midoux, N., Favier, M., and Charpentier, J. C., *J. Chem. Eng. Japan*, **9**, 350 (1976).

Minet, R. G., and Olesen, O., paper presented at Joint Meeting ACS and Chem. Soc. Japan, April 1–6, 1979, Honolulu.

Montagna, A. A., and Shah, Y. T., *Chem. Eng. J.*, **10**, 99 (1975a).

Montagna, A. A., and Shah, Y. T., *Ind. Eng. Chem. Process Des. Dev.*, **14**, 479 (1975b).

Montagna, A. A., Shah, Y. T., and Paraskos, J. A., *Ind. Eng. Chem. Process Des. Dev.*, **16**, 152 (1977).

Morsi, B. I., Midoux, N., and Charpentier, J. C., *AIChE J.*, **24**, 357 (1978).

Murase, A., Roberts, H. L., and Converse, A. O., *Ind. Eng. Chem. Process Des. Dev.*, **9**, 503 (1970).

Olson, K. E., Luss, D., and Amundson, N. R., *Ind. Eng. Chem. Process Des. Dev.*, **7**, 96 (1968).

Onda, K., Sada, E., and Okumoto, Y., *J. Chem. Eng. Japan*, **1**, 63 (1968).

Otake, K., and Okada, K., *Kagaku Kogaku*, **17**, 176 (1963).

Petersen, E. E., *Chemical Reaction Analysis*, Prentice-Hall, Englewood Cliffs, N.J., 1965.

Ponzi, P. R., and Kaye, L. A., *AIChE J.*, **25**, 100 (1979).

Puranik, S. S., and Vogelpohl, A., *Chem. Eng. Sci.*, **29**, 501 (1974).

Ramage, M. P., Graziani, K. R., and Krambeck, F. J., Chem. Eng. Sci., **35**, 41 (1980).

Ray, W. H., *Proc. 5th Eur. Symp. Chem. React. Engng.*, Amsterdam (1972).

Rajadhyaksha, R. A., and Vasudeva, K., *Chem. Eng. Sci.*, **30**, 1399 (1975).

Reiss, L. P., *Ind. Eng. Chem., Process Des. Dev.*, **6**, 486 (1967).

Roberts, S. M., *Dynamic Programming in Chemical Engineering and Process Control*, Academic Press, New York, 1964.

Ross, L. D., *Chem. Eng. Progr.*, **61**, 77 (1965).

Rückenstein, E., and Pulvermacher, B., *AIChE J.*, **19**, 456 (1973a).

Rückenstein, E., and Pulvermacher, B., *J. Catal.*, **29**, 224 (1973b).

Satterfield, C. N., Pelossof, A. A., and Sherwood, T. K., *AIChE J.*, **15**, 226 (1969).

Schlünder, E. U., in *Chem. React. Eng. Reviews*, *ACS Symp. Series*, **72**, D. Luss and V. W. Weekman, eds. (1978).

Schmalzer, D. K., and Hoelchser, H. E., "Stochastic Model of Packed Bed Mixing and Mass Transfer," *AIChE J.*, **17**(1), 104 (1971).

Shah, M. J., *Ind. Eng. Chem.*, **59**, 72 (1967).

Shah, Y. T., *Gas-Liquid-Solid Reactor Design*, McGraw Hill, New York, 1979.

Shipman, L. N., and Hickman, J. B., *Chem. Eng. Progr.*, **64**(5), 59 (1968).

Smith, T. G., and Carberry, J. J., *Adv. Chem. Series*, **133**, 362 (1974).

Smith, R. B., *Chem. Eng. Progr.*, **55**, 76 (1959).

Sundaresan, S., Amundson, N. R., and Aris, R., *AIChE J.*, **26**, 529 (1980).

Sweeney, D., *AIChE J.*, **13**, 633 (1967).

Trambouze, P., *Chem. Engng.*, **86**, 122 (1979).

Turpin, J. L., and Huntington, R. L., *AIChE J.*, **13**, 1196 (1967).

Van Krevelen, D. W., and Krekels, J. T. C., *Rec. Trav. Chim.*, *Pays-Bas*, **67**, 512 (1948).

Van Welsenaere, R. J., and Froment, G. F., *Chem. Eng. Sci.*, **25**, 1503 (1970).

Villadsen, J. V., and Michelsen, M. L., *Solution of Differential Equation Models by Polynomial Approximation*, Prentice-Hall, Englewood Cliffs, N. J., 1978.

Vortmeyer, D., *Chem. -Ing. -Techn.*, **51**, 839 (1979).

Weekman, V. W., *Proc. 4th Int. Symp. Chem. React. Engng.*, Heidelberg (1976).

Weekman, V. W., and Nace, D. M., *AIChE J.*, **16**, 397 (1970).

Weekman, V. W., *AIChE-Monograph Series 11*, **75** (1979).

Yagi, S., and Kunii, D., *AIChE J.*, **3**, 373 (1957).

Yagi, S., and Kunii, D., *AIChE J.*, **6**, 97 (1960).

Yagi, S., and Wakao, N., *AIChE J.*, **5**, 79 (1959).

Zehner, P., and Schlünder, E. U., *Chem. -Ing. -Techn.*, **42**, 933 (1970). Ibid, **44**, 1303 (1972). Ibid, **45**, 272 (1973).

6
FLUID – FLUID REACTORS

Y. T. SHAH AND W.-D. DECKWER

I.	Major Issues	202
	A. Practical Examples	204
II.	Scaleup Considerations	204
	A. Problem Areas	206
	B. Determination of Rate Controlling Step	207
	C. Packed Bed Absorbers	210
	D. Bubble Columns	217
	1. Variations of Nonadjustable Parameters in Scaleup	217
	a. Flow Regimes	217
	b. Gas Holdup	221
	c. Interfacial Area	222
	d. Volumetric Mass Transfer Coefficients, $k_L a$	223
	e. Gas-Side Mass Transfer Coefficients, k_G	224
	f. Liquid–Solid Mass Transfer	224
	g. Mixing Parameters	225
	h. Heat Transfer	226
III.	Applicability of Models to Scaleup	228
	A. Nonisothermal Operation	230
	B. Variation of Reactor Performance	230
	1. Fast Reaction Regime	231
	2. Slow Reaction Regime	233
	C. Model Simplification and Examples	235
	1. Role of Solids	243

IV. Uncertainties	245
A. Dispersion Coefficients	245
B. Gas Holdup and Interfacial Area	246
C. Flow Regime	246
D. Kinetic Data	246
E. Heat Transfer and Stability	246
Nomenclature	247
References	249

I. MAJOR ISSUES

Fluid–fluid reactions can be divided as liquid–liquid, gas–liquid, and gas–liquid–solid reactions. Among these, gas–liquid systems are more widely used in the chemical process industry than liquid–liquid systems. Gas–liquid–solid systems are beginning to be increasingly important, particularly in the petroleum industry and potentially in the synthetic fuel industry.

Scaleup of a fluid–fluid reactor is a rather complex problem since in most cases reactor performance depends quite significantly upon the prevailing hydrodynamics, transport, and mixing characteristics of the reactor. The variations in these characteristics with the reactor scaleup are not well understood. The applicability of the hydrodynamic and mixing models (e.g., dispersion model) used to correlate the reactor performance can depend on the scale of the reactor. Since the apparent reaction rate depends upon the various transport resistances (e.g., transport resistance at the fluid–fluid interface), the controlling resistance also can depend upon the scale of the reactor. This means that the reactor performance model used for a small-scale reactor may not be useful for the large-scale reactor.

It is often useful to distinguish fluid–fluid systems on the basis of the nature of the continuous phase and the dispersed phase, that is, the discontinuous phase. Liquid-in-gas dispersions where the gas is the continuous phase, as would be the case in spray columns, are seldom found. (Spray columns are commonly used in liquid–liquid reactions.) Of larger significance are gas-in-liquid dispersions which are predominantly processed in aerated and mechanically agitated tank reactors and in bubble column reactors. A third and widely used gas–liquid contactor is the packed bed chemical absorber in which gas (or lighter phase) can be either a continuous or a dispersed phase. Liquid (or the denser phase) often forms a film around the solid packing elements.

Some of the commonly used gas–liquid, liquid–liquid, and gas–liquid–solid reactors are described by Danckwerts (1970), Laddha and Degaleesan (1978), and Shah (1979), respectively. In a gas–liquid–solid system the solids can be inert material, catalyst, reactant, or reaction product. Examples of the function of solids in three-phase systems are summarized in Table 6-1. Inerts provide a

TABLE 6-1 Function of Solids in Three-Phase Systems

Solid Function	Practical Examples	References
Inert material	Noncatalytic gas–liquid reactions in packed bed Chemical absorbers Gas purification (CO_2, CO, COS, H_2S, SO_2, NO_x, HCl, Cl_2, $COCl_2$)	Alper and Danckwerts (1976), Astarita (1983). Danckwerts (1970), Danckwerts and Alper (1975), Danckwerts and Sharma (1966)
Reactant	Production of alumina alkyls	Albright (1967)
	Ca$(HSO_3)_2$ (for treatment of cellulose)	Volpolicelli and Massimilla (1970)
	SO_2 removal from flue gases with lime and limestone slurries	Engdahl and Rosendahl (1978), Rosenberg (1978)
	Coal hydrogenation	Kronig (1977), Wen and Tone (1978)
	Biotechnological processes	
	Production of primary and secondary metabolites	Smith and Greenshields (1974), Humphrey (1974)
	Production of single cell proteins (SCP)	Moo-Young (1975)
	Wastewater treatment	Leistner er al. (1979), Brauer and Sucker (1979)
Reaction product	Polymers from olefins	Reichert (1977), Gates et al. (1979)
	Oxamide by HCN oximation	Riemenschneider (1978)
	Production of biomass (SCP)	Moo-Young (1975), Rosenzweig and Ushio (1974)
Catalyst	Numerous hydrogenation and oxidation processes to produce organic and inorganic intermediates	
	Hydrogenation, hydrodesulfurization, and demetalization of residual oils	Weekman (1976), Satterfield (1975)
	Upgrading of coal oils, heavy oil fraction, etc.	vanDriesen and Steward (1964), Oestergaard (1971)
	Coal hydrogenation	Kronig (1977)
	Fischer–Tropsch synthesis	Kolbel and Ralek (1977) (1980)

uniform distribution of the phases over the whole cross-sectional area, a high interfacial area, high turbulence intensities, and reduce the longitudinal mixing. This results in a high driving force along the bed when operated countercurrently. The most practical gas–liquid–solid reactions are certainly those in which the solid acts as a catalyst. Gas–liquid reactions in the presence of a solid catalyst are mainly carried out in packed bed (trickle flow), slurry, and fluidized bed reactors.

In this chapter we restrict our discussion to the three most widely used fluid–fluid reactors, namely:

Packed bed absorber.
Gas–liquid bubble column.
Gas–liquid–solid slurry and fluidized bed reactors.

Attention is focused on gas–liquid bubble columns since they are widely used and their analysis forms the basis for analyzing gas–liquid–solids slurry and fluidized bed reactors. Trickle bed catalytic reactors have been discussed in Chapter 5. While the discussion is mainly centered on gas–liquid systems, many of the underlying principles are applicable to liquid–liquid systems.

A. Practical Examples

The number of fluid–fluid reactions (either with or without the presence of solids) carried out industrially is large. Gas–liquid and gas–liquid–solid reactions are used in hydrogenation, oxidation, hydration, sulfonation, chlorination, hydroformylation, polymerization, and amination processes.

All the reactants in liquid–liquid reactions are basically nonvolatile under the reaction conditions. Although liquid–liquid reactions are not as widespread and as widely explored as gas–liquid reactions, there are many examples of industrial importance such as nitration, alkylation, saponification (of fats), and oximation of cyclohexanone. Liquid–liquid reactions are also involved in solvent extraction processes applied in hydrometallurgy (for the recovery of zinc, copper, cobalt, nickel, uranium, aluminum, indium) and waste water treatment. Another important field of application is in the separation of fission products and the recovery of uranium and plutonium in nuclear fuel reprocessing plants. In recent years, liquid–liquid reactions have been studied in systems involving phase transfer catalysis and liquid membranes.

II. SCALEUP CONSIDERATIONS

The proper design and scaleup of a fluid–fluid reactor is facilitated by a good mathematical model of the reactor. A procedure normally followed for this purpose is illustrated in Figure 6-1. Application of the procedure requires the definition of throughput, nature of the reaction system, and the product yield structure desired.

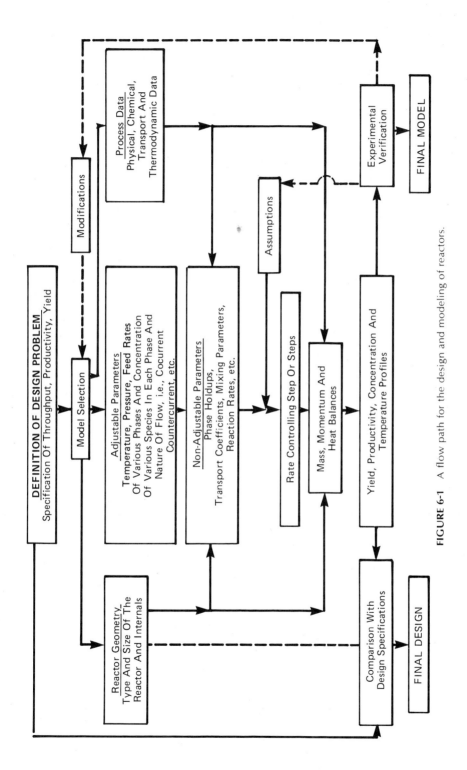

FIGURE 6-1 A flow path for the design and modeling of reactors.

To obtain the desired goal it is necessary to begin with specification of first level quantities such as reactor geometry, adjustable reactor operating conditions, and process data. As shown in Figure 6-1, these quantities are interrelated. For any reaction system, the intrinsic kinetic information (i.e., reaction paths, rate constants, and heats of reaction associated with reaction path) must either be estimated from the literature or measured. Knowledge of physical properties of the reaction mixture (i.e., densities, viscosities, and surface tension of various phases, heat and mass diffusivities, etc.) and thermodynamic data (i.e., phase equilibrium data such as solubilities of various reactants and products, phase and reaction equilibrium constants, etc.) are also required. These data permit a choice of the reactor geometry, dimensions, internals, and the nature of the phase distributors. The desired production rate, reactor geometry, and the process data also fix bounds on so-called adjustable operating conditions such as phase velocities, temperature, pressure, and the direction of flows (cocurrent, countercurrent, etc.). The reactor geometry, process data, and adjustable operating conditions together dictate the hydrodynamics, that is, the flow pattern within the reactor.

For single phase reactors this information would be sufficient to design and scale up the reactor given knowledge about residence time distribution. In the design of a multiphase reactor, there is, however, another group of important parameters. These are the "nonadjustable" quantities which are also dependent on the chosen reactor geometry, the adjustable operating conditions, and the process data. The "nonadjustable" quantities are the phase holdups, the interfacial areas, the heat and mass transfer properties, and the dispersion coefficients or the mixing parameters. The reactor geometry, the reaction parameters, and the adjustable and the nonadjustable parameters are then introduced in the fundamental reactor model equations derived on the basis of the physical and chemical phenomena suspected to take place within the reactor. Usually the sophisticated model equations are solved numerically as they contain strong nonlinearities (temperature dependency of reaction and solubilities, phase flow variation). Whenever possible, the model must be simplified with the knowledge of the rate controlling mechanism. The general scheme shown in Figure 6-1 would be iterated several times since a desired optimal reactor design cannot be obtained explicitly and in addition is subject to various economic choices. The model equations and the outlined scheme are usually processed through an optimization procedure. Such optimization techniques can suggest that one reactor is acceptable depending on the specific objectives. Furthermore, the usefulness of a given model for scaleup of a reactor depends on the validity of the assumptions made for different reactor sizes.

A. Problem Areas

The outlined procedure is not without problems. Process data estimation involves some inaccuracies. In multiphase systems these quantities can seldom be measured independently and they are always coupled with interphase transport of heat and mass. Isolation of intrinsic kinetics from the interphase transport requires either careful planning of experiments or rigorous analysis of experimental data as discussed by Danckwerts (1970), Laddha and Degaleesan (1978), Shah (1979), and Astarita (1983). This is essential in determining the rate controlling step (or steps) in the overall reaction and then deciphering how this controlling step changes with scaleup of the reactor.

Estimation of nonadjustable quantities presents major problems. Arbitrary variation of these parameters is not possible to a significant degree since they are determined by complex relationships and interactions from all the quantities given in the first level of Figure 6-1. For estimating the nonadjustable parameters, numerous correlations are proposed in the literature. However, the experimental data used in the correlations have been determined largely from measurements with water, aqueous solutions, or pure organic liquids at normal temperatures and pressure, conditions seldom encountered in industrial practice.

The majority of the data on nonadjustable parameters have been obtained in laboratory scale equipment. Effects can often be observed in this equipment which do not occur in industrial scale reactors, and vice versa. Also, the data used in correlations for the nonadjustable parameters represent only integral values for a reactor. Actually, the values can be space dependent. To what extent such local dependencies may modify the performance of a reactor is largely unknown. The problem, of course, is magnified as the reactor size becomes larger.

Most parameter values are determined for "dead" systems, that is, in equipment where no changes in molar flow rate occur as one phase rises through the other. In industry, however, chemical reactors can produce the largest possible changes in flow rates, since conversions are maximized. Often the parameter estimation is based on certain assumptions about the residence time distribution of the phases. In bubble columns, for instance, overall liquid–side mass transfer coefficients ($k_L a$) have been evaluated often from inlet–outlet concentration measurement by applying the NTU method as discussed by Metzner and Brown (1956), Houthton et al. (1957), Voyer and Miller (1968), Mashelkar and Sharma (1970), and Mashelkar (1970). Alternatively, one can assume the liquid phase to be well mixed as has been done by Eckenfelder and Barnhart (1961), Towell et al. (1965), Yosida and Akita (1965), and Akita and Yashida (1973, 1974). With the first approach, minimum $k_L a$ values are obtained while in the second $k_L a$ is at the upper limit. Neither of the assumptions underlying the model are completely correct. The impact of their inaccuracies are magnified with increased size of the equipment.

B. Determination of Rate Controlling Step

Fluid–fluid systems involving simultaneous mass transfer and chemical reaction can conform to various mechanisms as discussed by Danckwerts (1970),

Laddha and Degaleesan (1978), Shah (1979), and Astarita (1983). When the concentration profile of the solute in the fluid diffusion film is flat, the reaction is often called a "very slow" reaction, and the transfer process is said to be in the kinetically controlled regime. Under certain conditions, diffusion and reaction may conform to a "slow" mechanism. By this term, one means that a solute A diffuses through the film experiencing little reaction in the time of passage, and then reacts in the bulk fluid. In terms of classical film theory, the processes of chemical reaction and diffusion become two steps in series for a "slow" reaction. The solute transfer rate is then largely unaffected by the chemical reaction.

Interfacial mass transfer accompanied by a chemical reaction is considered to be in a "fast" reaction regime when the solute transfer rate is considerably affected by the chemical reaction. According to film theory, in a "fast" reaction regime, both reaction and difussion occur in parallel within the diffusion film. If concentrations of all nonvolatile species taking part in the reaction change significantly near the interface, the "fast" reaction is said to occur under "depletion" conditions.

A reaction between a transferring species and a nonvolatile reactant may, in an extreme case, be "instantaneous." The increase in transfer rate due to the chemical reaction is maximized under this situation. If the reaction is irreversible, the "instantaneous" reaction occurs when the reaction between the transferring solute and the nonvolatile reactant is so rapid that they cannot coexist in the liquid. Then a "reaction plane" is formed in the liquid, where the instantaneous reaction occurs, and both absorbed species and the nonvolatile reactant diffuse toward this "reaction plane," where they react. The reaction is, then, solely mass transfer controlled. However, even a reaction that may not be intrinsically instantaneous can become "instantaneous" under certain conditions. Then the rate of reaction is controlled by the rate of diffusion of the transferring species and the nonvolatile reactant to the reaction plane.

If the reaction between between the transferring species and the nonvolatile reactant is reversible, the term "instantaneous reaction" is synonymous with "equilibrium reaction." Both forward and backward reactions in this case are so fast that, at all times, the concentrations of the various reacting species at the reaction plane in the liquid are in equilibrium. The solute transfer rate in this situation would be independent of the reaction and solely determined by the diffusion of various reacting species.

Many reactions produce volatile species which are desorbed into a gas phase. The mechanism of desorption with chemical reaction follows the basic concepts described above. In many gas–liquid reactions, the mass transfer resistance of the gas-side film at the gas–liquid interface is negligible. In most liquid–liquid systems, however, the resistances at both sides of the interface are important.

For a second order reaction between two species A and B, the typical concentration profiles for slow, fast, and instantaneous reaction regimes are illustrated in Figure 6-2. The mathematics associated with the above concepts

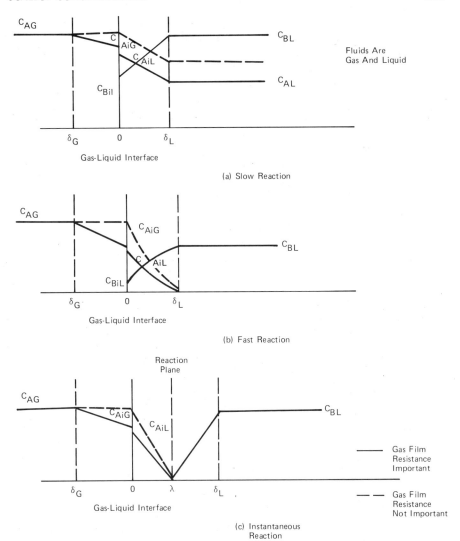

FIGURE 6-2 Concentration distributions for gas–liquid reaction based on film theory.

have been repeatedly described in the references previously cited. What is critical is that the apparent reaction rate depends on the phenomenological coefficients such as gas–liquid mass transfer coefficient, gas–liquid interfacial area, holdups of various phases, and so on. These parameters in turn depend on the reactor size and phase velocities.

The mathematical model required for the scaleup of the reactor can be considerably simplified if there is knowledge about the controlling reaction regime. Values of the intrinsic kinetic parameters do not depend on the shape, size, and the internal design of the reactor vessel or the velocities of the various phases. However, the phenomenological coefficients such as gas–liquid, and liquid–solid mass transfer coefficients and phase holdups, on the other hand, do depend on the scale of operation of the reaction process and nature and size of the reactor vessel.

During the process of scaleup of a reactor the relative importance of transport and kinetic resistances on the apparent reaction rate is almost always changed. These changes must be calculated for a proper prediction of the changes in the reactor performance with the scaleup. The mathematics involved in the calculation of the apparent reaction rate during fluid–fluid reaction not only depend upon the nature of the reaction regime as mentioned above but also upon the number of reactions, some of which may be slow, some fast, and others instantaneous. In many gas–liquid reactions involving highly soluble gases such as the absorption of NH_3, HCl, Cl_2, and SO_2 in water and hydrocarbon solvents, the gas film resistance may be very important. Kinetic influences in liquid–liquid reactions have also been evaluated in Laddha and Degaleesan (1978). The design and scaleup of pertinent reactors can be completely based on kinetics assuming pseudohomogeneous behavior even for reactions like aromatic nitrations which are known to be so rapid that their control is difficult as discussed by Hanson et al. (1974).

As mentioned previously, the solid can play a variety of roles in gas–liquid–solid reactions. Solids can be inert for the reaction and the mechanism described above can be used for mathematical calculations as shown by Danckwerts (1970). When the solid is a catalyst, the mass transfer resistance at the fluid–solid interface becomes important and the effect of this resistance on the overall kinetics is discussed by Shah (1979). If the solid is a reactant, its dissolution rate affects the controlling mechanism and the effects of solid concentration and particle size on the reaction regime are also discussed by Shah (1979) and Joshi et al. (1980).

C. Packed Bed Absorbers

The design and scaleup of gas–liquid reactors presents a difficult task because of uncertainties involved in the determination of the process data and the self-adjusting hydrodynamic parameters (k_L, a, k_G, holdup, mixing properties, etc.). Contrary to reactors that process gas-in-liquid dispersions like aerated vessels and bubble columns, however, the situation is not as complex with packed bed chemical absorbers since the hydrodynamic properties do not interfere as strongly with physicochemical properties and the operating conditions. In addition, there are some effective methods and procedures available which permit reliable scaleup of packed bed chemical absorbers. The subject has been reviewed by Charpentier (1978), Alper and Danckwerts (1976), and Alper (1979).

In countercurrent flow packed bed chemical absorbers, it is generally reasonable to neglect dispersion phenomena of both phases and also maldistri-

SCALEUP CONSIDERATIONS

bution of liquid. The governing liquid-phase balance equations for two species A (gaseous) and B (liquid) undergoing a reaction $A + zB \to$ product is given by Danckwerts and Alper (1975), Alper and Danckwerts (1976), and Alper (1979) as

$$\overline{R}(k_G, k_L, H, p_A, c_{AL}, c_{BL}) a\, dx = u_L\, dc_{AL} + \varepsilon_L r(c_{AL}, c_{BL})\, dx \quad (6\text{-}1)$$

and

$$-u_L\, dc_{BL} = z\varepsilon_L r(c_{AL}, c_{BL})\, dx \quad (6\text{-}2)$$

Here a is the gas–liquid interfacial area per unit volume of the reactor, z the stoichiometric coefficient, x the axial distance along the absorber, u_L the linear liquid velocity, ε_L the liquid holdup, $\overline{R}$ the mean local absorption rate, c_{AL} and c_{BL} the concentrations of A and B in bulk liquid phase, respectively, r the reaction rate, H the Henry's constant, and p_A the partial pressure of A in gas phase. By rearranging and integration one can obtain

$$\frac{aL}{u_L} = \int_{\text{in}}^{\text{out}} \frac{dc_{AL}}{\overline{R}(k_G, k_L, H, p_A, c_{AL}, c_{BL})} - \int_{\text{in}}^{\text{out}} \frac{dc_{BL}}{z\overline{R}(k_G, k_L, H, p_A, c_{AL}, c_{BL})} \quad (6\text{-}3)$$

and

$$\frac{\varepsilon_L L}{u_L} = \int_{\text{in}}^{\text{out}} \frac{dc_{BL}}{zr(c_{AL}, c_{BL})} \quad (6\text{-}4)$$

For the calculation of the desired packing length L the fluid dynamic parameters $(a, \varepsilon_L, k_G, k_L)$ and dependency of the local absorption rate must be known. The hydrodynamic properties have been measured and reviewed extensively by Shah (1979). Alper (1979) summarized the available data by proposing the following guidelines:

The interfacial area a is almost independent of the properties of the applied chemical system. For a given kind of packing a is only influenced by the liquid flow rate. This is demonstrated in Figure 6-3 where values of a measured by Sridharan and Sharma (1976), Jhaveri and Sharma (1968), and Vidwans and Sharma (1967) for various absorption systems are presented. Correlations and measured data of the interfacial area and the liquid holdup for various types of packing material can be found in Alper (1976, 1982), Buchanan (1967), Hoffman (1975), Kolar et al. (1970), Kolev (1976), Laurent and Charpentier (1974), Linet et al. (1955, 1974, 1977), Mohunta and Luddha (1965), Onda (1972), Ponter et al. (1976), Puranik and Vogelpahl (1974), Sharma and Danckwerts (1970), Shende and Sharma (1974), Shulman et al (1955a, b), and Ticky (1973), respectively.

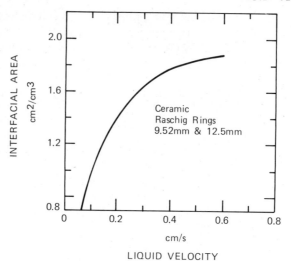

FIGURE 6-3 Dependence of interfacial area on liquid velocity absorption of CO_2 in amines.

It has been shown by Sahay and Sharma (1973) and Vidwans and Sharma (1967), that for a given packing the gas-side resistance is only dependent on the gas velocity. The gas-side mass transfer coefficient k_G is dependent on the nature of the gas and related to gas-phase diffusivity as $k_G \propto D_G^{1/2}$ as shown by Vidwans and Sharma (1967) and Yadav and Sharma (1978).

The liquid-side mass transfer coefficient k_L depends mainly on the liquid flow rate (for a certain packing material). The viscosity also has a strong influence on k_L and it can be correlated with sufficient accuracy by using $k_L \alpha D_L^{1/2}$ where D_L is a function of liquid viscosity. Data and correlations for k_L and $k_L a$ for various packings are reported by Sahay and Sharma (1973), Sharma and Danckwerts (1970), Sridharan and Sharma (1976), Laurent and Charpentier (1974), Reichelt and Blass (1974), and Onda (1972).

Following the guidelines of Alper (1979), the hydrodynamic properties in the packed bed can be estimated rather easily. The intrinsic rate expression $r(C_{AL}, C_{BL})$ for the given reaction system is either known or independently evaluated by the experimental methods outlined by Danckwerts (1970) and Astarita (1967). This leaves us then with the problem of determining the absorption rate R which varies over the entire column. All possible absorption-reaction regimes may be examined theoretically. The computations require essentially detailed knowledge of solubilities, diffusivities, and reaction rate constant. However, the complete theoretical-computational approach can be avoided by performing measurements in model absorbers of lab scale.

The Danckwerts-Gillham-Alper method (Danckwerts and Alper, 1975; Danckwerts and Gillham, 1966; Alper, 1971) presents a procedure where each

SCALEUP CONSIDERATIONS

point within the packed bed is simulated in a stirred cell which is constructed and operated under such conditions that:

The interfacial area is known.

The mass transfer coefficients (k_G, k_L) can be adjusted to those values which prevail in the packed bed.

With a knowledge of the measured absorption rates (at various concentrations of A and B) the integration of Equation 6-3 can be carried out when no reaction occurs in the bulk liquid phase. The Danckwerts–Gillham–Alper method is also called the point model or differential simulation of packed bed chemical absorbers. The usefulness of the stirred cell to measure absorption rates was shown by many authors. In particular, Alper (1971), Danckwerts and Alper (1975), and Laurent (1975) demonstrated the applicability of the point model for the design of a packed bed for absorbing CO_2 in sodium hydroxide solution.

The point model cannot be used to simulate the packed bed if the absorption process is accompanied by reaction in the bulk liquid phase or if two reactions occur. Also, if two gases are absorbed simultaneously, differential simulation with the stirred cell is not possible. Furthermore, the differential design equations (Equations 6-3 and 6-4) must be fulfilled by the experimental model absorber and the relationships for k_L and k_G for the packed bed absorber. Such an experimental model absorber, called a complete model, addresses not only the microscopic behavior, but also the integral macroscopic performance behavior of the absorber. Alper and Danckwerts (1976) demonstrated the validity of this experimental method for the design and scaleup of the reactors for the system outlined in Table 6-2. For all the systems studied, the absorption rates in the packed column were predicted from measurements in the complete model absorber with a deviation of 7 percent.

The potential of the Alper–Danckwerts method was emphasized by Charpentier (1978). In principle, such simulation methods should also be applicable to other gas–liquid reactors. However, the difficulty of estimating the hydrodynamic properties of the reactor which are strongly dependent on physical-chemical properties of the absorption system, along with the flow rates and the nature of packing material is limiting. Some other scaleup considerations for packed bed liquid–liquid reactors are outlined by Laddha and Degaleesan (1978) and they will not be repeated here.

Example 6.1 Application of Danckwerts–Gillham–Alper Method to Packed Bed Absorber. CO_2 is to be absorbed by NaOH solution in a countercurrent column packed with 1.27 cm ceramic Raschig rings. The inlet concentration of NaOH is 0.6 mol/L, the inlet mole fraction of CO_2 is 0.116 and the absorption process is carried out at atmospheric pressure. The inlet gas and liquid velocities are 15 and 0.28 cm/s, respectively. The interfacial area is 1.3 cm^{-1}

TABLE 6-2 Absorption–Reaction Systems Used to Check the Alper–Danckwerts Method

Absorption–Reaction System	Nature of the Reaction
CO_2 absorption in MEA containing As_2O_3	Fast reaction at interface followed by consecutive reaction in bulk
Simultaneous absorption of NH_3 and CO_2 in H_2O	Gases react with each other in solution, therefore the absorption of the one gas increases the absorption rate of the other gas
Simultaneous absorption of SO_2 and CO_2 in amine solution	Each gas reacts with the same liquid-phase reactant, hence, the absorption of the one decreases the absorption rate of the other gas
Absorption of CO_2 in solution of 2, 6-dimethyl morpholine	35 percent of the overall resistance is located at gas side
CO_2 absorption in solution which contain two amines	The absorption rates are such that different regimes occur
MEA and diisopropyl amine	
MEA and 2-methyl ethanol amine	

and the value of the liquid-side mass transfer coefficient is 0.0057 cm/s. For this value of k_L the absorption rate of CO_2 per unit interfacial area was measured by Danckwerts and Alper (1975) in a stirred cell with plane interface and various compositions ($NaOH/Na_2CO_3$). In Figure 6-4, the absorption rates are plotted versus the CO_2 mole fraction.

The required column height can be calculated using the Danckwerts–Gillham–Alper method on the basis of the absorption rates measured in stirred cell and the assumption that NaOH outlet concentration is 0.1 mol/L.

The CO_2 outlet mole fraction y is obtained from an overall material balance.

$$\frac{u_{G0}y_0 - u_G y}{V_M} = \frac{u_L}{z}(C_{BL0} - C_{BL}) \qquad (6\text{-}5)$$

where u_G and the u_L are in cm/sec, y is the CO_2 mole fraction, V_M the molar volume ($= 22.414$ L/mol), z the number of moles of NaOH needed per mole of $CO_2 (= 2)$, and C_{BL} the OH^- concentration in mol/L. An overall balance on the inerts gives

$$u_G = u_{G0}\left(\frac{1 - y_0}{1 - y}\right) \qquad (6\text{-}6)$$

SCALEUP CONSIDERATIONS

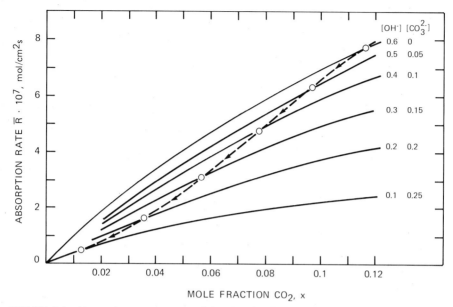

FIGURE 6-4 Absorption rates measured in stirred cell with plane interface, CO_2–NaOH system.

From Equations 6-5 and 6-6, one obtains

$$y = \frac{y_0 - \alpha(C_{BL0}C_{BL})}{1 - \alpha(C_{BL0} - C_{BL})} \qquad (6\text{-}7)$$

with

$$\alpha = \frac{V_M u_L}{u_{G0} z} = 0.2092 \text{ L/mol}$$

If $C_{BL} = 0.1$ mol/L, it follows

$$y = 0.0127$$

Hence, the CO_2 conversion is 89 percent. The column height can be estimated from Equation 6-3. For CO_2 absorption in NaOH solution the reaction takes place in the liquid film and C_{AL} (bulk concentration of CO_2) is essentially zero. Therefore Equation 6-3 reduces to

$$\frac{zaL}{u_L} = \int_{\text{out}}^{\text{in}} \frac{dC_{BL}}{\bar{R}} \qquad (6\text{-}8)$$

TABLE 6-3 C_{BL} as a function of y

C_{BL}, mol/L	y	$\bar{R} \times 10^7$, mol/cm² s	$10^{-7}/\bar{R}$, cm²/s/mol
0.6	0.116	7.75	0.129
0.5	0.097	6.30	0.159
0.4	0.077	4.75	0.210
0.3	0.057	3.15	0.317
0.2	0.035	1.65	0.607
0.1	0.013	0.50	2.000

Now, the corresponding values of C_{BL} and y can be calculated from Equation 6-7 and Figure 6-4. These data are summarized in Table 6-3.

Figure 6-5 presents a plot of $1/\bar{R}$ versus C_{BL}; the area under the curve is 2060 s/cm. Therefore, from Equation 6-8,

$$L = 2060 \frac{0.28}{1.3 \times 2} = 224 \text{ cm}$$

The column height required to absorb 89 percent of the CO_2 is 224 cm. The dotted line in Figure 6-4 shows how the absorption process proceeds within the column.

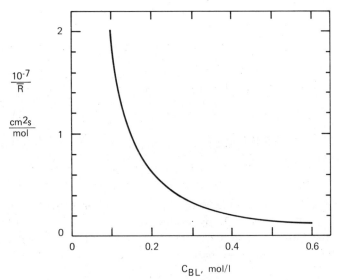

FIGURE 6-5 Plot of $1/\bar{R}$ versus c_{BL} for determination of the integral in Equation 6-8.

□

D. Bubble Columns

1. Variations of Nonadjustable Parameters in Scaleup

Reactor scaleup usually means increases in reactor diameter, length, and phase (gas and liquid or gas, liquid and solid) velocities. These changes will be accompanied by changes in flow regime, phase holdup characteristics, axial and radial mixing, and transport coefficients. Since these factors significantly affect the performance of a bubble column, the current state of the art is reviewed here. Where the definitive knowledge is available, the role of the solid on the bubble column dynamics can also be evaluated.

a. FLOW REGIMES

If a gas is distributed in a liquid by means of a specific sparger, the bubbles are of rather uniform size and equally distributed provided the gas velocities are low, say less than 5 cm/s. This regime is called bubbly or homogeneous flow; the bubble size distribution is narrow and the rise velocities of the bubbles in the swarm lie between 20 and 30 cm/s. Lockett and Kirkpatrick (1975) showed that the bubble flow regime can be realized up to gas holdups of 60 percent.

At higher gas velocities this pseudohomogeneous gas-in-liquid dispersion cannot be maintained. The flow becomes unstable, and coalescence sets in. The flow regime where large bubbles with high-rise velocities coexist in the presence of small bubbles is called the heterogeneous or churn–turbulent flow regime. The large bubbles are no longer spherical, but take the form of spherical caps of varying form with a very mobile and flexible interface. These large bubbles can grow up to diameters of about 10 cm.

A peculiar situation can occur in investigations of small diameter columns. At high gas flow rates, the larger bubbles are stabilized by the column wall and this leads to the formation of bubble slugs. In tall columns bubble slugs can be observed even if the column diameter is as large as 20 cm. The dependence of the flow regime on column diameter and gas velocity can be roughly estimated from Figure 6-6. However, other parameters like the behavior of the sparger, physico-chemical properties, liquid velocity, and the presence of solids can also affect the transitions from one flow regime to the other. In highly viscous solutions, for both Newtonian and non-Newtonian fluids, large bubbles can be formed even at gas velocities considerably less than 5 cm/s.

The nature of the gas sparger also significantly influences the flow regimes and the transition ranges between them. Porous spargers with mean pore sizes less than 150 μm lead commonly to bubbly flow up to gas velocities of about 5 to 8 cm/s. On the other hand, if perforated plates or single and multinozzle distributors with orifice diameter larger than 1 mm are used, homogeneous flow can only be realized at very low gas velocities. At larger orifice diameter, bubbly flow may not occur at all if pure liquids are aerated. With liquid mixtures the situation may change again. The rising bubbles can cause recircu-

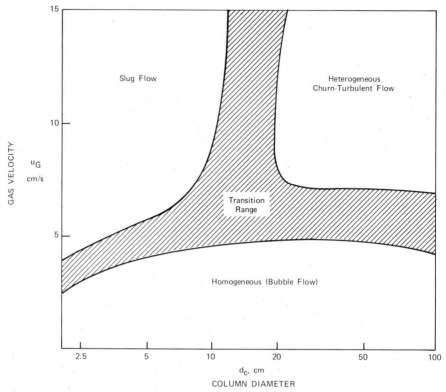

FIGURE 6-6 Flow regime as functions of column diameter and gas velocity (for low viscosity fluids).

lation of liquid within the column. Ueyama and Miyauchi (1979) have suggested that the heterogeneous flow regime be called the recirculating flow regime.

A knowledge of the transition from the bubble flow to the churn turbulent and slug flow regimes is important because in many cases of heterogeneous flow the achievable conversion decreases strongly with the increase in gas velocity. For example, Figure 6-7 shows the conversion for CO_2 absorption in alkali solution versus the gas velocity for various column heights (diameter of column 10.2 cm, sintered plate with pores of 150 μm mean diameter). At low gas velocities the conversion is complete. At gas velocities of about 5 cm/s the conversion decreases sharply which indicates transition to churn–turbulent flow where with increasing gas velocity an increasing amount of gas flows through the reactor in the form of large bubbles and slugs with high rise velocity, causing deterioration of reactor performance. For a given reaction system, results similar to the ones shown in Figure 6-7 are useful in making an approximate determination of the systems conditions for change in flow regime.

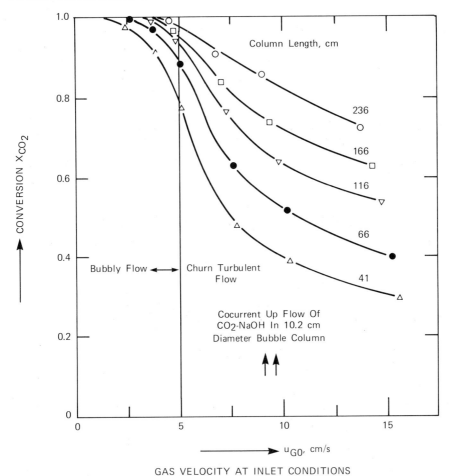

FIGURE 6-7 CO_2 converison drop at transition to churn–turbulent flow.

Beinhauer (1971) and Kölbel et al. (1972) measured the holdup of the large and small bubbles in a bubble column of 10 cm diameter using water as liquid phase and a sintered plate as gas sparger. Their results are shown in Figure 6-8. Although the holdup with small bubbles is rather large (i.e., about 20 percent of the reactor volume) for the gas velocities larger than 4 cm/s, the large bubbles rise up much more quickly as shown in Figure 6-9. (The constant rise velocity above $u_G = 8$ cm/s results from the fact that the velocity of the slugs is determined mainly by wall friction.) Nicklin (1962) showed that the transition from bubble flow to heterogeneous flow can usually be recognized easily by the sharp increase of bubble rise velocity, u_G^*. Although the holdup curves permit estimation of the formation of larger bubbles, the u_G^* versus u_G curve generally gives a more accurate estimate where the transition to churn–turbulent flow occurs.

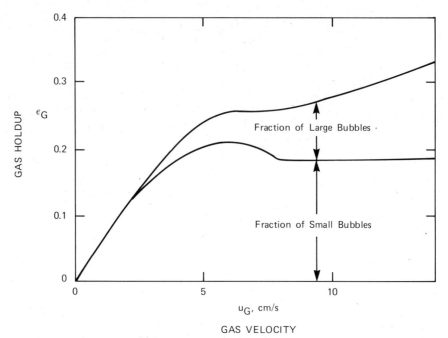

FIGURE 6-8 Fractional gas holdup of large and small bubbles.

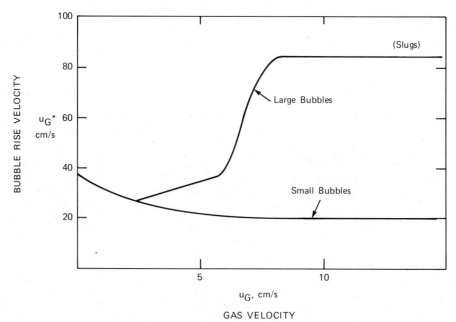

FIGURE 6-9 Rise velocity of large and small bubbles.

Flow transition and the bubble size is very important for the scaleup of bubble columns. Commercial vertical sparged reactors are usually operated at high gas velocities compared to laboratory scale reactors. If the flow regime and bubble dynamics in these two reactors are considerably different, their performance may vary significantly, particularly for the cases where the gas–liquid interface mass transfer plays an important role in the overall reaction process. Furthermore, because of the drastically different flow patterns that could prevail in the columns with different gas velocities, the same mathematical model (e.g., one-parameter dispersion model) may not be applicable to both large and small reactors.

b. Gas Holdup

Provided the ratio of the column to bubble diameter is large, say > 40, the column diameter does not significantly affect the holdup. Usually a column diameter of 10 cm is sufficient to yield holdup values which are close to the ones obtained in larger diameter columns under the same conditions. This is pointed out by Ueyama and Miyauchi (1979) whose analysis showed that the holdup is nearly proportional to $u_G^{0.5}$ and tends to decrease only slightly with increasing column diameter.

The dependence of the gas holdup on the gas velocity is generally of the form

$$\varepsilon_G \alpha u_G^n \qquad (6\text{-}9)$$

The values of n depend on the flow regime. For the bubble flow regime, values of n varying from 0.7 to 1.2 are reported in the literature. In the churn–turbulent flow the influence of u_G on ε_G is less pronounced. The log–log plots of gas holdup versus gas velocity measured by various authors for tap water are illustrated in Figure 6-10. Only data for single orifice and multinozzle spargers ($d \geq 1$ mm) are considered in this figure. Under such sparging conditions the bubble flow regime at low gas velocities is often not very pronounced and the transition to heterogeneous flow cannot be discerned from the ε_G versus u_G curve, that is, the flow is apparently heterogeneous over the entire range of u_G. Although the data shown in Figure 6-9 are obtained in columns with diameters varying from 7.5 to 550 cm, the different curves are surprisingly close together, indicating the relative unimportance of column diameter. The results for water given in Figure 6-9 lead to an exponent of 0.6 in Equation 6-9.

A large number of correlations for ε_G have been proposed in the literature. For many practical applications, empirical correlations often fail. In addition, the holdup may strongly depend on the liquid-phase composition (particularly if electrolytes and organics are present). For an accurate scaleup of a given fluid–fluid reactor it is therefore recommended that some measurements of holdup be carried out in small-scale equipment (but with a column diameter greater or equal to 10 cm). As diameter and length of the column generally

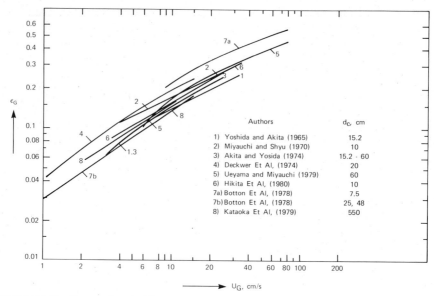

FIGURE 6-10 Reported gas holdup measurements in water for single and multinozzle spargers.

exert little effect on ε_G, such measurements yield a valuable check of the applicability of the above correlation for the problem at hand. This is particularly desirable for high-pressure fluid–fluid reactors as shown by Kölbel et al. (1961) and Neubauer and Pilhofer (1978). For the most practical range of liquid velocity (0–3 cm/sec), the gas holdup is not significantly affected by the liquid velocity.

The gas holdup in a three-phase slurry or fluidized bed column exhibits even more complex behavior. Some aspects are discussed by Shah (1979). More work is needed in this area.

c. Interfacial Area

A number of physical and chemical methods for measuring interfacial area have been examined in the literature. The chemical methods are generally preferred. Nagel and coworkers (1971, 1973, 1976, 1978, 1979) used the sulfite oxidation method to determine interfacial areas in a variety of gas–liquid contactors and correlated their measured data by using the energy dissipation rate per unit volume of reactor (E/V_R) (as calculated from the pressure drop) as the major correlating parameter. On the basis of Kolmogoroff's theory of isotropic turbulence Nagel and Kurten (1976) and Nagel et al. (1979) derived the following expression for the specific interfacial area

$$a = k \left(\frac{E}{V_R} \right)^{0.4} \varepsilon_G^n \qquad (6\text{-}10)$$

SCALEUP CONSIDERATIONS

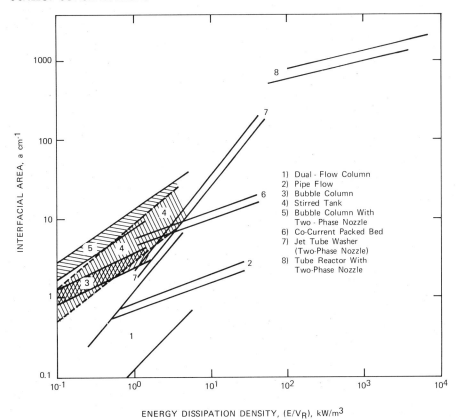

FIGURE 6-11 Correlation of interfacial area with energy dissipation density (sulfite – oxidation).

Equation 6-10 applies only if the condition of isotropic turbulence, and hence a spatially uniform energy dissipation, is fulfilled. Therefore, in practice, the exponent of the energy dissipation density will vary between 0.4 and 1. The correlation obtained by Nagel et al. (1978) is shown in Figure 6-11. In their articles Nagel et al. demonstrated the usefulness of Figure 6-11 for design and scaleup considerations of gas–liquid reactors. Oels et al. (1978) pointed out that the correlation shown in Figure 6-11 can result in errors for simulated fermentation media. In tall bubble columns, the value of a may decline along the height, due to bubble coalescence.

d. VOLUMETRIC MASS TRANSFER COEFFICIENTS, $k_L a$

Like the holdup and the interfacial area, the volumetric mass transfer coefficients depend on gas velocity, physico-chemical properties, and the gas sparger. Porous spargers, like sintered plates usually lead to higher $k_L a$ values. In addition, the $k_L a$ values are spatially varied, larger $k_L a$ values being observed often in the immediate vicinity of the sparger. Kastanek (1977) used Higbie's penetration theory along with Kolmogoroff's theory and derived

$$k_L a = b u_G^n \tag{6-11}$$

where the value of n depends on u_G. For water and electrolyte solution Kastanek proposed a value of $n = 0.82$ if the gas velocity is less than 25 cm/s. The liquid velocity is found to have no effect on $k_L a$. Also, the column diameter has little effect on $k_L a$ provided $d_c \geq 10$ cm. Kastanek also found n to be essentially unchanged by the sparger design. However, the constant b largely depended on the distributor and the liquid medium. The effect of the nature of the sparger on gas–liquid reactor performance and particularly on $k_L a$ has been discussed by many authors. This factor is particularly important for reactor scaleup because the nature of the gas sparger used in larger reactor is seldom the same as the one used in smaller reactor. As a rule of thumb, it can, however, be stated that in case of orifices spargers with diameters ≥ 1 mm no significant effect on $k_L a$ is found. Also the number of orifices does not give any influence on $k_L a$. In tall bubble columns a significant decline in $k_L a$ may occur from the bottom to the top of the column due to the bubble coalescence. For aqueous and nonaqueous systems, the works of Akita and Yoshida (1974), Schugerl et al. (1977), Lücke et al. (1976), Oels et al. (1978), Deckwer and coworkers (1974, 1978a, b, 1980a, b), and Nakanoh and Yoshida (1980) are particularly noteworthy.

e. Gas-Side Mass Transfer Coefficients, k_G

Gas-side resistances may be important during the absorption of highly soluble gases such as NH_3 in sulfuric acid, and HCl and SO_2 in alkalies. For the absorption of chlorine in benzene, van den Berg and Hoornstra (1977) observed significant gas-side resistances even for gas mixtures with about 50 percent by volume of chlorine. The reactor performance for these systems would strongly depend on the value of $k_G a$. Mehta and Sharma (1966) reported the following relationship for $k_G a$,

$$k_G a \alpha D_G^{0.5} u_G^{0.75} L^{0.33} \tag{6-12}$$

where D_G is the diffusivity, u_G the gas velocity, and L the dispersion height. $k_G a$ is believed to be essentially independent of column diameter and liquid velocity.

f. Liquid–Solid Mass Transfer

A thorough literature survey on liquid–solid mass transfer in three-phase slurry and fluidized bed reactors is given by Shah (1979). The literature indicates that the liquid–solid mass transfer coefficient $k_S a$ is a strong function of gas velocity, particle diameter, and fluid viscosity. It is a rather weak function of reactor diameter and length and of liquid velocity.

g. Mixing Parameters

Liquid Phase. Axial dispersion coefficients of the liquid phase in vertical gas–liquid cocurrent/countercurrent flow, that is, bubble columns, have been reviewed by Shah et al. (1978). Reported empirical correlations such as those of Deckwer, Burckhart, and Zoll (1974) indicate the dispersion coefficient to be dependent on the gas velocity and column diameter. A significant influence of the flow direction (i.e., cocurrent or countercurrent flow) has not been observed. Over the range of liquid velocities used in industrial operation, the liquid velocity appears to have no influence on liquid–phase dispersion.

It is usually assumed that the dispersion coefficient does not depend on the column height. There are, however, studies by Deckwer et al. (1973) and Badura et al. (1974) that indicate dispersion coefficients may increase along the column height. For reactor scaleup purposes, the most useful relation as shown by Baird and Rice (1975) for E_L is

$$E_L \alpha d_c^{4/3} u_G^{1/3} \tag{6-13}$$

The above described dependence of E_L on u_G and d_c is valid even when solids are present as long as solid particle size and its concentration are small.

Gas Phase. Gas-phase dispersion was measured by Kölbel et al. (1962), and Kölbel and Langeman (1964), Diboun and Schugerl (1977), Carleton et al. (1967), Towell and Ackerman (1972), Pilhofer et al. (1978), and Mangartz and Pilhofer (1980). Compared to liquid-phase dispersion, these measurements are sparse, and in general, the data reveal considerable scatter.

Using the data of Kölbel and coworkers (1962, 1964) and those of Carleton et al. (1967), Towell and Ackerman (1972) proposed the following empirical equation.

$$E_G = 0.2 d_c^2 u_G \tag{6-14}$$

whereas Mangartz and Pilhofer (1980) proposed the relation

$$E_G = 5 \times 10^{-4} u_G^{*3} d_c^{1.5} \tag{6-15}$$

where u_G^* is the bubble rise velocity in swarm, $u_G^* = u_G/\varepsilon_G$. E_G has been found to be independent of liquid velocity, column height and solids concentration as long as particle size is small.

Solid Phase. The dispersion coefficient for the solid phase is approximately the same as that for the liquid phase as long as solid particles are small. At low Froude numbers and large particle diameters Kato et al. (1972) presented the relation

$$\frac{u_G d_c}{E_s} = \text{Pe}_S = \left(1 + 0.009 \text{Re}_S \text{Fr}'^{-0.8}\right) \frac{13 \text{Fr}'}{1 + 8 \text{Fr}'^{0.85}} \tag{6-16}$$

where

$$\text{Re}_S = \frac{d_S u_{St}}{v_L} \quad \text{and} \quad \text{Fr}' = \frac{u_G}{\sqrt{g d_c}}$$

For the mean settling velocity of the particles in the swarm, they presented the following relation

$$u_S = 1.2 u_{St} \left(\frac{u_G}{u_{St}}\right)^{0.25} \left(\frac{1-\varepsilon_S}{1-\varepsilon_S^*}\right)^{2.5} \tag{6-17}$$

Here u_{St} is the settling velocity of the single particle which is calculated from the Stokes equation. ε_S^* represents the solids holdup of the gas free suspension for a particle concentration of 0.1 g solid/cm³.

If the solid suspension flows continuously through the reactor the solids concentrations within the reactor and feed or exit flow may be different. The ratio of these two concentrations is an important design and scaleup parameter, and Kato et al. expressed their results by

$$\frac{C_{SR}}{C_{SF}} = 1 + 0.5 \left(\frac{u_{St}}{u_G}\right)^{0.4} \tag{6-18}$$

h. Heat Transfer

The best up-to-date correlation for slurry-wall heat transfer coefficient is by Deckwer (1980); it can be expressed as

$$\text{St} = 0.1 (\text{ReFrPr}^2)^{-0.25} \tag{6-19}$$

where

$$\text{St} = \frac{h}{\rho c_p u_G}, \quad \text{Re} = \frac{u_G d_c}{v} \quad \text{Fr} = \frac{u_G^2}{g d_c} \quad \text{Pr} = \frac{\mu c_p}{k'} \tag{6-20}$$

Deckwer et al. (1980b) also showed that for slurries of small particles the suspension phase, that is, the liquid including the solid, can be regarded as a pseudohomogeneous phase. This view is supported also by mass transfer measurements into slurries of nonporous small diameter particles by Zaidi et al. (1979). However, if particles of larger diameter are suspended in a three-phase fluidized bed then the particle diameter becomes one of the major important parameters. Wall-to-bed heat transfer in three-phase fluidized beds was measured by Oestergaard (1964), Viswanathan et al. (1965), and Armstrong et al. (1976). The results of these authors are summarized and critically

TABLE 6-4 General Trends of Reactor Scaleup Variables on Bubble column Reactor Dynamic Variables

Effect on/Increase in	Reactor Length	Reactor Diameter	Gas Velocity	Liquid Velocity[a]	Solids Concentration
Flow regime	Generally no change except near transition boundary	May change	May change (around 5–10 cm/sec)	Most likely will not change under practical range of operation	May change at high solids concentration and for large particle size
Gas holdup	For very tall bubble columns holdup may decreases due to coalescence	Essentially no change	Increase	Essentially no change	May change-effect is complex and not well known yet
Gas–liquid interfacial area	May decrease in tall bubble column	Essentially no change if proper gas distributor is used	Increase	Essentially no effect unless liquid velocity is large	Effect not known—depends on particle size and concentration—decreases for larger particles and concentrations
Volumetric gas–liquid mass transfer coefficient ($k_L a$)	May decrease along column length	Essentially unchanged	Increase	No or small change	Effect is quite complex—depends on particle size and concentration usually decreases
Gas-phase dispersion	No observed effect	Increase	Increase	No observed effect	Not known—most probably only small effect under the practical range of operations
Liquid-phase dispersion	Observed increase along length	Increase	Increase	No observed effect	No effect under most practical conditions—may show effect for large particles
Solid-phase dispersion	No observed effect	Increase	Increase	No observed effect	Small effect at large concentrations
Slurry-wall heat	No observed effect	Increase	Increase	Some increase	Small increase may show optimum

[a] Generally liquid velocity range of 0–3 cm/sec is considered to be of most practical importance. Gas to liquid velocity ratio in most practical operations is of the order of 10 to 1.

reviewed by Shah (1979). Most recently, Joshi et al. (1980b) presented a unified model for the heat transfer in multiphase systems. This model should be very useful for scaleup purposes.

Based on the above discussions, the general trends of various nonadjustable parameters for two- (gas–liquid) and three-phase (gas–liquid–solid) bubble columns with variations in important scaleup variables are summarized in Table 6-4. The table shows the effect of increase in a number of independent scaleup variables such as reactor length, diameter, and so on, on the hydrodynamic, transport, and mixing characteristics of the bubble column.

III. APPLICABILITY OF MODELS TO SCALEUP

In a gas–liquid reactor, due to the difference in phase velocities, generally a large degree of backmixing prevails in the liquid phase. The effect of backmixing on the reactor performance is generally evaluated by the standard axial dispersion model. For first order slow liquid–liquid and gas–liquid reactions, such a model was evaluated by Pavlica and Olson (1970). The unified approach given by these authors, however, did not consider nonlinear kinetics and the axial variations in phase flow which are particularly important for the fast reactions. Several models for gas–liquid and gas–liquid–suspended solid reactors which consider other complex cases are presented by Schaftlein and Russel (1968), Cichy et al. (1969), Mhaskar (1974), Szeri et al. (1976), Juvekar and Sharma (1977), Schumpe et al. (1979), Parulekar and Shah (1980), and many by Deckwer and coworkers.

In general, the following points are essential when deriving a model for the gas–liquid or gas–liquid–solid bubble column reactor to be used for scaleup.

> Both liquid and gas phases should be assumed axially dispersed. Experimental evidence indicates that, generally, radial dispersion effects can be neglected.
>
> The axial distribution of solid (catalyst) is nonuniform unless the catalyst particles are very fine (approximately 100 μm or less). This distribution can be evaluated by the sedimentation–dispersion models of Cova (1966), Farkas and Leblond (1969), and Kato et al. (1972).
>
> Heat and mass axial dispersion coefficients in the liquid (or slurry) phase are equal and are related to each other by the expression

$$E_L = \left(\frac{k_{\text{eff}}}{C_p \rho} \right)_L \quad (6\text{-}21)$$

In low-pressure systems, the heat carried by the gas phase is usually small and can be neglected as long as reactor inlet gas and liquid temperatures are close to each other.

APPLICABILITY OF MODELS TO SCALEUP

In gas–liquid–solid systems, the mass transfer resistances may not only occur at the gas–liquid interface but also at the liquid–solid interface. In addition, Chaudhari and Ramachandran (1980) have shown that intraparticle diffusion limitations can be important for large catalyst particles.

If the reaction occurs at the catalyst surface, the absorption enhancement at the gas–liquid interface will not be observed except for very small particles, that is, with diameters in the micrometer range (Alper et al., 1980).

The reactor model must consider the variation in gas velocity along the reactor length when: (1) the absorption rate is high, (2) the hydrostatic head is significantly decreased, and (3) the reaction produces large amounts of volatile products with molecular weights different from that of the reactant.

Under these circumstances, Deckwer (1976) has shown that generally the overall gas balance and the balance over the inert gases are also required for the development of the model.

In tall bubble columns, particularly when the gas velocity varies along the reactor length, Deckwer et al. (1978b, 1980a) show that the axial variation in gas holdup and the gas–liquid mass transfer coefficient should also be considered. In fluid–fluid reactors, pressure decreases linearly along the reactor. The pressure profile is given by

$$P(z) = P_T[1 + \alpha(1 - z)] \qquad (6\text{-}22)$$

where P_T represents the pressure at column top and α is the ratio of the maximum hydrostatic head to P_T

$$\alpha = \frac{\rho_L g \varepsilon_L L}{P_T} \qquad (6\text{-}23)$$

For high-pressure operations $P(z)$ can be taken as constant.

Reactor models which consider the above points, at least in part, have been outlined by Deckwer (1976, 1977a, b, 1979), Deckwer and others (1978b, 1980a, 1981a), and Parulekar and Shah (1980). The models combined the hydrodynamic mixing and transport characteristics with the intrinsic kinetic rate expressions for the particular system at hand. Generally, the axial dispersion model gives a system of coupled differential equations which are subject to Danckwerts boundary conditions. In most cases, the equations are solved numerically on the computer. A completely generalized model encompassing all the features of chemical systems is generally very complex and not very meaningful. Some aspects of the generalized gas–liquid reactor models are given by Shah (1979) and Danckwerts (1970); they will not be repeated here. Instead, only some limiting cases of the general model are discussed below. Although these cases may contain some oversimplifications, they allow analytical solutions which can be used to provide some insight on how certain scaleup variables may modify reactor performance. The predicted results have also been found to be in general agreement with practical experience.

A. Nonisothermal Operation

Small-scale fluid–fluid reactors are generally operated isothermally. Large-scale reactors, particularly those involving three phases (e.g., hydrogenation, coal liquefaction, hydroprocessing) are often carried out adiabatically. The thermal behavior of large-scale reactors is not always well understood, particularly when complex reactions are involved. A standard technique used to model the steady-state behavior of hydroprocessing reactors is described by Parulekar and Shah (1982). Raghuram and Shah (1977), Raghuram et al. (1979), Singh et al. (1982), Ding et al. (1974), and Parulekar et al. (1980) have also examined the question of multiple temperature steady states in fluid–fluid reactions. During scaleup, the reactor can go from unique to multiple steady-state conditions or vice versa. The number and nature of steady states, in general, depend strongly on the nature of the reactor and the values of the reactor parameters and therefore each reactor system needs to be evaluated separately. The effect of reactor scaleup on the transient behaviors of the adiabatic three-phase reactors, particularly for high-pressure operations, is not currently well understood.

B. Variation of Reactor Performance

For the sake of simplicity the discussion is restricted to gas–liquid bubble column reactors operated under isothermal conditions. Most of the underlying principles are, however, applicable to slurry reactors. Scaleup of a gas–liquid reactor does not only imply increase in diameter and reactor length but it is also often accompanied by a change in devices used for phase distribution. The design of a gas sparger should be such that a uniform distribution of gas is guaranteed and this means that a different design may have to be implemented in large- and small-scale reactors.

The major effects of scaleup result from the variations in the hydrodynamic, mixing, and transport properties within the reactor.

Some of these are:

Changes in flow regime.

Increase in gas- and liquid-phase dispersion both of which strongly depend on reactor diameter.

Axial variation of gas velocity caused by high conversion which also changes holdup, gas–liquid mass transfer, and heat transfer rates.

Change in absorption-reaction regime along the reactor.

Assuming pseudo-first order kinetics, the influences of gas- and liquid-phase dispersions and of axial variations of gas flow rate on reactor performance have been studied by Deckwer (1976).

The change in gas flow rate depends on the absorption rate, the ratio α (see Equation 6-23), and the inlet mole fraction. Numerical simulations clearly indicated that use of simplified models which neglect gas flow rate changes is justified only at high operating pressures and small inlet mole fractions of absorbed gas. Otherwise, the decrease in gas volume that can occur at high solubilities and particularly in the fast reaction regime enlarges the gas residence time resulting in an appreciable increase in the overall amount of gas absorbed. This effect was denoted by Deckwer (1976) as "absorption enhancement by absorption," and should be considered during reactor scaleup. The importance of gas flow variations along with changes in gas holdup and volumetric mass transfer coefficients was demonstrated by Deckwer (1977b), and Deckwer et al. (1980a) for the absorption of CO_2 and isobutene in tall bubble column reactors. Gas flow variations are also important in large-scale reactors operated at atmospheric pressure, like fermentors and tower bioreactors for wastewater treatment. In these cases, the gas expands due to the reduced hydrostatic head. Isobaric reactor models which do not consider gas expansion have been shown to predict high conversions.

The importance of other effects depends on the nature of the chemical process and the desired conversion. In most cases, axial dispersion causes detrimental effects but its extent strongly depends on the nature of the chemical reaction. Also, a certain level of liquid dispersion is often desired. Examples are biological reactions like fermentations and wastewater treatment. In these cases mixing is needed to maintain the reaction and to prevent washout of biomass and therefore the reactors are often equipped with an internal or external recycle of biomass. However, reasonable guidelines for such processes taking place either in fast or slow reaction regimes can be obtained.

1. Fast Reaction Regime

For many reaction systems the transition from bubble to churn–turbulent flow is accompanied with marked changes in hydrodynamic properties. In churn–turbulent flow the interfacial area decreases. Thus, if the overall conversion depends mainly on interfacial area, that is, as in the fast reaction regime, a considerable drop in conversion can be observed with increasing gas velocity. For the absorption of CO_2 in alkali, Deckwer and Schumpe (1979) and Schumpe et al. (1979) determined the typical effects of change in flow regime (which can be caused by scaleup) on the conversion as illustrated in Figure 6-12. If for a given gas throughout (corresponding to superficial gas velocity = 7.5 cm/s and churn–turbulent flow) the reactor volume is enlarged by increasing the dispersion height, the increase of conversion is only moderate.

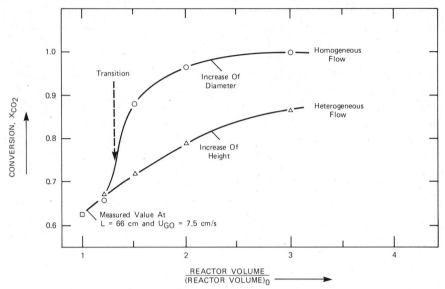

FIGURE 6-12 CO_2 absorption in alkali impact on conversion of enlarging reactor volume at constant throughput.

However, if the diameter is increased then transition to bubble flow takes place (as gas velocity decreases) which results in a large increase in conversion even for smaller reaction volumes.

Where the flow regime transition occurs also depends upon the nature of gas sparger. If the design of the gas sparger is changed during the reactor scaleup, its effect should be considered.

Along with axial variation in gas flow rate, in the "fast" reaction regime the axial variation in the gas–liquid mass transfer coefficient and the gas–liquid interfacial area also play important roles on the reactor performance. In tall bubble columns, both of these parameters can vary appreciably due to bubble coalescence and other hydrodynamic changes along the length of the reactor. These factors must be considered for the proper reactor scaleup.

Generally, in small-scale reactors, the gas phase is assumed to be moving in plug flow. This assumption may become questionable in large diameter bubble columns. The presence of gas-phase dispersion may affect the reactor performance for the "fast" reactions where a significant depletion in the gas-phase concentration occurs. Some examples are chlorination of bauxite in a molten salt bath, absorption of CO_2 in buffer solution, absorption of ammonia in water, and air oxidation of ethylene (to acetaldehyde). Generally the degree of gas-phase dispersion depends upon column diameter and gas velocity as shown by Mangartz and Pilhofer (1980). The extent of gas-phase dispersion also depends on the flow regime. Equation 6-9 indicates that in the bubble flow regime, the gas holdup is approximately proportional to u_G. In this flow

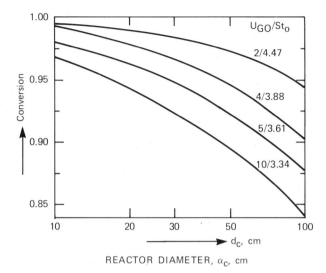

FIGURE 6-13 Dependence of conversion on reactor diameter and gas-phase dispersion.

regime, the dependence of E_G on u_G is small. However, if heterogeneous flow prevails, the parameter n is generally less than 1 and the effect of u_G on E_G may be rather pronounced, namely $E_G \propto u_G^{3(1-n)}$.

Using the correlation of Mangartz and Pilhofer (1980) for E_G, the effect of gas-phase dispersion on reactor performance was studied by Deckwer and Serpemen (1981b). Figure 6-13 shows loss in conversion resulting from an increase in reactor diameter under otherwise identical conditions. At high linear gas velocities, the decrease of conversion with increasing d_c is appreciable. A desired conversion level can be maintained by increasing the reactor length. An effective means for suppressing gas-phase dispersion is to build in additional plates giving staged bubble column.

In the "fast" reaction regime, liquid-phase dispersion is only important if along with a high gas-phase conversion a high conversion of a liquid-phase component (b) is desired as shown by Schumpe et al. (1979). For such cases, one must also consider the variation in absorption/reaction regime along the length of the column. The effect of gas-phase dispersion on required column length is discussed further in Examples 6.2 and 6.3.

2. Slow Reaction Regime

In fluid–fluid reactors it is usually desirable to generate high interfacial areas; this implies high values of gas holdup. However, in the slow reaction regime, it is also desirable to have a large liquid volume for the reaction to take place. These opposing effects cause a maximum in the space time yield (STY) versus gas velocity function. The space time yield is the amount of gas converted per unit time per unit reactor volume and presents a more appropriate quantity to characterize reactor performance than the conversion.

For the scaleup of a bubble column reactor operating in the slow reaction regime, an important task is to find out that gas velocity at which the STY has its maximum value. This can be done by the numerical computations for various design alternatives. By assuming a simplified (lumped) reactor model, Schumpe et al. (1979) derived a criterion for the optimum gas velocity for a pseudo-first order reaction. This can be expressed as

$$u_{G,\mathrm{opt}} = \frac{1}{\varepsilon_G^* \left(1 + \sqrt{\dfrac{k_L a^*}{k_1 \varepsilon_G^*}}\right)} \qquad (6\text{-}24)$$

where ε_G^* is the gas holdup and a^* the interfacial area at $u_G = 1$ cm/s. This criterion is often applicable to other reaction kinetics and slurry reactors. (See Example 6.3.)

The effect of the volumetric mass transfer coefficient on reactor performance in the slow reaction regime is less pronounced than in the fast reaction regime. Also, the detrimental effects of the gas-phase dispersion on the reactor performance are less significant in this flow regime. Schumpe et al. (1979) studied a second order reaction between gaseous and liquid species in the slow reaction regime and presented some useful conclusions for the reactor scaleup.

For fast and slow reaction regimes, Table 6-5 summarizes how various hydrodynamic mixing and transport parameters are affected by the reactor scaleup. The conclusions given in this table should be considered as guidelines that are applicable to many cases of practical importance.

TABLE 6-5 Scaleup Effects in Bubble Columns

	Fast Reaction Regime	Slow Reaction Regime
Flow regime changes	Drastic effects	Minor changes
Gas–liquid mass transfer, $k_L a$ coefficients	Most important, absorption rate and space time yield proportional	Important, but both $k_L a$ and reaction rate determine absorption rate, value of u_G which maximizes space time yield has to be found
Gas-phase dispersion	Very important	Important, if absorption rates are high
Liquid-phase dispersion	No influence on gas-phase conversion, but affects liquid-phase conversion	Important for conversion of liquid-phase component, affects gas-phase conversion only slightly

C. Model Simplification and Examples

A general mathematical model for fluid–fluid reactors comprises of a set of differential mass balance equations and a heat balance equation. As shown in Figure 6-14 the complete reactor model can often be simplified considerably. Large-scale reactors are frequently operated adiabatically. Alternatively in many large-scale reactors axial mixing is large. Hence, temperature gradients are small and then the temperature can be taken as constant over the entire reactor volume. Under these circumstances, the heat balance reduces to an algebraic equation and the reactor itself is isothermal.

In an isothermal reactor, the most complex case is that of the slow reaction regime since here all component mass balance equations have to be considered. This is generally the case for many gas–liquid reactions and for all gas–liquid–solid (catalyst) reactions. However, in many catalytic hydrogenations, oxygenations (including air oxidations and fermentations), and addition and substitution reactions like chlorinations and hydrochlorination, pure gases are frequently processed under gas recycle. Hence, the gas-phase concentration is constant and only the liquid (slurry) phase balance needs consideration.

On the other hand, if reaction in the bulk liquid phase can be neglected, that is, the process takes place in the fast reaction regime, the liquid-phase balances can be omitted if the concentration of the liquid-phase reactant B is fixed or is considered in an overall (lumped) balance equation. Then only solution of the gas-phase balances (entire and component balances because gas velocity may not be constant) is necessary. However, at small inlet mole fractions of gaseous reactant and high operational pressures, variation of gas velocity is negligible. Such conditions are encountered in many gas purification processes, and then

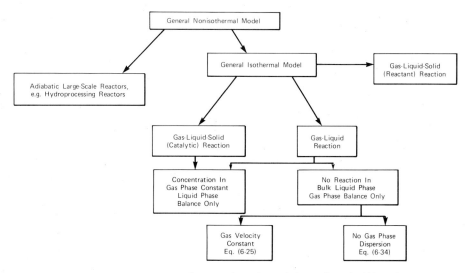

FIGURE 6-14 Model simplifications for two- and three-phase bubble columns.

the basic design equation reduces merely to

$$\frac{1}{\text{Pe}_G} \frac{d^2 C_{G,A}}{dz^2} - \frac{dC_{G,A}}{dz} - \text{St}_G C_{G,A} = 0 \qquad (6\text{-}25)$$

Here $C_{G,A}$ is the concentration of gaseous reactant (in gas phase), z is the dimensionless reactor length, and the dimensionless numbers are given by

$$\text{Pe}_G = \frac{u_G L}{E_G \varepsilon_G} \qquad (6\text{-}26)$$

and

$$\text{St}_G = k_L a \frac{L}{u_{G0}} \frac{RT}{H} \sqrt{1 + M} \qquad (6\text{-}27)$$

With consideration of the classical Danckwerts' boundary conditions, the solution of Equation 6-25 at the reactor outlet is

$$C_{G,A}(1) = \frac{4q \exp(\text{Pe}_G/2)}{(1+q)^2 \exp(q\,\text{Pe}_G/2) - (1-q)^2 \exp(-q\,\text{Pe}_G/2)} \qquad (6\text{-}28)$$

where

$$q = \sqrt{1 + 4\text{St}_G/\text{Pe}_G} \qquad (6\text{-}29)$$

If $\text{Pe}_G \geq 2$, the second term in the demoninator can be neglected with an error smaller than 10^{-3} in $C_{G,A}$. This gives a relation which explicitly predicts reactor length for a desired conversion:

$$L = \frac{2 E_G \varepsilon_G}{u_G(1-q)} \ln\left[\frac{1-X_A}{4q}(1+q)^2\right] \qquad (6\text{-}30)$$

q does not contain any elements of reactor length. Equation 6-30 is valuable for getting an estimate of reactor height if the above conditions are approximately fulfilled.

In bubble columns of smaller diameter, say $d_c \leq 40$ cm, which are operated at bubbly flow conditions, the effect of gas-phase dispersion can usually be neglected, particularly if conversion is less than 90 percent. Then the gas-phase balance on component A as shown by Deckwer (1976, 1977a, b, 1979) reduces to

$$\frac{d\bar{x}_A}{dz} + \frac{\text{St}_G \beta(z)}{(1+\alpha)(1-x_{A0})}(1 - x_{A0}\bar{x}_A)^2 \left(\bar{x}_A - \frac{1+\alpha}{\beta(z)}\bar{A}\right) = 0 \qquad (6\text{-}31)$$

where the variation of gas velocity is given by an inerts balance

$$\bar{u}_G = \frac{(1 + \alpha)(1 - x_{A0})}{[1 + \alpha(1 - z)](1 - x_{A0}\bar{x}_A)} \tag{6-32}$$

and $\beta(z)$ is given by

$$\beta(z) = 1 + \alpha(1 - z) \tag{6-33}$$

For the important case of negligible reaction in bulk liquid phase (diffusional and fast reaction regime) A is zero and the liquid-phase balance on A can be neglected.

Equation 6-31 simplifies to

$$\frac{d\bar{x}_A}{dz} + \frac{\mathrm{St}_{G,\mathrm{PF}}\beta(z)}{(1 + \alpha)(1 - x_{A0})}\bar{x}_A(1 - x_{A0}\bar{x}_A)^2 = 0 \tag{6-34}$$

The closed solution of this equation can be written as shown by Deckwer (1977a) as

$$x_{A0}\bar{X}_A - (1 - x_{A0})\ln(1 - \bar{x}_A) = \frac{1 + 0.5\alpha}{1 + \alpha}\mathrm{St}_{G,\mathrm{PF}} \tag{6-35}$$

The index PF on the Stanton number is to remind one that it applies to plug flow in gas phase.

Equation 6-35 permits calculation of the Stanton number required for a given conversion, and therefore the reactor length. It is also particularly suited for evaluating measurements in small-scale bubble columns, since gas-phase dispersion plays no role in those cases. For other orders of reaction Juvekar and Sharma (1977) have derived analytical solutions.

Dispersion in gas phase must be taken into account in industrial bubble columns. For the case of no reaction in the bulk liquid phase, the component balance and the entire balance of the gas phase must be solved numerically. From such conversions calculated for given parameter combinations (Pe_G, $\mathrm{St}_G, \alpha, x_{A0}$), the Stanton numbers for the case of negligible dispersion ($\mathrm{St}_{G,\mathrm{PF}}$) can then be calculated from Equation 6-35. The ratio of the Stanton numbers, $\mathrm{St}_G/\mathrm{St}_{G,\mathrm{PF}}$, gives direct information about the influence of gas-phase dispersion on the reactor volume and the reactor length, respectively. From systematic calculations of this kind, design charts as introduced by Levenspiel and Bischoff (1959, 1961) and Deckwer and Schumpe (1979) can be constructed.

An example for $x_{A0} = 0.3$ is shown in Figure 6-15. The curves are not very sensitive to inlet mole fraction x_{A0}, therefore Figure 6-15 can also be used for other x_{A0}. A relatively reliable design of bubble column reactors can be made from such a chart. First $\mathrm{St}_{G,\mathrm{PF}}$ is calculated from Equation 6-35 for a given conversion and negligible gas-phase dispersion. Therefrom, reactor length L

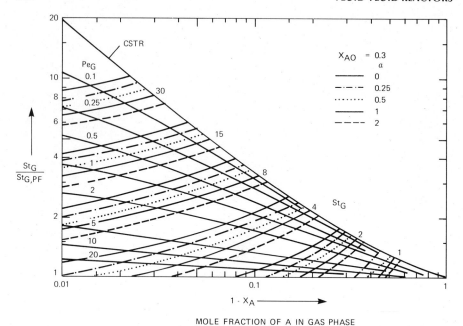

FIGURE 6-15 Influence of gas-phase mixing (Pe_G) on ratio of reactor volumes ($St_G/St_{G,PF}$).

for a plug flow reactor is obtained. Now Pe_G can be estimated, for instance, from correlations presented in this chapter. Then, from Figure 6-15 the ratio of $St_G/St_{G,PF}$ and hence L/L_{PF} can be obtained by interpolation. This is illustrated in Example 6.2.

Example 6.2 Design of an Isobutene Absorber. Isobutene is separated from C_4 cracked fractions by absorption and reaction in sulfuric acid as discussed by Kröper et al. (1969).

$$C_4H_8 + H_2O \rightarrow C_4H_9OH$$

The rate constant of the acid catalyzed hydration of isobutene to tertiary butanol is about 1000 times larger than that of other olefins present in the C_4 fraction. The reaction product tertiary butanol plays an important role as it increases the isobutene solubility in the liquid phase which appreciably improves reactor performance. Under industrial conditions 2 to 4 mol/L tertiary butanol are present in the liquid phase, and the sulfuric acid concentration varies from 35 to 45 wt%. This absorption reaction process was studied in various bubble columns by Popovic and Deckwer (1975) and Deckwer (1977b, c).

For a given liquid-phase composition the absorption–reaction parameters and the operation conditions are given in Table 6-6. The length of a 1.2-m-

APPLICABILITY OF MODELS TO SCALEUP

TABLE 6-6 Data for Isobutene Absorption in Sulfuric Acid – Butanol Mixtures, Example 6.2

Gas velocity, u_{G0}, cm/s	5
Gas holdup, ε_G	0.23
Interfacial area, a, cm^{-1}	12.0
Pressure at reactor top, P_T, kPa	200
Inlet mole fraction, x_{A0}	0.35
Reactor diameter, m	1.2
Density of liquid, g/cm^3	1.26
Henry's constant, kPa cm^3/mol	2×10^7
Liquid-side mass transfer coefficient, k_L, cm/s	0.01
Rate constant, k_1, s^{-1}	1.34
Temperature, °C	30.0
Diffusivity, cm^2/s	3×10^{-6}

diameter bubble column required for 90 percent conversion is desired. The length is obtained by calculating the reactor length assuming plug flow in the gas phase and Equation 6-35. Then the Pe$_G$ number is estimated, and from Figure 6-15 the ratio of St$_G$/St$_{G, \text{PF}}$ is obtained. This ratio corresponds to the ratio of reactor lengths as long as the other conditions are constant.

First, one must check whether Equation 6-35 can be applied. On the basis of a lumped model Danckwerts (1970) derived a criterion to be fulfilled for a reaction to take place in the diffusional regime. This criterion is

$$\frac{k_L^2(k_1 + 1/\tau_L)}{Dk_1(k_1 + k_L a + 1/\tau_L)} \gg 1 \gg \frac{k_L a}{k_L a + k_1 + 1/\tau_L} \quad (6\text{-}36)$$

As $1/\tau_L \ll k_1, k_L a$ we obtain for the present system

$$18.6 \gg 1 \gg 0.104$$

Since the reaction takes place in the slow reaction regime, Equation 6-35 is applicable.

The value of the enhancement factor

$$E = \left(1 + \frac{k_1 D}{k_L^2}\right)^{1/2} = 1.023 \quad (6\text{-}37)$$

is very close to unity. This also supports the above conclusion.

For $X = 0.9$ the LHS of Equation 6-35 is

$$\text{LHS} = 0.35 \times 0.9 - 0.65 \times \ln(0.1) = 1.81$$

Therefore, we have

$$1.81 = \frac{1 + 0.5\alpha}{1 + \alpha} \text{St}_{G,\text{PF}} \qquad (6\text{-}38)$$

Introducing

$$\alpha' = \frac{\alpha}{L} = \frac{\rho_L g \epsilon_L}{P_T} = 4.75 \times 10^{-4} \text{ cm}^{-1}$$

and

$$\text{St}'_{G,\text{PF}} = \frac{\text{St}_{G,\text{PF}}}{L} = k_L a \frac{RT}{Hu_{G0}} = 3.02 \times 10^{-3} \text{ cm}^{-1}$$

Equation 6-38 yields a quadratic expression in L

$$L^2 + \frac{2L}{\text{St}'_{G,\text{PF}}\alpha'}(\text{St}'_{G,\text{PF}} - 1.81\alpha') - \frac{2 \times 1.81}{\text{St}'_{G,\text{PF}}\alpha'} = 0 \qquad (6\text{-}39)$$

which gives the length of the plug flow reactors as

$$L = 683 \text{ cm} = L_{\text{PF}}$$

From the correlation of Mangartz and Pilhofer (1980), given as Equation 6-15, we obtain

$$E_G = 6.75 \times 10^3 \text{ cm}^2/\text{s}$$

and

$$\text{Pe}_G = \frac{u_{G0} L}{\epsilon_G E_G} = 2.2$$

For this Peclet number and $x_{A0} = 0.3$, Figure 6-15 indicates $\alpha < 0.3$; at

TABLE 6-7 Reactor Lengths Required for Plug Flow (PF) and Dispersed Plug Flow (DPF) at Various Conversion Levels

X	LHS, Equation 6-35	cm	$\text{Pe}_G{}^a$	α^a	$\dfrac{\text{St}_G}{\text{St}_{G,\text{PF}}}{}^b$	L_{DPF}, cm
0.75	1.1636	420	1.35	0.20	1.4	590
0.90	1.810	683	2.20	0.325	1.6	1093
0.95	2.28	886	2.85	0.422	1.8	1595
0.98	2.886	1162	3.74	0.553	1.7	1975

^{a}Calculated with L_{PF}.
^{b}Estimated from Figure 6-15.

APPLICABILITY OF MODELS TO SCALEUP

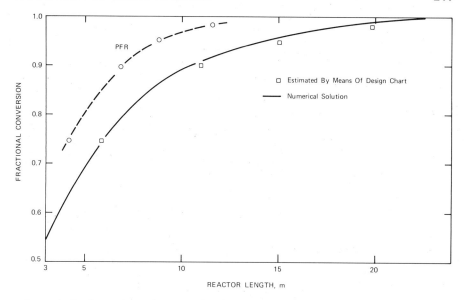

FIGURE 6-16 Comparison of numerical and approximate solutions. Example 6.2 — design of isobutene absorber.

$X = 0.9$ then the value of $St_G/St_{G,PF}$ is about 1.6. Therefore, the length of the dispersed reactor should be

$$L_{DPF} = 1.6 \times 683 = 1093 \text{ cm}$$

In Table 6-7 required reactor lengths for other conversions are presented.

The method predicts a conservative estimate, of course, as both the Peclet number and α are taken to be too small. In Figure 6-16 the predictions of the approximate method are compared with those by numerical solutions of the correct model equations. It is obvious that the differences are small. For instance, for the desired conversion of 0.9 the predictions of both methods differ by less than 50 cm. □

Example 6.3 Calculation of Absorber Height. CO_2 has to be scrubbed from synthesis gas into carbonate–bicarbonate buffer solution which contains arsenite as catalyst. The absorption is to be carried out in a 1-m-diameter unstaged bubble column at a temperature of 333 K and a pressure of 5.06 MPa. The superficial gas velocity is 5.5 cm/s; at this velocity the gas holdup is estimated to be 0.17 and the volumetric mass transfer coefficient $k_L a$ is 0.04 s^{-1}.

Under these operating conditions the enhancement factor $E = \sqrt{1 + M}$ is 3.2, and the CO_2 solubility at 1-atm pressure is 0.005 mol/L. The reactor length for an inlet mole fraction of CO_2 of 0.038 and a conversion of 90 percent is desired.

Here the pressure is high and the inlet mole fraction of gaseous reactant is small; then, gas flow variations can be neglected. The absorption process is taking place in the fast regime of absorption-reaction theory. Therefore, a liquid-phase balance equation need not be considered. For the chosen reactor diameter, one can suspect that axial dispersion considerably affects the reactor performance. Design of the absorber can be based on the one-phase axial dispersion model.

If we assume at first that $Pe_G > 2$, then the reactor length can be calculated from Equation 6-30 with sufficient accuracy. From the solubility of CO_2 in the reaction mixture the Henry constant can be calculated as

$$H = \frac{P}{C_L^*} = 200 \text{ atm L/mol}$$

The axial gas-phase dispersion coefficient E_G is computed from the correlation of Mangartz and Pilhofer (1980), Equation 6-15,

$$E_G \approx 1.7 \times 10^4 \text{ cm}^2/\text{s}$$

With this value and the other quantities given previously

$$\frac{St_G}{Pe_G} = \frac{0.04}{5.5^2} \frac{0.08206 \times 333}{200} \times 0.17 \times 1.7 \times 10^4 \times 3.2 = 1.671$$

$$q = 2.772$$

It follows from Equation 6-30 that

$$L = \frac{2 \times 1.7 \times 10^4 \times 0.17}{5.5 \times 1.772} \ln\left[\frac{0.1}{4 \times 2.772} 3.772^2\right] = 1218 \text{ cm}$$

A height of 12 m will give the required conversion. The largest uncertainty in the above calculation is the estimation of E_G as Mangartz and Pilhofer (1980) measured E_G only in smaller diameter columns. Towell and Ackerman (1972) provides some data for larger diameter columns. An E_G of 10,820 cm²/s for a gas velocity of 5.5 cm/s can be estimated. The column height based on this value of E_G, that is $L = 11.14$ m, is in rather good agreement with that value obtained by using Mangartz and Pilhofer's correlation. The Peclet number is

$$Pe_G = \frac{1218 \times 5.5}{17000 \times 0.17} = 2.32$$

and therefore application of Equation 6-30 is reasonable. □

1. Role of Solids

In a three-phase (gas–liquid–solid) reactor the solid can be either a catalyst or a reactant (or a product). The most important case is the one where the solid is catalyst. If the catalyst particle diameters are small (say less than 100 μm) and if its volume concentration is below about 10 percent, it is reasonable to assume the liquid–solid slurry to be pseudohomogeneous. This has been experimentally verified recently by Deckwer (1980b) for the Fischer–Tropsch synthesis in slurry phase.

Even where the diffusional resistances within the catalyst particles and at the liquid–solid interface are important, the liquid–solid slurry can be considered pseudohomogeneous by introducing appropriately defined effectiveness factors. Therefore, the guidelines given in Table 6-5 and the simplifications outlined in Figure 6-14 for the slow reaction regime can be applied directly to the slurry reactors.

On the basis of some reliable experimental data obtained in small-scale reactors, Deckwer et al. (1981a) have presented a sophisticated model for Fischer–Tropsch synthesis in a large slurry reactor. The model predictions are in excellent agreement with the facts observed in practice. It is therefore believed that the procedure applied by Deckwer may be useful for many other catalytic three-phase reactors. Other gas–liquid–solid processes were investigated by Parulekar and Shah (1980) and Joshi et al. (1981). These studies derive some specific conclusions on the scaleup of a number of reactors. The role of solids in gas–liquid–solid reactors has also been evaluated by Shah (1979).

An important point in the scaleup of a fluid–fluid reactor with a solid phase concerns the distribution of the solid phase. Due to higher mixing the solid distribution in large-scale reactors is much more uniform than in small-scale reactors. The small-scale reactors where catalyst settling is pronounced can give lower conversions than those obtained in large-scale reactors.

Example 6.4 Optimum Gas Velocity for Large-Scale Fischer–Tropsch Slurry Reactor. Kolbel and Ackerman (1956) and Kolbel and Ralek (1977, 1980) have reported on measurements with the Fischer–Tropsch slurry process in a 4.7-cm-diameter bubble column of 3.5-m height. The study was carried out at 539 K and 1.01 MPa, and with an inlet gas velocity of 3.5 cm/s. The catalyst concentration was 11 wt% Fe. From the reported conversion in this reactor Deckwer (1981c) evaluated a first order rate constant for synthetic gas conversion of 0.245 s^{-1}.

With a simplified model the inlet gas velocity giving a maximum value of the space time yield in a large-scale slurry reactor can be estimated for conditions where a catalyst concentration of 23 wt% Fe is used and the synthesis gas conversion (X_{CO+H_2}) is about 90 percent.

The rate constant can be assumed proportional to the catalyst concentration. Gas holdup and the volumetric mass transfer coefficient were measured in small-scale reactors by Deckwer (1980b) as follows:

$$\varepsilon_G = 0.053 u_G^{1.1} = a_1 u_G^{1.1} \tag{6-40}$$

$$k_L a = 0.045 u_G^{1.1} = a_2 u_G^{1.1} \tag{6-41}$$

where u_G is in cm/s.

Deckwer (1981a) has shown that these equations also apply to larger scale reactors operating at higher gas velocities. The volume contraction factor $\bar{\alpha}$ is defined by

$$\bar{\alpha} = \frac{Q(X_{CO+H_2} = 1) - Q_0}{Q_0} \tag{6-42}$$

is about -0.5 for the Fischer–Tropsch synthesis according to Deckwer (1981a, c).

Let us assume completely mixed phases for the large-scale reactor. Then the volumetric absorption rate, which is equivalent with the space time yield, is given by

$$R_A = \text{STY} = k_L a (C_{AL}^* - C_{AL}) = k_1 C_{AL} (1 - \varepsilon_G) \tag{6-43}$$

By eliminating C_{AL} one obtains

$$\text{STY} = \frac{k_L a k_1 (1 - \varepsilon_G)}{k_L a + k_1 (1 - \varepsilon_G)} C_{AL}^* \tag{6-44}$$

Introducing Equations 6-40 and 6-41, differentiating with respect to u_G and setting the result equal to zero gives

$$\bar{u}_{G,\text{opt}}^{2.2} \left(\frac{a_1 a_2}{k_1} - a_1^2 \right) + 2 a_1 \bar{u}_{G,\text{opt}}^{1.1} - 1 = 0 \tag{6-45}$$

Solution of this equation leads to

$$\bar{u}_{G,\text{opt}}^{1.1} = \frac{1}{a_1 \left(1 + \sqrt{\dfrac{A_2}{k_1 a_1}}\right)} \tag{6-46}$$

The derivation of this equation is the same as that given by Schumpe et al. (1979). They assumed $\varepsilon_G \sim u_G$ and $k_L a \sim u_G$ and obtained Equations 6-24. With $k_1 = 0.245 \times 23/11 = 0.512 \text{ s}^{-1}$, it follows from Equations 6-46

$$\bar{u}_{C,\text{opt}}^{1.1} = \left[0.053 \left(1 + \sqrt{\frac{0.045}{0.512 \times 0.053}} \right) \right]^{-1} = 8.25 \text{ cm/s}$$

$$\bar{u}_{G,\text{opt}} = 6.81 \text{ cm/s}$$

UNCERTAINTIES

In the simplified model, this gas velocity is assumed to be a mean value over the entire reactor volume. Considering the contraction factor the outlet gas velocity for a given conversion follows from

$$u_G = u_{G0}(1 - \bar{\alpha} X_{CO+H_2}) \quad (6\text{-}47)$$

Hence, the mean gas velocity is given by

$$\bar{u}_G = u_{G0}(1 + 0.5\bar{\alpha} X_{CO+H_2}) \quad (6\text{-}48)$$

Therefore, the optimum inlet gas velocity for maximizing the space time yield and the desired conversion is obtained from Equation 6-48.

$$u_{G0,opt} = \frac{6.81}{1 - 0.5 \times 0.5 \times 0.9} = 8.79 \text{ cm/s}$$

Kölbel and Ackerman (1956) and Kölbel and Ralek (1977, 1980) have reported that the Rheinpreussen Koppers demonstration plant (1.29-m effective diameter and 7.7-m height) was operated under optimal conditions with an inlet gas velocity of about 9.5 cm/s. Deckwer et al. (1981a) calculations based on a more complicated reactor model predicts optimum gas velocities between 9 and 11 cm/s. The value of $u_{G0,opt}$ estimated with the lumped model is therefore in good agreement with the practical experience.

This example demonstrates that the assumption of a pseudohomogeneous liquid-catalyst phase and the use of extremely simplified model may be a valuable tool for predicting optimum operation condition in large-scale slurry reactors.

IV. UNCERTAINTIES

Fluid–fluid reactors are carried out in a large number of chemical reactors. Only some major aspects of scaleup of such reactors have been covered in this chapter. Reasonable rules of thumb which apply to many reaction systems have been given. However, one should always note that for a particular system at hand, drastic differences may be possible. There remains a number of uncertainties in the scaleup of fluid–fluid reactors.

A. Dispersion Coefficients

These coefficients are strongly dependent on the reactor diameter. However, their influence on reactor performance is very pronounced at high conversions and for the standard reactor configuration rather reliable correlations are at hand. It is, however, not at all clear whether the correlations for the dispersion coefficients can be applied to other geometric arrangements or for the reactors equipped with baffles or other internals.

B. Gas Holdup and Interfacial Area

These quantities depend on sparger design and the coalescene promoting or hindering properties of the specific chemical system. In general, coalescene depends on the rise time of the bubbles and may be very important in taller reactors. In reaction media where bubbles tend to coalesce, the flow regime may change from bubbly flow close to the sparger to churn–turbulent flow at some distance away from the sparger. Holdup and interfacial area under these circumstances may be rather strong functions of reactor length. Of course, local dependencies of this kind are not considered in empirical correlations which largely describe overall integral values.

Occurrence of strong coalescene is often caused by trace impurities. In small-scale operations such impurities can be avoided but in continuous large-scale reactors they may be the result of undesired parallel and/or consecutive reactions which may often be unavoidable due to changes in residence time and temperature.

The accumulation of impurities or side products can also cause the undesirable effect, such as foaming. Foam formation is undesirable because it reduces the liquid volume required for the reaction to take place. Foaming is a common phenomenon in biological reactors and special foam destroyers have to be used to facilitate bubble disengagement in large-scale reactors.

C. Flow Regime

When designing a large-scale fluid–fluid reactor on the basis of the experimental data obtained in small-scale operation, the same flow regime in both units should be maintained. Otherwise, the scaleup will be uncertain. Above all, slug flow should be avoided in small reactors as this flow regime usually does not occur in large units.

D. Kinetic Data

Intrinsic kinetic data for fluid–fluid reactions are difficult to determine and are therefore often subject to serious errors. Such uncertainties in kinetic data and the possibility of side reactions may remain undetected in small-scale reactors but may become relevant in large-scale reactors and may lead to reduced selectivity. In addition, undesired side reactions may yield impurities which lead to drastic variations in gas holdup.

E. Heat Transfer and Stability

Limited information on the thermal behavior of large-scale fluid–fluid reactors with high heat generation is available. One can expect that under certain conditions instabilities may be encountered for highly exothermic oxidations, chlorinations, and hydrogenations including coal liquefaction processes. Investigations in this area for larger scale reactors are urgently needed.

NOMENCLATURE

a	Interfacial area
A	Species A
$\bar{A}$	Dimensionless concentration of A in a liquid phase
B	Species B
C^*	Solubility
C_A	Concentration of species A
C_{AL}, C_{BL}	Bulk liquid concentrations of species A and B, respectively
C_p	Specific heat
C_S	Solids concentration
CSTR	Continuous stirred tank reactor
d_c	Column diameter
D_G, D_L	Gas and liquid molecular diffusivities
E	Axial dispersion coefficient
E	Activation energy
(E/V_R)	Energy per unit volume of reactor
Fr'	Froude number as defined by Equation 6-16
g	Gravitational acceleration
h	Heat transfer coefficient
H	Henry's law constant
k'	Thermal conductivity
k'_{eff}	Effective heat conductivity, heat dispersion coefficient
k_1, etc.	Rate constants
$k_G a, k_G$	Volumetric and intrinsic gas film gas–liquid mass transfer
$k_L a, k_L$	Volumetric and intrinsic gas–liquid mass transfer coefficient
L	Length of column
M	Hatta number, $k_1 D_L / k_1^2$
n	Exponent in Equation 6-9
p_A	Partial pressure of species A
Pe	Peclet number
P	Total pressure
P_T	Total pressure at reactor top
Q	Volumetric gas flow rate
r	Reaction rate

R	Universal gas constant
$\bar{R}$	Mean absorption rate
Re_S	Reynolds number as defined by Equation 6-16
Sc	Schmidt number
St	Stanton number
T	Temperature
u_G, u_L	GAs and liquid velocities, respectively
$\bar{u}_G$	Mean gas velocity or dimensionless gas velocity u_G, u_{G0}
u_G^*	Bubble rise velocity
$u_{G,opt}$	Optimum gas velocity (with regard to space time yield)
u_s	Settling velocity of the particles in swarm
U_{st}	Settling velocity of a single particle
V_{SL}	Slurry volume
x	Axial distance
X	Conversion
$\bar{x}_A$	Dimensionless mole fraction of species A in gas phase, x_A/x_{A0}
y	Mole fraction
z	Dimensionless axial distance
z	Stoichiometric coefficient

Greek

α	Parameter defined by Equation 6-23
$\beta(z)$	A factor defined by Equation 6-33
$\varepsilon_L, \varepsilon_G$	Liquid and gas holdups respectively
μ	Viscosity
δ	Film thickness
ν	Kinematic viscosity
$\bar{\rho}$	Density
τ_L	Liquid residence time

Subscripts

G, l, s	Refers to gas, liquid, and solid phases
0	Refers to inlet or initial condition
PF	Plug flow condition
R	Refers to the reactor
F	Refers to the reactor feed

REFERENCES

Akita, K., and Yoshida, F., *I & EC Process Des. Dev.*, **12**, 76 (1973).

Akita, K., and Yoshida, F., *Ind. Eng. Chem. Proc. Des. Dev.*, **13**, 84 (1974).

Albright, L. F., *Chem. Eng. S.*, 179 (December 4, 1967).

Alper, E., Ph.D. Thesis, University of Cambridge (1971).

Alper, E., and Danckwerts, P. V., *Chem. Eng. Sci.*, **31**, 599 (1976).

Alper, E., *Proceedings of NATO Summer School on "Two-Phase Flow and Heat Transfer,"* Instanbul, 1976.

Alper, E., *Chem.-Ing.-Tech.*, **51**, 1136 (1979).

Alper, E., Wichtendahl, B., and Deckwer, W. D., *Chem. Eng. Sci.*, **35**, 217 (1980).

Alper, E., *Inst. Chem. Eng. Symp. Ser.*, **73**, D129–D143 (1982).

Armstrong, E. R., Baker, C. G. J., and Bergougnou, M. A., in *Fluidization Technology*, D. L. Keairns, ed., Hemisphere Publ. Corp., Washington, D.C., 1976, Vol. 1, p. 453.

Astarita, G., Savage, D. W., and Bisio, A., *Gas Treating with Chemical Solvents*, J. Wiley, New York, 1983.

Badura, R., Deckwer, W.-D., Warnecke, H. J., and Langemann, H., *Chem.-Ing.-Tech*, **46**, 399 (1974).

Baird, M. H. I., and Rice, R. G., *Chem. Eng. J.*, **9**, 17 (1975).

Beinhauer, R., Dr.-Ing.Thesis, TU Berlin (1971).

Botton, R., Cosserat, D., and Charpentier, J. C., *Chem. Eng. J.*, **16**, 107 (1978).

Brauer, H., and Sucker, D., *Ger. Chem. Eng.*, **2**, 77 (1979).

Buchanan, J. E., *Ind. Eng. Chem. Fund*, **6**, 400 (1976).

Carleton, A. J., Flain, R. J., Rennie, J., and Valentin, F. H. H., *Chem. Eng. Sci.*, **22**, 1839 (1967).

Charpentier, J. C., *ACS Symp.-Series*, **72**, 223 (1978).

Chaudhari, R. V., and Ramachandran, P. A., *AIChE J.*, **26** (2), 177 (1980).

Cichy, P. T., Ultman, J. S., and Russel, T. W. F., *Ind. Eng. Chem.*, **61**(8), 6 (1969).

Cova, D. R., *Ind. Eng. Chem. Proc. Des.*, **5**, 21 (1966).

Danckwerts, P. V., and Sharma, M. M., *Chem. Engr.* (London), CE 244 (Oct. 1966).

Danckwerts, P. V., and Gillham, A. J., *Trans. Instn. Chem. Engrs.*, **44**, T42 (1966).

Danckwerts, P. V., *Gas-Liquid Reactions*, McGraw-Hill, New York, 1970.

Danckwerts, P. V., and Alper, E., *Trans. Instn. Chem. Engrs.*, **53**, 34 (1975).

Deckwer, W.-D., Graeser, U., Langemann, H., and Serpemen, Y., *Chem. Eng. Sci.* **28**, 1223 (1973).

Deckwer, W.-D., Burckhart, R., and Zoll, G., *Chem. Eng. Sci.*, **29**, 2177 (1974).

Deckwer, W.-D., *Chem. Eng. Sci.*, **31**, 309 (1976).

Deckwer, W.-D., *Chem.-Ing.-Tech.*, **49**, 213 (1977a).

Deckwer, W.-D., *Chen. Eng. Sci.*, **32**, 51 (1977b).

Deckwer, W.-D., Allenbach, U., and Bretschneider, H., *Chem. Eng. Sci.*, **32**, 43 (1977c).

Deckwer, W.-D., Adler, I., and Zaidi, A., *ACS Symp. Ser.*, **65**, 359 (1978a).

Deckwer, W.-D., Adler, I., and Zaidi, A., *Can. J. Chem. Eng.*, **43** (1978b).

Deckwer, W.-D., *Int. Chem. Eng.*, **19**, 21 (1979).

Deckwer, W.-D., and Schumpe, A., in *Two-Phase Momentum, Heat and Mass Transfer in Chemical, Process, and Engineering Systems*, F. Durst, G. V. Tsiklauri, and N. H. Afgan, eds. Hemisphere Publ. Corp., Washington, D.C., 1979, Vol. 2, p. 1038.

Deckwer, W.-D., Hallensleben, J., and Popovic, M., *Can. J. Chem. Eng.*, **58**, 190 (1980a).

Deckwer, W.-D., Louisi, Y., Zaidi, A., and Ralek, M., *Ind. Eng. Chem. Proc. Des. and Dev.*, **19**, 699 (1980b).

Deckwer, W.-D., Serpemen, Y., Ralek, M., and Schmidt, B., *Chem. Eng. Sci.*, **36**, 765 (1981).

Deckwer, W.-D., *Chem. Eng. Sci.*, **35**, 1341 (1980).

Deckwer, W.-D., Serpemen, Y., Ralek, M., and Schmidt, B., *Ind. Eng. Chem. Proc. Des. Dev.*, **21**, 222–231 (1982).

Deckwer, W.-D., Serpemen, Y., Ralek, M., and Schmidt, B., *Ind. Eng. Chem. Proc. Des. Dev.*, **21**, 231–241 (1982).

Diboun, M., and Schügerl, K., *Chem. Eng. Sci.*, **22**, 147 (1967).

Ding, J. S. Y., Sharma, S., and Luss, D., *Ind. Eng. Chem. Fundam.*, **13**, 76 (1974).

Eckenfelder, W. W., and Barnhart, E. L., *AIChE J.*, **7** 631 (1961).

Engdahl, R. B., and Rosendahl, H. S., *Chemtech*, *S.*, 118 (February 1978).

Farkas, E. J., and Leblond, P. F., *Can. J. Chem. Eng.*, **47**, 215 (1969).

Gates, B. C., Katzer, J. R., and Schuit, G. C., *Chemistry of Ctalytic Processes*, McGraw-Hill, New York, 1979.

Hanson, C., Hughes, M. A., and Marsland, J. G., "Mass Transfer with Chemical Reaction," *Proc. Int. Solv. Extr. Conf.*, 2401, Society of Chemical Industry, London (1974).

Hikita, H., Asai, S., Tanigawa, K., Segawa, K., and Kiteo, M., *Chem. Eng. J.*, **20**, 59 (1980).

Hofmann, H., *Chem.-Ing.-Tech.*, **47**, 823 (1975).

Houghton, G., McLean, A. M., and Ritchie, P. D., *Chem. Eng. Sci.*, **7**, 26 (1957).

Humphrey, A. E., *Chem. Engng.*, **81**, 98 (1974).

Jhaveri, A. S., and Sharma, M. M., *Chem. Eng. Sci.*, **23**, 669 (1968).

Joshi, J. B., Shah, T. Y., Reuther, J. A., and Ritz, J. H., "Particle Size Effects in Oxidation of Pyrite in Air/Water Chemical Coal Cleaning," a paper presented at National AIChE Meeting, Chicago (1980a).

Joshi, J. B., Shah, Y. T., and Sharma, M. M., "Heat Transfer in Bubble Column," a paper presented at ACS Meeting, Las Vegas (August 1980b).

Joshi, J. B., Abichandari, J., Shah, Y. T., Reuther, J., and Ritz, H., "Modelling of Three Phase Reaction: A Case of Oxydesulfurization of Coal," *AIChE J.*, **27**, 937 (1981).

Juvekar, V. A., and Sharma, M. M., *Trans. Instn. Chem. Engrs.*, **55**, 77 (1977).

Kastanek, F., *Coll. Czechoslov. Chem. Commun.*, **42** 2491 (1977).

Kataoka, H., Takuchi, H., Nakao, K., Tadaki, T., Otake, T., Miyauchi, T., Washimi, K., Watanabe, K., Yoshida, F., *J. Chem. Eng.*, **12**, 105 (1979).

Kolar, V., Broz, Z., and Tichy, J., *Coll. Czech. Chem. Comm.*, **35**, 3344 (1970).

Kölbel, H., and Ackerman, P., *Chem.-Ing.-Tech.*, **28**, 381 (1956).

Kölbel, H., Borchers, E., an Langemann, H., *Chem.-Ing.-Tech.*, **33**, 668 (1961).

Kölbel, H., Langemann, H., and Platz, J., *Dechema-Monogr.*, **41**, 225 (1962).

Kölbel, H., and Langemann, H., *Dechema-Monogr.*, **49** 253 (1964).

Kölbel, H., Beinhauer R., and Langemann, H., *Chem.-Ing.-Tech.*, **44**, 697 (1972).

Kölbel, H., and Ralek, M., in *Chemierohstoffe aus Kohle*, J. Falbe, ed., Thieme Verlag, Stuttgart, 1977.

Kölbel, H., and Ralek, M., *Cat. Rev.-Sci. Eng.*, **21**, 225 (1980).

Kolev, N., *Chem.-Ing.-Tech.*, **48**, 1105 (1976).

Kronig, W., in *Chemierohstoffe aus Kohle*, J. Falbe, ed., Thieme Verlag, Stuttgart, 1977.

Kröper, H., Schlömer, K., and Weitz, H. M., *Hydrocarbon Proc.*, **48**, 195 (1969).

Laddha, G. S., and Degaleesan, T. E., *Transport Phenomena in Liquid Extraction*, McGraw-Hill, New York, 1978.

REFERENCES

Laurent, A., and Charpentier, J. C., *The Chem. Eng. J.*, **8**, 85 (1974).
Laurent, A., Thesis, University of Nancy (1975).
Leistner, G., Muller, G., Sell, G., and Bauer, A., *Chem.-Ing.-Tech.*, **51**, 288 (1979).
Levenspiel, O., and Bischoff, K. B., *Ind. Eng. Chem.*, **51** 1431 (1959).
Levenspiel, O., and Buschoff, K. B., *Ind. Eng. Chem.*, **53**, 313 (1961).
Linek, V., Stoy, V., Machon, V., and Krivsky, Z., *Chem. Eng. Sci.*, **29**, 1955 (1974).
Linek, V., Krivsky, Z., and Hudec, P., *Chem. Eng. Sci.*, **32**, 323 (1977).
Lockett, M. J., and Kirkpatrick, R. D., *Trans. Instn. Chem. Engrs.*, **53**, 267 (1975).
Lücke, J., Schügerl, K., and Todt, J., *Chem.-Ing.-Tech.*, **48**, 73 (1976).
Mangartz, K.-H., and Pilhofer, Th., *Verfahrenstechn.*, **14**, 40 (1980).
Mashelkar, R. A., and Sharma, M. M., *Trans. Inst. Chem. Engrs.*, **48**, 162 (1970).
Mashelkar, R. A., *Brit. Chem. Engng.*, **15**, 1297 (1970).
Mhaskar, R. D., *Chem. Eng. Sci.*, **29**, 897 (1974).
Mehta, V. D., and Sharma, M. M., *Chem. Eng. Sci.*, **21**, 361 (1966).
Metzner, A. B., and Brown, L. F., *I & EC*, **48**, 2041 (1956).
Miyauchi, T., and Shyu, C. N., *Kagaku Kogaku*, **34**, 958 (1970).
Mohunta, D. M., and Laddha, G. S., *Chem. Eng. Sci.*, **20**, 1069 (1965).
Moo-Young, M., *Can. J. Chem. Eng.*, **53**, 113 (1975).
Nagel, O., Kürten, H., and Sinn, R., *Chem.-Ing.-Tech.*, **44**, 367 and 899 (1972).
Nagel, O., and Kürten, H., and Hegner, B., *Chem.-Ing.-Tech.*, **45**, 913 (1973).
Nagel, O., and Kürten, H., *Chem.-Ing.-Tech.*, **48**, 513 (1976).
Nagel, O., Hegner, B., and Kürten, H., *Chem.-Ing.-Tech.*, **50**, 934 (1978).
Nagel, O., Kurten, H., and Hegner, B., in *Two-Phase Momentum, Heat, and Mass Transfer Systems in Chemical Process, and Engineering*, F. Durst, G. V. Tsiklauri, and N. H. Afgan, eds., Hemisphere Publ. Corp., Washington, D.C., 1979, Vol. 2, p. 834.
Nakonoh, M., and Yoshida, F., *Ind. Eng. Chem. Proc. Des. & Devel.*, **19** 190 (1980).
Neubauer, G., and Pilhofer, T., *Chem.-Ing.-Tech.*, **50**, (1978).
Nicklin, D. J., *Chem. Eng. Sci.*, **17**, 693 (1962).
Oels, U., Lucke, J., Bucholz, R., and Schugerl, K., *Ger. Chem. Eng.*, **1**, 115 (1978).
Oestergaard, K., in *Fluidization*, Society for the Chemical Industry, London, 1964, p. 50.
Ostergaard, K., Chapter 18 in *Fluidization*, J. F. Davidson and D. Harison, eds., Academic Press, New York, 1971.
Onda, K., *Memoirs of Faculty of Engng., Nagoys Univ.*, **24** (2), 165 (1972).
Parulekar, S., and Shah, Y. T., *Chem. Eng. J.*, **20** (1), 21 (1980).
Parulekar, S., Raghuram, S., and Shah, Y. T., *Chem. Eng. Sci.*, **35**, 745 (1980).
Parulekar, S. J., and Shah, Y. T., "Steady State and Adiabatic Three Phase Slurry Reactor—Coal Liquefaction Under Slow Hydrogen Consumption Regime," *The Chem. Eng. J.*, **23**, 15 (1982).
Pavlica, R. T., and Olson, J. H., *Ind. Eng. Chem.*, **62** (12), 45 (1970).
Pilhofer, Th., Bach. H. F., and Mangartz, K. H., *ACS-Symp. Series*, **65**, 372 (1978).
Ponter, A. B., Taymour, N., and Dankyi, S. O., *Chem.-Ing.-Tech.*, **48**, 636 (1976).
Popovic, M., and Deckwer, W.-D., *Chem. Eng. Sci.*, **30**, 913 (1975).
Puranik, S. S., and Vogelpohl, A., *Chem. Eng. Sci.*, **29**, 501 (1974).
Raghuram, S., Shah, Y. T., *Chem. Eng. J.*, **13**, 81 (1977).
Raghuram, S., Shah, Y. T., and Tierner, J. W., *Chem. Eng. J.*, **17**, 63 (1979).
Reichelt, W., and Blass F., *Chem.-Ing.-Tech.*, **46**, 171 (1974).

Reichert, H. H., *Chem.-Ing.-Tech.*, **49**, 626 (1977).
Riemenschneider, W., *Chem.-Ing.-Tech.*, **50**, 55 (1978).
Rosenberg, H. S., *Hydrocarbon Processing*, S. 132 (May 1978).
Rosenzweig, M., and Ushio, S., *Chem. Eng.*, **81**, 62 (1974).
Sahay, B. N., and Sharma, M. M., *Chem. Eng. Sci.*, **28**, 41 (1973).
Satterfield, C. N., *AIChE J.*, **21**, 209 (1975).
Schaftlein, T. W., and Russel, T. W. F., *Ind. Eng. Chem.*, **60**(5), 12 (1968).
Schügerl, K., Lücke, J., and Oels, U., *Adv. Biochem. Engng.*, T. K. Ghose, A. Fiechter and Blakeborogh, N. eds., **7**, 1 (1977).
Schumpe, A., Serpemen, Y., and Deckwer, W.-D., *Ger. Chem. Eng.*, **2**, 234, 267 (1979).
Shah, Y. T., Stiegel, G. J., and Sharma, M. M., *AIChE J.*, **24**, 369 (1978).
Shah, Y. T., *Gas-Liquid Solid Reactor Design*, McGraw-Hill, New York, 1979.
Sharma, M. M., and Danckwerts, P. V., *British Chem. Eng.*, **15**, 206 (1970).
Shende, B. W., and Sharma, M. M., *Chem. Eng. Sci.*, **29**, 1763 (1974).
Shulman, H. L., Ullrich, C. F., and Wells, N., *AIChE J.*, **1**, 247 (1955a).
Shulman, H. L., Ullrich, C. F., Wells, N., and Prolux, A. Z., *AIChE J.*, **1**, 259 (1955b).
Singh, C. P. P., Shah, Y. T., and Carr, N. L., *Chem. Eng. J.*, **23**, 101 (1982).
Smith, E. L., and Greenshields, R. N., *Chem. Engng.*, **81**, 28 (1974).
Sridharan, K, and Sharma, M. M., *Chem. Eng. Sci.*, **31**, 767 (1976).
Szeri, A., Shah, Y. T., and Madgavkar, A., *Chem. Eng. Sci.*, **31**, 225 (1976).
Tichy, J., *Chem. Eng. Sci.*, **28**, 655 (1973).
Towell, G. D., Strand, C. P., and Ackerman, G. H., in *Mixing-Theory Related to Practice*, P. A. Rottenburg, ed., 1965, p. 97.
Towell, G. D., and Ackerman, G. H., *Proc. 2nd Int. Symp. Chem. React. Engng.*, B3-1, Amsterdam (1972).
Ueyama, K., and Miyauchi, T., *AIChE J.*, **25**, 258 (1979).
van den Berg, H., and Hoornstra, R., *Chem. Eng. J.*, **13**, 191 (1977).
van Driesen, R. P., and Steward, N. C., *Oil and Gas J.*, S 100 (May 18, 1964).
Vidwans, A. D., and Sharma, M. M., *Chem. Eng. Sci.*, **22**, 673 (1967).
Viswanathan, S. A., Kakar, A. S., and Murti, P. S., *Chem. Eng. Sci.*, **20**, 903 (1965).
Volpolicelli, G., and Massimilla, L., *Chem. Eng. Sci.*, **25**, 1361 (1970).
Voyer, R. D., and Miller, A. I., *Can. J. Chem. Engng.*, **46**, 335 (1968).
Weekman, V. W., *Proc. ISCRE 4*, Vol. 2: Survey Papers, Dechema (1976).
Wen, C. Y., and Tone, S., *ACS Symp. Series*, **72**, 108 (1978).
Yadav, G. D., and Sharma, M. M., *Chem. Eng. Sci.*, **4**, 1423 (1978).
Yoshida, F., and Akita, K., *AIChE J.*, **11**, 9 (1965).
Zaidi, A., Louisi, Y., Ralek, M., and Deckwer, W.-D., *Ger. Chem. Eng.*, **2**, 94 (1979).

7
SELECTION OF REACTOR TYPES

R. L. KABEL

I.	Major Issues in Reactor Selection	254
II.	Reactor Types	255
	A. Vessels and Flows	255
	1. Batch Reactors	255
	2. Semibatch Reactors	256
	3. Continuous Stirred Tank Reactors (CSTRs)	257
	4. CSTRs in Series	257
	5. Tubular Reactors	257
	6. Recycle Reactors	257
	B. Classification by Phases Present in Reactor	258
	C. An Example of Reactor Selection	261
III.	Goals in Selection	263
	A. Flexibility	263
	B. Cost Minimization	263
	C. Product Selectivity	264
	D. Thermal Control	265
	E. Comparison of Reactor Types	266
	1. Solid-Catalyzed Vapor–Liquid Reactors	266
	2. Absorption in Reactive Solutions	268
IV.	Overview of Selection	270

V. Some Realities 270
    A. Mode of Operation 270
    B. Unpredictable Behavior and Safety 272
    C. Scaledown 272
    D. Influence of Experience 272
References 273

I. MAJOR ISSUES IN REACTOR SELECTION

Economics is the ultimate driving force in the selection of the type of reactor for a given application. When reactors are very large, as in the Synthol hydrocarbon synthesis process, or are so complex that they are available from only a small number of fabricators (e.g., large multitubular heat exchanger type reactors), then the cost of the reactor itself can be controlling. However, the major issue is much less often the cost of the reactor, which is frequently minor, than the guarantee that the reactor will make possible the efficient and dependable operation of the overall process. For example, the cost of being not able to make a product in a commercial unit can easily run from a million dollars a month to a million dollars a day. The economic impact of a reactor not operating, while huge, is remote. The main focus in scaleup is on the operating characteristics and the specific mission of the reactor.

One should not expect to identify an obvious best reactor for a job. Indeed, many different kinds, each with individual advantages and drawbacks, can suffice. An open mind and good ideas are more important here than any cookbook. You cannot design a reactor until you have selected its type, and you cannot know if your type selection was wise until you have designed it. Obviously, then, the selection-design optimization is an iterative procedure. It is evident from this statement that the data required to select a reactor type are the same as are used to design it.

Nishida et al. (1981) reviewed process synthesis, "the step in design where the chemical engineer selects the component parts and how to interconnect them to create his flowsheet." They considered in detail the following five topics: "chemical reaction paths, separation systems, heat exchanger networks, complete flow sheets, and control systems." They noted that most optimization studies have concerned simple reactions in ideal plug flow or continuous stirred tank reactors and commented on how little reactor network synthesis research has been done. This is not so surprising when the nearly infinite variety of chemical reactions and process contexts are contemplated. They observed that "the only effective approaches to synthesis have taken advantage of every possible special characteristic of the problem being considered."

Neither the author's experience nor a literature search has provided many generalizations on selection of reactor type for which one could not quickly think up a counterexample. For example, perhaps the most prominent consideration in reactor type selection is heat removal in exothermic reactions. And yet for autothermal reactions the heat buildup can be an asset. Accordingly, this chapter will discuss the major considerations and illustrate them with examples of productive approaches from the literature.

Hill (1977), in the first several pages of his Chapters 8 and 12, provides a good introduction to the attributes of the various types of homogeneous and heterogeneous reactors, respectively. Among the conventional textbooks, Walas (1959) provides the most insight into practical reactor issues. While being fundamentally sound, the book incorporates real problems, real data, and real solutions throughout. Further, it devotes 28 pages to the characteristics of industrial reactors. Another source of mechanical details and practical wisdom is Rase (1977). His two-volume set on *Chemical Reactor Design for Process Plants* is addressed specifically to practicing engineers and embodies a philosophy very similar to that of Chapters 3–7 of this book. For guidance in selection and scaleup of chemical reactors, the reader should find Rase's book a valuable complement to this one.

II. REACTOR TYPES

Reactors may be categorized in a variety of ways, each appropriate to a particular perspective. For example, Henglein (1969) chooses a breakdown based on the source of energy used to initiate the reaction (i.e., thermal, electrochemical, photochemical, nuclear). More common breakdowns are according to:

The types of vessels and flows that exist.

The number and types of phases present in the reacting mixture.

These are discussed in more detail below.

A. Vessels and Flows

Figure 4-1 on page 118 illustrates the most common types of homogeneous reactors: batch, semibatch, semiflow, continuous stirred tank, and tubular reactors. A qualitative indication of the applicability of these various reactor types is given in Figure 7-1. Of course, the dividing lines on the figure should be taken only as very wide bands.

1. Batch Reactors

The batch reactor is the almost universal choice in the chemist's laboratory where most chemical processes originate. The reason is the simplicity and versatility of the batch reactor, whether it be a test tube, a three-neck flask, an autoclave, or a cell in a spectroscopic instrument. Regardless of the rate of the reaction, these are clearly low production rate devices. As scaleup is desired,

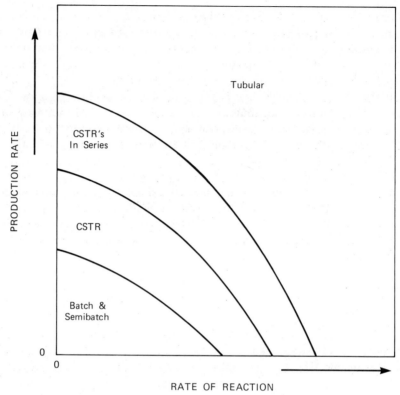

FIGURE 7-1   Relative utility of various reactor types.

the most straightforward approach is to move to a larger batch reactor such as a large vat or tank.

Commercial batch reactors can be huge, 100,000 gal or more. The cycle time, often a day or more, typically becomes longer as reactor volume increases in order to achieve a substantial production rate with an inherently slow reaction. Fabrication, shipping, or other factors place a limit on the scale of all kinds of reactors. For example, transportation capacity can limit the size of a batch reactor for which shop, as opposed to on-site, fabrication of the heat exchange surface is required. This limits the production rates for which batch reactors may be economically utilized. Also, batch reactors must be filled, emptied, and cleaned. For fast reactions these unproductive operations consume far more time than the reaction itself and continuous processes can become more attractive.

2. Semibatch Reactors

Some reactions may yield a product in a different phase from the reaction mixture. Examples would be the liberation of a gas from a liquid-phase reaction or the formation of a precipitate in a fluid-phase reaction. To drive the reaction to completion, it may be desirable to continuously separate the raw product phase. A semibatch operation may result as well from differing modes of feeding the individual reactants. For reasons we will discuss later, it may be desirable to charge one reactant to the reactor at the outset and bleed a second reactant in continuously over time. Such reactors have both a batch and a flow character and, like batch reactors, are useful for slow reactions and low production rates.

3. Continuous Stirred Tank Reactors (CSTRs)

It is a small step from the batch reactor to the CSTR. The same stirred vessel may be used with only the addition of piping and storage tanks to provide for the continuous in- and outflow. Faster reactions can be accommodated and larger production rates can be achieved because of the uninterrupted operation. CSTRs are most often used for liquid-phase reactions, such as nitration and hydrolysis, and multiphase reactions involving liquids with gases and/or solids. Examples would be chlorination and hydrogenation. The effectiveness of mixing comes into play in a CSTR and this is addressed in Chapters 4 and 8.

4. CSTRs in Series

It was shown in Chapter 4 that considerable gains in production rate and economics can be achieved by passing the reacting mixture through a series of CSTRs. Again, we see how easy it is to achieve a gradual scaleup, say for a specialty chemical for which demand is increasing. CSTRs in series are usually used for liquid-phase reactions.

5. Tubular Reactors

As the production rate requirement increases, batteries of CSTRs become increasingly complex and tubular reactors become attractive. With the transition to tubular reactors, some versatility is lost and more process integration is required. Nevertheless, tubular reactors find extensive application in liquid-phase reactions, for example, polymerization, and are almost always the continuous reactor of choice for gas-phase reactions, for example, pyrolysis. Exceedingly high production rates can be achieved with tubular reactors either by increasing the diameter of the tube or more commonly by using a sufficient number of tubes in parallel.

6. Recycle Reactors

Recycle reactors can be batch, CSTR, tubular, and so on in nature with the purpose of the recycle varying from one case to the next. Many large-scale commercial processes incorporate the recycle of one or more streams back to an earlier point in the process to conserve raw materials. This practice often

TABLE 7-1 Reactors Characterized by Phases Present

Reaction Type	Suitable Commercial Reactors
Homogeneous: gas phase	Empty tube, continuous
Homogeneous: liquid phase	Empty tube or stirred vessel, continuous
	Stirred vessel, batch
Heterogeneous: liquid–liquid	Stirred vessel, batch or continuous
	Plate and/or agitated column, continuous
Heterogeneous: liquid–gas	Stirred vessel, semibatch or continuous
	Absorption column, continuous
Heterogeneous: liquid–solid	Stirred vessel, batch or continuous
	Fixed bed, continuous
Heterogeneous: gas–solid	Fixed bed, continuous
	Moving bed, continuous
	Fluidized bed, continuous
Heterogeneous: gas–liquid–solid	Stirred vessel, semibatch or continuous (slurry reactor)
	Packed column or fixed bed, continuous (trickle bed reactor)

results in the accumulation of impurities, which in turn requires separation. Usually it is not simply the reactor outlet stream that is recycled back to the reactor inlet, but it can be. For example, in a batch reactor the reacting mixture can be recycled, or pumped around, through a heat exchanger to provide thermal control.

Recycle reactors have also found valuable application in the laboratory and pilot plant because of their special characteristics. At one extreme, in which all of the product is recycled (no net flow), the reactor is the exact equivalent of the well-stirred batch reactor. At the other extreme of no recycle, the reactor is simply the tubular variety. If there is some net flow but the recycle rate is high, the overall reactor performs like a CSTR. Yet the reaction tube itself behaves like differential tubular reactor. This versatility of the recycle reactor can be exploited to great advantage in research and development.

B. Classification by Phases Present in Reactor

The classification of reactors by vessel type and flow regime is what is found in most chemical reaction engineering textbooks. They deal mainly with homogeneous (single phase) reactions. Many commercial processes involve heterogeneous (multiphase) systems. Happel and Jordan (1975) argue that it is more convenient to classify reactors "according to the number and kind of phases present." Table 7-1 follows their organization with a number of additions. It

REACTOR TYPES

will be noted that the entries in Table 7-1 are consistent with the preceding discussion on homogeneous reactors.

The hydrodynamics associated with various types of reactors are discussed in the chapters which deal with those reactors. The design of homogeneous reactors has been treated in Chapter 4. The most thorough treatment of liquid–liquid systems is given by Rase (1977), who concludes that this is an especially difficult reactor scaleup problem. This is not to say that scaleup of other multiphase systems is easy. For gas–liquid reactions, Chapter 6 deals with packed bed absorbers and bubble columns. A gas–liquid semibatch reactor is analyzed in depth in Chapter 4. Another excellent reference is Astarita et al. (1982).

The design of fixed-bed, fluid–solid reactors is the subject of Chapter 5. The moving bed reactor is only different mechanically. Gas–liquid–solid reactors are treated in Chapter 6 and also by Shah (1979). Fluidized beds are treated in Chapter 10.

	FIXED BED		FLUID BED		TRANSPORT
TYPE OF REACTOR	Over Flowed 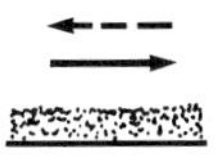	Permeated 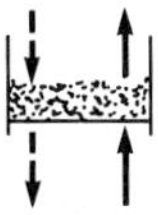	Fluidized Bed 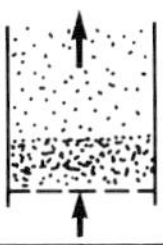	Expanded Fluid Bed 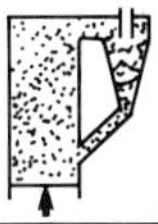	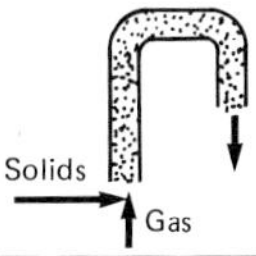
Typical Reaction Devices	Multiple Hearth Rotary Kiln Belt Dryer	Shaft Furnace Traveling Grate Grate Firing Boiler	Fluid Cracker Fluid Bed Roaster Multistage Fluidized Bed Furnace	Circulating Fluid Bed Venturi Fluid Bed	Flash Dryer Cyclone Preheater Melting Cyclone Burner
Solids Motion By	Mechanics	Gravitation Mechanics	Gravitation Drag Force		Gravitation Drag Force
Gas/Solids Flow	Cocurrent Countercurrent Crosscurrent		Stirred Flow Countercurrent } In Stages Crosscurrent }		Cocurrent, Stirred Flow By Recycle; Counter Current Stages
Particle Diameter	Small-Very Large	Medium-Very Large	Small-Medium	Very Small-Small	Very Small
Solids Retention Time	Hours-Days		Hours	Minutes	Seconds And Less
Gas Retention Time	Seconds		Seconds		Fractions Of Seconds
Heat + Mass Transfer Rates	Very Low	Low-Medium	High	Very High	Very High
Temperature Control	Medium-Good	Poor-Medium	Good	Very Good	Medium-Good
Volume/Time Yield	Very Low-Medium	Medium	Medium-High	High	Very High

FIGURE 7-2 Types of reactors classified according to the state of motion of the solids.

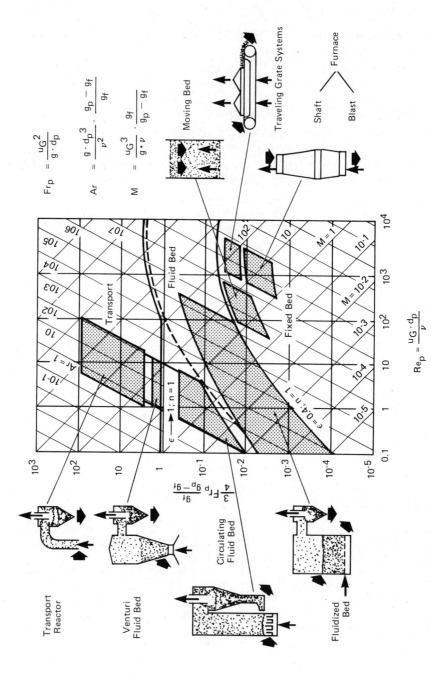

FIGURE 7-3 Status chart for gas/solid systems with antigravitational gas flow including operational regimes of typical gas/solid reactors.

REACTOR TYPES

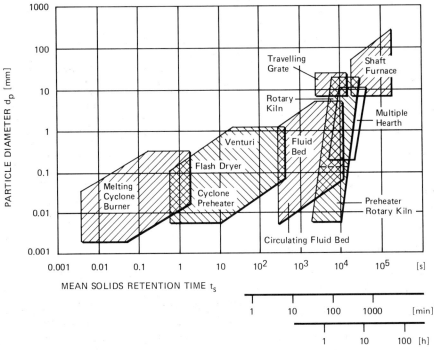

FIGURE 7-4 Regimes of application for industrial gas/solids high-temperature reactors as a function of mean particle diameter and mean solids retention time.

C. An Example of Reactor Selection

It has been pointed out that scaleup is especially difficult when solids handling is involved. A case in point would be the calcination of one solid ($CaCO_3$) to another (CaO) where heat for the endothermic reaction is provided by burning finely pulverized coal, a third solid. An article by Reh (1978) on "Selection Criteria for Noncatalytic Gas/Solid High Temperature Reactors" provides excellent guidance for the scaleup engineer faced with such a problem. He points out that his article cannot answer the question of which type of reactor represents the optimal technical and economic solution for a given application, but he does make an effort "to deduce criteria and classification principles common to all gas/solid high-temperature reactors in order to facilitate the selection of the optimal reactor for a specific purpose."

Having restricted his attention to gas–solid reactors, Reh classifies the different "types of reactors according to the state of motion of the solids" as shown in Figure 7-2. He discusses the attributes of the various reactor types describing the greater ease in building fixed-bed reactors for high production capacities, explaining that the main obstacles to the use of fluid bed and

TABLE 7-2 Temperature Control of Gas/Solid Reactors

Exothermic Reaction Cooling by	Endothermic Reaction Heating by
Heating of cool reaction gas or circulating gas	Cooling of preheated reaction gas or circulating gas
Addition of cool solids, circulation of cooled reaction product or inert heat carrier	Addition of preheated solids, circulation of heated reaction product or inert heat carrier
Cooled wall or tube surfaces	Heated wall or tube surfaces
Latent heat, e.g., evaporation of injected water	Latent heat, e.g., recrystallization effects
Simultaneous endothermic reaction	Simultaneous exothermic reaction, such as combustion
Power production, e.g., MHD	Electric heating

transport reactors are scaleup problems. Reh also considers the fluid mechanics as shown in Figure 7-3, that exist in a variety of industrially important gas–solid reactors. This graph characterizes the operating regimes of the various reactor types in terms of a Reynolds number (Re_p), a Froude number (Fr_p), an Archimedes number (Ar), a dimensionless velocity number (M), an acceleration factor (n), and the void fraction (ε).

Having treated the fluid mechanics in a rather fundamental and comprehensive way, Reh goes on to consider the reaction and transport processes. "Gas/solids reactions take place inside the individual particles of solid. Heterogeneous reactions such as roasting or reduction of ores are affected by a complicated interplay of heat transfer, heat conduction, mass transfer, pore diffusion, and conversion according to the laws of chemical kinetics at the site of reaction." He quantifies these phenomena emphasizing the overwhelming importance of particle size. This treatment makes possible the classification of "the operational regimes of industrial gas/solids high temperature reactors with regard to particle diameters d and solids retention time," shown in Figure 7-4. In terms of limestone calcination, which is usually carried out in a rotary kiln, a wide range of limestone and lime particle sizes may be accommodated and a solids retention time on the order of hours is required. In contrast, for the concomitant combustion of finely pulverized coal, the same rotary kiln functions as a transport reactor. Very small particles are used and the solid is retained (exists) for less than a second. Solid fuel burners could be located in the "melting cyclone burner" block on Figure 7-4.

It will be shown momentarily that thermal control is often the dominant consideration in reactor scaleup. This aspect for gas/solid reactions is summarized in Table 7-2. For the limestone calcination simultaneous endothermic and exothermic reactions, along with other factors, combined to provide the requisite thermal environment. Beyond what has already been described here, Reh provides an analysis of the fundamental processes occurring in some actual operating reactors. He concludes with: "An optimal combination of theory and experiment concentrating on specific objectives will probably long remain the only realistic approach to work in this field, due to the complexity of the problems involved." His article is well worth reading by anyone dealing with scaleup of fluid–solid reactors. Many of his ideas are applicable to catalytic reactions as well. The scaleup engineer should find Reh's approach to be a valuable template in the definition and solution of problems involving many other kinds of reaction systems.

III. GOALS IN SELECTION

To select a particular type of reactor at any stage in a scaleup, it is important to identify your purpose, or goal. As mentioned earlier, completely different reactor types may be optimal for bench scale research, pilot plant application, and commercial operation. For example, the spinning basket reactor or one of its descendents [see Carberry (1976)], may be the best possible choice for catalyst evaluation in gas-phase reactions. But such a device would be ludicrous on a commercial scale or when one of the products was a liquid or, worse still, a solid. A variety of considerations are illustrated below according to different goals.

A. Flexibility

For the small-scale chemical manufacturer, few criteria are as consequential as flexibility with respect to operating conditions and product demand. This product market seldom justifies a massive research program, thus optimal operating conditions are not likely to be well defined. Further, a series of campaigns producing appropriate amounts of several products in the same equipment may be the normal mode of operation. Both of these circumstances require easily adaptable equipment. Cycle time may be quite variable. Contamination must be quickly eradicable. The advantage of stirred tanks, operated continuously or batchwise, is obvious.

B. Cost Minimization

The economics of the overall process are often little influenced by the cost of the reactor. For example, use of a more expensive catalyst may pay for itself by obviating costly pretreatment of the reactor feed. Use of a larger reactor to approach equilibrium more closely may greatly reduce the costs of product separation and reactant recycle. The minimization of reactor volume, and therefore reactor cost, for single reactions is treated at length in Chapter 4 and in many textbooks. Wirges and Shah (1976) present a number of tables and graphs purporting to make the correct selection of the optimum reactor for homogeneous reactions easy. Their approach is to allow selection among CSTRs, CSTRs in series, and plug flow reactors (PFR) to minimize the total reactor volume, given rate equations and stoichiometry. The telling statement appears at the end of their article, however: "Effects of factors, such as volume changes accompanying the reaction, heat transfer effects, mixing patterns, material problems, maintenance and safety considerations are not easy to quantify in the framework of this article." It just goes to show that there are no routine solutions. The reader is referred back to the quote of Nishida et al. (1981) at the beginning of this chapter.

C. Product Selectivity

The ability of a chosen reactor to optimize product selectivity, or maximize productivity, is far more important in practice than the minimization of the reactor cost. Carberry (1976) is the one author to emphasize consistently this aspect throughout his book. No doubt this awareness and emphasis resulted from his industrial experience before taking up an academic career. He discusses the impact on selectivity of types of reactors or combinations of reactors, the character of flow and state of mixing, thermal factors, mass transfer, catalyst decay, reaction kinetics, and so on. A simple illustration will be given here to make the point and the reader can check the references for greater depth and variety.

Consider the simultaneous isomerization of A to R and to S and let the rates of formation of R and S be first and second order, respectively.

$$A \to R \quad r_{A \to R} = k_1 C_A$$

$$A \to S \quad r_{A \to S} = k_2 C_A^2$$

The point selectivity is defined as the ratio of the rates of formation at any point in time and space, that is, in the momentary, local reaction environment. Note that this point selectivity may differ quantitatively from the overall selectivity achieved in an integral reactor because of the variation of the rates of formation with reaction progress.

$$\text{selectivity} = \frac{r_{A \to R}}{r_{A \to S}} = \frac{k_1 C_A}{k_2 C_A^2} = \frac{k_1}{k_2}\left(\frac{1}{C_A}\right) \qquad (7\text{-}1)$$

From this relationship, it can be seen that large values of C_A keep the selectivity to R down and hence favor the formation of S. On the other hand, low values of C_A favor the formation of R.

These relationships can be used to choose between a CSTR and a tubular reactor. For example, all of the liquid in a well-mixed CSTR is at the con-

GOALS IN SELECTION

centration of the product stream, that is, at low C_A. In contrast, the reaction near the inlet of a tubular reactor occurs at high C_A. Only at the outlet, for the same conversion, does C_A become as low as for the CSTR. Thus a single CSTR should be used to maximize selectivity to R, whereas use of a tubular reactor or CSTRs in series will enhance the formation of S. The impact of such strategy can be tremendous.

In a slightly different circumstance, imagine parallel reactions of A with B to form R (desired) and S (undesired). Also assume that the formation of S increases more strongly with the concentration of B than does the formation of R. A semibatch reactor with A charged initially and B bled in slowly can be used to advantage. By keeping the concentration of B low, the selectivity to the formation of the desired component, R, will be enhanced. Rase (1977) provides analyses similar to these for a good many other cases.

In either of the above two examples, there exists the possibility of R and S reacting to produce impurity, Y. A detailed kinetic study of the reaction environment which promotes the formation of Y may be infeasible. Nevertheless, the conditions that increase Y to an intolerable level should be explored.

Denbigh (1965) offers a very nice presentation of "chemical factors affecting the choice of reactor." He considers parallel reactions, as in the preceding illustration, as well as sequential reactions like degradation, polymerization, and crystallization. He focuses on point yield, which in the context of the previous example would be the ratio of the rate of formation of R to the total rate of reaction of A. In most of his examples the kinetics are known. He treats in considerable detail, however, one wartime example where the reactor selection is based on experimental studies of point yields while the reaction rate equation remains undetermined. This circumstance may have considerable relevance to engineers looking for a shortcut to effective scaleup.

In the nitration of hexamethylene-tetramine ("hexamine") to form the explosive cyclonite (R.D.X.), small crystals of hexamine were fed into concentrated nitric acid. As the reaction proceeded the acid became diluted by the water formed. At first this provided a modest increase in the rate of reaction, but further dilution strongly inhibited the reaction. In this context Denbigh points out how an appropriately chosen, single CSTR may be preferred to maximize the yield. However, if the excess acid is to be recovered and recycled, then a batch or tubular reactor may be the better choice. If the CSTR is to be used for reason of thermal control, then Denbigh suggests using a series of them with hexamine crystals fed independently to each. Finally, he explains how a CSTR followed by a tubular reactor could be superior to any of the aforementioned possibilities.

D. Thermal Control

Control of temperature is of such overwhelming importance to reactor scaleup that many reactors take on the appearance of heat exchangers. The rates and equilibria of reactions are profoundly affected by temperature. Accordingly, so are side reactions, by-product formation, yield, selectivity, and so on. In multiphase systems, the temperature dependencies of phase equilibria and rates of mass transfer also come into play. Catalyst performance can deteriorate rapidly in a suboptimal thermal environment. All of these factors mandate effective temperature control and therefore impact on selection of reactor type.

Figure 7-2 and Table 7-3 show that heat transfer and therefore temperature control in fixed-bed reactors is rather poor. The problem is exaggerated especially in large sizes, where the reactor volume exceeds twice the catalyst volume. Nevertheless, the advantages of fixed-bed reactors are so numerous and significant that these reactors find widespread application. Much design consideration is addressed to the thermal characteristics of such reactors as shown in Figure 3-3 and Table 7-2. In Chapter 5 extensive attention is given to the avoidance of runaway in adiabatic and other nonisothermal fixed-bed reactors.

Fluidized bed reactors have excellent heat transfer characteristics. It is possible to use inert solids simply to provide better heat exchange, but most often the fluidized solids are catalyst particles. In fact, continuous regeneration of the catalyst is one of the main reasons for choosing a moving or fluidized bed reactor.

The preceding discussion has centered on high throughput equipment, but thermal control is equally crucial on a smaller scale. If a large production rate is not required, the single CSTR can be used to ameliorate thermal effects. Because the reaction occurs at the final reactant concentration, the rate of reaction and hence of heat release or uptake is usually quite small. Thus, a larger reactor is required to achieve the desired conversion, but overall savings may accrue by elimination of heating or cooling coils, reduction of power consumption for agitation, and improved product quality.

A similar reduction in reaction rate and hence temperature buildup or decline can be realized by converting from batch to semibatch operation. If only one reactant is charged to a vessel, the other can be bled in at such a rate as to control any temperature excursions. Even in the case of the isothermal semibatch reactor design of Chapter 4, it was necessary to show by calculation that the requisite heat transfer was available to maintain isothermality.

E. Comparison of Reactor Types

The great variety in reactor types and duties and the fact that no individual type is suitable for all purposes makes it necessary to establish criteria for selection. This chapter has dealt with the matter of selection largely by specific illustrations. To provide a broader perspective on the problem, two more comprehensive examples are presented.

1. Solid-Catalyzed Vapor–Liquid Reactions

Weekman (1974) considered the problem of selecting the most suitable laboratory reactor to study a typically complex, industrially important reaction. The

GOALS IN SELECTION

feed had a wide boiling range giving both vapor and liquid phases at reaction conditions. There were many, highly endothermic reaction paths. The catalyst was a powdered solid; highly active but rapidly decaying at reaction conditions. The goal was to obtain kinetic rate data accurate enough for reactor design. Weekman's paper provides useful insight into the characteristics and limitations of a variety of laboratory reactors, to some extent in general but especially in the context of this reaction system. More importantly, he provides a logical framework that the scaleup engineer can follow to solve his own particular problem.

Weekman established the following five attributes as being the most crucial. Of course, different attributes would be specified for commercial reactors:

Sampling and analysis of product composition.

Isothermality.

Residence-contact time measurement.

Selectivity-time disguise decay.

Construction difficulty and cost.

The importance of each of these attributes and an analysis for eight different reactor types are given in Table 7-3.

Every reactor has good and bad features and none is clearly superior to the rest. Since a single reactor has to be chosen, a compromise among objectives is required. A better approach, where the issues are critical and both time and resources allow, is to conduct scaleup studies with two very different reactors. Not only can you enjoy the most favorable attributes of each, but more importantly you can obtain independent confirmation of results. More likely

**TABLE 7-3 Summary of Reactor Ratings
Gas – Liquid, Powdered Catalyst, Decaying Catalyst System**

Reactor Type	Sampling and Analysis	Isothermality	Residence-Contact Time	Selectivity Decay	Construction Problems
Differential	P–F	F–G	F	P	G
Fixed bed	G	P–F	F	P	G
Stirred batch	F	G	G	P	G
Stirred contained solids	G	G	F–G	P	F–G
Continuous stirred tank	F	G	F–G	G	P–F
Straight-through transport	F–G	P–F	F–G	G	F–G
Recirculating transport	F–G	G	G	G	P–F
Pulse	G	F–G	P	F–G	G

G = good, F = fair, P = poor.

you will get apparent inconsistencies which, when properly interpreted, will lead you to greater insight.

Weekman provides tables similar to Table 7-3 reevaluating the reactor choices for less stringent conditions. For example, if the requirements are relaxed, in turn, to allow for a nondecaying catalyst, a single reacting phase, no diffusional limitations, and low heats of reaction, the result, not surprisingly, is a considerable reordering of reactor suitabilities.

Although the focus is on laboratory reactors, Weekman's methodology is valid for commercial reactors. The procedure for selecting the most suitable commercial reactor would follow the same format with the major difference being the development of a new collection of attributes. For example, one might specify in addition:

Time between shutdowns.

Ease of maintenance.

Provision for additional heat exchange surface.

Product selectivity.

Production rate.

Versatility of performance.

Elapsed time to commercialization.

Initial and continuing costs.

2. Absorption in Reactive Solutions

The author went through a Weekman-like exercise to evaluate laboratory reactors for the study of equilibrium and rate of reaction for the absorption of acid gases (CO_2, H_2S) in diethanolamine-promoted, aqueous K_2CO_3 solutions. The important operating variables and characteristics in such a system are:

Pressure and temperature.

Gas and liquid flow rates.

Gas and liquid compositions.

Absorption and desorption capability.

Thermal effects are mild but important.

An evaluation summary table analogous to Weekman's is given in Table 7-4. Reactors suitable only for reactions with an active solid phase were deleted and reactors that have been used for gas–liquid reactions were added. An overall evaluation column was added.

**TABLE 7-4 Summary of Reactor Ratings
Acid – Gas Absorption in Reactive Solutions**

Reactor Type	Sampling and Analysis	Isothermality	Residence-Contact Time	Selectivity Disguise Decay	Construction Problems	Overall
Differential	P	G	P	G	G	F
Stirred batch or semibatch	G	G	G	P	G	F–G
Continuous stirred tank	F	G	G	F	G	F–G
Straight-through transport	P	F	F	F	F	F
Recirculating transport	P	G	F	F	F	F
Pulse	P	F	P	G	F	P–F
Inert packed column	G	F	F	P	G	F
Wetted wall, chain of spheres	G	G	G	G	F	G
Pendent drop	F	F	G	G	F	F

G = good, F = fair, P = poor.

These ideas on laboratory reactors can be extended to reactors for any purpose. It has already been noted that different reactor types serve best at different stages in a scaleup. In an actual research and development program at Exxon both a stirred semibatch reactor and a single wetted-sphere reactor were selected for use. For a subsequent pilot plant investigation, a column packed with inert material was used. In commercial practice packed and plate columns are the common choices.

IV. OVERVIEW OF SELECTION

An attempt is made here to establish a sequence of steps leading to the selection of a type of reactor for any particular purpose. It is a summary of the elements of the methods presented in this chapter.

State the goals of the reaction process under consideration. The statement should include any criteria, constraints, and special characteristics.

List all possible types of reactors that might be suitable for the intended application. Characterize the hydrodynamics and other fundamental phenomena associated with each type of reactor.

Gather pertinent kinetic data on rates, selectivity, impurity formation, and so on. Gather also the relevant physical and thermodynamic property data. Correlate the data as far as possible.

Explore heat and mass transfer effects and other complicating factors. Seek qualitative and quantitative generalizations.

Design, compare, and cost out the most attractive reactor possibilities.

Consider timetable to commercialization and integrate findings with broader economic analyses. If commercialization is not the momentary goal, consider how your findings fit in with the immediate goal and management plan.

With new insight iterate through the preceding six steps.

V. SOME REALITIES

A. Mode of Operation

Batch and semibatch reactors are unsteady-state devices, of course. Almost all continuous flow reactors are designed for steady-state operation. Too little attention is paid to reactor dynamics and various transient operations. These impact both on the economics of operation and especially on safety. Before a process can operate at steady state, it must be started up. Some scaleups have failed because startup was not achievable. During normal operation perturba-

SOME REALITIES

tions in feedstock, process requirements, or process conditions occur. Finally, all processes must be shut down. Even barring unexpected happenings, the dynamics of any selected reactor must be understood. Lack of understanding of the dynamics of one reactor type may be grounds for choosing a better understood type for actual use.

Two kinds of instability are particularly troublesome: fluid mechanical and thermal. We have already seen that laminar and turbulent flows can produce quite different results in reactor performance. So can surging behavior and channeling. Good mechanical design such as flow distributors in packed beds, avoidance or proper location of abrupt constrictions, and availability of effective mechanical agitation can help avoid such problems.

Similarly, certain thermal regimes can produce multiple steady states, some of which are unstable and some of which are only locally stable. This matter was discussed in some detail in Chapter 4 for a homogeneous CSTR. Multiple steady states and the accompanying potential for instability can also occur in catalyst beds and, indeed, within catalyst particles. Many papers on this subject have been published by Luss and others, for example, Balakotaiah and Luss (1982). While the results of such instabilities can be catastrophic, they can usually be avoided by judicious choice of reactor type and/or operating conditions. However, first one must be aware that they are possible. Supplementary heating or cooling capability can be provided to aid in startup and to provide for control of temperature excursions.

More troublesome is the potential for coupling of fluid mechanical and thermal instabilities. For example, a rise in temperature can result in the initiation of two-phase flow. A drop in temperature can result in the transition from turbulent to laminar flow. Conversely, a shift from turbulent to laminar flow accompanying polymerization could alter not only the residence time distribution but also wall heat exchange. With these alterations could come a change in molecular weight distribution.

All of these factors and others should be the subject of scrutiny during research, development, and testing. Theoretical analyses of reactor dynamics should be performed so far as possible. But these are complex and experiment will have to be resorted to. Data taking should not be restricted to steady-state regimes alone but should continue during intended and unintended transients. At least qualitative interpretation of all transient behavior should be attempted. Quantitative models should be developed if possible. Attention to unsteady state conditions early will facilitate design for process control, planning for startup and shutdown, and response to unexpected trouble as discussed by Graboski et al. (1982).

The exploitation of reactor dynamics needs not be directed only at problem avoidance. It can be the basis for superior modes of operation, for example, periodic processing. Examples of effective use of periodic processing are occuring with greater frequency now. Different reactor types have different dynamic characteristics, such as inertia and controllability, which can be expected to influence reactor type selection.

B. Unpredictable Behavior and Safety

The objective of scaleup studies is the preclusion of unpredictable behavior. As we have seen, different reactors can be used to advantage at different stages of scaleup. While this means that the reactor selected for commerical application may not have been tested previously, there is a lot of security provided by the variety of reactors that have been explored. Most reactors can be made to work and the question in the context of the overall process is "How well?" Common problems are premature catalyst decay due to buildup of impurities, by-product formation because of inadequate temperature control, and suboptimal separation circumstances resulting from off-specification reactor performance. While costly, these deficiencies are not catastrophic.

Unpredictable behavior becomes catastrophic, though, when safety is the issue. Reactors, where streams are mixed at severe conditions, are prime targets for explosions. Rase (1977) discusses this issue in some detail. Reactors are also a potential source of toxic materials and other wastes. This potential should be explored at the earliest stages of a project.

C. Scaledown

The concept of scaledown involves acknowledging your full-scale goal and then retreating to a smaller scale for preliminary testing, as invoked in Chapter 17. A corollary is to recognize the largest size your equipment might be and optimize downward from there. For example, it was mentioned earlier that fabrication and transportation ultimately limit the realizable volume of process vessels. Bisio (1982) suggested that the above factors and the supporting systems would limit the largest reactor that could be built to a production rate of 10^4 barrels per day. One look at the seemingly endless succession of Lurgi gasifiers in the SASOL coal to hydrocarbon liquid projects as shown in any photograph of SASOL II and III (*Chem. Eng. Prog.*, 1980) seems to confirm this notion.

D. Influence of Experience

Maybe more than we acknowledge, we allow experience rather than creativity to determine our choice of reactors. Subconscious adherence to such a policy inhibits the deliberate consideration of alternatives. On the other hand, experience represents our cumulative store of knowledge. Coupled with creativity, experience can lead to ever-improving levels of performance. The benefits of this latter approach are evident in the competitive world of the chemical process industries.

To illustrate the importance of experience in scaleup and selection of reactor type, the "Mobil Methanol to Gasoline Process" as reviewed by Liederman et al. (1980) is particularly instructive. The process is a solid-catalyzed, vapor-phase reaction to convert methanol directly to gasoline. Fixed-bed and fluid bed reactors are under consideration. Both fixed- and fluid bed units have been successfully tested at the bench scale and in four barrel per day pilot plants. The fixed-bed operation is cyclic because of the need to regenerate the catalyst.

Mobil believes that the fixed-bed process is more suitable than the fluid bed process for small-scale operation and that it has been "ready for immediate commercialization for some time." They have begun construction of a 14,000 barrel per day plant in New Zealand [*Wall Street Journal* (1982)]. This implies a scaleup ratio of 3500. Comparison of this ratio with typical values presented in Chapter 1 shows the benefit of the confidence that comes with experience. Not only are intermediate steps eliminated, but commercial operation can be achieved much earlier by using familiar technology such as the fixed-bed reactor.

For large-scale operation, the fluid bed process may be more attractive. Liederman et al. (1980) see the following advantages for fluid bed operation:

"The reaction heat removal is simplified by using the superior heat transfer characteristics of a fluid bed."

"When coupled with alkylation the fluid bed gives a higher gasoline yield than the fixed bed."

"Constant catalyst activity, gasoline selectivity, and quality can be maintained with the fluid bed operation."

With an eye to the future, Mobil proceeded with a 100 barrel per day fluid bed pilot plant in West Germany. The much smaller scaleup ratio, 25 in this case, is due in part to the greater difficulty in scaleup when solids must be handled. The fluid bed pilot plant program, through conceptual design of a commercial plant, is scheduled for completion in 1985 which is the same year that the fixed-bed commercial plant is expected to come onstream.

REFERENCES

Astarita, G., Savage, D. W., and Bisio, A. L., *Gas Treating with Chemical Solvents*, Wiley, New York, 1982.

Balakotaiah, V., and Luss, D., "A Novel Method for Determining the Multiplicity Features of Multireaction Systems," ACS Symposium Series, No. 196, Chemical Reaction Engineering, Boston, American Chemical Society, Washington, D.C., 65–75 (1982).

Bisio, A. L., personal communication (May 13, 1982).

Carberry, J. J., *Chemical and Catalytic Reaction Engineering*, McGraw-Hill, New York, 1976.

Denbigh, K. G., *Chemical Reactor Theory*, Cambridge University Press, Cambridge, England, 1965.

Graboski, M. S., Kabel, R. L., Danner, R. P., and AlAmeeri, R. S., "Process Input Analysis," *Chem. Eng. Communications*, **17**, 137–149 (1982).

Happel, J., and Jordan, D. G., *Chemical Process Economics*, 2nd ed., Dekker, New York, 1975.

Henglein, F. A., *Chemical Technology*, translated by R. F. Lang, Pergamon Press, New York, 1969.

Hill, C. G., *An Introduction to Chemical Engineering Kinetics & Reactor Design*, Wiley, New York, 1977.

Liederman, D., Yurchak, S., Kuo, J. C. W., and Lee, W., "Mobil Methanol-to-Gasoline Process," 15th Intersociety Energy Conversion Engineering Conference, Seattle, WA (August 18–22, 1980).

Nishida, N., Stephanopoulos, G., and Westerberg, A. W., "A Review of Process Synthesis, *AIChE J*. **27**, 321–351 (1981).

"Picture Tour of Sasol-II Coal Liquifaction Plant," *Chem. Eng. Progr.* **76**, 85–88 (March 1980).

Rase H. F., *Chemical Reactor Design for Process Plants*, Vol. I, Wiley, New York, 1977.

Reh, L., "Selection Criteria for Noncatalytic Gas/Solid High-Temperature Reactors," *Ger. Chem. Eng.* **1**, 319–329 (1978).

Shah, Y. T., "Gas-Liquid-Solid Reactor Design," McGraw-Hill, New York, 1979.

Staff Reporter, "Synthesis Gasoline Loan of $1.7 Billion Set for New Zealand," *Wall Street Journal* (June 8, 1982).

Walas, S. M., *Reaction Kinetics for Chemical Engineers*, McGraw-Hill, New York, 1959.

Weekman, V. W., Jr., "Laboratory Reactors and Their Limitations," *AIChE Journal*, **20**, 833–840 (1974).

Wirges, H-P., and Shah, S. R., "For a Given Kinetic Duty... Select Optimum Reactor Quickly," *Hydrocarbon Proc.*, **55**(4), 135–138 (1976).

8

FLOW PATTERNS AND RESIDENCE TIME DISTRIBUTIONS

E. B. NAUMAN

I.	Introduction	276
II.	Residence Time Distributions	277
	A. Distribution Functions	278
	B. Moments of the Distribution	278
	C. Normalized Distributions	279
	D. Special Distribution Functions	280
III.	Direct Applications	281
	A. Surge Damping	281
	B. First Order Reactions	282
	C. Other Simple Reactions	283
	D. Complex Reactions	285
IV.	Measurement Techniques	286
	A. Closed Systems	286
	B. Open Systems	288
V.	Analytical Distribution Functions	290
	A. Piston Flow and Ideal Mixing	290
	B. Tanks in Series	292
	C. Laminar Flow Models	293
	D. Axial Dispersion	295
VI.	Scaleup Considerations	298
	A. The Almost Delta Distribution	298
	B. Laminar Flow Systems	299
	C. The Almost Exponential Distribution	300

VII. Complications and Extensions 303
    A. Intermediate Distributions 303
    B. Unmixed Feed Streams 304
    C. Nonisothermal or Nonhomogeneous Reactors 305
Nomenclature 305
References 307

I. INTRODUCTION

This chapter considers scaleup from the viewpoint of flow patterns and residence time distributions in processing equipment. Primary emphasis is naturally on continuous, steady-state processes although it is sometimes possible to treat startups, shutdowns, and cyclic operations, including batch systems, within the conceptual framework of residence time theory.

Consider a process operating at some volumetric flow rate, Q_1. Suppose that the process is conducted in pilot scale equipment and that the product properties associated with Q_1 are judged to be "good." Then the classical scaleup problem is to design a larger process with flow rate $Q_2 \gg Q_1$ which produces material having the same product properties. The scaleup factor is

$$S = \frac{Q_2}{Q_1} \quad (8\text{-}1)$$

and in the simplest possible world, everything about the process would scale by this factor. For example, we would normally want the system volume to be given by

$$S = \frac{V_2}{V_1} \quad (8\text{-}2)$$

In processes where heat transfer to the environment is important, it would also be desirable if the external surface area of the process followed the same scaling law as in Equation 8-2. Unfortunately, this is rarely possible. Indeed, it is impossible if one wants to maintain geometric similarity between the pilot plant and production unit because surface area would then scale as $S^{2/3}$. As emphasized throughout this book, one can never maintain all forms of similarity (geometric, dynamic, thermal, etc.) during a scaleup. The art of scaleup is knowing which similarities to keep and which to sacrifice. A constraint on the scaleup is to maintain product properties at least as good as in the pilot plant and it is often desired and even possible to improve certain properties during the scaleup.

Equation 8-1 represents the definition of S and is thus automatically satisfied. The use of volumetric flow rate is, of course, arbitrary and was chosen for convenience in this chapter. Equation 8-2 may or may not be satisfied by the scaleup. However, it is a theme of this chapter that it is often desirable to satisfy Equation 8-3 since this gives the same mean residence time in the production unit as in the pilot plant

$$\bar{t} = \frac{V_1}{Q_1} = \frac{V_2}{Q_2} \qquad (8\text{-}3)$$

The above expression is restricted to a conserved component—for example, one that is not created or destroyed by the process. It may also seem restricted to constant density systems but, in fact, V can be replaced by any measure of the inventory or holdup of a conserved component. Equation 8-3 will also apply when Q is replaced by the mass flow rate of that material (measured in the same mass units as V). A recent text by Nauman and Buffham (1983) discussed Equation 8-3 and its generalizations at greater length. Their book provides a detailed treatment of mixing and residence time theory and can be regarded as a general supplement to the ideas presented below which emphasize special applications to scaleup. Another general reference is Nauman (1981a).

Constancy of $\bar{t}$ during scaleup is natural for chemical reactors since any alteration will change the reaction yield and perhaps the selectivity. For heat and mass transfer devices, the need to hold $\bar{t}$ constant is less clear. However, $\bar{t}_2 > \bar{t}_1$ causes the volume to increase faster than S which leads to large in-process inventories and doubtful economies of scale. The case of $\bar{t}_2 < \bar{t}_1$ is sometimes possible when the pilot unit is overdesigned for a particular operation. One can also achieve $\bar{t}_2 < \bar{t}_1$ by using higher driving forces in the production unit, but the required change in temperature, pressure, or composition between the pilot and production units usually entails great risk.

Given that $\bar{t}$ is to remain constant, we next question whether the distribution of residence times about this mean should also be constant. For chemical reactors, the answer is usually yes since any change in the distribution will alter yields and selectivities. However, the distribution might be allowed to change in a direction known to be favorable; and very occasionally such a shift might enable scaleup with a reduced $\bar{t}$. For heat and mass transfer devices, the residence time distribution has a less direct bearing on performance than for reactors; but any change in the distribution reflects a change in the underlying flow patterns which at least must be understood.

II. RESIDENCE TIME DISTRIBUTIONS

Consider any steady-state flow system which has some conserved entity passing through it. Usually these entities will be molecules of a conserved species but they could be atoms if no molecules are conserved or they could be larger agglomerates such as Brownian particles. Each entity flowing through the system will have had some first entrance into the system and some final exit from it. The time the particle spends inside the system boundaries is called the residence time, t. Most process systems are *closed* which means that a particle which once enters will not leave on temporary excursions but will leave only once, never to return. In closed systems, the residence time is equal to the elapsed time between entering and leaving. *Open* systems arise with recycle or when there is significant particle diffusion across a system boundary. Temporary excursions outside the system boundaries are then possible, but the time spent on such excursions does not contribute to t.

A. Distribution Functions

There are three common functions used to describe a residence time distribution: the cumulative distribution function, the washout function, and the differential distribution function. Any of these may be regarded as fundamental and the others derived it. Here, we begin with the *washout function*:

$W(t)$ = fraction of particles leaving the system never to return,

which experienced a residence time greater than t (8-4)

The washout function is defined over $(0, \infty)$, is nonincreasing, and has $W(0) = 1$ and $W(\infty) = 0$. Closely related to it is the *cumulative distribution function*,

$$F(t) = 1 - W(t) \tag{8-5}$$

which gives the fraction of particles leaving the system with a residence time less than t. The differential distribution function is defined by

$$f(t) = -\frac{dW}{dt} = \frac{dF}{dt} \tag{8-6}$$

so that $f(t)\,dt$ gives the fraction of exiting particles with residence times between t and $t + dt$.

In closed systems, the functions $W(t)$, $F(t)$, and $f(t)$ are readily measured using transient experiments with nonreactive tracers. Analytical expressions for these functions are also known for all simple system models and for many complex models. Before getting into the actual determination of these functions, however, we will first explore some general properties and applications.

B. Moments of the Distribution

The nth moment about the origin is defined as

$$\mu_n = \int_0^\infty t^n f(t)\,dt \tag{8-7}$$

RESIDENCE TIME DISTRIBUTIONS

where $n = 0, 1, 2, \ldots$. The zeroth and first moments have special values

$$\mu_0 = \int_0^\infty f(t)\, dt = 1 \tag{8-8}$$

and

$$\mu_1 = \int_0^\infty tf(t)\, dt = \bar{t} \tag{8-9}$$

where $\bar{t}$ may be found by dividing the holdup of a conserved species by its throughput, see Buffham (1978) or Nauman (1981a). Equations 8-8 and 8-9 must be satisfied for any residence time distribution. Failure to do so represents an experimental or conceptual error.

A useful alternative to Equation 8-7 is

$$\mu_n = n \int_0^\infty t^{n-1} W(t)\, dt \tag{8-10}$$

The use of the washout function to calculate the moments usually gives simpler analytic integrations and more accurate numerical ones.

Moments about the mean are defined as

$$\mu'_n = \int_0^\infty (t - \bar{t})^n f(t)\, dt \tag{8-11}$$

of which the case of $n = 2$ is the most important since this gives the *variance* of the distribution,

$$\sigma^2 = \mu_2 = \mu'_2 - (\bar{t})^2 \tag{8-12}$$

Similar but more complex equations relate μ'_3 to μ_3 and so on.

C. Normalized Distributions

It is sometimes desirable to compare the "shapes" of residence time distributions having different values of $\bar{t}$. This may be accomplished through normalization:

$$\mathscr{W}(\tau) = W(\bar{t}\tau) \tag{8-13}$$

or

$$\mathscr{f}(\tau) = \bar{t} f(\bar{t}\tau) \tag{8-14}$$

where $\tau = t/\bar{t}$ is the dimensionless residence time. The moments of the normalized distribution are related to those of $f(t)$ by

$$\nu_n = \frac{\mu_n}{(\bar{t})^n} \tag{8-15}$$

and $\nu_1 = 1$ so that all normalized distributions have unit means. The variance of the normalized distribution is

$$\sigma_\tau^2 = \nu_2' = \nu_2 - 1 = \frac{\sigma^2}{(\bar{t})^2} \tag{8-16}$$

This parameter is often used to characterize distributions and to fit models to experimental data. It is usually considered to have a range from 0 to 1.0 but $\sigma_\tau^2 > 1$ is possible in systems with bypassing or in laminar flow systems with low particle diffusivities.

D. Special Distribution Functions

Two special distributions merit introduction at this point. One is the delta function distribution corresponding to piston flow:

$$f(t) = \delta(t - \bar{t}) \tag{8-17}$$

or

$$W(t) = 1, \quad 0 < t < \bar{t}$$
$$= 0, \quad t > \bar{t} \tag{8-18}$$

All particles leaving a piston flow reactor have the same residence time, and thus $\sigma^2 = 0$.

The other special distribution is the exponential distribution

$$f(t) = \frac{1}{\bar{t}} e^{-t/\bar{t}} \tag{8-19}$$

or

$$W(t) = e^{-t/\bar{t}} \tag{8-20}$$

for which $\sigma^2 = (\bar{t})^2$ and $\sigma_\tau^2 = 1.0$. It happens that a perfectly mixed stirred tank has an exponential distribution of residence times, but an observed exponential distribution certainly does not imply perfect mixing on a molecular scale.

The delta function and exponential distributions often—but not always—place limits on the performance of a real system. For example, the delta distribution is the worst possible distribution with respect to surge damping and is the best possible distribution with respect to the yield of a first order reaction. The exponential distribution is the best possible with respect to surge damping and is poor but not the worst for reaction yields (zero yield is possible in systems with extreme bypassing).

III. DIRECT APPLICATIONS

This section considers some ways in which the residence time distribution directly determines the performance of a process unit. Equivalent performance between pilot plant and production units will thus require identical $W(t)$. Alternatively, if $W(t)$ changes during scaleup, the effects of this change may be calculated or at least closely bounded.

A. Surge Damping

Mixing tanks are sometimes installed to damp out surges in the composition or flow rate of feed streams to downstream equipment. The simplest case is where only the composition changes at constant flow rate. Then the outlet concentration of some key component is related to the inlet concentration via convolution

$$C_{\text{out}}(\theta) = \int_0^\infty C_{\text{in}}(\theta - t) f(t) \, dt = \int_{-\infty}^\theta C_{\text{in}}(t) f(\theta - t) \, dt \quad (8\text{-}21)$$

This equation was first derived by Danckwerts and Sellers (1951) who give some sample applications. Its generalization to systems where both the input composition and flow rate vary with time requires unsteady residence time theory [Nauman (1969)]. However, either situation is easily scaled up using the approach for almost exponential distributions which is discussed subsequently. Briefly stated, exact scaleup of a mixing tank requires the same residence time distribution in the production unit as in the pilot plant. Conservative scaleup results when $\bar{t}_2 > \bar{t}_1$ with a constant normalized distribution or with $\bar{t}_2 = \bar{t}_1$ but $W_2(t)$ being more nearly exponential than $W_1(t)$.

As a simple example of surge damping, suppose the inlet concentration has a sinusoidal ripple

$$C_{\text{in}}(\theta) = \bar{C}_{\text{in}} + \alpha \sin \beta \theta \quad (8\text{-}22)$$

where $\alpha < \bar{C}_{\text{in}}$ is the amplitude of the ripple. Then the (maximum) amplitude of the outlet ripple is unchanged if the mixing tank has the delta distribution of a piston flow element but is substantially damped if the mixing tank has an exponential residence time distribution:

$$C_{\text{out}}(\theta) = \bar{C}_{\text{in}} + \frac{\alpha[\beta \bar{t} \cos \beta \theta - \sin \beta \theta]}{1 + \beta^2 \bar{t}^2} \quad (8\text{-}23)$$

Figure 8-1 illustrates the damping effect of an exponential distribution for the case $\alpha = 0.5 \bar{C}_{\text{in}}$ and $\beta = 2\pi/\bar{t}$.

Residence time distributions intermediate between these extremes will naturally give intermediate levels of damping. Exact scaleup of surge damping

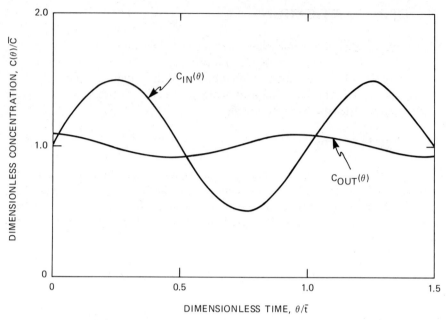

FIGURE 8-1 Typical smoothing of concentration fluctuations in a stirred tank.

phenomena is rarely necessary; but if it were, the pilot plant and production units would need the same $W(t)$.

B. First Order Reactions

Many industrially important reactions are first order or at least pseudo-first order. Few industrial reactors are isothermal and homogeneous; but when they are, the yield of a first order reaction is uniquely determined by the residence time distribution:

$$\frac{C_{\text{out}}}{C_{\text{in}}} = \int_0^\infty e^{-kt} f(t)\, dt \qquad (8\text{-}24)$$

Reactors that are isothermal and homogeneous tend to be liquid-phase stirred tanks with exponential or almost exponential residence time distributions. Equation 8-24 then integrates to give

$$\frac{C_{\text{out}}}{C_{\text{in}}} = \frac{1}{1 + k\bar{t}} \qquad (8\text{-}25)$$

From the yield viewpoint, scaleup with $\bar{t}_2 \geq \bar{t}_1$ is conservative. It also appears conservative to allow the residence time distribution to shift away from the exponential distribution toward the delta function distribution. The yield is then given by

$$\frac{C_{\text{out}}}{C_{\text{in}}} = e^{-k\bar{t}} \qquad (8\text{-}26)$$

which always predicts higher conversions than Equation 8-25. However, poorly conceived departures from residence time similarity are apt to give channeling or bypassing of the feed through the stirred tank. Then, of course, the yield will be lower than had the distribution remained exponential.

C. Other Simple Reactions

A reaction is called *simple* when the rate of reaction of a key component depends only on the concentration of that component

$$\text{rate} = \mathcal{R}(C) \qquad (8\text{-}27)$$

Unimolecular reactions are thus simple. Multimolecular reactions may be treated as simple if the reactants have similar flow and mixing patterns in the vessel so that stoichiometry is locally preserved. This requirement is usually satisfied in single-phase, premixed feed systems where the various reactants have similar diffusivities.

Given a simple, isothermal, and homogeneous reaction, suppose also that the rate expression is either a concave upward or concave downward function of concentration:

$$\mathcal{R}''(C) = \frac{d^2\mathcal{R}}{dC^2} < 0, \qquad \text{concave downward}$$

$$\mathcal{R}''(C) = \frac{d^2\mathcal{R}}{dC^2} > 0, \qquad \text{concave upward} \qquad (8\text{-}28)$$

We suppose one of these conditions is satisfied for all C in the range $0 \leq C \leq C_{\text{in}}$ (assume the component is consumed by the reaction). Then residence time theory may be used to place limits on the yield of the reaction (Chauhan et al., 1972). These limits correspond to the theoretical extremes of molecular level mixing (usually called *micromixing*) which are possible with a given residence time distribution.

One extreme of micromixing is called *complete segregation* [Danckwerts (1958); Zwietering (1959)]. At this extreme, fluid flows in small, segregated packets. The packets have different residence times so that an arbitrary $f(t)$ is possible but there is no mixing between packets. The yield for this case is given by

$$C_{\text{out}} = \int_0^\infty C_{\text{batch}}(C_{\text{in}}, t) f(t)\, dt \qquad (8\text{-}29)$$

where C_{batch} is the concentration in a batch reactor at time t given an initial concentration of C_{in}. The fluid packets in a completely segregated system behave as small batch reactors, and Equation 8-29 is just the weighted average of the individual outlet concentration.

The opposite micromixing extreme is known as *maximum mixedness* [Zwietering (1959)]. At this extreme, there is the most possible mixing at the molecular level compatible with a given $f(t)$. If $f(t)$ is the exponential distribution, this most possible mixing corresponds to complete homogeneity at the molecular scale so that the reactor is a perfect mixer. For other residence time distributions, some compositional variations must exist between different points in the reactor. In a sense, the condition of maximum mixedness approaches that of complete segregation as $f(t)$ tends toward the delta distribution. With an exponential distribution, the maximum mixedness yield is found by solving the algebraic equation

$$C_{out} - \bar{t}\mathscr{R}(C_{out}) = C_{in} \tag{8-30}$$

which is just the steady-state material balance on a perfect mixer. With the delta function distribution,

$$C_{out} = C_{batch}(C_{in}, \bar{t}) \tag{8-31}$$

which is the same result as for complete segregation in a piston flow reactor. For distributions other than the two special cases, the outlet concentration is found by solving Zwietering's differential equation

$$\frac{dC}{d\lambda} = \mathscr{R}(C) + \frac{(C - C_{in})f(\lambda)}{W(\lambda)} \tag{8-32}$$

where λ is the residual lifetime and $C_{out} = C(\lambda = 0)$. The boundary condition associated with Equation 8-32 is that $C(\lambda)$ is bounded and positive for all λ. Equations 8-29 and 8-32 provide upper and lower limits for the yield of simple, isothermal, homogeneous reactions provided one of the inequalities in Equation 8-28 is satisfied. If $\mathscr{R}''(C) < 0$, Equation 8-29 predicts the highest possible conversion for a given $f(t)$ and Equation 8-32 gives the lowest possible conversion. If $\mathscr{R}''(C) > 0$, the converse is true with Equation 8-29 providing the lower limit on conversion and Equation 8-32 the upper limit.

As an example of simple kinetics, consider the nth order reaction

$$\mathscr{R}(C) = -kC^n \tag{8-33}$$

then

$$\mathscr{R}''(C) = -n(n-1)kC^{n-2} \tag{8-34}$$

which is negative for $n > 1$ and positive for $n < 1$. Thus reactions of order

DIRECT APPLICATIONS 285

higher than first have greater conversions in completely segregated systems than in maximum mixedness systems. Normally, the bounds on the yield provided by Equations 8-29 and 8-32 are quite close. For a second order reaction, the maximum possible difference in conversion is only 7 percent [Methot and Roy (1971)]. This maximum difference applies in the case of an exponential residence time distribution and is smaller for other $f(t)$. Novosad and Thyn (1966) have calculated the complete segregation and maximum mixedness yields for a variety of reaction orders and residence time distributions. The fractional conversion seldom differs by more than a few percent in absolute difference. Thus micromixing theory provides tight bounds on conversion, and scaling up with a constant residence time distribution will ensure very close to constant conversion regardless of possible changes in micromixing.

A partial exception to the above generalization applies when $C_{out}/C_{in} \to 0$. The absolute conversion difference between complete segregation and maximum mixedness remains small (indeed it approaches zero), but the relative difference becomes large. Thus if one is concerned about residual amounts of an unreacted component, it may be important to maintain a constant level of micromixing during scaleup as well as a constant residence time distribution. Fortunately, this is often possible.

D. Complex Reactions

The previous section used a somewhat unconventional definition of a simple reaction, and any reaction not following that definition is called *complex*. Consider the following reaction network

$$2A \xrightarrow{k_1} B \xrightarrow{k_2} C$$
$$A \xrightarrow{k_3} D \qquad (8\text{-}35)$$

where reactant A can either dimerize to form B or decompose into D. A plausible rate expression might be

$$\mathcal{R}(A) = -k_1 C_A^2 - k_3 C_A \qquad (8\text{-}36)$$

which has the form of Equation 8-27. Thus this reaction scheme is simple with respect to component A, and the conversion of A will be maximized in a completely segregated reactor.

The same reaction is complex with respect to components B, C, and D; and the theory of Chauhan, Bell, and Adler does not apply to the yields of these components. Thus there is no mathematical assurance that the micromixing extremes of complete segregation and maximum mixedness provide limits on the yield of these components. Confident scaleup of such a system thus requires that both the residence time distribution and the detailed state of micromixing be maintained.

IV. MEASUREMENT TECHNIQUES

A. Closed Systems

Measurement of the residence time distribution in a closed system is easily accomplished through inert tracer experiments. The theoretical basis for this statement rests on Equation 8-21. The system is disturbed with a known $C_{in}(\theta)$ and the outlet response, $C_{out}(\theta)$, is measured. Deconvolution of Equation 8-21 then gives $f(t)$. In principle, this deconvolution is possible with nearly any $C_{in}(\theta)$. In practice, the mathematics are much simpler for certain $C_{in}(\theta)$ and these simple C_{in} are normally used experimentally.

Suppose the system is disturbed by a negative step change with corresponds to turning off the supply of nonreactive tracer:

$$C_{in}(\theta) = 1, \quad \theta < 0$$
$$C_{in}(\theta) = 0, \quad \theta > 0 \tag{8-37}$$

Equation 8-37 describes a washout experiment, and the outlet response gives the washout function directly, $W(\theta) = C_{out}(\theta)$. Table 8-1 illustrates the data reduction technique for the case where there is a small, background concentration of tracer even as $t \to \infty$. The experimental range of concentrations was 0.412 wt% (at $t = 0$) to 0.004 wt% (as $t \to \infty$). These experimental results were scaled to give a dimensionless concentration with range 1.0 to 0 which directly defines the washout function, $W(t)$. The mean residence time ($\bar{t} = 1.55$ min in the example) was then found from Equation 8-10 with $n = 1$, and this allowed the dimensionless times τ to be calculated.

Suppose the inlet step change is positive:

$$C_{in}(\theta) = 0, \quad \theta < 0$$
$$C_{in}(\theta) = 1, \quad \theta > 0 \tag{8-38}$$

The cumulative distribution function, $F(\theta)$, is obtained as the outlet response.

The last, commonly used form of simple input disturbance is the Dirac delta function

$$C_{in}(\theta) = \delta(\theta), \quad \theta \geq 0 \tag{8-39}$$

This function can be approximated by the rapid injection of tracer particles at the inlet to the system. The outlet response must be integrated to determine the total amount of injected tracer. Define

$$I = \int_0^\infty C_{out}(\theta)\, d\theta \tag{8-40}$$

MEASUREMENT TECHNIQUES

TABLE 8-1 Sample Residence Time Measurement Using a Negative Step Change

Time, min	Measured Concentration, wt %	$W(t)$ Dimensionless Concentration (with Zero Suppression)	$\tau = t/\bar{t}$ Dimensionless Time, Based on $\bar{t} = 1.55$ min
0	0.412	1.000	0
0.25	0.411	0.998	0.16
0.50	0.412	1.000	0.32
0.75	0.397	0.963	0.48
1.00	0.298	0.721	0.64
1.25	0.221	0.532	0.81
1.50	0.164	0.392	0.97
1.75	0.123	0.292	1.13
2.00	0.090	0.211	1.29
2.25	0.068	0.157	1.45
2.50	0.051	0.115	1.61
2.75	0.038	0.093	1.77
3.00	0.031	0.066	1.94
3.25	0.024	0.049	2.10
3.50	0.018	0.034	2.26
3.75	0.013	0.022	2.42
4.00	0.013	0.022	2.58
4.5	0.009	0.012	2.90
5.0	0.006	0.005	3.23
6.0	0.005	0.002	3.87
7.0	0.005	0.002	4.52
8.0	0.004	0	5.16
10.0	0.004	0	6.45

Then $f(\theta) = C_{out}(\theta)/I$ so that the delta function input gives the differential distribution as the output.

The best form of tracer experiment depends on the intended purpose of the data. Accurate determination of $\bar{t}$ and higher moments is best done using step change inputs since this avoids the scaling of Equation 8-40 and allows integration of the experimental data with a more favorable weighting function, for example, t^{n-1} in Equation 8-10 rather than t^n as in Equation 8-7. Of the step change experiments, the washout determination is preferred when background amounts of the tracer are negligible since the baseline response is then known, $W(\infty) = 0$. The positive step change normally requires experimental determination of $F(\infty) = C_{out}(\infty) = 1.0$.

The chemical engineering literature has put considerable emphasis on the estimation of model parameters from experimental residence time data, and particularly from measured values of the dimensionless variance, σ_τ^2. As suggested above, step change experiments are best for calculating higher moments and thus are best for fitting models in this manner. However, even second moments are very sensitive to the tail of the distribution and accurate calculations are difficult. This difficulty is compounded by the fact that second moments are theoretically infinite in the absence of molecular diffusion. This fact follows from the "zero slip" boundary conditions of hydrodynamics [Nauman (1977a)], and means in practice that variances can become very large in liquid-phase, laminar flow systems. The best approach to parameter estimation is to fit the entire experimental response curve, $C_{out}(t)$, directly to a model using nonlinear least squares. When done in this way, an experimental $f(t)$ curve is usually preferred since the differential distribution tends to be more sensitive to the model parameters.

There are, of course, many possible sources of error in experimental residence time measurements. White (1962) reviews most of the errors possible in inert tracer experiments. Perhaps the most serious possibility is a change in fluid properties due to the addition or deletion of the tracer. The whole approach to measurement assumes the tagged fluid behaves identically to the untagged fluid, but even small changes in density or viscosity can give rise to major changes in flow patterns. For example, salinity differences of 0.1 part per thousand (out of 35 parts per thousand total) help drive major currents in the Pacific Ocean [Kerr (1981)]. The experimental approach is to perform step changes of both the positive and negative variety. When the responses agree, the tracer artifact is probably insignificant. The responses are apt to disagree in unagitated, low velocity systems.

Tracer experiments provide a fairly direct means for measuring flow patterns. They can be used in the diagnostic mode to detect problems such as bypassing and stagnancy [Levenspiel (1972); Woodrow (1978)]. If all streamlines are the same length, as in pipes and other closed conduits, they can be used to predict average velocities.

$$u = \frac{L}{t} \qquad (8\text{-}41)$$

This u combines the effects of eddy and molecular diffusivity and hydrodynamic velocity. In low-diffusivity, laminar flow systems, u represents the hydrodynamic velocity alone. A suitable mathematical transformation may then allow calculation of the geometric velocity profile, for example, $u(r)$ in a circular pipe. In the author's experience this approach has been used to measure the highly elongated velocity profiles sometimes encountered in tubular polymerizations [Nauman (1974)]. The calculation of $u(r)$ is unique under the assumption that $u(r)$ is a monotone decreasing function of r.

B. Open Systems

Figure 8-2 attempts to illustrate the differences between open and closed systems. The transfer lines are small and have high velocities in closed systems

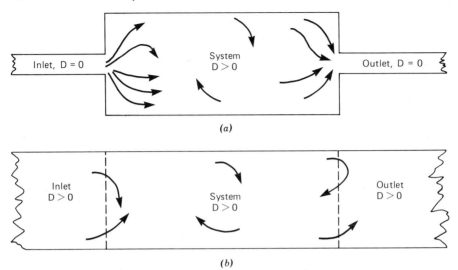

FIGURE 8-2 (a) A closed system. (b) An open system.

so that any backward flow of material in the transfer lines is negligible. Large transfer lines and fluid velocities similar to those in the system as a whole characterize open systems. In terms of the axial dispersion model discussed below, closed systems have the dispersion coefficient, $D = 0$ in the inlet and outlet transfer lines, while open systems have $D > 0$ in these lines.

Material that once enters a closed systems stays in it until it finally exits, never to return. The residence time distribution in such systems is easily determined by inert tracer experiments.

In open systems as in closed systems, particles have some initial entrance into the system and some final exit from it. In between these times, however, open systems allow temporary excursions outside the system boundaries while closed systems do not. Since residence stops outside the system boundaries, time spent outside does not contribute to the residence time distribution. However, it does still contribute to the transient response of inert tracers since nonreactive tracers have no way of knowing whether or not they are inside the (arbitrarily defined) system boundaries. Equation 8-21 still governs the transient response of an open system but the kernel, $f(t)$, is no longer the differential residence time distribution. This situation has been discussed extensively elsewhere [Nauman and Buffham (1983); Nauman (1981b, 1981c)]. The true residence time distribution can be deduced mathematically from a model of the open system or experimentally using reactive tracers. Such determinations are awkward at best, and it is fortunate that most chemical engineering systems are closed or nearly closed. Scaleup of open systems should be avoided if possible. When impossible to avoid, accurate modeling using some form of the convective diffusion equation is necessary.

V. ANALYTICAL DISTRIBUTION FUNCTIONS

This section is concerned with the residence time distribution functions which correspond to mathematical models of continuous flow systems. Such distribution functions are known analytically (as opposed to numerically). All have at least one parameter, the mean residence time, which is a scalable variable. Most contain at least one other parameter which may or may not be readily scalable.

A. Piston Flow and Ideal Mixing

We have already referred to the delta function distribution, Equation 8-17, which gives the residence time distribution in a piston flow reactor, and to the exponential distribution, Equation 8-19, which applies to the well-stirred (but not necessarily perfectly mixed) vessel. Most well-designed flow systems have residence time distributions intermediate between these two; and it is thus natural to approximate a real system as a series combination of piston flow and ideal mixing elements. The washout function for this model is given by

$$\mathcal{W}(\tau) = \exp\left[\frac{-(\tau - \tau_p)}{(1 - \tau_p)}\right], \quad \tau > \tau_p \tag{8-42}$$

where $\tau = t/\bar{t}$ and τ_p is a dimensionless parameter known as the fractional tubularity. This model is valid only for $0 \leq \tau_p \leq 1$. The model represents piston flow when $\tau_p = 1$ and ideal mixing when $\tau_p = 0$. The model may be fit to real systems by setting $\bar{t}$ to be experimentally determined value and by setting $\tau_p = 1 - \sigma_\tau$ where σ_τ is the dimensionless standard deviation, also experimentally determined. The parameter τ_p can also be found by regression analysis or just by observing that τ_p is a (dimensionless) first appearance time which corresponds to the fastest moving fluid in the system. The exact value for τ_p will naturally depend on the specific method used to find it. However, the various methods should give approximately the same value for τ_p or else the fractional tubularity model is a poor choice to fit the system.

Figure 8-3 illustrates $\mathcal{W}(\tau)$ for the fractional tubularity model. The real system must have two characteristics for this model to provide a relatively good fit to the experimental data:

> The system must have a fairly sharply defined first appearance time.
>
> The tail of the experimental response curve must decay rapidly, indeed exponentially.

Liquid-phase (low-diffusivity) laminar flow systems satisfy the first of these requirements but not the second. Gas fluidized beds satisfy both requirements, and Gilliland and Mason (1952) used the fractional tubularity model to fit their experimental data. The data of Table 8-1 show a reasonably good fit with

ANALYTICAL DISTRIBUTION FUNCTIONS

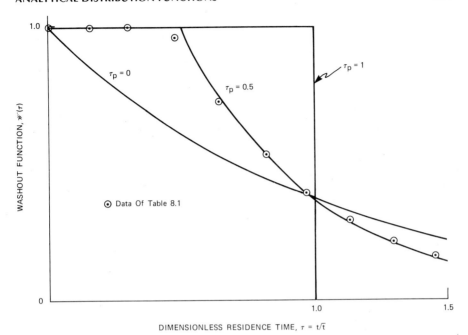

FIGURE 8-3 Washout function for fractional tubularity model.

the fractional tubularity model. This is illustrated in Figure 8-3 for the case of $\tau_p = 0.5$, although close scrutiny of the data shows that $\tau_p = 0.47$ would give an improved fit.

Once a model has been fit to experimental data on the pilot scale, it remains to estimate model parameters for the plant scale equipment. This is easy for $\bar{t}_2 = V_2/Q_2$, but there is no clear way of estimating τ_p even assuming the functional form of Equation 8-42 still applies at the plant scale. The difficulty is common to all models of the "black box" variety. Obtaining a good fit at the pilot scale and even obtaining dimensionless correlations at that scale (e.g., τ_p versus Re) will not assure a good scaleup. The difficulty becomes more pronounced as the number of parameters grows, and many of the complex residence time models that have been proposed in the literature should be rejected on the grounds that they are unscalable. At least, the person utilizing them should be prepared for relatively elaborate scaleup studies. This is the case for the fractional tubularity model when used in the "black box" mode.

It is often possible to avoid scaleup complications by using simple models with relatively scalable parameters. The fractional tubularity model will be one of these when the piston flow and ideal mixing elements in the model have physical counterparts in the real equipment. In other words, the real system must consist of a piston flow element in series with a stirred tank. Then

$$\tau_p = \frac{V_P}{V_P + V_S} \qquad (8\text{-}43)$$

where V_P and V_S represent the piston flow and stirred tank volumes, respectively. Now, τ_p is determined as a physical parameter. It remains, however, to ensure that the (almost) piston flow and (almost) ideal mixing elements retain these idealized residence time characteristics during scaleup.

B. Tanks in Series

The residence time distribution for J equal sized tanks in series is given by

$$\mathscr{W}(\tau) = e^{-J\tau} \sum_{i=0}^{J-1} \frac{J^i \tau^i}{i!} \tag{8-44}$$

or

$$f(\tau) = \frac{J^J \tau^{J-1} e^{-J\tau}}{(J-1)!} \tag{8-45}$$

These results assume that the tanks individually have the exponential distribution of Equation 8-19 with $\bar{t}_j = \bar{t}/J$. They also assume J is an integer. Figure 8-4 displays the washout function for various values of J. Two characteristics of the model are:

> There is no sharp time of first appearance; some molecules move through the system in zero time.
>
> The tail of the distribution is exponential.

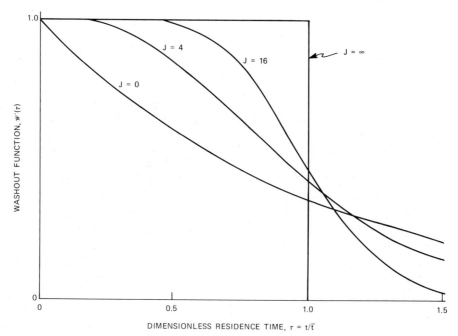

FIGURE 8-4 Washout function for stirred tanks in series.

ANALYTICAL DISTRIBUTION FUNCTIONS

Accurate fitting of experimental data to the tanks in series model requires noninteger values for J. Two extensions to Equations 8-44 and 8-45 are available for this purpose. The gamma function extension of Buffham and Gibilaro (1968) is purely mathematical in nature. It replaces the factoral in Equation 8-45 with the gamma function $\Gamma(J)$. Then the finite sum in Equation 8-44 becomes an incomplete gamma function. The fractional tank extension of Stokes and Nauman (1970) retains a physical interpretation for noninteger J by defining a fraction $0 < v < 1$ such that

$$J = j + v \qquad (8\text{-}46)$$

The model then consists of $j + 1$ tanks in series, j of which have the common value, V/J, and one of which has the smaller volume, vV/J. Equations for $\mathscr{W}(\tau)$ and $\mathscr{f}(\tau)$ are given elsewhere (Nauman and Buffham, 1983).

Where the physical system actually consists of J stirred tanks in series, it is easy to scale up the residence time distribution, the only requirement being that the individual tanks retain their exponential distributions. This literal interpretation is suitable for some continuous styrenic polymerizations where $J = 2$ or 3 or even for $J = 25$ to 40 as in some continuous emulsion processes. Scaleup from pilot plant to production unit would normally retain the same J although it is doubtful that this occurred in the emulsion case. When J is used as an adjustable parameter to fit arbitrary data, scaleup is less certain and requires a correlation of J with the appropriate dimensionless groups. An intermediate case occurs when several radial flow impellers (see Chapter 9) are installed in a single vessel. Residence time measurements on such vessels tend to fit the tanks in series model with J being the number of impellers. Retaining geometric similarity should give the same J although this is obviously less certain than when the individual "tanks" are physically separated.

C. Laminar Flow Models

In the absence of molecular (or eddy) diffusion, knowledge of the velocity profile within a piece of processing equipment allows the residence time distribution to be calculated. A general case of three-dimensional flow would require rather tortuous exercises in geometry and vector algebra before $W(t)$ is determined, but the method of approach is conceptually straightforward [Wein and Ulbrecht (1972); Nauman (1981a)].

The simple case of a one-dimensional, monotonic, velocity profile in a circular tube will serve to illustrate the residence time characteristics of laminar flow systems. By one dimensional, we mean that there is a single axial velocity component, $u(r)$, which is a function of r alone. By monotonic, we mean that $du(r)/dr$ does not change sign in the region $0 < r < R$. Given these assumptions, the washout function is given by

$$W(r) = \frac{\int_r^R u r \, dr}{\int_0^R u r \, dr} \qquad (8\text{-}47)$$

This gives W as a function of radial position r. To find it as a function of t, note that

$$t = \frac{L}{u} \tag{8-48}$$

implicitly gives r as a function of t. Substitution gives $W(t)$. For a Newtonian fluid with constant viscosity

$$u = 2\bar{u}\left(1 - \frac{r^2}{R^2}\right) \tag{8-49}$$

and

$$W(t) = \frac{(\bar{t})^2}{4t^2}, \quad t \geq \frac{\bar{t}}{2} \tag{8-50}$$

In common with all other diffusion-free flow systems, this residence time distribution has two characteristics:

There is a sharp first appearance time which corresponds to the fastest moving fluid.

The distribution has a very long tail, so long in fact that σ^2 and all higher moments are infinite.

Laminar flow, diffusion-free residence time distributions are not particularly desirable in terms of reaction uniformity; but they are scalable. One need only keep geometric similarity and an adequately low Reynolds number to obtain the same $W(t)$. If diffusion was negligible on the small scale, it will be doubly so upon scaleup. This scalability does not depend on the analytical form for $W(t)$ being predictable from first principles. As one example, the residence time distribution in a motionless mixer should be independent of scale provided geometric similarity and creeping flow (Re < 10) are maintained. One should be able to base design calculations either directly on the experimental data [Tung (1976); Nigam and Vasudeva (1980)] or on the semiempirical model of Nauman (1982). This model treats the Kenics Static Mixer as a segmented circular pipe. Within a segment, the residence time distribution has the same form as that for a parabolic velocity distribution. Between segments, complete radial mixing is assumed to occur within a region of vanishingly small length. The number, M, of such mixing points is the single parameter in the model other than $\bar{t}$. As a reasonable approximation, it takes four mixing elements of the Kenics variety to equal one point of complete radial mixing, for example, $M = 4$ for a 16-element Kenics mixer. Figure 8-5 illustrates this model for various M, the curve for $M = 0$ corresponding to an uninterrupted pipe, Equation 8-50.

ANALYTICAL DISTRIBUTION FUNCTIONS

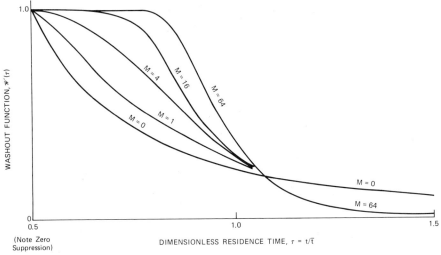

FIGURE 8-5 Washout functions for laminar flow in a tube with intermediate mixing points.

D. Axial Dispersion

Probably the best known and most useful model in chemical engineering is that for axial dispersion. The performance of many, more-or-less tubular flow systems can be approximated by the one-dimensional, convective diffusion equation:

$$\frac{\partial C}{\partial \theta} + \bar{u}\frac{\partial C}{\partial z} = D\frac{\partial^2 C}{\partial z^2} + \mathcal{R}(C) \tag{8-51}$$

where $C = C(\theta, z)$. In this formulation, $\bar{u}$ represents the average fluid velocity while D is a kind of effective diffusivity known as the axial dispersion coefficient. This coefficient lumps the effects of a nonuniform velocity profile, molecular diffusivity, and eddy diffusivity into a single parameter. The model provides a reasonably good fit for the transient and steady-state response of packed beds, turbulent flow in tubes of arbitrary cross section, and even laminar flow with molecular diffusion.

Equation 8-51 is usually expressed with independent variables in dimensionless form as

$$\frac{\partial C}{\partial \tau} + \frac{\partial C}{\partial x} = \frac{1}{\text{Pe}}\frac{\partial^2 C}{\partial x^2} + \bar{t}\mathcal{R} \tag{8-52}$$

where $\tau = \theta\bar{u}/L$ and $x = z/L$. The parameter $\text{Pe} = \bar{u}L/D$ is known as the axial Peclet number, and it is this dimensionless group that determines the residence time distribution in a region of length L:

$$\mathscr{W}(\tau) = e^{Pe/2} \sum_{i=1}^{\infty} \frac{8\omega_i \sin\omega_i \exp\left[-\left(\frac{Pe^2 + 4\omega_i^2}{4Pe}\right)\tau\right]}{Pe^2 + 4Pe + 4\omega_i^2} \tag{8-53}$$

where the ω_i are the positive roots of

$$\tan\omega_i = \frac{4\omega_i Pe}{4\omega_i^2 - Pe^2} \tag{8-54}$$

This series converges rapidly for small Pe but an asymptotic solution is necessary to avoid convergence problems for large Pe. For Pe > 16, the following result provides an excellent approximation

$$\mathscr{W}(\tau) = 1 - \int_0^\tau \frac{Pe}{4\pi\theta^3} \exp\left[\frac{-Pe(1-\theta)^2}{4\theta}\right] d\theta \tag{8-55}$$

Figure 8-6 illustrates the behavior of $\mathscr{W}(\tau)$ for various values of Pe.

Residence time distributions for the axial dispersion model have the following characteristics:

> There is no first appearance, the fastest particles move with infinite speed. The tail of the distribution is exponential and moments of all positive orders exist.

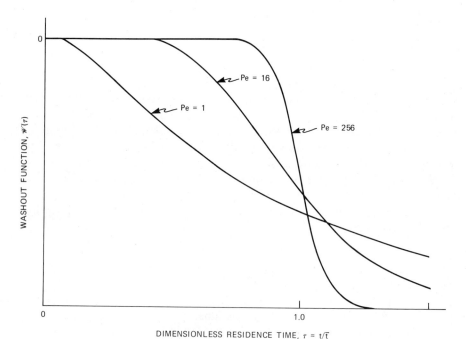

FIGURE 8-6 Washout function for the axial dispersion model.

ANALYTICAL DISTRIBUTION FUNCTIONS

Many real systems have these characteristics, and it is often possible to correlate Pe = $\bar{u}L/D$ with dimensionless groups such as Re which allows scaleup with relative certainty. Standard texts such as those by Levenspiel (1972), Himmelblau and Bischoff (1968), and Wen and Fan (1975) give correlations for packed beds and turbulent pipe flows, the Wen and Fan book also giving extensive references to the experimental literature. Figure 8-7 reproduces Levenspiel's (1958) correlation for $D/\bar{u}d_t$ in a circular tube of diameter d_t. For laminar flow, the Peclet number is given by Taylor (1953) and Aris (1956):

$$\text{Pe} = \frac{\bar{u}L}{D} = \frac{192\,\text{ReSc}(L/d_t)}{192 + \text{Re}^2\text{Sc}^2} \tag{8-56}$$

provided

$$\frac{L}{d_t} > 0.04\frac{\bar{u}d_t}{\mathscr{D}} = 0.2\sqrt{\frac{\bar{u}L}{\mathscr{D}}} \tag{8-57}$$

Similar results are available for non-Newtonian fluids and for tubes of noncircular cross section. See Wen and Fan (1975) for a summary.

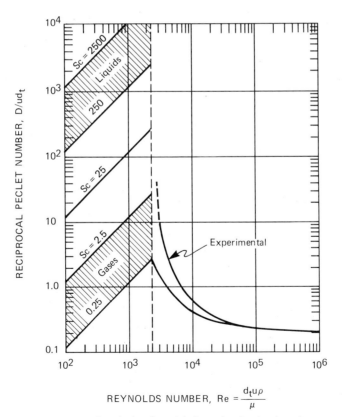

FIGURE 8-7 Correlation for axial dispersion in circular tubes.

VI. SCALEUP CONSIDERATIONS

There are three basic situations where scaleup of residence time distributions can be done with a high degree of confidence:

The pilot system is an open tube or packed bed with a residence time distribution which approximates piston flow. The desired scaleup maintains $\bar{t}$ and allows the distribution to approximate piston flow even more closely.

The pilot system is in laminar flow and molecular diffusivity is negligible. The desired scaleup maintains $\bar{t}$ and $\mathscr{W}(\tau)$.

The pilot system is a stirred tank with an exponential distribution of residence times. The desired scaleup maintains $\bar{t}$ and the exponential distribution.

These three situations encompass many, perhaps most, scaleup cases. The reason for this is that almost all processing equipment operates best at one of the mixing extremes, either piston flow or the well-stirred tank. Optima are rarely found at intermediate states of macromixing. Thus the pilot plant will have operated close to one of these mixing limits, and the desire for the production unit is that it operate at least as closely to the limit. Laminar flow systems rarely represent optima in the above sense, but the intractable nature of the fluid (typically a polymer) prevents any closer approach to the optimal residence time distribution. The pilot plant engineer has found a system that works. He would be satisfied to replicate this performance upon scaleup.

The following sections outline scaleup methods which are usually suitable in the above situations. We will always regard fluid properties as being independent of scale (e.g., the scaleup is done at constant Schmidt number.) Usually, we will also maintain geometric similarity.

A. The Almost Delta Distribution

When fluid properties are constant, the dimensionless residence time distribution in open tubes or packed beds is a function only of geometry and Reynolds number. The Reynolds number dependence becomes weak in open tubes with Re > 10,000 or in packed beds with Re ≥ 10 (N_{Re} based on packing diameters, d_p). For turbulent flows, scaleup with geometric similarity and constant $\bar{t}$ is slightly conservative with respect to narrowness of the residence time distribution. In this case, $\bar{u}$, L, and d_t (or d_p) all increase as $S^{1/3}$ while Re increases as $S^{2/3}$. The radial Peclet number, $\bar{u}d_t/D$ (or $\bar{u}d_p/D$), will slightly increase, for example, see Figure 8-7. Since the aspect ratio, L/d_t (or L/d_p), is constant, this means that the axial Peclet number, $\bar{u}L/D$, will slightly increase and that the residence time distribution will narrow at constant $\bar{t}$.

The disadvantage of scaling up with geometric similarity and constant $\bar{t}$ is that the pressure drop will increase when going from the pilot plant to the production unit. The magnitude of the increase is easily estimated based on the

Blasius correlation of friction factor with Reynolds number:

$$\text{Fa} = \frac{1}{2}\frac{\Delta P}{L}\frac{d_t}{\rho \bar{u}^2} = 0.0791 \text{Re}^{-1/4} \tag{8-58}$$

The case of constant L/d_t and $\bar{t}$ gives ΔP increasing as $S^{1/2}$. This increase may be reasonable in liquid-phase systems, particularly if the pilot scale experimentation is done at pressure drops which will remain reasonable when scaled up.

In gas-phase systems, Equation 8-3 should be interpreted in mass units rather than volumetrically since the gas density will be a function of pressure. Scaleup with L, d_t, and $\bar{u}$ all proportional to $S^{1/3}$ will not give constant $\bar{t}$ but instead $\bar{t}_2 > \bar{t}_1$. This is probably undesirable, and one would ordinarily want to operate the pilot plant and production units at the same absolute pressure drop. The conventional way of achieving this is to scale in parallel using multiple tubes, particularly since this also allows the heat transfer area to increase as S^1 rather than as $S^{2/3}$. For adiabatic systems, however, one has the option of holding a constant pressure drop by violating geometric similarity. Again using Equation 8-58 gives ΔP = constant when R increases as $S^{11/27}$ while L and $\bar{u}$ increase only as $S^{5/7}$. The Reynolds number will increase as $S^{16/27}$ but the aspect ratio, L/d_t, will decrease as $S^{-2/9}$. The net effect of these changes will probably be a decrease in Peclet number, but the decrease may be slight enough to be acceptable for scaleup. By operating the pilot plant in the transitional region, one could probably devise a scheme where the scaleup increase in Re exactly compensates for the decrease in L/d_t, but existing correlations of $\bar{u}d_t/D$ versus Re are not accurate enough to allow this with high precision.

Laminar flow systems can approach piston flow when the Reynolds number is low, when L/d_t is high, and when molecular diffusivity is high. Alternatively stated, $\bar{t}$ must be long enough so that there is time for diffusion to remove concentration gradients in the radial direction. Scaleup can be attempted based on geometric similarity and constancy of $\bar{t}$, but this scaleup is nonconservative with the production unit having a broader distribution of residence times than the pilot plant. This may be seen from Figure 8-7 since $\bar{u}d_t/D$ is a decreasing function of Re in the laminar region. The time available for diffusion is unchanged but the pathlength for diffusion increases as $S^{1/3}$. For sufficiently large S, the residence time distribution for laminar flow in an open tube will approach Equation 8-50 even though it was almost a delta distribution at $S = 1$. Molecular diffusion is a benefit which decreases with scale, and this fact can cause difficulty in the scaleup of laminar systems.

B. Laminar Flow Systems

Scaleup with geometric similarity and constant $\bar{t}$ will preserve the exact functional form for $\mathcal{W}(\tau)$ provided:

The Reynolds number is sufficiently low.
Molecular diffusion is negligible.

As for the case of turbulent flow, the Reynolds number will increase as $S^{2/3}$. However, the friction factor is now given exactly as

$$\text{Fa} = \frac{1}{2}\frac{\Delta P}{L}\frac{d_t}{\rho \bar{u}^2} = 16\text{Re}^{-1} \qquad (8\text{-}59)$$

which has the happy result that ΔP is independent of S.

The restriction on Re will be most critical at the production scale since Re increases with scale. To ensure constancy of $\mathscr{W}(\tau)$, the production unit must operate with Re below about 200 or else there is a risk that entrance effects or flow disturbance will be propagated downstream. Even when the flow is stable, the dimensionless entrance length increases with Re

$$\left(\frac{L}{d_t}\right)_{\text{ent}} \approx 0.04\text{Re} \qquad (8\text{-}60)$$

Since L/d_t is constant, this means that the entrance region will constitute a progressively larger fraction of the system length as the scale is increased. Constance of $\mathscr{W}(\tau)$ thus imposes a second restriction: $(L/d_t)_{\text{ent}}$ must be small compared to L/d_t.

The diffusivity restriction is most critical at the pilot scale. Merrill and Hamrin (1970) have shown that molecular diffusivity can be ignored provided

$$\frac{\mathscr{D}\bar{t}}{R^2} < 3 \times 10^{-3} \qquad (8\text{-}61)$$

This requirement effectively imposes a lower limit on the size of the pilot plant. Below that size, molecular diffusivity will give a narrower residence time distribution than will be possible in the production unit. This restriction rarely arises in practice since fluids with high enough $\mathscr{D}$ usually have such low viscosities that laminar flow is unlikely. For a typical solvent-in-polymer solution diffusivity of 10^{-6} cm^2/sec, Equation 8-61 indicates diffusion will be negligible in a 1-in. ID tube if $t < 1.3$ hr.

C. The Almost Exponential Distribution

Figure 8-8 depicts a recycle system. The net throughput for the system is Q, and the mean residence time remains V/Q. Within the system, however, flow rates are higher since an amount q of the fluid leaving the flow element is recycled. An interesting aspect of this system is its residence time distribution. As the ratio q/Q is increased, $W(t)$ tends to the exponential distribution, Equation 8-20, for any type of flow element. There are exceptions to this statement, but they tend to be pathological situations of more interest to the theoretician than to the practical engineer. If one wants to achieve the exponential distribution of an ideal mixer, simply increase q/Q and nature (or mathematics) will take care of the rest.

SCALEUP CONSIDERATIONS

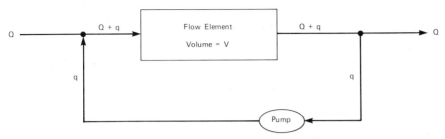

FIGURE 8-8 A recycle system.

To scale up the residence time distribution of a recycle system, geometric similarity and constant values for $\bar{t}$ and q/Q are maintained. This approach is not completely conservative since the Reynolds number in the flow element will increase by a factor of $S^{2/3}$ which will mean a slightly higher value for $\bar{u}L/D$. Thus the "per pass" residence time distribution will be somewhat closer to piston flow so the distribution for the recycle system will be somewhat farther from ideal mixing. However, Equation 8-20 is obtained in the limit of high q/Q, even when the per pass $f(t)$ is a perfect delta function. Thus if q/Q remains high, say $q/Q > 16$ [Nauman (1974)] the recycle system will remain well stirred upon scaleup. In this scaleup, Q and q will both increase as S^1 while the pressure drop across the flow element will increase approximately as $S^{1/2}$. Total power input to the system will thus increase as $S^{3/2}$ while power per unit volume will increase as $S^{1/2}$.

The most important industrial example of a recycle system is the stirred tank reactor where the impeller acts as the pump and the tank acts as the flow element. Detailed scaleup considerations for agitated vessels are given in Chapter 9. The present section outlines the treatment for the special case of scaling up residence time distribution. Pertinent literature studies include those by Berresford et al. (1970), Connolly and Winter (1969), Bowen (1969), and Gosling and Hubbard (1980).

Consider scaleup of a stirred tank reactor assuming geometric similarity, constant $\bar{t}$, and constant q/Q. For Newtonian fluids in the turbulent regime, pumping capacity, q, scales as ND_I^3 where N is the rotational speed and D_I is the impeller diameter, while Q scales with S and thus with D_I^3. This implies that maintaining a constant q/Q requires a constant rotational velocity for the impeller. Total power input scales as $N^3 D_I^5$ or, with geometric similarity and constant N, as $S^{5/3}$. Power per unit volume thus increases as $S^{2/3}$. Within the accuracy of the analysis, these scale factors are the same as those for the pipe flow model of a recycle loop which was developed previously.

Agitation experts are loath to increase power per unit volume as the vessel is increased in size. The usual approach is to maintain constant power per unit volume and sometimes even to decrease it. We shall see that constant power per unit volume is reasonable from the viewpoint of micromixing. It is also reasonable from the viewpoint of macromixing if q/Q remains sufficiently high in the production unit, say $q/Q > 16$. Most pilot plant units operate at considerably higher values of q/Q so that some reduction is possible upon scaleup. Indeed Gosling and Hubbard (1980) have observed that keeping ND_I constant (constant impeller tip speed) is sometimes appropriate for maintaining a constant $W(t)$. This scaling rule allows power per unit volume to decrease as $S^{-1/3}$.

A nearly exponential distribution can be maintained in the scaleup of a stirred tank reactor if q/Q does not drop below some critical value. This will determine the performance of the system with respect to such simple characteristics as surge damping and the yield of a first order reaction. For more complex circumstances, one wishes to keep both the same residence time distribution (constant macromixing) and the same degree of segregation (constant micromixing). This appears possible by maintaining constant power per unit volume. To see this, note that the degree of segregation can be modeled using a single parameter known as the segregation number [Nauman (1975)],

$$\text{Sg} = \frac{\eta^2}{\pi^2 \mathscr{D} \bar{t}} \qquad (8\text{-}62)$$

where η is some characteristic dimension to which fluid is dispersed by the agitator. The model envisions molecular level mixing by diffusion between spheres of radius η. For turbulent systems, η is postulated as being proportional to the Kolmogoroff scale of turbulence [see, for example, Hinze (1975)]:

$$\eta = \left(\frac{\mu^3}{\rho \varepsilon} \right)^{1/4} \qquad (8\text{-}63)$$

where ε is the (local) rate of energy dissipation per unit volume of fluid. Thus constant power per unit volume will, on average at least, give the same η and Sg.

The above analysis has been aimed at turbulent, Newtonian fluids. The general concept of scaling up with constant (or at least adequate) q/Q should also apply in non-Newtonian and/or laminar flow situations. However, the pumping capacity of the impeller will be a different function of the operating variables than it is in the turbulent, Newtonian case. This function must be known for confident scaleup. NonNewtonian fluids sometimes show unusual behavior of q. Even flow reversals are possible with elastic fluids and may account for some of the unusual scaleup results reported by Gosling and Hubbard (1980). Caution is indicated when pilot scale experiments show the Weissenberg effect (1947).

Micromixing is accomplished by molecular diffusion. In turbulent flow fields, the distance scale over which this diffusion must occur can be kept constant (though use of constant power per unit volume) and the time scale can also be held constant (by maintaining $\bar{t}$ = constant). Thus the extent of micromixing is scalable in turbulent flow fields. In laminar flow, however, the distance scale in Equation 8-62 can be identified as a striation thickness, and this striation thickness probably scales as the characteristic length, that is, as $S^{1/3}$. Since $\bar{t}$ is constant, one can expect micromixing to decrease as scale increases. Laminar flow systems will tend to become more segregated as the vessel size becomes larger. This is the same phenomenon found when scaling up tubular reactors in laminar flow. A truly constant $W(t)$ is achievable only when the vessel, whether tubular or a stirred tank, is completely segregated on the pilot scale.

VII. COMPLICATIONS AND EXTENSIONS

A. Intermediate Distributions

It is rarely desirable or practical to scale up a residence time distribution which is intermediate between the extremes of piston flow and ideal mixing. Laminar flow without diffusion is the major exception to this generalization. Other exceptions are stirred tank reactors in series, mixer/settlers, agitated vessels with multiple impellers, and similar cases of physical compartmentalization.

When attempting to scale up an intermediate distribution, it is often necessary to sacrifice geometric similarity. We illustrate the general problem by considering laminar flow in a tube with significant molecular diffusion. Suppose that the tube is sufficiently long so that the axial dispersion model holds. Then an exact scaleup of $W(t)$ requires that $\bar{t}$ and $\text{Pe} = \bar{u}L/D$ remain constant. This may be possible for a limited range of S. As a first example, suppose that the product ReSc is fairly large. This is a reasonable possibility for the laminar flow of liquids. Then Equation 8-56 can be approximated as

$$\text{Pe} \approx \frac{192(L/d_t)}{\text{ReSc}} \qquad (8\text{-}64)$$

The conditions of constant Pe and $\bar{t}$ force d_t also to be constant but $\bar{u}$ and L are free to vary as S^1. Thus the tube may be lengthened and the flow rate increased proportionately while maintaining a constant $\bar{t}$. This is a strange form of scaleup, but it does illustrate the theoretical possibility of scaling with constant $W(t)$.

As another example, suppose the product ReSe to be small compared to 192. This situation is possible, although not very practical, with gases. Then Equation 8-56 becomes

$$\text{Pe} \approx \text{ReSc}\left(\frac{L}{d_t}\right) \qquad (8\text{-}65)$$

Exact scaleup of $W(t)$ is now possible at constant L and $\bar{u}$ but with R increasing as $S^{1/2}$.

The foregoing examples should illustrate that scaleup of an intermediate residence time distribution is very demanding in terms of the accuracy of the system model; and even when the scaleup is possible, the allowable range of S may be rather small.

B. Unmixed Feed Streams

The application of residence time theory to chemical reactor design is primarily limited to the case of mixed feed streams [but see Ritchie and Tobgy (1978)]. Mixed feed streams are generally desirable from the viewpoint of reaction engineering, but it is occasionally necessary to combine the operations of reactant mixing and reaction into a single step. Scaleup of the combined operation is clearly more difficult than when the operations are separated, but it is sometimes possible to do so using the same approach as for scaling up systems with premixed feed. Residence time theory treats mixing between fluid elements which were initially identical but which have become different due to the difference in time they have been in the system. Typically, this type of mixing proceeds by exactly the same mechanism as that between components which were initially different in composition.

Consider the mixing and subsequent reaction of two components in turbulent, pipeline flow. Had these components been premixed, the conditions of geometric similarity and constant $\bar{t}$ should yield a conservative scaleup of the almost delta distribution. The case of mixing between initially separated components which do not react has been studied experimentally by Beek and Miller (1959) and theoretically by them and Brodkey (1967) among others. It turns out that geometric similarity and constant $\bar{t}$ is almost always a conservative approach to scaling up this form of mixing.

The analysis is rooted in the statistical theory of turbulence and begins with the idea that the rate of mixing is proportional to the local fluctuating velocity, u'_x. Specifically, Beek and Miller found that the extent of mixing depended on a dimensionless mixing time which can be given as

$$\theta = k_0 \frac{u'_x L}{\bar{u}} \tag{8-66}$$

where k_0 is the wave number of the largest turbulent eddy in the system. Beek and Miller took this as

$$k_0 = \frac{4}{d_t} \tag{8-67}$$

so that the size of the largest eddy is assumed to one-quarter of the pipe diameter. Thus,

$$\theta = 4 \frac{u'_x}{\bar{u}} \frac{L}{d_t} \tag{8-68}$$

so that a geometrically similar scaleup with constant $u'_x/\bar{u}$ should give a constant extent of mixing. The ratio $u'_x/\bar{u}$ is a weakly increasing function of Reynolds number so that scaleup with a constant Re should be adequate and scaleup with an increasing Re should be conservative. Residence time distributions scale in the same way since geometric similarity and constant Re will give the same $\mathscr{W}(\tau)$ but with $\bar{t}$ increased by a factor of S^2. Scaleup with constant $\bar{t}$ increases Re. The reactants will mix somewhat more quickly (quickness here being measured in terms of fractional length down the tube) and will react in an environment which is closer to piston flow. One would expect this combination to give superior performance in terms of reaction yield and selectivity. If it does not, a tubular reactor was probably the wrong design choice for the reaction.

C. Nonisothermal or Nonhomogeneous Reactions

Residence time theory is adequate for yield prediction only in the case of isothermal and homogeneous reactions. Nonisothermal reactors can be brought into a unified conceptual framework by replacing the residence time distribution with the thermal time distribution [Nauman (1977b)]; but actual applications of this idea have so far been limited to interpretation of results rather than to scaleup. Nonisothermal reactors are readily scaled only when adiabatic and tubular. Then, the methods advocated for scaling up residence time distributions should be satisfactory. Note, however, that keeping the reactor adiabatic may be difficult at the pilot scale. The best approach is to heat the reactor walls, using an axial temperature profile, $T(z)$, which corresponds to a theoretical adiabatic profile. Tubular reactors with radial temperature gradients are difficult to scale since the pathlength for diffusion of either heat or mass increases with scale but the time available for diffusion does not. A typical approach is to scale in parallel using multiple tubes.

Fluidized bed reactors and other nearly isothermal gas–solid contactors can be analyzed using contact time distributions which are the heterogeneous analog of residence time distributions. Measurement of these distributions is fairly easy using nonreactive tracers which are adsorbed on the surface of the solid particles. A number of such experiments have been reported [Nauman and Collinge (1968), Orth and Schuegerl (1972), Yates and Constans (1973)] but these measurements have not yet provided a basis for scaleup.

NOMENCLATURE

C Concentration of key species
d_t Diameter of tube
d_p Diameter of packing

D	Axial dispersion coefficient
D_I	Diameter of impeller
$\mathscr{D}$	Molecular diffusivity
f	Differential residence time distribution
f	Dimensionless differential distribution
F	Cummulative distribution function
Fa	Friction factor
i	Index of summation
I	Amount of injected tracer
j	Integral number of tanks
J	Number of tanks in series
k	Reaction rate constant
k_0	Wavenumber of largest eddy
L	Length of tubular reactor
M	Number of static mixing elements
n	Order of moments
N	Rotational velocity of impeller
Pe	Peclet number
Q	Flow rate through system
r	Radial coordinate
R	Rube Radius
$\mathscr{R}$	Reaction rate
Re	Reynolds number
S	Scale factor based on throughput
Sc	Schmidt number
Sg	Segregation number
t	Residence time
$\bar{t}$	Mean residence time
T	Temperature
u	Local velocity
$\bar{u}$	Average axial velocity
u'_x	Fluctuating velocity component
v	Volume of fractional tank
V	Volume of the system
W	Washout function
$\mathscr{W}$	Dimensionless washout function
x	Dimensionless axial coordinate
z	Axial coordinate

Greek

- α Amplitude of input signal
- β Frequency of input signal
- δ Dirac delta function
- ΔP Pressure drop
- ε Energy dissipation per unit volume
- λ Residual lifetime
- η Size of Kolmogoroff eddy
- θ Time
- μ Viscosity
- μ_n Moment about the origin
- ν_n Moment about the mean
- ρ Density of fluid
- τ Dimensionless residence time
- ω_i Eigenvalue

Subscripts

- 1   Refers to pilot scale system
- 2   Refers to product scale system
- p Refers to piston flow element
- s Refers to stirred tank element
- τ Denotes dimensionless moment

REFERENCES

Aris, R., *Proc. Roy. Soc.*, **A235**, 67 (1956).

Beek, J., Jr., and Miller, R. S., *Chem. Eng. Proc. Symp. Ser. No. 25*, **55**, 23 (1959).

Berresford, H. I., et al., *Trans. Inst. Chem. Eng.*, **48**, 21 (1970).

Bowen, R. L., *Chem. Eng. Prog.*, **65**(11), 14 (1969).

Brodkey, R. S., *The Phenomena of Fluid Motions*, Addison-Wesley, 1967.

Buffham, B. A., and Gibilaro, L. G., *AIChE J.*, **14**, 805 (1968).

Buffham, B. A., *Nature*, **274**, 879 (1978).

Chauhan, S. P., Bell, J. P., and Adler, R. J., *Chem. Eng. Sci.*, **27**, 585 (1972).

Connolly, J. R., and Winter, R. L., *Chem. Eng. Proc.*, **65**(b), 70 (1969).

Danckwerts, P. V., and Sellers, E. S., *The Industrial Chemist*, 395, September, 1951.

Danckwerts, P. V., *Chem. Eng. Sci*, **2**, 1 (1953).

Danckwerts, P. V., *Chem. Eng. Sci.*, **8**, 93 (1958).

Gilliland, E. R., and Mason, E. A., *Ind. Eng. Chem.*, **44**, 218 (1952).

Gosling, C. D., and Hubbard, D. W., "Residence Time Distribution and Scaleup in Non-Newtonian Mixing," Paper 66e, 73rd Annual Meeting AIChE, Chicago, November 1980.

Himmelblau, D. M., and Bischoff, K. B., *Process Analysis and Simulation*, Wiley, 1968.

Hinze, J. O., *Turbulence*, 2nd ed., McGraw-Hill, New York, 1975.

Kerr, R. A., *Science*, **213**, 632 (1981).

Levenspiel, O., *Ind. Eng. Chem.*, **50**, 343 (1958).

Levenspiel, O., *Chemical Reaction Engineering*, 2nd ed., Wiley, New York, 1972.

Merrill, L. S., Jr., and Hamrin, C. E., Jr., *AIChE J.*, **16**, 194 (1970).

Methot, J. C., and Roy, P. H., *Chem. Eng. Sci.*, **26**, 569 (1971).

Nauman, E. B., and Collinge, C. N., *Chem. Eng. Sci.*, **23**, 1317 (1968).

Nauman, E. B., *Chem. Eng. Sci.*, **24**, 1461 (1969).

Nauman, E. B., *J. Macromol. Sci.—Rev. Macromol. Chem.*, **C10**, 75 (1974).

Nauman, E. B., *Chem. Eng. Sci.*, **30**, 1136 (1975).

Nauman, E. B., *Chem. Eng. Sci.*, **32**, 287 (1977a).

Nauman, E. B., *Chem. Eng. Sci.*, **32**, 359 (1977b).

Nauman, E. B., *Chem. Eng. Commun.*, **8**, 53 (1981a).

Nauman, E. B., *Chem. Eng. Sci.*, **36**, 957 (1981b).

Nauman, E. B., *AIChE Symp. Ser. No. 202*, **77**, 87 (1981c).

Nauman, E. B., *Can. J. Chem. Eng.*, **60**, 136 (1982).

Nauman, E. B., and Buffham, B. A., *Mixing in Continuous Flow Systems*, Wiley, New York, 1983.

Nauman, E. B., *Chem. Eng. Sci.*, **39**, 173 (1984).

Nigam, K. D. P., and Vasudeva, K., *Can. J. Chem. Eng.*, **58**, 543 (1980).

Novosad, Z., and Thyn, J., *Collect. Czech. Chem. Commun.*, **31**, 710 (1966).

Orth, P., and Schuegerl, K., *Chem. Eng. Sci.*, **27**, 497 (1972).

Ritchie, B. W., and Tobgy, A. H., *Chem. Eng. Commun.*, **2**, 249 (1978).

Stokes, R. L., and Nauman, E. B., *Can. J. Chem. Eng.*, **48**, 723 (1970).

Taylor, G. I., *Proc. Roy. Soc*, **A219**, 186 (1953).

Tung, T. T., "Low Reynolds Number Entrance Flows: A Study of a Motionless Mixer," Ph.D. Thesis, Univ. of Mass., 1976.

Wein, O., and Ulbrecht, J., *Coll. Czech. Chem. Commun.*, **37**, 412 (1972).

Weissenberg, K., *Nature*, **159**, 310 (1947).

Wen, C. Y., and Fan, L. T., *Models for Flow Systems and Chemical Reactions*, Dekker, New York, 1975.

White, E. T., *Imp. Coll. Chem. Eng. Soc. J.*, **14**, 72 (1962).

Woodrow, P. T., 5th Int. Symp. of Chem. React. Eng., *ACS Symp. Ser. No. 65*, pp. 571–581, 1978.

Yates, J. G., and Costans, J. A. P., *Chem. Eng. Sci.*, **2B**, 1341 (1973).

Zwietering, T. N., *Chem. Eng. Sci.*, **11**, 1 (1959).

9

MIXING PROCESSES

J. Y. OLDSHUE

I.	Fluid Mixing with Impellers	310
	A. Principles	310
	B. Impeller Fluid Mechanics Principles	312
	C. Fluid Shear Rates	314
	1. Dynamic Similarity	315
	2. Shear Rate Relationships	319
II.	Scaleup Relationships	322
	A. Scaleup Parameters	322
	B. Maximum and Average Shear Rates	325
	C. Need for Pilot Plant Studies	327
	D. Scaleup Tools	328
	E. Power Measurements	331
	F. Viscosity	333
III.	Some Guidelines for Scaleup	334
	A. General Guidelines	334
	B. Example—Batch Process for a Chemical Specialty	334
	C. Example—Polymerization	336
	D. Example—Continuous Staged Gas–Liquid Slurry Process	339
	E. Example—Liquid–Liquid Emulsions	341
IV.	User–Contractor–Vendor Relationships	342
	Nomenclature	344
	References	345

I. FLUID MIXING WITH IMPELLERS

Translation of mixing processes from pilot scale to full scale commercial operation is fascinating and challenging. Significant differences can exist both in various processes and the mixing parameters as the scale of operation is increased. Qualitative and quantitative concepts of how these differences can be evaluated in specific areas is essential to good scaleup. Inherent in the scaleup of mixing is a good definition, and most desirably, a quantitative measure of the process result expected from the mixing system. At a minimum a qualitative description which can be used to evaluate and discuss mixing parameters is required. A mutual understanding of the process objectives and both the quantitative and qualitative observations and measurements made and to be made is essential since there will be extensive discussion within the company carrying out the process between production, research, engineering, and purchasing personnel, and often outside the company with vendors.

Several thousand different mixing applications are designed each year and each one may have major or minor differences in process performance and desired results. Finding an exact scaleup correlation for each one of these several thousand processes from the existing literature is not possible. Reference can only be made to typical relationships on similar processes, and very careful study must be made of published papers and correlations to see how closely the process objective corresponds to the project at hand. There is no shortage of proposed scaleup correlations. However, there is a real potential for problems in applying a suggested scaleup correlation to a different mixing requirement, particularly one outside the limits of the proposed correlation.

A. Principles

A large tank is not identical to a small tank in most of the mixing and fluid mechanics parameters that may be involved in determining process results. To scale up the production rate from a small mixing vessel, either a large number of small tanks can be used or a few larger vessels can be used. Figure 9-1 schematically illustrates that a large number of small tanks identical to the original tank will keep all the mixing parameters the same. However, as we go to a larger tank, as shown in Figure 9-2, there are greater distances to traverse. Some scale ratios go up linearly; some go up with the square and some with the cube of linear dimension. A combination of all the fluid mechanics in Figure 9-2 is not the same as the large number of small tanks in Figure 9-1.

A general background to describe the ways in which a large tank differs from a small tank is needed. This requires an understanding of some of the basic principles of fluid mixing and their roles in determining process performance. Ultimately, a mixer for the large tank must be selected. This means we must specify the power, speed of the impeller, type of the impeller, diameter of

FLUID MIXING WITH IMPELLERS

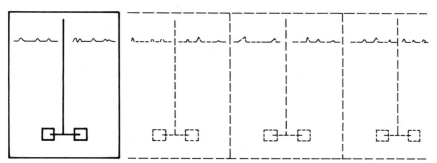

FIGURE 9-1 Schematic of pilot scale tank illustrating identical mixing parameters when using a large number of small tanks.

the impeller, and many other parameters which will affect the process. Normally more than one mixer will handle the process requirement; however, the mixers will differ in capital cost and operating cost relationship. The process performance may be expressed in very quantitative nonambiguous terms, such as a mass transfer rate in pound moles of component A per hour to be

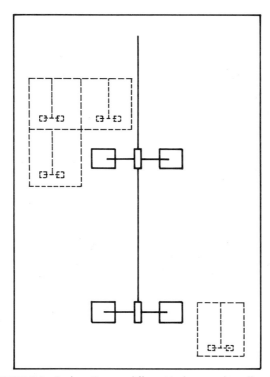

FIGURE 9-2 Scale ratios are different in large and small tanks.

transferred in the tank under specified operating conditions. On the other hand, the process may be affected by many different aspects of the fluid mixer, some of which may not be quantitatively definable. For example, making a successful polymer product may require meeting criteria of tensile strength, brittleness, dielectric constant, yield stress, and particle size distribution in the powder form. The role of mixing in its effect on all these various processes may not be completely understood. Moreover, the process description may be fairly general, with respect to uniformity of various components or the physical dispersion and suspension of some of the chemical components in the system. This makes a qualitative understanding of the differences between the pilot and the full scale plants and their possible effects much more important to understand.

B. Impeller Fluid Mechanics Principles

All of the power supplied to the mixer shaft goes into the batch and appears as heat at the rate of 1 J/s/W. The power appearing in the fluid stream can be expressed as the product of the pumping capacity and the velocity head, $u^2/2g$. The relationship between pumping capacity and power is shown below:

	Units	
	English	SI
Pumping capacity	lb/hr	kg/s
×		
Velocity head $\dfrac{u^2}{2g}$	ft	N m/kg or J/kg
=		
Power to fluid	ft lb/hr	N m/s or J/s

The height of a fluid of the same density, having the same potential energy as the velocity of this flowing stream, can therefore be directly calculated.

Since fluid in a mixing tank is not moving in a confined channel, there is considerable arbitrariness in determining the flow from the impeller, and in calculating the velocity head from the measured power. Therefore, the concepts are more useful in a relative form to predict how things change from one size to another or from one geometry to another. Rarely is there a need to calculate with high precision the actual pumping capacity and velocity head in the streams coming from the impeller.

The power to the fluid from an impeller is given by

$$P = QH \tag{9-1}$$

where Q is the pumping capacity of the impeller in kg/s, H is the velocity of

FLUID MIXING WITH IMPELLERS

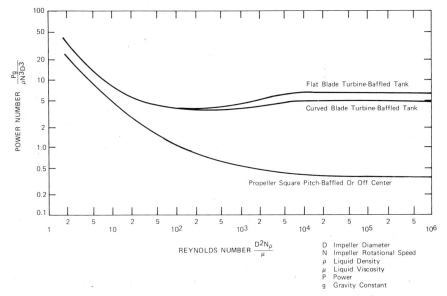

FIGURE 9-3 Power number – Reynolds numbers relationships.

fluid flowing in N m/kg, and P is the power appearing in the fluid stream in N m/s.

There are two other relationships that are helpful in a general discussion of fluid mechanics in the mixing tank. For medium and low viscosity fluids, flow varies with the first power of the impeller speed and the cube of the impeller diameter as shown in the following equation

$$Q = KND^3 \qquad (9\text{-}2)$$

where K is a constant for each particular impeller type.

Figure 9-3 shows the relationship between the power number and the Reynolds number. In the turbulent region the power number is constant. The power consumption of any impeller can be expressed as

$$P = k'N^3D^5 \qquad (9\text{-}3)$$

The exponent on N can be anywhere from the square to the 3.2 power, and the exponent on D can be anywhere from the cube to the 5.2 power. From Equations 9-1, 9-2, and 9-3, the flow-to-head ratio at constant power is proportional to the 8/3 exponent on the impeller diameter

$$\left(\frac{Q}{H}\right)_P \alpha (D)^{8/3} \qquad (9\text{-}4)$$

In the transition and viscous areas, various combinations of exponents are possible and for Q/H at constant power can vary between 7/3 and 9/3. The

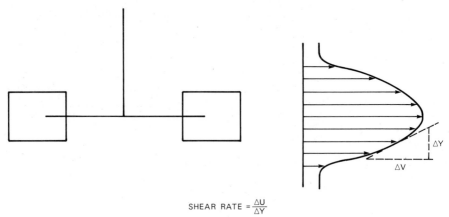

FIGURE 9-4 Typical velocity pattern coming from the blades of a radial flow turbine showing calculation of the shear rate, $\Delta V/\Delta Y$.

exponent 8/3 can be used as a reasonable approximation for a wide variety of fluid viscosities.

If large pumping capacities at relatively low velocity heads are desired then Equation 9-4 shows that large diameter impellers should be used at a given power level. This requires a low speed. Of course, the converse is required if high impeller heads are needed for the process result.

The velocity head is related to the fluid shear rate of the impeller in a general way in that as velocity head increases, fluid shear rate also increases. The functional form of the relationship is somewhat complicated, so it is better to think of Equation 9-4 as showing that the flow-to-fluid shear rate is proportional to the D/T to a positive exponent.

All of these equations are true for a given impeller type. However, they become much more complicated and approximate when trying to compare different impeller types such as radial and axial flow turbines or propellers.

C. Fluid Shear Rates

If the velocity profile coming off a radial flow turbine is measured, the velocity profile shown in Figure 9-4 is obtained. The slope at any point is a measure of fluid shear rate at that point. Fluid shear rate is a velocity gradient with units of reciprocal time. For example, a velocity of meters per second divided by distance in meters gives us a shear rate of reciprocal seconds. The shear rate from the impeller must be multiplied by the viscosity in the fluid to get the shear stress which will carry out the mixing process. For example, if the viscosity is changed from .001 to .005 Pa s, the shear stress from a given impeller will be five times greater in the more viscous fluid.

At any point in the fluid in the mixing vessel, there are velocity fluctuations. If the flow is fully turbulent, the velocity at a point in the vessel as a function of time is schematically represented in Figure 9-5. The average velocity can be

FLUID MIXING WITH IMPELLERS

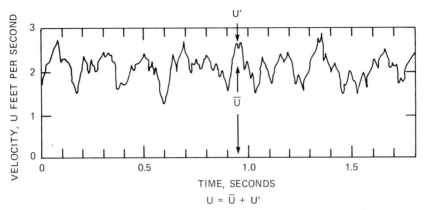

FIGURE 9-5 Velocity measurement at a point with time showing velocity fluctuation and average velocity.

used to calculate the shear rate between adjacent layers of fluid, or the velocity fluctuations can be used to obtain an estimate of the intensity and scale of the high frequency shear rates. The shear rate between the average velocities shown in Figure 9-4 is an important factor in macroscale processes involving large particles. By comparison, shear rates corresponding to the velocity fluctuations are involved with microscale processes. The velocity fluctuations are often expressed as in root-mean-square (RMS) values.

There are several different interpretations as to what determines whether macroscale or microscale mixing is controlling. Various estimates suggest that somewhere in the vicinity of 100–500 μm, the transition between macroscale and microscale mixing occurs. Probably 70 to 80 percent of industrial mixing applications are more sensitive to the pumping capacity of the impeller than to the various fluid shear rates. In these cases one needs not worry about shear rates or their distribution since this introduces unnecessary complications. However, some processes require an understanding of the fluid shear rate effects on the process. To evaluate macroscale shear rates it is necessary to consider at least four different points in the tank, as shown in Table 9-1. For microscale mixing, the RMS value at several different areas in the mixing tank must be determined.

1. Dynamic Similarity

To attain dynamic similitude, the four fluid forces in a mixing tank that are given in Table 9-2 must be considered. One is the inertia force put in by the mixer, F_I. The other three are opposing forces resisting mixing. These are the viscous force, F_v, gravitational force, F_g, and surface tension force, F_σ. The ratio of the inertia force put in by the mixer to the opposing force of:

Viscosity is the Reynolds number.
Gravity is the Froude number.
Surface tension is the Weber number.

TABLE 9-1 Definition of Shear Rates Around Impeller Zone and Root Mean Square Velocity Fluctuation

Average Point Velocity

Maximum impeller zone shear rate
Average impeller zone shear rate
Average tank zone shear rate
Minimum tank zone shear rate

RMS Velocity Fluctuations

$$\sqrt{(\bar{u}')^2}$$

TABLE 9-2 Dimensionless Groups Relating Fluid Forces: F_i, Inertia, F_v, Viscosity, F_g, Gravity, and F_σ, Surface Tension

Force Ratios

$$\frac{F_i}{F_v} = N_{Re} = \frac{ND^2\rho}{\mu}$$

$$\frac{F_i}{F_g} = N_{Fr} = \frac{N^2D}{g}$$

$$\frac{F_i}{F_\sigma} = N_{We} = \frac{N^2D^3\rho}{\sigma}$$

TABLE 9-3 Definition Hydraulic and Dynamic Similitude for Model (*M*) and Prototype (*P*)

Geometric

$$\frac{X_M}{X_P} = X_R$$

Dynamic

$$\frac{(F_I)_M}{(F_I)_P} = \frac{(F_V)_M}{(F_V)_P} = \frac{(F_G)_M}{(F_G)_P} = \frac{(F_\sigma)_M}{(F_\sigma)_P} = F_R$$

FLUID MIXING WITH IMPELLERS

TABLE 9-4 Relationship of Impeller Power Characteristics on Scaleup

Impeller Power Characteristics

$$P = f[D, N, \rho, \mu, g, T,]$$

$$\frac{\text{applied force}}{\text{fluid acceleration}} = f\left[\frac{\text{applied force}}{\text{resisting force}}\right]$$

$$\frac{Pg}{\rho N^3 D^5} = \left[\frac{ND^2\rho}{\mu}\right]^x \left[\frac{N^2 D}{g}\right]^y \left[\frac{D}{T}\right]^z$$

For dynamic similarity to be attained the inertia forces of both the model and prototype must be equal to the ratios of the viscous forces, the gravitational forces, and the surface tension forces as shown in Table 9-3. Using the same liquid in both pilot plant and full scale unit satisfies only two of these relationships. If the inertia force of the mixer is used as one of the relationships, then one can work with either the opposing force of viscosity, gravity, or surface tension.

In many civil and mechanical engineering problems, calculations of the fluid forces is the ultimate goal of the experimentation. Then one can use different fluids to model several of these dimensionless ratios. However, in a mixing system that luxury is not available and generally the same process fluid must be used in both the pilot plant and the commercial unit.

The use of dimensionless groups can give excellent correlation in certain cases. For example, the power drawn by a mixer is a function of the variables shown in Table 9-4. Then dimensional analysis would suggest that the inertia force divided by the fluid acceleration, (the Power number), should be a function of the ratio of inertia force to viscous forces, which is the Reynolds number. When the data are plotted, shown as smoothed curves on Figure 9-3, an excellent correlation is obtained for many mixing impellers, tank sizes, and fluid properties. One can also through dimensional analysis relate the heat

TABLE 9-5 Application of Hydraulic Similarity to Heat Transfer

$$h = f[N, D, \rho, \mu, C_P, k, d]$$

$$\frac{\text{result}}{\text{system conductivity}} = f\left[\frac{\text{applied force}}{\text{resisting force}}\right]$$

$$\frac{hD}{k} = \left[\frac{ND^2\rho}{\mu}\right]^x \left[\frac{C_P\mu}{k}\right]^y \left[\frac{D}{d}\right]^z$$

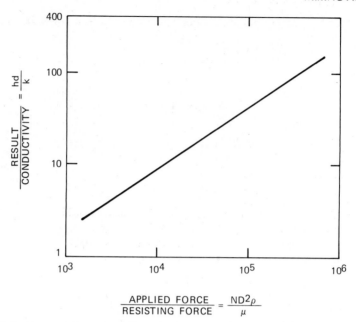

FIGURE 9-6 Heat transfer process dimensionless group, hd/k, as a function of the Reynolds number.

transfer coefficient to the thermal conductivity of the fluid as shown in Table 9-5. A length term is required, however, to complete the dimensionless number and then a good correlation can be obtained with the Reynolds number as shown in Figure 9-6. A similar analysis can be done for blend time, as shown in Table 9-6. A dimensionless blend number, θN, can be shown to be a function of Reynolds number as in Figure 9-7.

Unfortunately, there is seldom a good way to express the process result obtained in a mixing process in the form of a dimensionless group. Thus,

TABLE 9-6 Application of Hydraulic Similarity to Blending

$$\theta = f[N, D, \rho, \mu, T]$$

$$\frac{\text{result}}{\text{system conductivity}} = f\left[\frac{\text{applied force}}{\text{resisting force}}\right]$$

$$\theta N \alpha \left[\frac{ND^2\rho}{\mu}\right]^x \left[\frac{D}{T}\right]^z$$

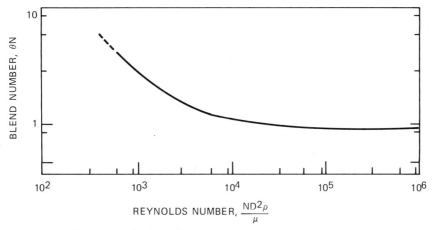

FIGURE 9-7 Blend number θN, as a function of Reynolds number.

dynamic similarity and dimensional analysis do not work except for the simplest mixing process.

The procedure used is to first get some data on a small scale. This study should be extensive enough to give some idea of the controlling factor or factors in the process so a particular relationship can be ratioed to the full scale performance expected. Shear rate is often found to be a controlling factor.

2. Shear Rate Relationships

Typical velocity fluctuation measurements are shown in Figure 9-8. Here, the mean velocities at each point are calculated and plotted as a function of the probe position above and below the impeller centerline for a radial flow, flat blade turbine, resulting in the curves shown in Figure 9-8. The slope at any point is the shear rate, and both the average and maximum shear rates in the impeller zone can be calculated. If data are taken at different impeller speeds, the maximum and average shear rates can be related to impeller speed as shown in Figure 9-9.

A series of studies on different sizes of impellers, summarized in Figure 9-10, showed that, in the impeller zone, average shear rate does not change with impeller diameter at the same RPM, but the maximum shear rate does. This leads, then, to the concept in Figure 9-11 that the maximum shear rate will tend to increase on scaleup to large equipment while the average shear rate tends to decrease. Therefore, there will be a greater variation of shear rates in large tanks than in small tanks. This difference often accounts for differences in the overall performance.

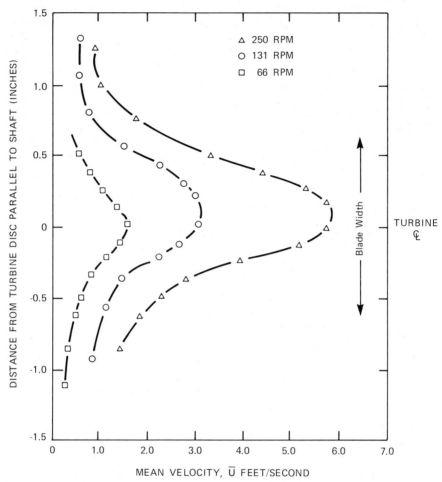

FIGURE 9-8 Experimental data points showing mean velocity at various positions above and below the impeller centerline.

A plot of the root-mean-square velocity fluctuations (RMS) as a function of position above and below the impeller centerline can be made as shown in Figure 9-12. The ratio of the RMS velocity value to the average velocity at a point shown in Figure 9-13 indicates that the turbulent intensity is about 50 percent of the mean velocity in the impeller zone. This is a very high level of velocity fluctuations since pipeline fluctuations are on the order of 5 to 15 percent. Data taken in other parts of the tank as shown in Figure 9-14 indicate that the velocity fluctuations in the rest of the tank are indeed on the order of 5 to 15 percent. Figure 9-11 is of importance for macroscale shearing effects which impact particularly on particle sizes and fluid clumps while Figure 9-13

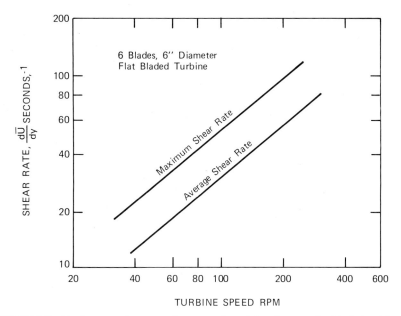

FIGURE 9-9 Maximum and average shear rates for a 6-in. diameter flat bladed turbine.

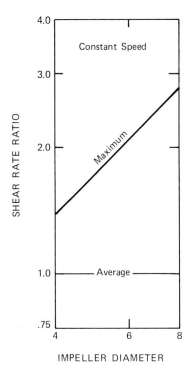

FIGURE 9-10 Shear rate dependence on impeller diameter.

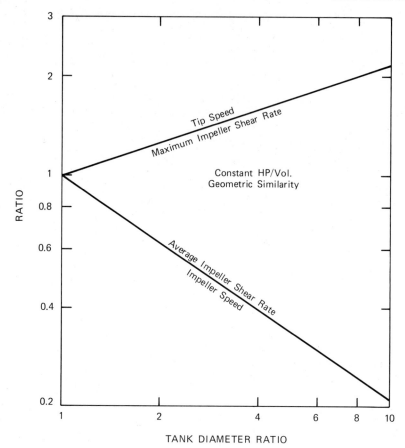

FIGURE 9-11 Scaleup of shear rates with impeller diameter.

describes the microscale shearing effects which are mainly a function of power dissipation/unit volume.

II. SCALEUP RELATIONSHIPS

A. Scaleup Parameters

In Table 9-7 the scaleup factors are shown for a change from a pilot scale vessel of 20 L capacity to a full scale vessel of 2500 L, a five-fold change in linear dimension. These calculated values assume vessels with a constant D/T ratio, a constant Z/T ratio, and the same liquid in both tanks. On the left side of Table 9-7 are many of the parameters involved in mixing processes: power, power/volume, speed, diameter, flow (Q), flow/volume, tip speed (ND), and Reynolds number. All of the parameters have been assigned a value of 1 in

SCALEUP RELATIONSHIPS

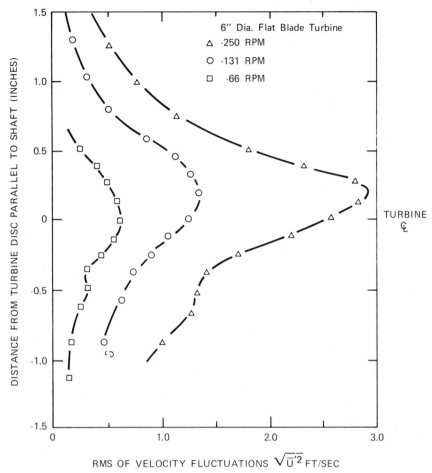

FIGURE 9-12 Velocity fluctuations and centerline depth.

column 2 of Table 9-7. In columns 3, 4, 5, and 6 the parameters have been changed, depending on what scaleup procedure is used:

Constant P/volume.
Constant Q/volume.
Constant tip speed.
Constant Reynolds number.

The general observation is that as the ratios of these different parameters change, there is no way to keep all the factors constant, regardless of what scaleup parameter is held constant.

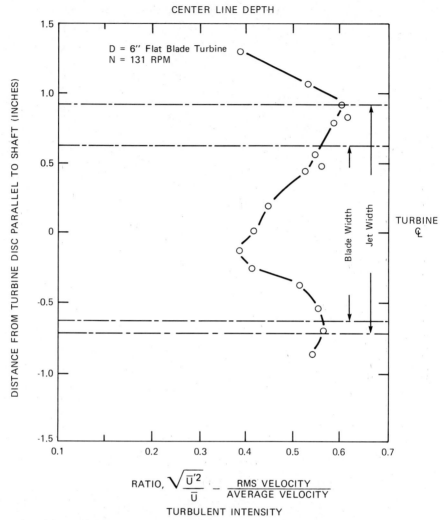

FIGURE 9-13 Experimental data for the ratio of velocity fluctuation to average velocity as a function of position around impeller discharge zone.

A more basic question must be asked—is it possible that for every conceivable mixing process there is a constant scaleup parameter that can be used for design? The more practical concept is that these parameters, P/volume, Q/volume, tip speed, and Reynolds number, are correlating parameters. Then it is necessary to find out how they change on moving from a smaller to a larger tank to scale up successfully. There might be a constant scaleup parameter for a given mixing process but even so one might not find it. For example, constant impeller tip speed could be a constant for scaleup to larger

SCALEUP RELATIONSHIPS

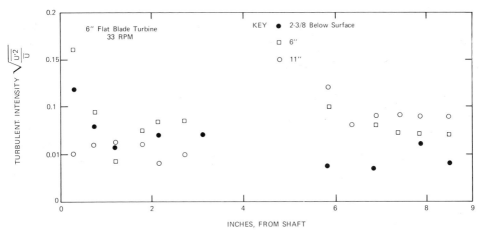

FIGURE 9-14 Ratio of velocity fluctuation to average velocity at different locations in mixing tank, excluding the immediate impeller zone.

TABLE 9-7 Properties of a Fluid Mixer on Scaleup

Property	Pilot Scale, 20 L	Plant Scale 2500 L Scaleup Procedure			
		P/Volume	Q/Volume	Tip Speed	ND^2l/μ
P	1.0	125	3125	25	0.2
P/volume	1.0	1.0	25	0.2	0.0016
N	1.0	0.34	1.0	0.2	0.04
D	1.0	5.0	5.0	5.0	5.0
Q	1.0	42.5	125	25	5.0
Q/volume	1.0	0.34	1.0	0.2	0.04
ND	1.0	1.7	5.0	1.0	0.2
$ND^2\rho/\mu$	1.0	8.5	25.0	5.0	1.0

sizes. This would then determine all other variables. In general, a practical approach to scaleup is usually between constant P/volume as a conservative estimate and constant tip speed as a very unconservative estimate.

B. Maximum and Average Shear Rates

Radial velocities for a small radial flow pilot turbine are shown in Figure 9-15. The average shear rate is that occurring across one-half the impeller blade

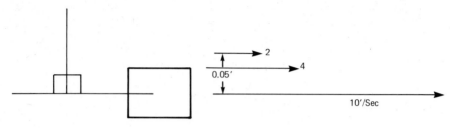

Average Shear Rate = $\frac{10 - 2}{.05}$ = 160 Sec^{-1}

Maximum Shear Rate = $\frac{4 - 2}{.00625}$ = 320 Sec^{-1}

FIGURE 9-15 Typical radial velocity produced by radial flow impeller of a pilot scale size.

width, and in this case it is 160 s^{-1}. The maximum shear rate occurs across a very small experimentally observed boundary layer 0.16 mm thick and is 320 s^{-1}.

If the impeller diameter is doubled as in Figure 9-16, twice the radial velocities are attained at the same RPM. However, since the blade width is also doubled, that leads to the same average shear rate for the larger diameter impeller. Since the maximum shear rate occurs across the same boundary layer, which has remained constant, the maximum shear rate is 640 s^{-1}. Maximum shear rates go up with impeller diameter while average shear rates do not. This is caused primarily by the fact that some dimension ratios scale up in proportion while others do not.

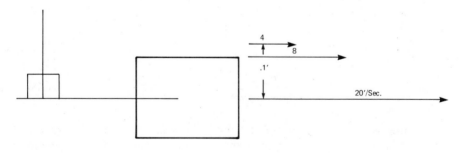

Average Shear Rate = $\frac{20 - 4}{.1}$ = 160 Sec^{-1}

Maximum Shear Rate = $\frac{8 - 4}{.00625}$ = 640 Sec^{-1}

FIGURE 9-16 Typical radial velocity pattern produced by an impeller twice the diameter of the impeller in Figure 9-15.

SCALEUP RELATIONSHIPS

TABLE 9-8   Methods Used in Designing Mixers

Design Status	Pilot Sizes Required
Complete correlation	0
Known scaleup effect	1
No previous data	2

C. Need for Pilot Plant Studies

The three situations that commonly exist in a mixing application are summarized in Table 9-8. In 95 percent of the applications there is sufficient data or experience so that the equipment supplier or equipment user can design the equipment for the required process result.

For the other 5 percent of the applications, some test work will be needed. About 80 to 90 percent of this remaining 5 percent are processes where the scaleup relationships are known or can be determined by testing in one tank size. The test work or pilot plant work done in one tank size must be directed at determining and confirming the major controlling steps in the process. An absolute value of performance must be determined on a small scale, so that scaleup can be related to this measured performance.

In the event that pilot plant studies on only one scale do not yield the controlling factor, or in the event that these studies show that available techniques may not be reliable, then the next step is to establish data in a second tank size. However, the experimental program need not be extensive; usually it will be sufficient to determine an absolute value of the performance result at one or two mixing conditions. Most of the detailed experimental work will have been done in the first series of tests.

Generally, it is sufficient for the second tank diameter to be on the order of $1\frac{1}{2}$ to 2 times larger than the first tank. However, the size ratio needed to establish the scaleup slope needs to be looked at for each application since some processes may be somewhat unusual.

The history of mixing process research and application is a continual one of establishing process mechanisms and scaleup relationships in two or three different tank sizes. Then, as application data are obtained in commercial units, these applications move into the category where complete data are available for sizing. There is a continual migration of process know-how from the need to study two tank sizes, to the need for only one tank size, to the use of proven selection methods.

However, in most applications involving gas–liquid, liquid–solid, or liquid–liquid mass transfer, some pilot plant or plant scale data are always necessary. These establish the absolute level of performance on some scale for the particular chemical system being utilized. Usually there are enough differences in chemical composition, trace elements, and unknown ingredients

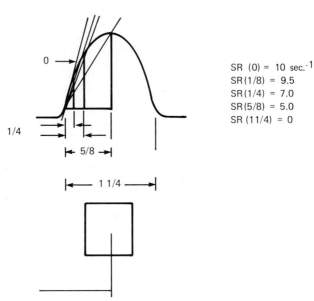

FIGURE 9-17 Schematic drawing showing shear rate as a function of distance intervals across impeller.

and their effects on surface or interfacial tension that an experimental run or data from a plant scale run is necessary to establish absolute performance details.

D. Scaleup Tools

There is a minimum size pilot plant for mixing studies. In heterogeneous processes the impeller blade cannot be smaller in physical dimension than the size of the largest particle, bubble, droplet, or fluid clump in which we are interested. For the reason behind this, look at a typical flow diagram coming from a flat blade turbine as shown in Figure 9-17. At the boundary of the jet, assume the shear rate has a maximum value of 10 s^{-1}. The blade width for this impeller is 1 cm and the width of the jet is approximately $1\frac{1}{4}$ cm.

Across the $\frac{1}{8}$ cm, the shear rate is 9.5 s^{-1}, across $\frac{1}{4}$ cm the shear rate is 7 s^{-1}; and across one-half of the blade width ($\frac{5}{8}$ cm) the shear rate is 5 s^{-1}. Across the entire jet width of $1\frac{1}{4}$ cm the shear rate is zero, since the velocity is the same on both sides of the impeller. This means that a particle $1\frac{1}{4}$ cm in diameter will experience essentially zero shear rate while a particle of micrometer dimensions will see a shear rate of 0 to about 10 as it circulates through the tank.

SCALEUP RELATIONSHIPS

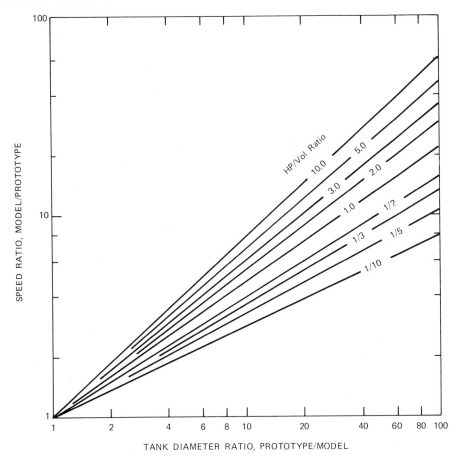

FIGURE 9-18 Relative speed in pilot plant and prototype as function of tank diameter of prototype, relative to pilot tank.

In the plant the impeller is bound to be much bigger than the particles. If we want pilot plant results to be proportional to full scale results, we cannot scale the impeller blade, and therefore the tank, down too far.

Power/volume as a constant is not a universal criterion for scaleup. But it is desirable to have some idea of the power levels in mind when carrying out experiments since you will want to relate the power level from small-scale tanks to large-scale tanks. In Figure 9-18 are shown the relationship between various speed ratios in the model and prototype for a number of scale ratios. For example, if the tank to diameter ratio, T, is 10:1, then the speed ratio between the model and prototype for constant power per unit volume must be five times greater in the pilot plant than it is going to be in the full scale tank.

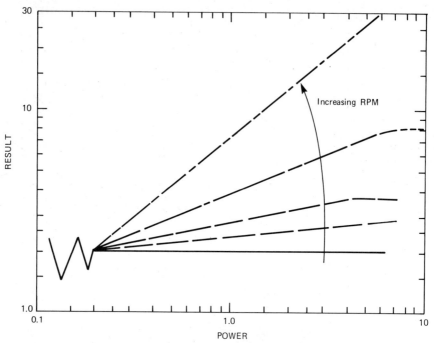

FIGURE 9-19 Possible experimental results obtained from experiments where power is varied by varying speed of a constant diameter impeller.

In any pilot plant program, the first thing to vary is the impeller speed. This changes everything; pumping capacity, blend time, as well as all the various shear rates. The results should be plotted as shown in Figure 9-19; if the results are quantitative, then a log–log plot is appropriate. High slopes indicate usually a gas–liquid mass transfer controlled process; slopes to zero indicate chemical reaction controlled processes, and intermediate slopes have a variety of interpretations.

If one wishes to determine the effect of the overall pumping capacity to fluid shear relationship, then varying the impeller diameter to tank diameter is suitable. At constant power, large impellers give more pumping capacity and lower fluid shear rates than do small impellers. This is one way to examine the effectiveness of these two quantities. If it is desired to evaluate the relationship between the macroscale and the microscale mixing processes, then it is normally necessary to vary blade width to blade diameter ratio in various steps. This is not readily obvious, but the relationships that lead to this conclusion are discussed in Oldshue (1970).

Processes involving gas, liquids, and solids are a special case; the influence of gas rate can be used to advantage in interpretation. If the gas rate is varied and a large change is observed in the process result as shown in Figure 9-20, this indicates that the gas–liquid mass transfer step is controlling. Gas rate has

SCALEUP RELATIONSHIPS

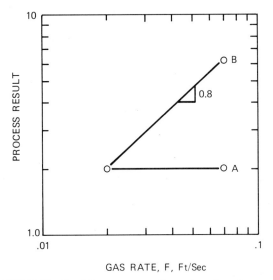

FIGURE 9-20 Possible process result as a function of gas rate.

been found experimentally to have a large effect on gas–liquid mass transfer rates, and a very small effect on liquid–solid mass transfer rates.

Whenever quantitative process results can be used as a criterion, plotting the results against the various mixing parameters can yield quantitative measures of mixer variables and their effect. On the other hand, if only qualitative results are available, they can also be plotted or listed in a table, but then the slopes will not have any quantitative meaning. When several criteria are important, each must be correlated against the mixing parameters. The final judgment then will be a compromise of the effect of various mixing parameters on each process result and some judgment as to their importance in the total evaluation of the plant performance.

The absorption of carbon dioxide into a caustic solution to make sodium carbonate can easily be quantified. The process result is described as moles of carbon dioxide absorbed per second per cubic meter. Quantitative measurements on both pilot and full scale tanks can then be made to evaluate the process result. However, the flavor and taste of a beverage as a function of mixer variables in an anaerobic fermentation step can only be described in qualitative terms. As different mixing parameters are investigated, a taste panel evaluates the particular result in comparison to the present product as "poor," "acceptable," and "excellent."

E. Power Measurements

Whether it be on a full scale tank or pilot scale equipment, the measurement of power is best done with a recording wattmeter. Clamp-on ammeters at best must be ratioed to the full load, name plate amperage, which varies with voltage, power factor, and motor type.

In general, it is better to use the wattmeter readings with motor curves prepared for the class of motor being tested and estimate motor output rather than using no-load amperage or watt readings. A motor curve presents power, amperes, power factor, and efficiency plotted against the percent of full load output. No-load readings can be very misleading as to the actual extent of losses in the motor efficiency and other mechanical elements in the train. In addition to motor curves, one should have an equation or a curve giving the losses to be expected in the speed reducer and some estimate of losses in the mechanical seal and stuffing box, if present. However, the inaccuracy in estimating these losses is normally far less in error than using a blanket no-load reading as a subtractor from the power reading.

For a fractional horsepower motor used in a pilot plant, it is almost impossible to use electrical readings to get accurate data, since no-load and full load amperage will be very similar. It is better to have an impeller calibrated on a dynamometer and to use the proper power number/Reynolds number curve shown in Figure 9-3, along with estimates of fluid properties to calculate

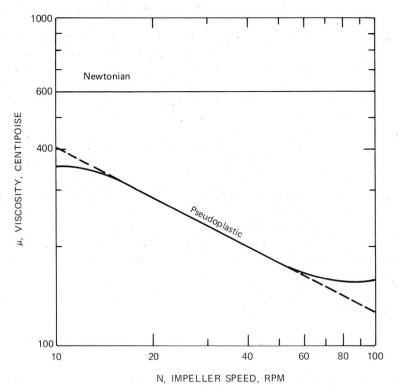

FIGURE 9-21 Viscosity/shear rate relationships.

F. Viscosity

To estimate the viscosity of complicated heterogeneous materials is difficult. The general principle is that we must measure the viscosity while the material is being mixed in a somewhat similar manner as in the commercial units and at similar shear rates to the plant.

For a non-Newtonian fluid, as shown in Figure 9-21, the curve of viscosity versus shear rate may have a complicated functional shape. The power law, Equation 9-5, is a very powerful and practical tool.

$$\text{shear stress} = k(\text{shear rate})^n \qquad (9\text{-}5)$$

However, it must be determined that the power law is accurate over the range of shear rates to be used in the mixing tanks. Viscosities obtained with a disk or cylindrical bob viscosimeter, where the material is not mixed in a manner similar to that in the plant, may not give the proper results. Also, if the disk is slipping by the interstices between the particles rather than developing the

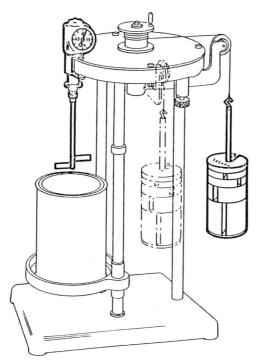

FIGURE 9-22 Typical mixing type viscosimeter, commonly called Stormer or Krebs viscosimeter.

response of an impeller to this two-phase mixture, different results will be obtained.

In Figure 9-22 is shown schematically a mixing type of viscosimeter which has been used in the paint industry and in many other areas to get reliable viscosity measurements. Oldshue and Sprague (1974) discuss the problems in using this equipment to obtain reliable data.

III. SOME GUIDELINES FOR SCALEUP

A. General Guidelines

While guidelines are best developed for specific application areas, as will be done in the examples, there are a few general ones that should be kept in mind:

For gas–liquid processes, the superficial gas velocity and power per unit volume are effective in correlating the gas–liquid mass transfer coefficient, $K_L a$. These relationships are relatively independent of scale. The mass transfer coefficient should be expressed as moles per second per cubic meter per Pascal second driving force.

When blending tanks are operated so one fluid is in motion by means of an impeller as the other material to be blended is injected into the tank, a blend time is obtained which is related to the total circulation time in the tank. It is usually found to be proportional to the impeller speed. With this type of blending process, to maintain pumping capacity from the impeller per unit tank volume constant on scaleup, the power per unit volume must increase proportionally to the square of the tank diameter. This is normally not practical. Therefore, the circulation time tends to increase with increases in tank diameter.

Free-settling solids can have a process description such as on-bottom, off-bottom suspension or complete uniformity. Constant power per unit volume is an approximate indicator of similarity of these definitions on scaleup to larger tank diameter.

When the slurries are very viscous, have high percent solids, on the order of 40 percent or more, and are pseudoplastic in their flow characteristic, it is usually found that power per unit volume will decrease with an increase in tank diameter.

Batch-versus-continuous operation is an issue of consequence. While a pilot plant may be run batchwise, the commercial plant may be batch or continuous. The following examples address both cases.

B. Example — Batch Process for a Chemical Specialty

There are many different kinds of chemical specialties manufactured in batch processes. These include cosmetic products, shampoo, face creams, floor polishes, viscosity-increasing gums, and all manner of unusual products. What

SOME GUIDELINES FOR SCALEUP

can be done in the pilot plant to predict performance in the plant with this tremendous variety of process material and process objectives?

The first step is to consider the possible mixing effects on the particular product under consideration. In general, concepts that will be of help are:

Large-scale tanks tend to swirl and produce a deeper vortex than do small-scale tanks. On some processes, it may be that a certain amount of swirl and liquid vortexing is essential to bringing in materials and producing the overall flow pattern. This may not require any baffles in the small-scale equipment. However, as we scale up to the large vessel, we often have to add baffles which may be narrower or fewer than the so-called standard baffles (standard baffles are four in number, $\frac{1}{12}$ the tank diameter in width) to obtain the proper combination of flow pattern on full scale.

Component addition time, blend time, and heat transfer time, will probably increase in the plant size unit. Realistic estimates must be made of this increase so production cycles may be estimated.

There will be higher maximum shear rates and lower average shear rates in the big tank. If the ultimate particle size in a solid or liquid phase is determined by the maximum shear rate, then this number must be carefully evaluated during the scaleup studies. In addition, the frequency with which product particles will be in contact with this high shear zone must be estimated.

The process result may be governed not by the magnitude of the shear rate, but by the product of shear rate multiplied by the period of time the shear rate is applied. This yields a concept of shear work, which is a dimensionless number and can be thought of as the meters of difference in movement between adjacent layers over a period of time. If this is true, then it is possible that applying one-half shear rate for double the time will give the same process result as obtained in the initial condition. On the other hand, if the absolute magnitude of the shear rate is important, it may be that one-half the shear rate will never give the process result no matter how much time is allowed.

By examining some of these differences between circulation time and shear rates in large and small tanks, it is usually possible to design suitable experiments in a pilot plant to indicate possible effects on the commercial scale, even though these indications may only be qualitative in nature. However, if the differences between the small- and large-scale tanks are ignored, then the small-scale tests can actually be very misleading and any time saved in making a "quick test" in the laboratory will be lost in time and money on full-scale "fix up."

In any process where shear rates are important, it really is the shear stress which governs the effect of shear rate on the process. Therefore, in working with a pseudoplastic material, the shear rate around the tank must be estimated; then it is necessary to multiply the shear rate(s) by the appropriate viscosity at that shear rate to obtain the shear stress distribution around the tank. It is the shear stress distribution which ultimately governs the overall process result.

In pilot planting a chemical specialty, proceed as follows:

Adjust whatever mixer variables are required in the way of speed, diameter, and other parameters to achieve the process result in terms of quantitative and qualitative process result specifications.

Estimate the importance of pumping capacity as measured by circulation time or blend time, and the importance of shear rate from the impeller as indicated by the effect of shear stress on particle sizes or product viscosity.

Use the relationships in a quantitative or qualititative way to predict performance in a larger tank. Remember that blend time and circulation time will increase on scaleup and that there will be a greater divergence of maximum and average shear rates around the impeller in the tank.

The critical nature of these changes on potential process result changes will usually indicate the need for further test work on a different scale, or possibly moving right ahead to full scale design and production.

C. Example — Polymerization

For emulsion and suspension polymerizations, mixer shear rates and the resulting shear stresses have an effect on the particle size distribution of the product polymer. In Figure 9-23 a typical distribution of shear rates at two different D/T ratios in the pilot plant is shown. These curves show that at steady state, there would be a similar distribution of shear rates. Since particles are circulating in and out of the various shear rate regions, the resultant polymer particle size distribution will be a function of where the particles are when conditions are reached that stabilize their particle size. Obviously, the precise particle size distribution of the polymer is a function of the steady-state shear rate distribution produced by the mixer, as well as a very complex phenomenon of polymerization time and process cycle.

Typically there are greater shear rate variations with small D/T ratios than with large D/T ratios as shown in Figures 9-23 and 9-24. Comparison of these figures shows that in a full scale tank there is a greater variety of shear rates at a given D/T ratio than in small tanks. At constant D/T ratio, therefore, a large-scale mixing tank will have a higher maximum impeller zone shear rate; however, the average impeller zone shear rate will decrease with the average shear rate in the large-scale tank also.

Particle size distributions in polymerization processes are also a function of the chemistry and processing techniques used. Many times as the process is scaled up from 400 L to 4000–20,000 L, changes in the shear rate distribution from the mixer can be compensated for quite adequately by changes either in

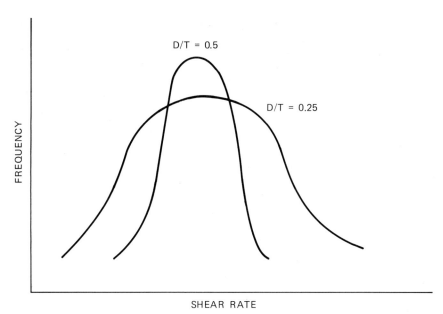

FIGURE 9-23 Pilot scale – shear rate as a function of impeller (D)/tank (T) diameter.

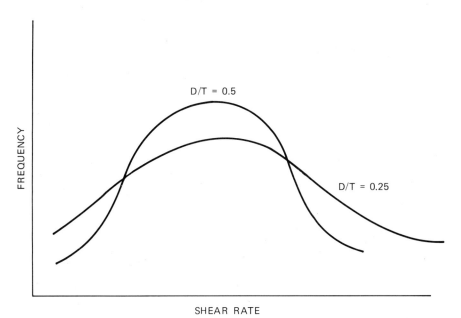

FIGURE 9-24 Plant scale – shear rate as a function of impeller (D)/tank (T) diameter.

the chemical formulations or the various chemical stabilizers used in the process.

However, for large-size units, say 40,000 to 80,000 L, or sometimes even larger, the changes in the resulting shear rate distribution cannot be compensated for adequately by chemical variables. Then it is necessary to consider just what is involved in changing the geometry of the impeller and the impeller-to-tank size ratio to modify the full-scale process particle size distribution.

An installed mixer in the plant usually has an installed horsepower, a fixed operating speed, and a fixed impeller diameter. At constant speed and horsepower, it is normally not practical to change the pumping capacity of the impeller and the maximum shear rate around the impeller more than 5 or 10 percent. Such a small change would seldom markedly affect a polymerization process. In order to decide whether marked change in D/T ratio will result in suitable improvement in the process results, it is usually necessary to demonstrate potential improvements in a pilot scale run.

It is instructive to examine what is involved in changing the D/T ratio on a large-size mixer. To increase the D/T ratio, the impeller speed must normally decrease. At lower operating speed the mixer horsepower rating of the drive will be lower, since most mixer drives are rated at essentially constant torque. Mixer drives have a horsepower rating approximately proportional to output speed, so the reduced speed required will be accompanied by a reduced power output rating for the equipment. If the lower horsepower is acceptable, then this may be an acceptable solution. However, even this change may require extensive modifications to the gearing specifications of the mixer and the actual drive. If the reduced power rating is not acceptable, then a larger torque capacity mixer drive would usually be considered.

If the desired change is to a smaller D/T ratio, then it is usually necessary to increase the speed of the impeller. There is always the cost of modification, which can be considerable, for changing the internal gearing, reducer, or putting in other speed increasers to the mixer drive chain. However, the higher operating speed brings the shaft closer to its natural critical frequency. Most often this will require a larger diameter shaft, a change that has many adverse cost ramifications for bearings, stuffing boxes, and other mechanical considerations.

An essential point in Figures 9-23 and 9-24 is the realization that if the shear rate distribution of a 0.25 D/T tank in the plant is to be duplicated, it would be necessary to run in a tank with a D/T of 0.1 or 0.15 in the pilot plant. This may or may not be practical in the pilot plant, depending upon the overall requirement for pumping capacity and fluid shear rates. However, to duplicate the shear rate distribution for a tank with a D/T ratio of 0.5 in the plant, either which exists now or is being proposed, we would need to run a tank with a D/T ratio in the pilot plant of around 0.2 or 0.3. This illustrates again that geometric similarity does not necessarily control individual mixing

SOME GUIDELINES FOR SCALEUP

relationships. In fact, it may not be possible to duplicate all of the process variables in any one run.

It may first be necessary to look at the effect of maximum shear rate, leaving the minimum shear rate to go where it will. Separately, one would look at average or minimum shear rates, letting the maximum shear rate go in a different direction. It is necessary before starting a pilot plant study to see what is involved in carrying out the process, and to establish if one will proceed on an overall or a stepwise basis to develop the design information.

In polymerization processes, it is usually possible by running a tank with a D/T of 0.15 and then with a D/T ratio of 0.25 to project the process result that will come about with a larger diameter, lower speed impeller in a full-scale plant. Ideally, this study should be made in the pilot plant as a part of scaling studies before a full scale plant size unit is built. These studies would indicate the type of geometry and mixer drive required for the full scale plant. A commercial plant size design is rarely based on a process duplicating D/T ratios in both the pilot plant and commercial unit.

D. Example — Continuous Staged Gas – Liquid Slurry Process

Assume there is continuous flow of a slurry containing 5 percent of a valuable metal to be oxidized and brought into solution in a number of mixing tanks. The equipment is shown schematically in Figure 9-25; the significant process variables are the volume of air, the mole fraction of oxygen in the entrance and exit streams, the mixer power level in the system, and the slurry flow rate.

In order to calculate the mass transfer driving force for scaleup, one must know the equilibrium partial pressure, P^*, of oxygen in solution. This is not usually available for gas–liquid slurry systems. However, a reasonable assumption can be made about the equilibrium P^* value at various reaction conditions in the pilot plant and a mass transfer coefficient $K_G a$ for the pilot plant can be calculated. This $K_G a$ can be scaled up to larger systems, using the same

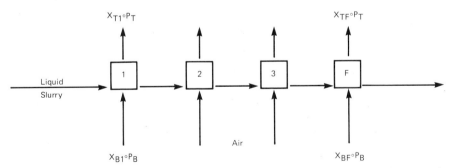

FIGURE 9-25 Four-stage continuous flow system, liquid, slurry passes from tank 1 through tank 4, with fresh air supply going into each tank. Symbols indicate Mol fraction and total pressure at bottom and top of each tank.

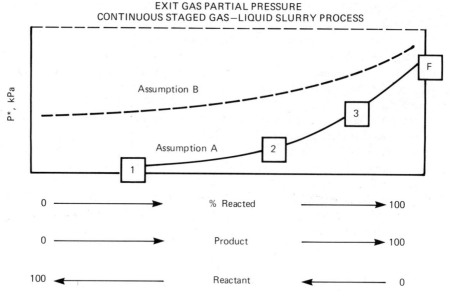

FIGURE 9-26 Partial pressure of oxygen in equilibrium with experimental dissolved oxygen levels in various tanks of the four tanks in series in Figure 9-24.

equilibrium P^* relationship. The procedure is reliable with some error, since the variation of the true equilibrium parital pressure P^* with percent reacted material is not known.

Consider the first tank in the train of slurry reactors. This is the tank where the reaction rate normally is the highest. A reasonable first assumption is that at that reaction level the P^* value in the exit of tank 1 is zero. This then allows the calculation of a $K_G a$ for tank 1 at the power level and superficial velocity of that tank. For tank 2, there may be a different power level and gas rate from those in tank 1. Typical correlations [Oldshue (1969)] for the effect of horsepower and gas rate on $K_G a$ can be used to estimate to $K_G a$ which would exist in tank 2. Then the P^* value can be calculated since the mass transfer rate and gas-phase composition are known. This gives a P^* value at the reaction level of tank 2 based on the assumption of P^* in tank 1. The calculation can be repeated for tanks 3 and 4, resulting in a curve shown as curve A in Figure 9-26.

The values of $K_G a$ are then plotted as a function of mixer power level and gas velocity to see whether the correlations as used for changing the $K_G a$ values for tanks 2, 3, and 4 are confirmed. If not, the coefficients are adjusted and the calculations repeated for tanks 2, 3, and 4. An alternative assumption, for example, is that P^* in tank 1 is approximately 30 percent of saturation. This allows the calculation of another set of results for tanks 2, 3, and 4, shown as curve B in Figure 9-26.

SOME GUIDELINES FOR SCALEUP   341

TABLE 9-9 Typical Full-Scale Mixer Selections for Oxidation
of Slurry Based on Two Different Assumptions for P^* Value

	Tank Number			
	1	2	3	4
Horsepower for assumption A	100	80	50	40
Horsepower for assumption B	80	20	50	35
Horsepower to use for design	100	80	50	40

Scaleup of $K_G a$ to a full size system depends on whether the flow regime is gas controlled or mixer controlled. One will also take into account the available relationships for this particular type of gas–liquid–slurry system. The full scale mixers required for cases comparable to curve A and curve B in Figure 9-26 can be developed and an estimate made of the possible variance in the required mixer process specification. As a rule, the mixers will differ within reasonable design and process factors. A suitable series of mixers for this entire full scale process can be quite quickly determined. Table 9-9 shows a typical example of full scale results.

E. Example — Liquid – Liquid Emulsions

An emulsion is being produced in a 1 m diameter tank. One of the key characteristics is the minimum particle size of the immiscible liquid phase. However, the heat transfer to jacket, the blend time, and cycle time of various ingredients added to the emulsion system as well as stabilizers for the final product are also important. It is desired to scale this up to a tank 2.3 m in diameter. Consider some of the decisions that must be made about the design of the full scale mixer.

The viscosity of the continuous phase at normal operating temperatures is 0.01 Pa s. In the full scale plant the temperature may be lowered and the viscosity would then be about 0.015 Pa s.

The maximum impeller zone shear rate is the usual variable governing the minimum particle size produced. If a constant D/T ratio is used for the first scaleup, maintaining the same maximum impeller zone shear rate would require constant impeller peripheral speed. This means that the operating speed of the mixer would be 43 percent of the RPM for the 1 m tank. This would result in a power per unit volume in the 2.3 m diameter tank of 43 percent of that in the 1 m tank as well as a pumping capacity per unit volume of 43 percent.

This establishes the mixing conditions required for the same minimum particle size if the chemistry is such that this particle size could be maintained at the resulting blend time. However, it will take much longer for the material to circulate through the maximum impeller zone shear rate, and also there would be marked increase in blend time for the larger unit.

For the 2.3-m tank, the impeller diameter can also be reduced to 50 percent of that in the 1 m tank. This will increase the operating speed and give a power per unit volume and pumping capacity per unit volume of 11 percent of the smaller tank. Alternatively, impeller diameter can be increased 50 percent which will give a pumping capacity per unit volume and power per unit volume of 98 percent of the smaller tank. A qualitative decision will have to be made as to which of the three mixers to use in the full-scale unit. None of them duplicate exactly all the parameters found in the 1 m tank.

If the viscosity increases to 0.015 Pa s for the geometrically similar mixer in a larger tank, then the operating speed must be decreased 50 percent to reduce the maximum impeller shear rate back to the original value. This reduces the power per unit volume to 13 percent and the flow per unit volume to 29 percent. Here a further complexity is added as to what direction to go with different geometries in the full-scale system. Seldom are all parameters equally represented in a full-scale unit and approximate engineering decisions must be made every step of the way.

Obviously, mixers are scaled up on processes for all kinds of operations. Sometimes the change in pumping capacity and fluid shear rates counterbalance each other and the process result is the same on full scale. Other times the process result may be better or may be somewhat worse. The cycles are similar to or different, but surprises can be prevented by considering in what direction mixer variables go on scaleup, making some allowance in either expectation or actually changing designs to modify these scaleup differences.

IV. USER – CONTRACTOR – VENDOR RELATIONSHIPS

Methods for specifying mixing equipment can be divided into three broad categories:

Issuance of a process specification.
Issuance of a mechanical specification.
Setting up a mutual development program.

The process specification may be based on a description of the process, using existing mixing equipment, or general process requirements. Alternatively, the process specification may summarize the process pilot plant studies to be used for developing the required commercial size equipment.

In many case there may be a contractor involved in the design, specification, and purchase of the equipment. Therefore, the engineering responsibility can either be joined or shared by the contractor and user or may be retained by

USER-CONTRACTOR-VENDOR RELATIONSHIPS

one or the other. Issuance of a process specification requires a good definition of the process requirement, and a good understanding between the user and equipment manufacturer of the definitions and interpretations of the process descriptions. This allows the equipment manufacturer to use his knowledge about mixing equipment in various kinds of processes. Then, the resulting installation will take advantage of knowledge of both the user and manufacturer. The manufacturer will usually extend a process guarantee based on mutual understanding and interpretation of the process requirements.

The user can get a wide variety of quotations in response to his specification which may differ considerably in power, speed, diameter, and type of impeller. He will have to satisfy himself as to the background knowledge behind the quotations, especially if some seem unusually attractive in terms of power or cost. However, a variety of quotations allows the user to take advantage of the latest technology in optimizing the mixer for his particular process based on current experience and data.

Where pilot plant data are available, a request for a quotation can be based upon the process result defined in terms of a particular pilot plant condition with stipulation that pilot plant data should be used in the scaleup to the full size unit. This allows the manufacturer to use his knowledge and experience in designing an optimum full size mixer based on mixing technology. The customer will, however, have to sort out from a variety of quotations which one is most suitable to his particular requirements.

The issuance of mechanical specifications means that the user and/or the contractor have determined from the process requirements what type of mixer is required and also that all the elements of the mixer have been described. Any responsibility for process results then normally lie with the user; the manufacturer guarantees only the mechanical design of the equipment. The mechanical design depends upon an accurate statement of the conditions under which the mixers are to operate. These must be carefully spelled out when the quotation request is issued.

Working from a mechanical quotation request does not allow the equipment manufacturer to use his knowledge and experience in selecting the optimum mixer for the process. When the details of the mechanical specifications are too minute, this may also limit unduly the options the equipment manufacturer would suggest. It may not allow him to use the latest technology and mechanical design to arrive at an impeller selection and shaft design based on the latest technology.

Generally mixer costs are on the order of 3 to 5 percent of the total cost of a chemical processing plant. Since mixing may be the key to attaining the desired process result, it is well to consider elaborate specifications and extensive consultations with the manufacturer during the course of a mixing program. Since it is not generally feasible to work with many different suppliers in great detail, the typical procedure is to prepare a base process and mechanical specification and after an evaluation of the responses, choose a mixer manufacturer to work with on a noncompetitive basis during the pilot plant development and design program. Contractual assurances can be developed so that the manufacturer will use and demonstrate current "Data Book" prices.

The user will generally have to agree that he will not use the process data developed during the program to obtain competitive equipment bids. Rather he will have to agree to purchase the equipment from the selected manufacturer if the backup technical data and discussion support the analysis and suggested design. The manufacturer must discuss the process design factors and any mechanical design factors used in the final equipment selection with the user.

Both sides have advantages in this arrangement since either can terminate at any point. The user gets access to the research and engineering data of the manufacturer and the equipment can therefore be closely tailored to the actual process design conditions. He also has access to the technology of the manufacturer during design, experimentation, data interpretation, and even equipment startup. There is very close contact and review for both must stay current with the data as developed. The price for the equipment will be fair, but not discounted. To protect both, suitable confidentiality agreements can be executed, spelling out the nature of the data to be shared, and the terms and conditions of any restrictions on its use.

NOMENCLATURE

℄	Centerline
C_P	Specific heat
D	Diameter of impeller
D/T	Impeller diameter to tank diameter ratio
d	Heat transfer tube diameter
F_I	Inertia force
F_G	Gravity force
F_v	Vicosity force
F_σ	Surface tension
F_A	Applied force
Fr	N^2D/g
g	Gravity, hd/k
kPa	Kilo Pascals
k	Thermal conductivity
N	Impeller rotational speed
ND	Impeller peripheral speed
n	Power law exponent
$P_g/\rho N^3 D^3$	Power number

P	Horsepower
P/V	Power per unit volume
Re	$ND^2\rho/\mu$
T	Tank diameter
$\bar{u}'$	Average velocity fluctuation
u	Velocity
$\bar{u}'$	Mean or average velocity
u''	Velocity fluctuation
$\sqrt{(\bar{u}'')^2}$	Root-mean-square (RMS) velocity fluctuation
X	Dimension
x	Mol fraction
Z	Liquid level
We	N^2D^3/σ

Subscripts

M	Model
P	Prototype
T	Top
B	Bottom
i	Tank number i
F	Final tank

Greek

$\Delta u/\Delta Y$	Shear rate
ρ	Density
μ	Viscosity
θ	Blend time
θN	Blend number
σ	Surface tension

REFERENCES

Chemineer, "CE Refresher—Liquid Agitation," *Chemical Engineering*, **82**, 110–114 (1975); **83**, 139–145 (1976); **83**, 93–100 (1976); **83**, 102–110 (1976); **83**, 144–150 (1976); **83**, 89–94 (1976); **83**, 101–108 (1976); **83**, 109–112 (1976); **83**, 119–126 (1976); **83**, 127–133 (1976); **83**, 165–170 (1976). Reprints of these articles can be obtained from Chemineer-Kenics, Daytown, Ohio 45401.

Corpstein, R. R., Dove, R. A., and Dickey, D. S., "Stirred Tank Reactor Design," *Chem. Eng. Prog.*, **75**(2), 66–74 (1979).

Johnstone, R. E., and Thring, M., *Pilot Plant Models and Scaleup Methods in Chemical Engineering*, McGraw-Hill, New York, 1957.

Nagata, S., *Mixing—Principles and Applications*, Wiley-Halsted, New York, 1975.

Oldshue, J. Y., "Fermentation Mixing Scaleup Techniques," *Biotech. and Bioeng.*, **VIII**(1), 3–24 (1966).

Oldshue, J. Y., "Suspending Solids and Dispersing Gases Mixing Vessels," *Ind. Eng. Chem.*, **61**(9), 79–89 (1969).

Oldshue, J. Y., "Spectrum of Fluid Shear in a Mixing Vessel," CHEMECA '70 Australia, Butterworth, London, 1970.

Oldshue, J. Y., and Sprague, J., "Theory of Mixing," *Paint and Varnish Production*, **3**, 19–28 (1974).

10

FLUIDIZED BEDS

J. M. MATSEN

I.	Major Scaleup Issues in Fluidization		348
II.	Fundamental Concepts		353
	A.	Minimum Fluidization Velocity	353
	B.	Bubble Behavior	354
		1. Single Bubble Rise Velocity	354
		2. Bubble Velocity During Continuous Fluidization	355
		3. Bed Expansion	356
		4. Gas Flow Through Bubbles	358
	C.	Terminal Velocity and Entrainment	359
	D.	Reaction and Gas–Solids Contacting	361
	E.	Heat Transfer	365
	F.	Significance of Slug Flow	369
III.	Prediction of Performance in Large Equipment		371
	A.	Bubble Size	371
	B.	Mixing Rates	375
	C.	Grid Design	377
	D.	Entrainment	380
	E.	Cyclones	382
IV.	Practical Commercial Experience		385
	A.	Fluid Coking	385
	B.	Fluid Hydroforming	387
	C.	The Shell Chlorine Process	392
V.	Problem Areas in Scaleup		396

A.	Particle Size Balances	396
	1. Reaction Effects	397
	2. Attrition	397
	3. Agglomeration	398
B.	Erosion	398
C.	Choice of Reactor Models and Model Parameters	399
D.	Uncertainties in Performance	401
Nomenclature		402
References		404

I. MAJOR SCALEUP ISSUES IN FLUIDIZATION

When a gas or liquid flows upward through a bed of solid particles, the pressure difference across the bed increases approximately linearly with velocity. At a sufficiently high velocity (called minimum fluidizing velocity, U_{mf}) the drag force of the fluid on the particles becomes equal to their weight, and the particles become neutrally buoyant. In this state, the particles are said to be fluidized, and the particle bed takes on many characteristics of a liquid. The particles move freely if stirred, and waves can be easily formed. The bed surface will remain horizontal when the vessel is tilted, and the solid will stream freely through a hole punched in the vessel wall in accordance with Bernoulli's theorem.

As velocity is increased beyond the minimum fluidizing velocity, the bed begins to expand and the pressure difference across the bed remains essentially constant, equal to the weight of the bed per unit cross-sectional area. (Note that the pressure drop may deviate slightly from this ideal. In small diameter beds with high fluid velocity, the friction of particles against the wall can add slightly to pressure drop. In large beds, solids circulation patterns may cause density and pressure gradients to be greater near the wall than at the center of the bed. In that case a single density would be misleading.) The behavior of bed height and differential pressure as a function of fluid velocity is shown in Figure 10-1.

During the last 40 years, fluidized beds have become a much favored type of chemical reactor. The first major use of fluidization technology was in the catalytic cracking of petroleum, which became commercial in 1942 and now accounts for some 350 units and a daily product rate of 1,600,000 t worldwide. Some subsequent major applications have included limestone calcination, ore roasting and reduction, drying, coking, Fischer–Tropsch hydrocarbon synthesis, and phthalic anhydride manufacture. The applications under most intensive development today are coal combustion in fluid beds of limestone and

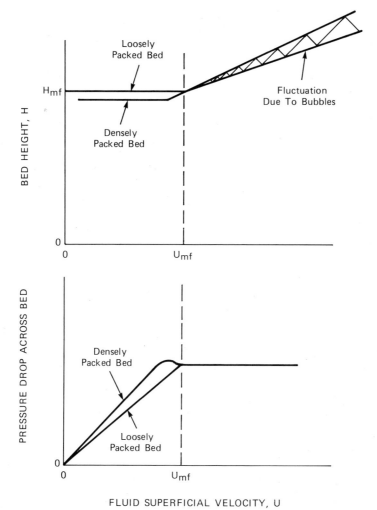

FIGURE 10-1 Bed expansion and pressure drop as a function of fluid velocity U. Bed exists in packed state below minimum fluidizing velocity U_{mf} and in fluidized state above U_{mf}.

various synthetic fuels processes such as coal gasification and oil shale processing.

Fluidized beds have a number of characteristics which make them attractive for certain types of process application. By far the most important of these is the suitability of fluid beds when fine solid particles are a reactant or a product of the reaction. Fluid beds greatly facilitate the continuous addition and withdrawal of solids from a process. The alternatives for solid reactants or products are (a) fixed beds, which would be operated batchwise, (b) moving beds, which have generated only limited interest, and (c) mechanical devices such as rotating kilns and moving grates, which achieve straightforward process operation at the expense of complex and expensive machinery. In some fluid bed processes, the solids may nominally be a catalyst which remains in the system, but such solids may also be a stoichiometric reactant and heat transfer medium flowing between a reaction and a regeneration vessel. In fluid catalytic cracking 5 to 10 kg of catalyst are circulated to the reactor for each kilogram of oil feed. It is safe to say that if solid reactants or products are involved, the process is a likely candidate for fluidization, while if the solids are captive, fluidization must confer critical advantages elsewhere in order to be attractive.

A second advantage of fluid beds is their characteristic of accommodating high gas flow rates through a bed of fine particles at much lower pressure drop than is possible with packed beds. A typical packed bed catalyst would be a 2×6 mm extrudate cylinder, while a typical fluid bed catalyst would be a microsphere of 0.06 mm diameter. In a packed bed at a superficial velocity of 1.3 m/s the extrudate might have a pressure drop of 10.4 kPa/m while the microspheres would have a pressure drop of 7400 kPa/m. In a fluid bed, either catalyst would have a pressure drop of 10 kPa/m or less.

A third advantage of fluid beds is outstanding heat transfer behavior. Because of rapid solids mixing, temperatures in a fluid bed are generally quite uniform even in the presence of highly exothermic reactions. Even within a large vessel, temperature differences in excess of 20°C are seldom seen. Temperature uniformity is of utmost importance in the synthesis of phthalic anhydride:

$$C_{10}H_8 + 4.5O_2 \rightarrow C_8H_4O_3 + 2CO_2 + 2H_2O$$

This reaction is highly exothermic, and if the temperature is slightly greater than optimum complete combustion to CO_2 and H_2O is greatly favored. Packed-bed phthalic anhydride reactors, consisting of many small catalyst-filled tubes immersed in a heat transfer medium, are subject to hot spots and temperature runaways which are successfully avoided in fluid beds. Besides uniformity of bed temperatures, fluid beds exhibit good fluid to surface heat transfer coefficients. A typical coefficient might be 300 W/m^2 K, compared to 500 W/m^2 K for water in turbulent flow and less than 50 W/m^2 K for a gas in turbulent flow. A high heat transfer coefficient permits a fluid bed boiler to be much more compact than its conventional counterpart. A final important heat transfer property is the very rapid quenching of gases injected into fluid beds. Quench rates 50,000,000°C/s have been measured for plasma jets in fluidized beds. High quench rates are important in fluid bed drying of heat sensitive materials, and it is practical to introduce very high temperature gases into a bed of cool particles with little concern for overheating.

Balancing their significant advantages, fluid beds also have definite disadvantages when compared to other types of reactors. The disadvantages invari-

MAJOR SCALEUP ISSUES IN FLUIDIZATION

ably become more onerous as the size of a process increases, and therein lurk the scaleup issues for fluid beds. Disadvantages may be classed as poor gas flow patterns, undesirable solids mixing patterns, and potential physical operating problems.

In a fluid bed operated at commercially significant velocities, most of the gas flows as bubbles. Bubbles rise through the bed at velocities much higher than the superficial gas velocity, and they have a relatively short contact time. Bubbles are also essentially empty of solid particles, and gas must usually percolate or diffuse from the bubbles into the emulsion or particulate phase of the bed before any reaction or other interaction can take place. As a result of the short residence time of most gas and the poor contacting of gas with bed particles, a fluid bed invariably gives a lower conversion of gaseous reactant than a fixed bed operated at the same space velocity. An early step in many scaleup efforts is to relate kinetic performance obtained in a thermogravimetric analyzer or a small packed bed to that seen in a bench scale fluid bed. On this scale the bubble-contacting limitations can be estimated fairly accurately, and other complications seldom need treatment. Subsequent scaleup often involves a simultaneous increase in bed diameter, bed depth, and gas velocity and a change in gas distributor design. These changes all tend to increase the size of bubbles present and the contacting becomes correspondingly worse. A major effort in fluid bed scaleup lies in estimating what the gas–solids contacting debits will be and in designing the next reactor to give the desired conversion despite these debits. Figure 10-2 is an example of the contacting debits experienced in fluid bed scaleup as noted by Frye et al. (1958).

While they may cause contacting debits, bubbles cause rapid solids mixing, which is responsible for the usually desirable uniform temperature observed in fluid beds. With respect to solids residence time distribution, most large fluid beds must be regarded as well-mixed reactors. As explained in Chapters 4 and 7, a well-mixed flow reactor is not suitable for high conversions of the well-mixed reactant. If fluid beds are to be used for such reactions, staging must be achieved by employing several beds in series or, in cases of very rapid reactions, by using a high velocity transfer line reactor. The alternative for high conversions of solids is to use a batch reactor.

A second effect of solids mixing is that the interstitial gas in the emulsion phase is backmixed in concert with the solid particles. Gaseous reactant conversion is accordingly reduced by solids mixing, usually a more severe debit than the effect on solid conversion. Mixing rate is very much increased with reactor size. Bench scale beds of small diameter and high length to diameter ratio exhibit much smaller solids diffusivities than larger beds, and costly errors have been made when the importance of solids mixing has not been appreciated.

The final critical issue in fluid bed scaleup has to do with the physical operability of the plant rather than the performance of the bed as a chemical reactor. While most processes experience a few operability problems during

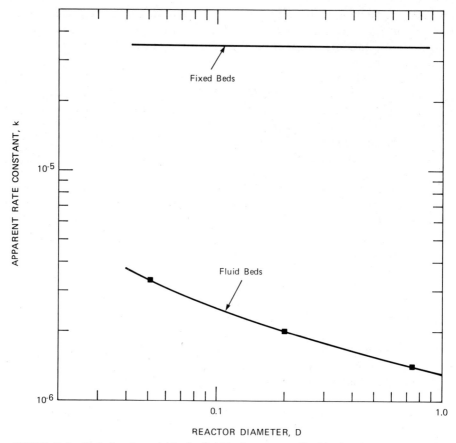

FIGURE 10-2 Typical scaleup debits for fluid bed reactors. A fluid bed is always less effective than a fixed bed and becomes progressively worse as size is increased.

scaleup, the problems can be expected to be more numerous and more serious with a fluid bed process than with stirred tanks, packed beds, sparged columns, and so on. A fundamental difficulty arises in trying to handle solid particles as fluids. The analogy between fluidized solids and true fluids is limited and does not hold in many critical areas. For example:

1. Flowing liquids or gases cause virtually no wear on exposed surfaces, whereas fluidized solids often have a very abrasive action.
2. Fluidized particles are often subject to substantial attrition.
3. Fluids flow easily at any rate through pipes of any size, while fluidized particles may settle out, bridge, or otherwise refuse to flow properly unless operating conditions are exactly right.

FUNDAMENTAL CONCEPTS

4. Particle cleanup devices such as cyclones or electrostatic precipitators are normally necessary.
5. Reactive solids may agglomerate or form slags detrimental to the process.
6. Large vessels and transfer lines are often subject to severe mechanical shocks or vibration.

Almost every new fluid bed process reveals a novel operability problem, and an important aspect of scaleup is to demonstrate operability of all mechanical aspects of the process or to anticipate problems and provide fallback positions for the first commercial unit.

II. FUNDAMENTAL CONCEPTS

A. Minimum Fluidization Velocity

The minimum gas velocity at which a bed can be fluidized, U_{mf}, sets a lower bound to the gas throughput for a fluidized system. It is also an important factor in the gas–solids contacting performance of a fluid bed.

Minimum fluidization velocity is simple to measure. Experimental determination eliminates the need for extensive data (on particle size distribution, shape factors, and packing density) which is needed for accurate calculation of U_{mf}, and may also provide useful insight into behavior of the subject particles. If direct measurement is not feasible, an approximate equation is

$$\frac{d_p U_{mf} \rho_g}{\mu} = \left[33.7^2 + \frac{0.0408 d_p^3 \rho_g (\rho_p - \rho_g) g}{\mu^2} \right]^{1/2} - 33.7 \qquad (10\text{-}1)$$

For particles of diameter, d_p, less than 500 μm the equation becomes approximately

$$U_{mf} = \frac{d_p^2 (\rho_p - \rho_g) g}{1650 \mu} \qquad (10\text{-}2)$$

For a bed having a distribution of particle sizes the correct value of d_p is the surface-volume mean diameter

$$d_p = \frac{\sum n_i d_i^3}{\sum n_i d_i^2} = \frac{1}{\sum W_i / d_i} \qquad (10\text{-}3)$$

The concept of minimum fluidization velocity usually implies a well-mixed bed if a range of particle sizes is present. When a bed of mixed sizes is operated at minimum fluidization velocity, however, the larger particles will soon settle out into an unfluidized packed-bed layer. The lowest operating velocity for a fluid bed is therefore usually higher than the well-mixed value of U_{mf}. While the mechanics of particle segregation are not fully understood it is safe to say that a bed should probably be operated at a velocity greater than the minimum fluidization velocity for the ninety-fifth percentile (by weight) particle size.

B. Bubble Behavior

Above the minimum fluidization velocity, a bed can expand in two distinct ways. In particulate fluidization the bed expands uniformly, with the particles maintaining a somewhat regular spacing as they move apart. Fixed-bed laws for pressure drop (e.g., the Carman–Kozeny or the Ergun equation) are applicable, with the provision that bed voidage, ε, increases with fluid velocity in order to maintain constant pressure drop, rather than being independent of velocity as for a fixed bed. In aggregative fluidization, on the other hand, the bed is very nonuniform. Most of the gas in excess of U_{mf} passes through the bed in fast-moving bubbles which are nearly empty of particles, while the rest of the bed, the emulsion phase, remains close to minimum fluidization conditions. Aggregative fluidization may become quite rough or violent at high fluid velocities, and the bed structure and fluid flow patterns are much more complex than in particulate fluidization. In general, liquid fluidized systems display particulate behavior while gas fluidized systems are aggregative. Stewart (1965) has shown, however, that very dense, coarse particles can fluidize aggregatively with liquids while fine particles in a dense gas may exhibit particulate fluidization tendencies. Only gas fluidized systems have achieved commercial significance and particulate systems are of mainly academic interest. This chapter will not deal further with liquid systems.

Perhaps the most important feature of an aggregatively fluidized bed is the existence of much of the fluidizing gas in the form of large pockets or bubbles. Much of the fundamental work in fluidization has been directed toward characterizing the properties of bubbles and deducing fluid bed behavior from those properties. Most large commercial fluid beds exhibit gas flow patterns and behavior far more complex than existing bubble theories can describe, but bubbling-bed models are nevertheless useful in making engineering judgments from limited data.

1. Single Bubble Rise Velocity

Gas bubbles in fluid beds have many of the properties of gas bubbles in inviscid liquids. Thus Davidson et al. (1959) discovered that fluid bed bubbles have the same spherical cap shape as those in liquids. Rise velocity of single bubbles in incipiently fluidized beds is given by the relationship

$$U_B = 0.71\sqrt{gD_B} \tag{10-4}$$

FUNDAMENTAL CONCEPTS

where the rise velocity, U_B, the acceleration of gravity, g, and the bubble diameter, D_B, may be expressed in any set of consistent units. This is the same relationship as had been found 9 years before by Davies and Taylor (1950) for gas/liquid systems. The equation may be derived theoretically, assuming conditions of potential flow of an inviscid fluid around a sphere. When the frontal diameter of the bubble D_B becomes greater than 0.3 times the bed diameter D the bubble is termed a slug, and the relationship becomes

$$U_B = 0.35\sqrt{gD} \tag{10-5}$$

This is also applicable to liquid systems and is theoretically derivable. A slug instead of being centered (axisymmetric) may rise along one side of the bed, traveling at a greater velocity. This equation then becomes

$$U_B = 0.49\sqrt{gD} \tag{10-6}$$

A popular apparatus for studying bubble behavior has been the two-dimensional fluid bed, in which the front-to-back thickness of the bed is small compared to the side-to-side width of the bubble. Such an apparatus provides a simple means to observe bubbles within the bed. Again the similarity to liquid system holds. The rise velocity becomes

$$U_B = 0.35\sqrt{gD_B} \tag{10-7}$$

Finally, in the case of a two-dimensional bed in which D approaches the width of the bed, a two-dimensional slug exists and the rise velocity is given by

$$U_B = 0.23\sqrt{gD} \tag{10-8}$$

where D is the width (not the thickness) of the bed. The five types of bubbles are shown in Figure 10-3.

2. Bubble Velocity During Continuous Fluidization

When a bed is fluidized by a substantial flow of gas, bubbles are generated continuously, and the bubble rise velocity is greater than when only a single bubble is present. Imagine that a single bubble is injected, and when it has risen well away from the bottom of the bed, high gas flow is started, as shown in Figure 10-4. Since no bubbles will break the surface of the bed for a while, the surface must rise at a velocity of $U - U_{mf}$ which is the volumetric rate of generation of bubbles at the bottom of the bed. Any other control surface drawn through the emulsion phase and moving so that no particles cross the surface must likewise rise at $U - U_{mf}$. The bubble will rise at U_B relative to these moving control surfaces, or at $U_A = U_B + U - U_{mf}$ relative to a stationary observer.

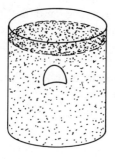
a. Sphere-Cap Bubble

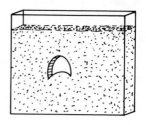
b. Two-Dimensional Bubble

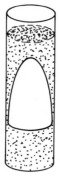
c. Symmetric Slug

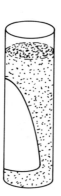
d. Wall Slug

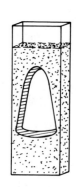
e. Two-Dimensional Slugs

FIGURE 10-3 Types of bubbles in fluid beds. Type a is typical of large commerical beds. The other types, especially c, are often found in small beds. The existence of different types of bubbles in small and large beds is a factor contributing to scaleup effects.

The above derivation is for an idealized case in which the time average gas flow is uniform across the bed cross section, in which there are no persistent solids circulation patterns, and in which negligible bubble coalescence occurs. In many large fluid beds coalescence occurs, and gulf/stream flow exists in which gas flow is channeled through the center of the bed and solids circulate down at the walls and up in the center. Such nonideal flow invalidates the above expression for bubble velocity. In the case of slug flow, however, the expression for U_A has been proven to be generally applicable.

3. Bed Expansion

The density of a bubbling fluidized bed may be derived from the equation for the absolute bubble rise velocity U_A, the two-phase theory of fluidization which

FUNDAMENTAL CONCEPTS

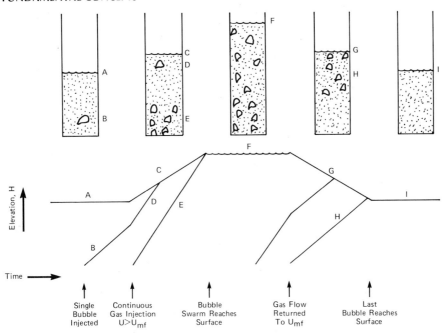

A Surface of bed.
B Single bubble rising at U_B. $\Big\}$ Bed at minimum fluidization $U = U_{mf}$
C Surface of bed when continuous bubble gas injection $U > U_{mf}$ has started and before bubble swarm reaches bed surface. $dH/dT = U - U_{mf}$.
D Bubble track B as altered by increase in U. $U_A = U - U_{mf} + U_B$
E Top of bubble swarm formed by increase in U. $U_A = U - U_{mf} + U_B$
F Surface of bed at steady state bubbling.
G Surface of bed immediately after gas velocity has been returned to U_{mf} $dH/DT = U - U_{mf}$
H Last bubble in swarm after gas velocity.
I Surface of bed at minimum fluidization.

FIGURE 10-4 Motion of bed surface and of bubbles during fluidization.

says that bubble gas flow rate is $U - U_{mf}$, and from the knowledge that bed expansion is due to the presence of bubbles. The resulting equation is

$$\frac{H}{H_{mf}} = \frac{\rho_{mf}}{\rho} = 1 + \frac{U - U_{mf}}{U_B} \qquad (10\text{-}9)$$

In laboratory scale beds with $(U - U_{mf}) > 0.2 U_B$ and with $H_{mf} > 5D$, slug flow will usually result, and then

$$\frac{H_{max}}{H_{mf}} = 1 + \frac{U - U_{mf}}{0.35\sqrt{gD}} \qquad (10\text{-}10)$$

The top surface of a bed in slug flow moves up and down quite makedly as bubbles rise in the bed and then break the surface. Matsen et al. (1969) showed that the maximum bed height or minimum bed density (rather than the average value) is the appropriate parameter for correlating expansion data of a slugging bed by the above equation.

The equation may also be used in the absence of actual data to predict the density of a non slugging bed. In such a case the bubble velocity will be given by $U_B = 0.7\sqrt{gD_B}$ and bubble size can be estimated according to the section on scaleup. Nonideal behavior such as gulf stream flow will often cause actual bed density to be somewhat greater than predicted by the equation, but no general method exists to quantify the effect.

4. Gas Flow Through Bubbles

The gas within a bubble is at uniform pressure from top to bottom, while the gas in the emulsion phase has a pressure gradient and has less than bubble pressure at the top and greater pressure at the bottom. This pressure imbalance causes gas to flow from the top of the bubble into the emulsion and from the emulsion into the bottom of the bubble. Davidson and Harrison (1963) have analyzed this flow in detail and have shown that the volumetric flow rate may be expressed as

$$Q_F = \frac{3\pi}{4} U_{mf} D_B^2 \qquad (10\text{-}11)$$

For slugs the corresponding equation is

$$Q_F = \frac{\pi}{4} U_{mf} D^2 \qquad (10\text{-}12)$$

This flow is an important mechanism by which bubble gas may come into contact with the reactant or catalyst particles in a fluid bed. Rowe (1961) has demonstrated that the flow out of the top of the bubble creates enough drag on the particles at that interface to keep them from falling into the bubble. Davidson (1961) has calculated streamlines for gas flow through a bubble. In the case where U_B is greater than U_{mf}/ε_{mf}, the bubble gas streamlines form a closed envelope or cloud as shown in Figure 10-5. The diameter of this closed envelope may be expressed as $(D_C/D_B)^3 = (U_B + 2U_{mf}/\varepsilon_{mf})/(U_B - U_{mf}/\varepsilon_{mf})$. The existence of bubble clouds becomes significant in calculations of gas–solid reactions in fluid beds, because the cloud severely restricts the number of particles accessible to bubble gas.

Diffusion offers a second mechanism for gas exchange between bubble and emulsion. For a sphere-cap bubble the solution has been derived by Baird and Davidson (1962) as

$$Q_D = 3.06 \delta_g^{1/2} \left(\frac{g}{D_b}\right)^{1/4} D_b^2 \qquad (10\text{-}13)$$

FUNDAMENTAL CONCEPTS

Conditions: $U_B = 25.9$ cm/s; $U_{mf}/\varepsilon_{mf} = 19.9$ cm/s

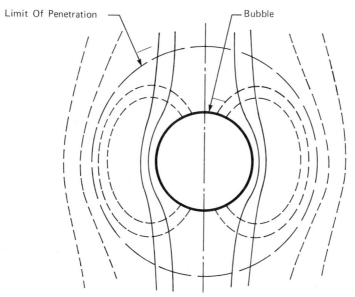

FIGURE 10-5 Limit of penetration of gas circulating from a bubble into the emulsion phase. The dashed lines are gas streamlines, while the solid lines are particle flow.

For slugs, a corresponding solution is given by Hovmand and Davidson (1971):

$$Q_D = 4(\pi \delta_g)^{1/2} \left(\frac{g}{D}\right)^{1/4} D^2 I \qquad (10\text{-}14)$$

The surface integral I relates the surface area of the slug to the diameter squared. It depends on the length of the slug as shown in Figure 10-6.

Throughflow and diffusion make essentially additive contributions to gas exchange from a bubble or slug. Throughflow usually predominates for low temperatures and coarse particles, while diffusion becomes more important for high temperatures and fine particles.

C. Terminal Velocity and Entrainment

Another important reference velocity in fluidization is the particle terminal velocity, V_T. This is the free fall velocity that a single particle in isolation would achieve. Experimental determination of V_T, particularly in gases, is often inconvenient experimentally. Terminal velocity may be calculated from

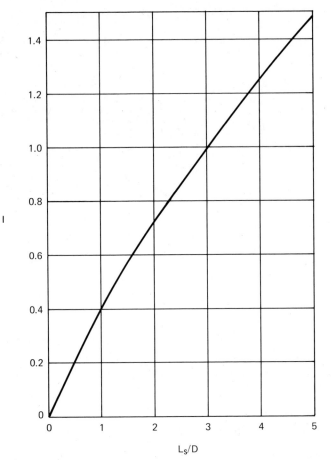

FIGURE 10-6 Surface integral I for diffusion from a slug.

the equations

$$V_T = (\rho_p + \rho_g)gd_p^2/18\mu \qquad \text{Re} < 2$$

$$V_T = 0.152 d_p^{0.114} g^{0.714}(\rho_p - \rho_g)^{0.114}/\rho_g^{0.285}\mu^{0.428} \qquad 2 < \text{Re} < 500$$

$$V_T = \left[3gd_p(\rho_p - \rho_g)/\rho_g\right]^{1/2} \qquad 500 < \text{Re}$$

(10-15)

Particles will be entrained and carried overhead from a bed fluidized at a superficial velocity greater than V_T. Entrainment often remains at acceptable levels even at velocities many times greater than V_T, however. When uncon-

FUNDAMENTAL CONCEPTS

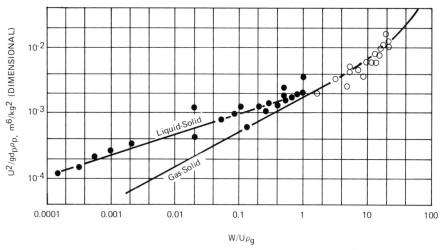

FIGURE 10-7 Correlation for mass flux of particles, W, entrained from a fluid bed. Note that Y axis is dimensional.

trolled entrainment becomes unacceptable, cyclones, which return entrained particles to the bed, can prevent excessive losses of particles from the process. Terminal velocity places no inherent upper limit on bed superficial velocity, and processes such as fluid catalytic cracking may operate at 5 to 10 times V_T while still maintaining an identifiable dense bed.

A comprehensive quantitative understanding of entrainment is not presently available. Perhaps the most reliable of the present empirical correlations is that of Zenz and Weil (1958), shown in Figure 10-7. To calculate entrainment, the term $U^2/gd_p\rho_p^2$ (having dimension m^6/kg^2) is calculated and the dimensionless entrainment $W/U\rho_g$, is read from the figure. If a range of particle sizes is present in the bed, the entrainment rate of each size fraction is its weight fraction in the bed times the entrainment rate calculated from Figure 10-7 for that size fraction. Total entrainment rate is the sum of the fractional entrainments. The value so calculated is entrainment far above the bed, while near the bed surface entrainment is greater and is affected by scaleup considerations.

D. Reaction and Gas – Solids Contacting

The aspect of fluid bed design receiving the greatest attention in the literature in recent years is the study and modeling of gas–solids contacting and its effect on the extent of chemical reaction in fluid beds. Very comprehensive treatments of such contacting have been published by Kunii and Levenspiel (1969) and by Davidson and Harrison (1971). The reader is referred to those works for a detailed account, and only a brief summary will be given here.

Bubbles of gas in a fluid bed account for substantial debits and also credits in the performance of fluid bed reactors when compared to fixed-bed reactors.

The bubbles produce relatively good mixing of solid particles within a bed, so that temperature gradients and solids concentration gradients within a bed are usually negligible and are almost always small compared to such gradients in fixed beds. At the same time bubble gas moves rapidly through a bed in poor contact with most of the bed solids, and the extent of chemical reaction may be severly debited by gas flow patterns.

A wide variety of assumptions have been used in the modeling of interaction of kinetics and gas–solids contacting in fluid bed reactors. Considering the computational complexity of a sophisticated model and the difficulty of obtaining independent data on many model parameters, a simple model is to be preferred unless special characteristics of the system under study dictate otherwise. The approach summarized here assumes the following:

The reaction is first order in gas reactant concentration.

The bubbles are essentially empty of solids, and no reaction takes place therein.

Bubble gas is exchanged with emulsion gas by a combination of convective gas throughflow and diffusion.

The bed is isothermal.

Emulsion gas composition is uniform in a horizontal plane, and no separate cloud zone is postulated.

Bubble gas is in plug flow.

Emulsion gas may be in plug flow or may be backmixed.

The number of reaction units is defined as

$$N_R = \frac{kH_{mf}}{U} \qquad (10\text{-}16)$$

which relates the amount of catalyst present to the flow rate and kinetic rate constant. The number of crossflow units, that is, the number of times that bubble gas is exchanged with emulsion gas during the time the bubble is rising through the bed, may be written as

$$N_C = \frac{QH}{U_A V_B} \qquad (10\text{-}17)$$

For sphere cap bubbles, the gas exchange rate Q is the sum of Q_F and Q_D given in Equations 10-11 plus 10-13, and equation becomes

$$N_C = \frac{6.3 H_{mf}}{D_B \sqrt{g D_B}} \left[U_{mf} + \frac{1.30 \delta_g^{1/2} g^{1/4}}{D_B^{1/4}} \right] \qquad (10\text{-}18)$$

For slugs, the exchange rate Q is the sum of Q_F and Q_D from Equation 10-12

FUNDAMENTAL CONCEPTS

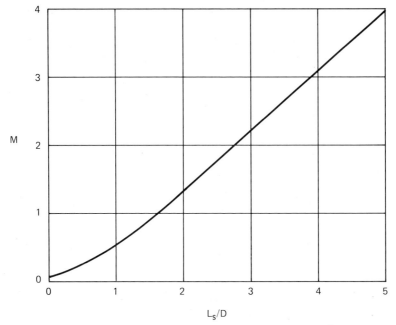

FIGURE 10-8 Shape factor for slug volume. $M = 4V_s/\pi D^3$.

and 10-14:

$$N_C = \frac{H_{mf}}{0.35\sqrt{gD}\; DM}\left[U_{mf} + \frac{16\varepsilon_{mf} I}{1 + \varepsilon_{mf}}\left(\frac{\delta_g}{\pi}\right)^{1/2}\left(\frac{g}{D}\right)^{1/4}\right] \quad (10\text{-}19)$$

The slug shape factor relating slug volume to bed diameter is defined by $M = 4V_s/\pi D^3$ and its dependence on slug length is shown in Figure 10-8. The surface integral I is given as a function of L_s/D in Figure 10-6, and the dependence of L_s/D on gas velocity is given in Figure 10-9. The factors M, I, and L_s/D are all from Hovmand and Davidson (1971).

The equations for chemical conversion under these circumstances become for plug flow in the emulsion phase

$$\frac{C_o}{C_i} = \frac{1}{M_1 - M_2}\left[M_1 e^{-M_2}\left(1 - \frac{M_2 U_{mf}}{N_c U}\right) - M_2 e^{-M_1}\left(1 - \frac{M_1 U_{mf}}{N_c U}\right)\right]$$
$$(10\text{-}20)$$

where

$$M_i = \frac{N_C + N_R \pm \sqrt{(N_C + N_R)^2 - 4N_C N_R U_{mf}/U}}{2U_{mf}/U} \quad (10\text{-}21)$$

and $M_i = M_1$ with the + sign and M_2 with the − sign.

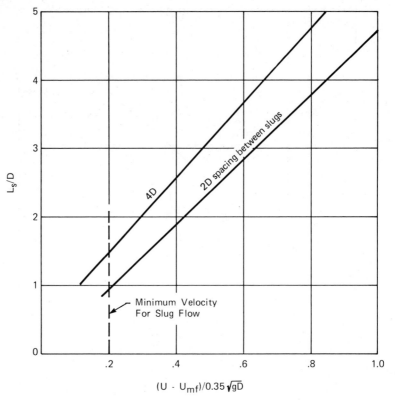

FIGURE 10-9 Length-to-diameter ratio for slugs as a function of fluidizing velocity.

For perfectly backmixed emulsion phase the proper expression is

$$\frac{C_o}{C_i} = \left(1 - \frac{U_{mf}}{U}\right)e^{-N_C} + \frac{\left[1 - \left(1 - \frac{U_{mf}}{U}\right)e^{-N_C}\right]^2}{N_R + 1 - \left(1 - \frac{U_{mf}}{U}\right)e^{-N_C}} \quad (10\text{-}22)$$

Several limiting cases are worthy of note:
For $U \gg U_{mf}$, Equation 10-20 becomes

$$\frac{C_o}{C_i} = \exp\left(-\frac{N_R N_C}{N_R + N_C}\right) \quad (10\text{-}23)$$

and then for $N_R \rightarrow \infty$,

$$\frac{C_o}{C_i} = e^{-N_C} \quad \text{for both cases} \quad (10\text{-}24)$$

FUNDAMENTAL CONCEPTS

while for $N_C \to \infty$,

$$\frac{C_o}{C_i} = e^{-N_R} \quad \text{for Equation 10-20} \quad (10\text{-}25)$$

$$\frac{C_o}{C_i} = \frac{1}{1 + N_R} \quad \text{for Equation 10-22} \quad (10\text{-}26)$$

Experience has shown that for a slug flow reactor, the emulsion gas is essentially in plug flow, while for a freely bubbling bed the mixing is somewhat between plug flow and perfect backmixing.

The formulae have been developed for cases where bubble diameter is constant throughout the bed. For the more general case where bubble coalescence occurs, either an average bubble size may be used or the bed may be broken down into small height increments for numerical solution.

E. Heat Transfer

The heat transfer properties that are of benefit in fluid bed operation are of three types: gas to particle heat transfer; immersed surface to bed heat transfer; and thermal conductivity within the bed. As with other fluid bed properties, theory and correlation are somewhat fragmentary, but enough information is available to permit rational decisions in most cases.

The state of the art for gas-to-particle heat transfer measurement according to Barker (1965) is indicated in Figure 10-10. This indicates a range of five orders of magnitude in experimental Colburn j factor for gas–particle heat transfer at a given Reynolds number. The great variation in measured heat transfer rates is the result of four factors: use of several definitions of the correlating temperature difference; use of different assumptions of relevant heat transfer area; use of different gas mixing assumptions; and great experimental difficulty in obtaining meaningful independent measurements of gas temperature and particle temperature within a fluid bed. As a practical matter, although the heat transfer coefficient is small, the gas rapidly approaches thermal equilibrium with the particles, because of the very large heat transfer surface available and the comparatively low volumetric heat capacity of the gas. Quench rates of plasma jets in fluid beds have been measured at 5×10^7 °C/s, and for most purposes it is safe to assume that local gas and particle temperatures are equal. When thermal equilibrium is not a useful assumption, the equation of Kothari (1967) may be used:

$$\text{Nu}_p = 0.03 \, \text{Re}_p^{1.3} \quad (10\text{-}27)$$

This correlation was developed for a plug flow model for the gas.

Apparent bed thermal conductivity is a second important heat transfer property in fluidization. No generalized correlations exist, but extensive mea-

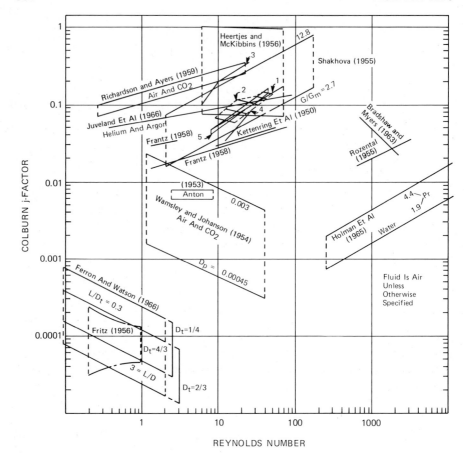

FIGURE 10-10 Summary of heat transfer coefficients reported in the literature.

The numbered lines refer to the work of the following:
1) Frantz correlation for "true" coefficients (1958)
2) Heertjes and McKibbins (1956)
3) Donnadieu (1961)
4) Sunkoori and Kaparthi (1960)
5) Walton Et Al, (1952)
The lower line labeled Frantz (1958) represents his correlation for "apparent" coefficients.

surements by Lewis et al. (1962) are shown in Figure 10-11. These data are from a 7.5-cm-diameter bed, and experience suggests that the thermal diffusivity will increase proportionally to bed diameter. For most cases, beds may be assumed to be isothermal.

The most critical heat transfer property for most fluid bed applications is the transfer coefficient from the bed to an immersed surface. Perhaps the most widely used expressions are due to Vreedenberg (1958), (1960). For horizontal

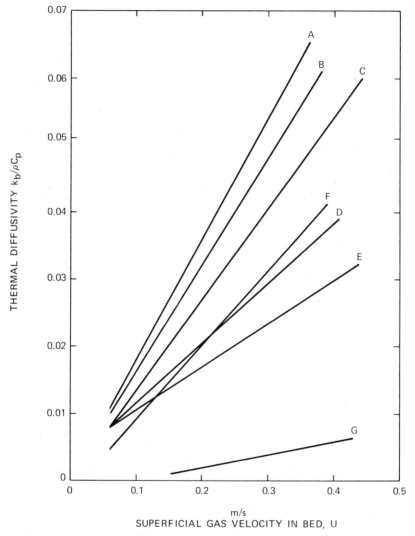

FIGURE 10-11 Thermal diffusivities measured in a 7.5-cm-diameter air fluidized bed.

tubes he proposed (1958) $(Ud_p\rho_p/\mu) < 2050$

$$\left(\frac{hd_t}{k_g}\right)\left(\frac{C_p\mu}{k_g}\right)^{-0.3} = 0.66\left(\frac{Ud_t\rho_p(1-\varepsilon)}{\mu\varepsilon}\right)^{0.44} \quad (10\text{-}28)$$

and for $(Ud_p\rho_p/\mu) > 2050$

$$\left(\frac{hd_t}{k_g}\right)\left(\frac{C_p\mu}{k_g}\right)^{-0.3} = 420\left(\frac{Ud_t\rho_p}{\mu} \times \frac{\mu^2}{d_p^3\rho_p^2 g}\right)^{0.3} \quad (10\text{-}29)$$

For vertical tubes the situation is more complicated with $(Ud_p\rho_p/\mu) < 2050$ and $U(D - d_t)\rho_p/\mu < 0.237 \times 10^6$,

$$\frac{h(D-d_t)}{k_g}\left(\frac{d_t}{D}\right)^{1/3}\left(\frac{k_g}{C_p\mu}\right)^{1/2} = 0.27 \times 10^{-15}\left(\frac{U(D-d_t)\rho_p}{\mu}\right)^{3.4}$$

$$(10\text{-}30)$$

With $Ud_p\rho_p/\mu < 2050$ and $U(D - d_t)\rho_p/\mu > 0.237 \times 10^6$,

$$\frac{h(D-d_t)}{k_g}\left(\frac{d_t}{D}\right)^{1/3}\left(\frac{k_g}{C_p\mu}\right)^{1/2} = 2.2\left(\frac{U(D-d_t)\rho_p}{\mu}\right)^{0.44} \quad (10\text{-}31)$$

With $Ud_p\rho_p/\mu > 2550$ and $U(D - d_t)/d_p^{3/2}g^{1/2} < 1070$,

$$\frac{h(D-d_t)}{k_g}\left(\frac{d_p}{D}\frac{d_t}{D-d_t}\right)^{1/3}\left(\frac{k_g}{C_p\mu}\right)^{1/3} = 0.105 \times 10^{-3}\left(\frac{U(D-d_t)}{d_p^{3/2}g^{1/2}}\right)^2$$

$$(10\text{-}32)$$

With $Ud_p\rho_p/\mu > 2550$ and $U(D - d_t)/d_p^{3/2}g^{1/2} > 1070$,

$$\frac{h(D-d_t)}{k_g}\left(\frac{d_p}{D}\frac{D_t}{(D-d_t)}\right)^{1/3}\left(\frac{k_g}{C_p\mu}\right)^{1/3} = 240\left(\frac{U(D-d_t)}{d_p^{3/2}g^{1/2}}\right)^{-0.1}$$

$$(10\text{-}33)$$

Needless to say, these equations are highly empirical and should be used with caution. The Equations 10-30 through 10-33 for vertical tubes predict a very substantial scaleup dependence on vessel diameter D.

In situations of commercial interest, typical bed-to-tube coefficients are usually in the range 200–400 W/m² °C.

FUNDAMENTAL CONCEPTS

F. Significance of Slug Flow

Physical properties of fluid beds change as bed diameter and depth are increased. One of the clearest examples is that of bed density or expansion. Figure 10-12 presents bed density reported by Volk (1962) which shows a marked scaleup effect. For a given velocity, density of a large diameter bed is noticeably greater than that of a small diameter bed. Matsen et al. (1969) showed that this density data could be explained quantitatively when one realized that the data for Figure 10-12 were taken on beds in slug flow. Equation 10-10 may be used to predict bed density or expansion. Using $U_B = \sqrt{0.35gD}$ and taking maximum bed height or minimum bed density, as is proper for slug flow, the data of Figure 10-12 can plotted in Figure 10-13. The scaleup effect of bed density is entirely due to the predictable effect of bed diameter D on slug velocity, U_B.

Recognition of the existence of slug flow is critical to the understanding of most laboratory scale fluid bed units, and slugging must be regarded as the natural flow regime in small beds. The minimum superficial gas velocity required for slugging is given by

$$U_{\min \text{slug}} - U_{\text{mf}} = 0.07\sqrt{gD} \tag{10-34}$$

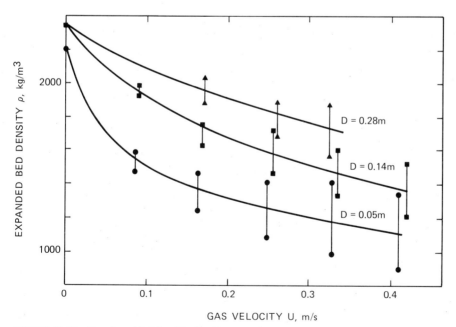

FIGURE 10-12 Density of fluidized beds of iron are showing scaleup effect. Points correspond to minimum and maximum bed heights during slugging.

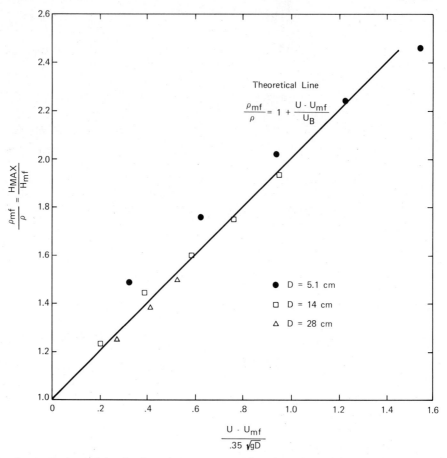

FIGURE 10-13 Bed density data of Figure 10-12 replotted according to slug flow theory. The correct density for slug flow corresponds to maximum bed height rather than average bed height.

Bed depth to diameter ratios of 2 or 3 are usually necessary to permit enough bubble coalescence for slugs to form, and this ratio is affected by distributor design. Maximum bed diameters for slugging are not well defined, but vessels as large as 0.6 m in diameter have been observed to slug even with fairly fine particles. Some investigators regard slug flow as "bad" fluidization because of the large bubbles and dramatic oscillation of bed height, and they make concerted efforts to avoid slugging by using low gas velocities, short beds, porous plate gas distributors, and baffles to break up bubbles. This approach seems misguided, however, and it is recommended that slug flow actually be sought and promoted in laboratory scale equipment for two important reasons. First slug flow represents a worst attainable case of gas–solids contacting in the lab unit, so that differences between small and large units are kept as small as possible. Second, slug flow is very well defined physically with regard to

III. PREDICTION OF PERFORMANCE IN LARGE EQUIPMENT

A. Bubble Size

Large fluid beds are generally not in slug flow, and it becomes important to estimate what size bubbles will be present. Harrison and Leung (1961) studied the initial size bubbles formed from a freely bubbling orifice. The relationship between gas flow rate and bubble size is shown in Figure 10-14 and may be

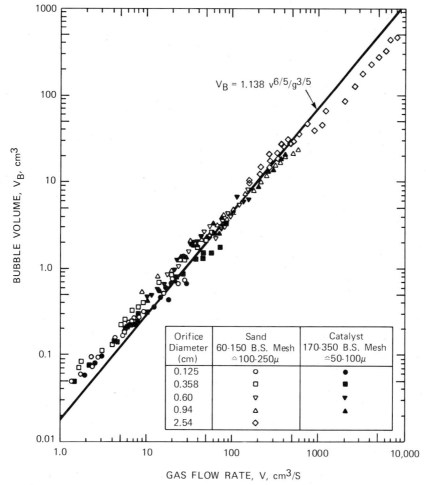

FIGURE 10-14 Volume of bubbles formed at an orifice as a function of gas flow rate. Volume is independent of orifice diameter as long as orifice operates in bubbling regime.

expressed as

$$V_B = 1.14 v^{6/5}/g^{3/5} \tag{10-35}$$

The relationship holds for gas bubbles in both fluidized solids and liquids and is derivable theoretically.

Bubbles formed at the distributor plate begin to coalesce rapidly as they rise through the bed. A mechanistic model of coalescence was proposed by Harrison and Leung (1962). They reasoned that a bubble rising in a fluid bed carries a wake of particles which extends about 1.2 bubble diameters behind it. A trailing bubble within this wake will exhibit a rise velocity, U_B, relative not to the average emulsion in the bed but to the wake of the leading bubble. If the absolute rise velocity of the leading bubble is given by $U_A = U - U_{mf} + U_B$, then the trailing bubble rises at $U_A = U - U_{mf} + 2U_B$, and coalescence will take place when the nose of the trailing bubble catches up with the base of the leading bubble. The resulting bubble will be of larger diameter and will have a correspondingly higher U_B. Successive coalescence and growth can be calculated in a stepwise manner. The calculation is not rigorous, being subject to a great many rather arbitrary assumptions of initial conditions. One finds, however, that the number of coalescence events which a bubble can encounter before reaching the surface of the bed is limited and not especially sensitive to initial condition assumptions.

Several empirical correlations of bubble coalescence have also been proposed. These are undoubtedly more accurate for the conditions for which they developed than the Harrison and Leung model, and they are easier to apply. In general, however, the empirical correlations were developed for relatively low gas velocities, shallow beds, and small diameter vessels and must be used cautiously for scaling to conditions of a typical commercial fluid bed. One widely used equation is that of Mori and Wen (1975)

$$\frac{D_{BM} - D_B}{D_{BM} - D_{BO}} = \exp\left(-0.3\frac{H}{D}\right) \tag{10-36}$$

$$D_{BM} = 1.63\left(\frac{\pi D^2}{4}(U - U_{mf})\right)^{0.4} \tag{10-37}$$

$$D_{BO} = 0.872\left(\frac{\pi D^2}{4n}(U - U_{mf})\right)^{0.4} \tag{10-38}$$

where x is height above the distributor with the correct units being meters and seconds. Note that Equations 10-37 and 10-38 are dimensional. A second such correlation is due to Rowe (1976),

$$D_B = (U - U_{mf})^{1/2}(H + H_o)^{3/4}/g^{1/4} \tag{10-39}$$

PREDICTION OF PERFORMANCE IN LARGE EQUIPMENT

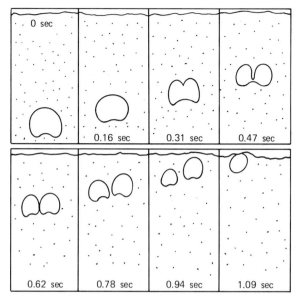

FIGURE 10-15 Cine sequence showing the splitting of a bubble in a 35-cm-deep two-dimensional bed of 230-m Ballotine.

where H_o is zero for porous plate distributors and is on the order of 1 m when large bubbles are formed at the grid.

Bubbles not only coalesce and grow, they also break up, as illustrated in Figure 10-15 from Rowe (1971). Although an exact mechanism has not been proven, theory holds that when a bubble becomes too large it becomes hydrodynamically unstable and breaks up. One theory due to Davidson and Harrison (1963) notes that toroidal gas circulation currents are set up within a bubble due to the relative downward motion of solids around the bubble, and this circulation velocity must be on the order of bubble velocity, $U_B = 0.7\sqrt{gD_B}$. When this circulation velocity exceeds the free fall velocity of the bed particle, these particles will be picked up from the bubble wake, filling the center of the bubble and causing it to break. A graphical presentation of stable bubble size based on this theory is given in Figure 10-16. A second theory due to Clift et al. (1974) holds that a Taylor instability causes the roof of the bubble to collapse when the bubble becomes too large. This seems to be the operative mechanism in Figure 10-15.

Matsen (1973) has reported several instances in which a stable bubble size was reached. Table 10-1 shows some of the data and compares it with the theory from Figure 10-16.

Figure 10-16 is for a single uniform particle size. The theory does not account for mixed particle sizes.

None of the approaches for calculating bubble coalescence or maximum stable size can be regarded as being definitive at this time. The designer should

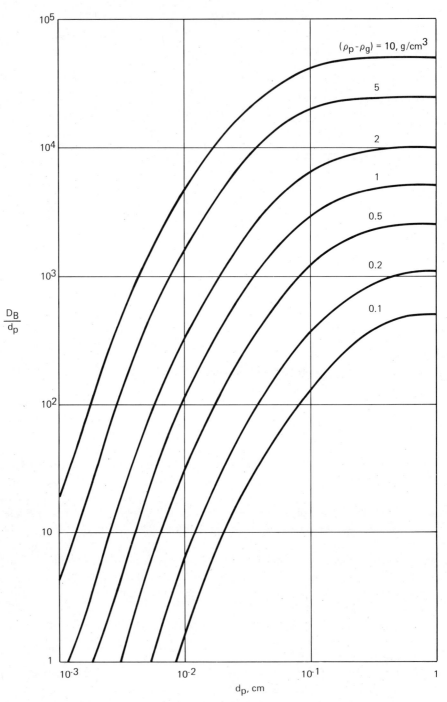

FIGURE 10-16 Theoretical predictions of maximum stable bubble size.

TABLE 10-1 Stable Bubble Size Observation

| | Particle Size, μm | | | Particle Size, μm, |
10%	50%	90%	Bubble Size	According to Theory
8	26	40	2.5 cm	86
12	70	108	6.2 cm	115
16	70	112	12.5 cm	148
30	85	112	15 cm	160

calculate bubble size by two or three methods to reach a consensus value. He should check this value against that from any available large scale expansion data. Finally he should make a size-sensitivity calculation to see how critical bubble size is to reactor performance. Very often a satisfactory margin of safety can be achieved with a slightly deeper bed or a somewhat lower gas velocity.

B. Mixing Rates

It is a common assumption that solids in a fluid bed are well mixed, with uniform properties throughout. That is usually a good assumption, and any other leads to a highly complex model of bed behavior. In some cases, however, solids properties are not constant throughout the bed, and significant scaleup effects can then be expected in:

a. Deep, small-diameter beds in slug flow. Top-to-bottom solids mixing is typically slow in such beds, and tens of minutes may be required to mix particles thoroughly.
b. Beds with high throughput of solids. Some processes have solids residence times of only a few minutes and may exhibit significant plug flow behavior.
c. Beds with very fast or very exothermic reactions. Hot spots near the gas distributor and near the solids introduction point are characteristic of such beds.

A unit volume of emulsion phase solids will carry with it a volume ε_{mf} of gas. Solids mixing is, therefore, the mechanism of emulsion phase gas backmixing, and hence it can greatly affect gas flow patterns and residence time distributions.

The rates of solids mixing and the accompanying gas mixing are very strongly affected by the size of the fluid bed reactor at hand. In large beds solids transport in wakes of bubbles makes only a small contribution. Much greater mixing is caused by large-scale "gulf stream" circulation patterns, in which solids flow rapidly down at the wall and up through the center of the

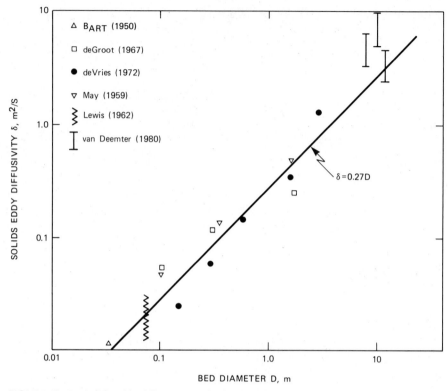

FIGURE 10-17 Solids eddy diffusivities in beds of cracking catalysts and similar fine particles.

bed. A comprehensive theory of large-scale mixing is yet to be developed, but empirical information is available. Solids mixing is often characterized by an effective diffusivity, and Figure 10-17 shows axial diffusivity measurements from several sources as a function of bed diameter. The large scaleup effect can be seen.

The directional effects of variables other than diameter on solids diffusivity are as follows:

1. Most of the data are for relatively tall narrow beds with height to diameter ratios of 2.5 and greater. It is expected that diffusivities will not change for taller beds but will decrease as the height to diameter ratio decreases below 1.
2. The data are for catalyst having a mass median diameter of about 80 μm and having a fines content (less than 44-μm particle diameter) of 15 to 25 percent. Lower or higher fines content as well as coarser average particle size will cause a decrease in diffusivity.

PREDICTION OF PERFORMANCE IN LARGE EQUIPMENT

3. The data are for gas velocities of 0.2 to 0.3 m/s. Most investigators find diffusivity proportional to gas velocity, but de Groot (1967) found little effect of velocity in large beds.
4. Diffusivities in the horizontal direction are expected to be somewhat less than the vertical diffusivities shown.
5. For calculations of thermal conductivity of the bed, it may be assumed that the effective diffusivity of the figure will be numerically the same as the thermal diffusivity, $k_b/\rho C_{ps}$ where k_b is the bed thermal conductivity, ρ is the bed density, and C_{ps} is the solids heat capacity.

In practice it is sometimes found that a simple diffusion model does not accurately describe the dispersion of tracer particles in a bed. However, no other general description of the dispersion process is presently available.

C. Grid Design

A fritted disk gas distributor (Figure 10-18a) seems to be the most common type in small-scale laboratory fluid beds. In small sizes such grids are easy to fabricate. They are easy to start up and shut down and are immune to solids backflow. They give good gas distribution and produce small bubbles, often desired by the investigators. Despite apparent advantages, the fritted disk grid can be a poor choice in process development work. An important objective in operating many small fluidization units is to obtain not the very best reactor performance but the performance most useful in scaling up the process. The very small bubbles produced by a fritted disk give deceptively good contacting and chemical conversion near the grid. This is not representative of the gas–solids contacting in larger units, and the exact nature of the contacting is difficult to characterize. It is, therefore, often preferable in small beds to promote slug flow deliberately at the very bottom of the bed, as explained in the section on fundamental concepts. A 60° cone with a single inlet nozzle at the vertex, as in Figure 10-18b, is an effective inlet geometry for promoting slugging.

An important consideration in the design of a large grid is to ensure sufficient pressure drop for good gas distribution and stable bed operation. One approach to calculating grid pressure drop necessary for stability is shown in Figure 10-19. The pressure drop through a grid hole will increase with gas velocity. Pressure drop through the bed immediately above that hole will decrease as gas preferentially channels through the hole; however, because the density of the bed decreases as gas flow increases. In order for the grid to be stable, total pressure drop through a grid hole plus the bed above that hole must increase as gas flow starts to channel preferentially there, so as to halt the channeling. A simplified case is instructive.

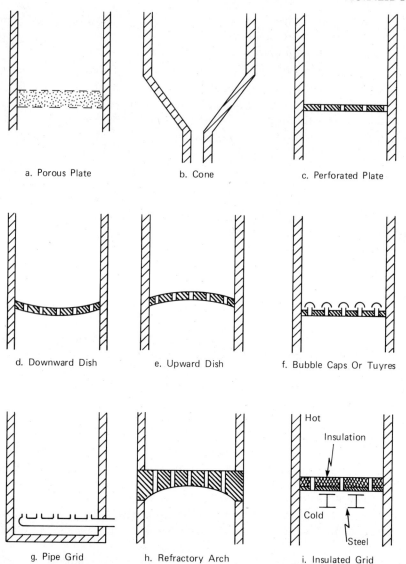

FIGURE 10-18 Typical types of grids for fluid bed applications.

Assuming that the bed is bubbling freely with a density given by Equation 10-9, pressure drop through the bed is given by

$$\Delta P_b = \rho g H = \frac{\rho_{mf} g H}{1 + \dfrac{U - U_{mf}}{U_B}} \qquad (10\text{-}40)$$

PREDICTION OF PERFORMANCE IN LARGE EQUIPMENT

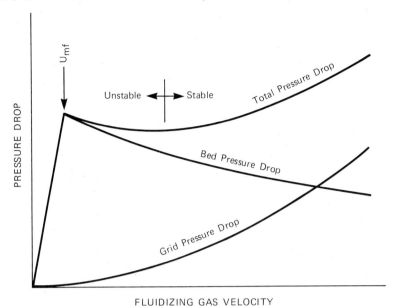

FIGURE 10-19 Criterion for grid stability.

and grid pressure drop is

$$\Delta P_g = CU^2 \qquad (10\text{-}41)$$

where C is an orifice coefficient based on superficial velocity. Solving for the conditions that $d(\Delta P_g + \Delta P_g)/dU > 0$ gives the criterion

$$\frac{\Delta P_g}{\Delta P_B} > \frac{U}{2(U + U_B - U_{mf})} \qquad (10\text{-}42)$$

for stable grid operation. A grid operated at too low a pressure drop will redistribute the gas flow so that some of the holes are inoperative and the remainder satisfy the criterion. In practice, grid pressure drop of 10 to 30 percent of bed pressure drop are usually sufficient to keep all holes active.

Gas entering a fluid bed from an orifice usually seems to form a jet, with jet penetration distances being determined by hole velocity and diameter. These jets allow gas to bypass the bed and so should not penetrate to the bed surface or occupy an inordinate fraction of bed height. Jet impingement can also cause severe erosion of surfaces within the bed and such submerged sufaces as heat transfer tubes and cyclone diplegs should be located outside any jet penetration zones. An empirical formula for jet penetration X is given by Merry (1975):

$$\frac{X}{d_o} = 5.2 \left(\frac{\rho_f d_o}{\rho_p d_p} \right)^{0.3} \left[1.3 \left(\frac{V_o}{gd_o} \right)^2 - 1 \right] \qquad (10\text{-}43)$$

Having selected a grid hole velocity and pressure drop sufficient for good gas distribution, significantly higher velocities should be avoided. Attrition of bed particles increases rapidly as grid hole velocity increases, and erosion of bed internals in the vicinity of the grid jet can also become serious. High grid pressure drop is, of course, wasteful of compressor power. In some multibed processes, high grid pressure drop adversely affects the pressure balance necessary for solids circulation.

The design of a commercial grid is often dictated by specific process considerations. One of the simplest and most widely used grid types is the simple perforated flat plate. One to three centimeter diameter holes on a square or triangular pitch are common. In cases such as some catalytic cracking reactor designs, erosive solids are carried upward through the grid by the fluidizing gas. Annular hard-surfaced inserts may then be used to improve erosion resistance of the grid holes. A perforated plate grid is shown in Figure 10-18c. A dished grid is inherently stronger than a flat one and is sometimes specified. A downward dish, Figure 10-18d, is usually preferable to an upward dish, Figure 10-18e, because more of the gas will be directed to the walls of the vessels, counteracting the natural tendency for gas to channel up through the center of the bed.

Many varieties of bubble caps or tuyeres have also been used, Figure 10-18f. These are often specified where backflow of solids through the grid must be avoided.

Pipe grids (Figure 10-18g) are simply networks of perforated pipes crisscrossing the bottom of the fluid bed. For large beds they are often cheaper than perforated plates, and they eliminate the need for a grid plenum or elaborate grid support structure. A pipe grid design must minimize the tendency for solids to fill up portions of the grid pipes during shutdown, since a filled section of pipe often cannot be blown clear when gas flow is resumed. Pipe grids with downward jetting holes are reported to produce stable gas distribution at very low grid pressure drop.

For very high temperature operation, metallic grids become unsuitable, and arches of refractory brickwork have been used in such cases (Figure 10-18h). If only the top surface of the grid is hot, a metal grid covered by a layer of insulating refractory (Figure 10-18i) is often preferable, particularly for large diameter vessels.

D. Entrainment

The entrainment rate calculated in Section II-C is a limiting value reached far above a fluid bed. Just above the bed surface, entrainment is much greater, but it decreases with height above the bed and becomes essentially constant at the "Transport Disengaging Height" or TDH. The value of TDH depends on bed diameter, so entrainment from a bed will, in general, depend on both the

PREDICTION OF PERFORMANCE IN LARGE EQUIPMENT

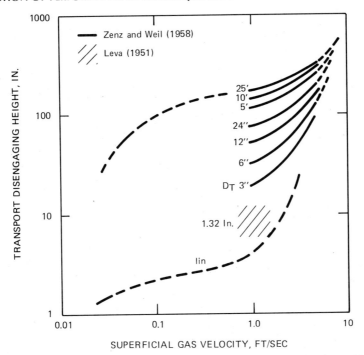

FIGURE 10-20 Empirical correlation for transport disengaging height, TDH.

height of the freeboard section and the vessel diameter. The principal work on determining TDH and the increase of entrainment below that level is presented by Zenz and Othmer (1960). The TDH correlation is shown in Figure 10-20. The data embodied in the figure were gathered for other purposes than TDH determination, and the correlation should not be regarded as highly accurate. Nevertheless, it represents real trends and data on six units having diameters from 5 cm to 5 m.

The usual explanation for the fact that entrainment increases as the bed surface is approached is that gas velocity at the surface of the bed is very nonuniform in time and space due to eruption of bubbles. Particles are entrained in the high velocity "jets" leaving the bed, and as these jets dissipate with height, their particle carrying capacity decreases. Defining the jet velocity as V_j and superficial velocity as U, Figure 10-21 shows how V_j/U decreases as (height above bed)/(TDH) approaches 1.

In order to calculate entrainment below TDH proceed as follows:

1. Estimate TDH from Figure 10-20.
2. Estimate V_j from Figure 10-21.
3. Using V_j in place of U, calculate entrainment from Figure 10-7 as outlined in Section II-C.

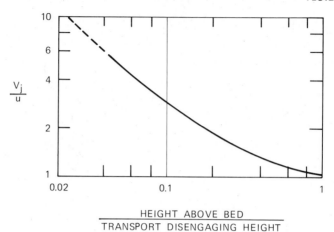

FIGURE 10-21 Dissipation of entraining jet below TDH.

E. Cyclones

Cyclone dust collectors are an essential element of many commercial fluid bed processes. By collecting and returning entrained particles to the bed, they permit bed operation with much finer particles and higher gas velocities than could possibly be tolerated if entrained material were lost from the process. As an example, entrainment from an FCC regenerator operating at 0.7 m/s superficial gas velocity would be about 10 kg catalyst/m³ flue gas, equivalent to 80 t/min in the largest units. Two sets of internal cyclones in series operate at about 99.997 percent collection efficiency, so that actual catalyst losses from the regenerator would be only 0.3 g/m³. This is usually an acceptable loss rate as far as process considerations are concerned, although additional high efficiency dust collectors may be needed to meet pollution regulations.

Cyclones are one of the cheaper and simpler dust collectors available but are normally less efficient than precipitators, scrubbers, or filters. A basic cyclone is shown in Figure 10-22. Dusty gas enters the tangential inlet, typically at velocities of 15–30 m/s. Gas flows in a helical path, first downward in an annulus and then upward in the center, passing out the top of the cyclone through the outlet pipe. Particles are driven to the cyclone walls by centrifugal accelerations on the order of 100 times gravity and exit from the dust outlet at the bottom of the cyclone cone.

The first step in calculating cyclone efficiency is to find the cut point diameter, d_{50}^0, the diameter of particle which is collected at 50 percent efficiency. A semiempirical expression for this is

$$d_{50}^0 = 1.16 \sqrt{\frac{\mu w_i D}{\rho_p U^2 t}} \qquad (10\text{-}44)$$

PREDICTION OF PERFORMANCE IN LARGE EQUIPMENT

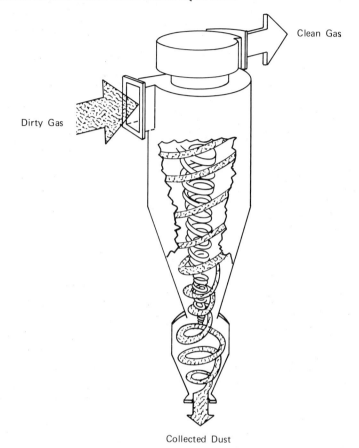

FIGURE 10-22 Typical cyclone separator used in commercial catalytic cracking units.

The average gas residence time in the cyclone can be approximated by

$$t = \frac{V}{h_i w_i U} \qquad (10\text{-}45)$$

This d_{50}^0 applies at an inlet dust loading of about 11 g/m³ (5 grains/ft³) which is a standard test value. Cyclone efficiency increases with dust loading, and the adjustment is

$$\frac{d_{50}^0}{d_{50}} = \left(\frac{L}{L^0}\right)^{0.2} \qquad (10\text{-}46)$$

where L^0 is the standard dust loading. After d_{50} is found, the collection efficiency for any other particle diameter d can be found from Figure 10-23.

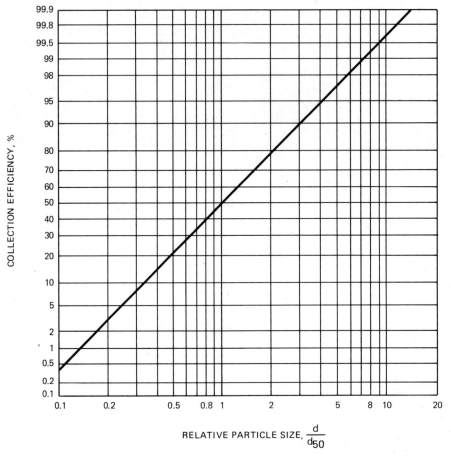

FIGURE 10-23 Normalized grade efficiency curve for cyclones.

Equation 10-44 can give directional effects for many geometric changes in design; but if data are available on the actual geometry of interest, it is much more satisfactory to use such data to modify the numerical constant in the equation. It is especially important to note that modifications to outlet pipe diameter do not enter into Equation 10-44, and the numerical constant is for cyclones with normal ratios of outlet to inlet area of 1.3 to 1.5. For certain applications area ratios well below 1.0 are now being offered. Such cyclones have notably increased efficiency at the expense of much higher pressure drop. Unfortunately, suitable published data are not available on the effect to permit quantitative prediction of effect on efficiency.

The obvious scaleup factor for cyclones arises from Equation 10-44, which shows how the diameter of collectable particles increases with cyclone size. A less apparent factor is that cyclones on small pilot plant and laboratory equipment often perform poorly and fail to achieve predicted efficiencies.

Among the reasons for this are:

1. Small cyclones are often an afterthought, poorly designed and crudely fabricated.
2. Small cyclones are frequently subject to plugging of the solids discharge line or to absence of a proper seal of the discharge line returning to the fluid bed.
3. In small units inadequate dipleg length may prevent return of solids to the bed. For geometrically similar cyclones operating at the same inlet velocity, the same dipleg length is required in the small pilot plant as in a commercial unit.
4. Effects of particle reentrainment are probably more severe in small cyclones.

For these reasons, direct use of pilot plant cyclone performance data may be risky, and a performance calculation rather than a scaleup from data is probably preferable.

IV. PRACTICAL COMMERCIAL EXPERIENCE

A. Fluid Coking

Coking is the thermal decomposition of residual petroleum fractions to useful distillates, gas, and coke. The batch process of delayed coking has been in refinery use for some six decades. The continuous fluid bed process was developed in the early 1950s by Exxon Research and Engineering Company, with the first commercial unit becoming operational in 1955 at Billings, Montana. Thirteen commercial units have been built, in sizes up to 72,000 barrels/day (11,000 t/day).

The process is shown in Figure 10-24. The vessels contain fluid beds of coke particles having an average diameter of 150 to 200 μm. Residuum feed is injected directly into the reactor vessel operating at 500–600°C where it cracks, depositing a thin layer of coke on the bed particles. Vapor products leave the bed as gas bubbles, passing out through cyclones and into a scrubber, where they are quenched and fractionated. Particles from the reactor pass down through a stripper, in which a countercurrent flow of stream strips out any interstitial hydrocarbons. The stripped coke then passes to the heater vessel, in which a portion is burned with air. Hot coke is circulated back to the reactor to supply the heat of reaction.

Early development of the fluid coking process was described by Krebs (1956). Initial screening studies for fluid coking were done in $\frac{1}{10}$ barrel/day bench scale equipment to obtain preliminary estimates of product yields and quality. Larger scale, more accurate laboratory work was then undertaken

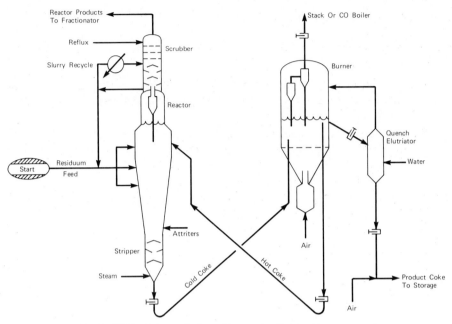

FIGURE 10-24 Schematic flow diagram for fluid coking process.

under better controlled conditions, showing favorable yields from even the poorest quality feedstocks and giving good quality gas oils for further refining. The next operation was in a 100 barrel/day pilot plant, which had originally been built for catalytic cracking. The pilot plant vessels were of small diameter but the height of the structure approached full commercial scale (about 40 m). The first commercial plant had a throughput of 3800 barrels/day, a 38-fold scaleup, and a further 19-fold increase in scale has now been achieved in commercial units.

From the standpoint of reactor engineering, the development of fluid coking was very straightforward. The residuum feed remains as a thin liquid film on coke particles within the reactor until the desired degree of cracking has occurred. At that point the molecular weight has become low enough for the oil product to vaporize, and it then rapidly leaves the bed as gas bubbles. While the yields of coke, various liquid fractions, and gases are a strong function of feed properties and reaction conditions, they are virtually independent of contacting parameters. Commercial yields are well predicted by bench scale data. In the heater vessel, oxygen consumption of 98 or 99 percent is routinely achieved, and if any scaleup effects are present they are of no practical consequence to heater operation. The major benefits of large pilot plant operation were not in classical reactor design but in (a) development of process operability, (b) generation of engineering data, and (c) manufacture of large supplies of liquid and coke products for treating and market development studies.

Development of oil feed nozzles was a major operability task. Several designs were tested to find the one giving the best distribution of oil to the bed while remaining relatively free of obstructing coke deposits. Nozzle designs were developed to permit cleaning during a run. A second problem was to develop operating conditions and engineering solutions to minimize formation of coke deposits within the reactor, particularly in the reactor cyclones. A series of "bogging" tests was conducted to determine at each temperature the maximum oil feed rate which could be attained without defluidizing the bed. Finally, because a coker is an especially complex and difficult fluid bed unit, careful development and demonstration was required for startup, shutdown, and emergency operating procedures.

Engineering studies included estimation of entrainment rates and cyclone performance, measurement of bed densities, and tests of coke standpipe and transfer line operation. Steam jet attrition was used to generate fine seed coke particles, and correlations of attriter operation were developed. Design criteria were developed for the quench elutriator, which cools product coke and selectively removes coarse particles while returning fines to the process.

As in most refining processes, liquid products from a fluid coker are a complex mixture of hydrocarbons, and as part of scaleup it was desired to demonstrate further processing, for example, fractionation or catalytic cracking. This required large amounts of product for a refinery run. Of even greater interest for fluid coking was production of large quantities of fluid coke, an entirely new product. Tests were made on coke burning characteristics in furnaces of several sizes. Development work on grinding and addition of auxiliary fuels was done to improve burning characteristics where necessary. Large-scale tests were also done on such specialty uses as carbon electrodes for the aluminum industry, and manufacture of calcium carbide, phosphorous, and carbon disulfide.

B. Fluid Hydroforming

Fluid Hydroforming is a catalytic process developed by Exxon Research and Engineering Co. for reforming heavy naphtha into a highly aromatic high octane gasoline. A flow diagram is given in Figure 10-25. In many important respects, the process is similar to fluid catalytic cracking, and the need for large-scale pilot plant work to demonstrate operability and to produce engineering design data was felt to be unimportant. The nature of the gasoline product was well known and no large quantities for processing or product tests seemed necessary.

When commercial units came on stream starting in 1954, an unanticipated debit in reactor performance became evident. While the process could meet either design throughput or design product octane levels, it could not meet

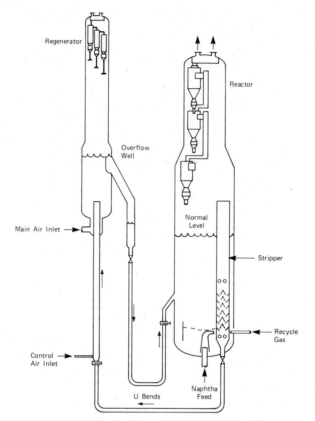

FIGURE 10-25 Schematic diagram for fluid hydroforming process.

both specifications simultaneously. This led to the extensive studies of W. G. May (1959), done in fluid beds of 0.076, 0.38, and 1.52 m in diameter. The reactor model used in his analysis is shown in Figure 10-26. Most gas passes through the reactor in plug flow in the bubble phase. Bubble gas is continuously exchanged with emulsion gas, with the crossflow ratio N_c denoting the number of times the bubble gas is exchanged before the bubble leaves the bed. All reaction occurs in the emulsion phase, which is backmixed with a solids eddy diffusivity δ_e.

Solids diffusivity was measured by injecting a small amount of radioactive solids at the top of the reactor and measuring concentration versus time of the tracer at selected locations within the bed. The solution to the unsteady-state diffusion equation is shown in Figure 10-27, which gives the ratio of instantaneous to steady-state concentrations C_o/C_∞ as a function of depth of sampling point, height H of the bed, time t, and eddy diffusivity δ_e. Figure 10-17 includes May's measured diffusivities. Crossflow was estimated by measuring

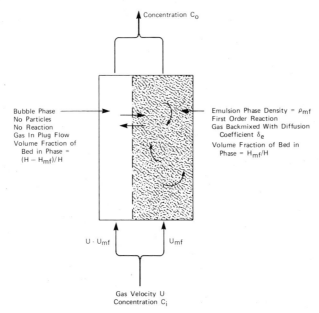

FIGURE 10-26 Contacting model for fluid bed with diffusive backmixing in emulsion phase.

the decay of outlet concentration of a tracer gas with time, when flow of tracer to the bed inlet was suddenly stopped. The slope of the curve $\ln C_o/C_i$ versus t/θ is related to diffusivity δ_e and crossflow N_C, with typical solutions shown in Figure 10-28. Crossflow depended on bed diameter, particularly below a diameter of 1.5 m.

Using May's terminology and letting

$$a = \frac{N_C}{(1 - U_{mf}/U)}, b = N_D\left(\frac{U_{mf}}{U}\right), d = N_D\left(1 - \frac{U_{mf}}{U}\right) \text{ and } f = N_R$$

the solution for chemical conversion in a partially backmixed fluid bed is

$$\frac{C_o}{C_i} = \frac{\dfrac{M_1}{P}e^{\alpha_1}\left[\dfrac{ad}{a+\alpha_1}+b\right] + \dfrac{M_2}{P}e^{\alpha_2}\left[\dfrac{ad}{a+\alpha_2}+b\right] + \dfrac{M_3}{P}e^{\alpha_3}\left[\dfrac{ad}{a+\alpha_3}+b\right]}{b+d}$$

(10-47)

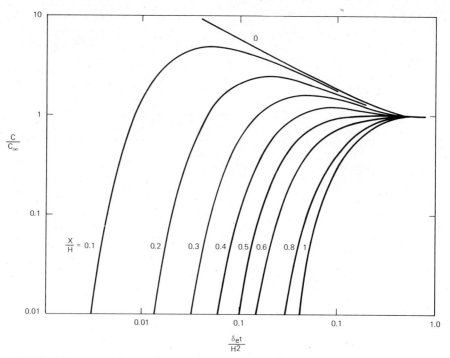

FIGURE 10-27 Unsteady-state dispersion of a tracer pulse injected into a captive fluid bed.

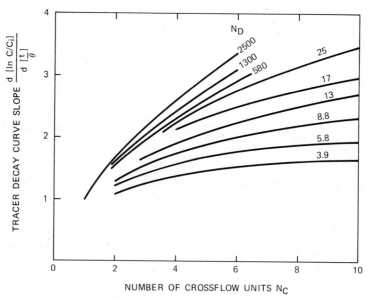

FIGURE 10-28 The effect of number of crossflow units and number of diffusion units on the slope of a gaseous tracer delay curve. Steady state flow of tracer at Concentration C_i in gas entering bottom of fluid bed is shut off at $t = 0$.

where α_1, α_2, and α_3 are the roots of the cubic equation:

$$R^3 + (a-b)R^2 - (ab + ad + fd)R - adf = 0 \qquad (10\text{-}48)$$

$$M_1 = (\alpha_2 - b)\alpha_3 e^{\alpha_3} - (\alpha_3 - b)\alpha_2 e^{\alpha_2} + \frac{ab}{a + \alpha_2}\alpha_3 e^{\alpha_3} - \frac{ab}{a + \alpha_3}\alpha_2 e^{\alpha_2}$$

$$(10\text{-}49)$$

$$M_2 = (\alpha_3 - b)\alpha_1 e^{\alpha_1} - (\alpha_1 - b)\alpha_3 e^{\alpha_3} + \frac{ab}{a + \alpha_3}\alpha_1 e^{\alpha_1} - \frac{ab}{a + \alpha_1}\alpha_3 e^{\alpha_3}$$

$$(10\text{-}50)$$

$$M_3 = (\alpha_1 - b)\alpha_2 e^{\alpha_2} - (\alpha_2 - b)\alpha_1 e^{\alpha_1} + \frac{ab}{a + \alpha_1}\alpha_2 e^{\alpha_2} - \frac{ab}{a + \alpha_2}\alpha_1 e^{\alpha_1}$$

$$(10\text{-}51)$$

$$P = \alpha_1 e^{\alpha_1}\left[\frac{a}{a + \alpha_2}(\alpha_3 - b) - \frac{a}{a + \alpha_3}(\alpha_2 - b)\right]$$

$$+ \alpha_2 e^{\alpha_2}\left[\frac{a}{a + \alpha_1}(\alpha_2 - b) - \frac{a}{a + \alpha_2}(\alpha_1 - b)\right]$$

$$+ \alpha_3 e^{\alpha_3}\left[\frac{a}{a + \alpha_1}(\alpha_2 - b) - \frac{a}{a + \alpha_2}(\alpha_1 - b)\right] \qquad (10\text{-}52)$$

Predicted conversions in both laboratory and commercial units using measured values of crossflow and diffusivity agreed well with actual reaction data. The model gave good insight into the limiting factors for the reaction, and the tracer techniques provided a quick and convenient way to measure pertinent parameters in laboratory and commercial units.

The experimental program demonstrated conclusively that the reason fluid hydroformers were performing below expectation was that crossflow was much lower and backmixing was much higher in commercial units than in the laboratory units which had been the basis of design. An extensive program, not documented in May's paper, was carried out to correct the deficiencies. The 1.52-m-diameter unit was used to test various improvements, with verification being obtained from tracer tests and reaction performance on commercial units. The importance of high levels of fine particles in the catalyst was clearly demonstrated. A new gas distributor was developed which greatly improved feed dispersion across the bed cross section and reduced "gulf stream" circulation patterns, while still meeting criteria of easy startup and mechanical strength. Finally proprietary baffles were developed which reduced backmixing and improved crossflow with only a minor debit to bed catalyst inventory. With these improvements, commercial units were able to operate properly and meet performance guarantees, marking a successful conclusion to the scaleup effort. Although the operation was a success, the patient died anyway. Fixed-bed reforming processes using platinum based catalysts were being developed simultaneously. The platinum catalyst (which could not be economically used in a fluid bed reactor because attrition losses were much too costly) gave superior yields and selectivities, and the fixed-bed process became clearly preferred. No fluid hydroformers were built after 1956, but several units remained in service for many years and one was still operating as of 1980.

C. The Shell Chlorine Process (SCP)

Perhaps the most thoroughly documented example of fluid bed scaleup is that of the Shell process for oxidation of hydrogen chloride to chlorine. Process development details have been given by Quant et al. (1963) and Fleurke (1968). Van Deemter (1961) presented the mathematical framework for reactor analysis. De Groot (1967), de Vries et al. (1972), and Van Swaaij and Zuiderwag (1972, 1973) gave details of solids mixing and gas residence time tests on several experimental reactors and a commercial unit.

The basic reaction is given by

$$4HCl + O_2 \rightarrow 2Cl_2 + 2H_2O$$

The reaction is exothermic and is equilibrium limited, with 79 percent HCl conversion being possible at 350°C. A fluid bed process was attractive because it permitted adiabatic operation with cold gas feed and prevented hot spots which would limit conversion and possibly cause catalyst deactivation.

On a laboratory scale more than 300 catalysts were tested. Factors in evaluation included:

a. High activities to permit reasonable operation at low temperatures where equilibrium conversion is highest.
b. Effect of chemical purity, specific surface area, and pore diameter of porous catalyst supports.
c. Effect of composition, concentration, and purity of numerous metal chlorides as the catalytic species.
d. Effect of impurities and possible catalyst poisons in the HCl feed stream.
e. Stability of catalyst to flue gases, which would be used commercially for initial heatup of the fluid bed.
f. Resistance of the catalyst to fluid bed attrition.

The copper based catalyst finally selected demonstrated insensitivity to impurities, good stability to 400°C, and maintenance of high activity in a life test of 2000 hr.

Process optimization studies were next conducted to determine the most desirable reaction conditions. Temperature, pressure, throughput, oxygen versus air, HCl to air ratio, and water vapor in the feed were all studied. Although pressure, pure oxygen, and dry gas feed all increased the conversion, improvement over near-atmospheric operation with air and wet feed did not justify the expense. A temperature of about 360°C and near stoichiometric HCl/air were optimum. Final choices in the process studies were the result of a careful balancing of economic and yield factors.

Product gases from the reactor pass through towers for absorption of HCl by water, drying by concentrated H_2SO_4, and absorption of Cl_2 by CCl_4. These steps also were tested for operability and design information, with special attention being given to demonstration of corrosion resistant materials of construction.

Studies of contacting for reactor design and scaleup were carried out separately from the process development studies. Tracer tests in unreacting systems and generally at ambient conditions were conducted in vessels of 0.1-, 0.3-, 0.6-, and 1.5-m diameter. Finally, tracer tests were performed in a commercial 3-m-diameter reactor and comparison was made with actual reactor performance.

The model used for reactor analysis was developed by Van Deemter (1961) and was similar to the earlier model of May (1959) discussed in the preceding example. Van Deemter developed his mathematics for gas exchange tests in which a short pulse of tracer gas is injected at the bottom of a fluid bed, and the standard deviation of the outlet pulse is measured. May analyzed the slope of the outlet concentration decay curve that resulted from suddenly stopping an inlet tracer which had been flowing continuously.

The model assumes gas in plug flow in a bubble phase exchanging at a finite rate with emulsion gas which is backmixed by solids diffusivity. The essential parameters are the number of reaction units, N_R, the number of crossflow units N_C, and the number of mixing units, $N_D = UH^2/\delta_e H_{mf}\varepsilon_{mf}$. In principle, these parameters could have been predicted from knowledge of bubble size, which in turn could be estimated from coalescence correlations or a theory of stable bubble size. In reality, as pointed out by de Groot (1967) and Van Swaaij and Zuiderweg (1972), available bubble theories did not give an adequate or consistent picture and the empirical measurements of N_C and N_D were necessary. Much of the Shell data on solids diffusivity δ_e is included in Figure

TABLE 10-2 Effect of Gas Velocity on Particle Diffusivity in a 3-m-Diameter Bed [van Swaaij and Zuiderweg (1973)]

Superficial velocity, m/s	0.1	0.2	0.3
Diffusivity relative to 0.2 m/s	0.77	1.0	1.3

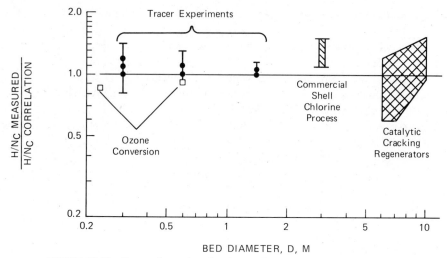

FIGURE 10-29 Comparison of H/N_c data with correlation of Equation 10-53.

10-17. These diffusivities are the maximum ones obtained at an optimum level of "fines" (particles less than 44 μm). The optimum fines level was about 24 percent in large beds and somewhat less in smaller ones. Diffusivity was a weak function of superficial gas velocity as shown in Table 10-2. Although diffusivity depended strongly on bed diameter, it was essentially independent of height.

Measurement of gas exchange rate showed a marked dependence on both bed diameter and depth. Van Swaaij and Zuiderweg (1973) found the height of a crossflow unit, H/N_C, to obey the equation

$$\frac{H}{N_C} = \left(1.8 - \frac{1.06}{D^{0.25}}\right)\left(3.5 - \frac{2.5}{H^{0.25}}\right) \qquad (10\text{-}53)$$

where bed diameter D and height H are in meters. The fit of this equation to laboratory and commercial data is shown in Figure 10-29. The effect of pressure on H/N_C was negligible, and an increase of temperature from ambient to 300°C decreased H/N_C by about 20 percent. The effect of fines concentration is given in Table 10-3.

Operating conditions for the commercial SCP reactor, 3 m diameter and 10 m deep gave $N_R = 20$, $N_c = 3.5$, and $N_D = 2.8$. As is shown in Figure 10-30,

TABLE 10-3 Effect of Fines on Height of a Crossflow Unit [van Swaaij and Zuiderweg (1973)]

Fines content, %	10	15	20
Correction factor to Equation 10-53	1.2	1.0	0.8

PRACTICAL COMMERCIAL EXPERIENCE

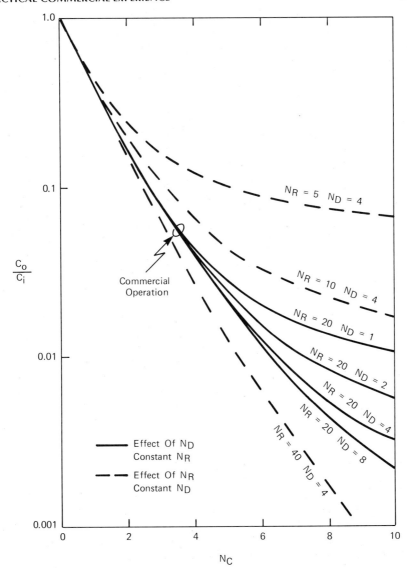

FIGURE 10-30 Effect of number of diffusion units N_D, number of reaction units N_R, and number of crossflow units N_C on reactor performance.

the reaction under these conditions is limited by crossflow and is essentially independent of backmixing.

During initial testing of the commercial reactor it became apparent that catalyst attrition was occurring at a high rate. The trouble was traced to the design of the grid nozzles. The design, shown in Figure 10-31a, had been selected so as to have a low velocity gas jet entering the bed so as to minimize

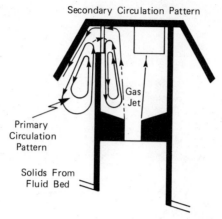

A. GAS FLOW PATTERN IN ORIGINAL SCP NOZZLE CAUSED EXCESSIVE EROSION AND ATTRITION

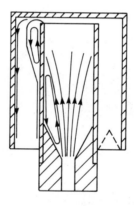

B. FLOW PATTERNS IN IMPROVED SCP NOZZLES

FIGURE 10-31 Grid nozzle flow patterns in the shell chlorine process.

attrition. However, the shape of the nozzle cap caused a circulation of bed particles back under the cap, causing severe attrition and erosive failure. Pilot plant tests with that nozzle had apparently been too short to reveal the erosion/attrition problem. The improved nozzle design in Figure 10-31*b* completely solved the difficulty.

V. PROBLEM AREAS IN SCALEUP

A. Particle Size Balances

In any fluid bed there will be some change with time in the size of individual particles. Some changes may be inherent in the process, for example, particle

PROBLEM AREAS IN SCALEUP 397

shrinkage due to combustion or growth due to a deposition reaction. Others are incidental, such as attrition of friable particles or agglomeration of sticky ones. Any feed and withdrawal or loss of particles will act to limit the change in average bed particle size in a continuous operation. Particle size is an extremely important parameter in fluid bed operation, and a critical aspect of scaleup is to assure an acceptable steady-state size distribution and possibly to manipulate the process so as to achieve that distribution.

The construction of mathematical models for determining equilibrium bed particle size distribution is extensively covered in Kunii and Levenspiel (1969), Chapter 11. The ability to construct and solve mathematical models for particle size balances considerably exceeds the knowledge of rate constants and mechanisms necessary for most design calculations. A model may be useful, however, for predicting changes in particle size due to changes in operating variables such as feed rate and cyclone efficiency.

1. Reaction Effects

The size of particles that are reactants or products of a reaction will change according to the nature of the reaction. In a few processes such as fluid coking and fluid bed calcination and drying, bed particles will grow as layers of solid material are formed on the surface. Particles are usually thought to increase in diameter at a constant rate, although experimental data to prove this are not readily available. More often particles are consumed, for example, in fluid bed combustion, and here again the change in diameter is usually presumed to occur at a constant rate. When the particle contains an unreactive residue such as ash, change of size is usually much more complex, and several ash particles usually result from combustion of a single fuel particle.

2. Attrition

Attrition of particles in a fluid bed occurs in many places and with many mechanisms. When first fed to a reactor, particles may fracture from thermal shock or from the rapid internal evolution of vapors. Attrition often continues for a time at a high rate as rough corners are knocked off and the weaker particles fracture, but the rate may later slow down when particles have become more spherical.

Within a bed, particle size is usually reduced by autogenous grinding. Abrasion slowly wears away the particle surface, causing shrinkage of the identifiable particle and generating very fine dust from the abraded material. The rate for this type of attrition is usually thought to be proportional to particle surface area, that is

$$\frac{dw}{dt} = k_0 A \rho_p \quad \text{or} \quad \frac{d(d_p)}{dt} = k_0 \qquad (10\text{-}54)$$

Impact attrition occurs in cyclones and where high velocity gas jets enter a bed. In this mode the original particle usually fractures into several large fragments. Larger particles, having a greater mass per unit cross section, are usually more subject to impact attrition than smaller particles.

There are no general correlations of attrition rates in fluid bed processes. A catalytic cracker may lose 1 percent of its catalyst inventory per day due to attrition, but losses may be temporarily much higher. A single missing restriction orifice on an aeration tap can double attrition. Likewise a new rough refractory lining in cyclones can cause a several fold increase in losses for a few weeks, until the lining is worn smooth.

3. Agglomeration

Agglomeration of particles can be thought of as the reverse of attrition. Particles may fuse together either due to an inherent stickiness, as with iron particles above 800°C, or due to a "glue" such as a thick aqueous salt solution being sprayed into a hot bed. Agglomeration can have a more sudden and serious effect on ability to operate a fluid bed than attrition. Particles can fuse together in a large mass so as to defluidize or "bog" a bed. In a less extreme case a handful of particles can agglomerate to pea sized to fist sized chunks which will settle out in low spots in the bed or in circulation lines.

While attrition seems to be an inherent characteristic of a fluid bed process, agglomeration is much more sensitive to specific operating conditions and mechanical details. High fluidizing velocity and elimination of stagnant zones are always helpful in reducing agglomeration. When a liquid feed is being introduced, high bed temperature, low feed rate, and attention to feed nozzle design are important in minimizing the problem. When the particles themselves are close to fusion temperature, a reduction of temperature is effective. Provision for on-stream chunk removal permits operation despite agglomerate formation.

B. Erosion

Erosion is an important design consideration for commercial fluid bed units handling abrasive particles. Because erosion is a strong function of specific mechanical design and because it may require many thousands of hours of operation to become apparent, erosion problems are seldom seen or solved in bench scale and pilot plant units. Erosion is minimized by careful mechanical design, conservative choice of operating variables, and use of erosion resistant materials.

In catalytic crackers, erosion of cyclones is probably the greatest single factor limiting run length. Cyclone inlet velocity is generally kept below 30 m/s, although low velocities result directionally in low efficiency and a more expensive cyclone installation. Cyclones are often lined with an extremely hard refractory such as Resco AA 22, and the refractory must be applied with meticulous attention to detail. It is common to size cyclones to permit easy access for making repairs to linings during turnarounds. Reactor manways should be sized to facilitate removal and replacement of cyclones when necessary.

Slide valves in solids transfer lines and in overhead lines handling dusty gas are also in critical erosion service. A slide valve is basically a gate valve constructed with large clearances and gas purges, to permit the gate to slide freely without jamming in the presence of solids. Additional gas purges to the valve bonnet keep erosive particles out of the valve stem packing gland annulus. Valve slides are often Stellite hard-surfaced to reduce wear. It is common practice to use two slide valves in series so as to have a spare when the first one becomes damaged. Pressure drop through slide valves is usually limited to 50 kPa to reduce erosion.

Gas jets in fluid beds, from grid holes or aeration connections, can be serious sources of erosion. Internals in a bed, such as cyclone diplegs or heat transfer tubes are usually placed a meter or more above an upward pointing grid hole. Alternatively, grid holes may be omitted immediately below internals. Aeration taps should be located and aimed so as not to impinge on nearby surfaces, and exit velocities should be kept below 30 m/s.

C. Choice of Reactor Models and Model Parameters

A heated debate exists between proponents of "fundamental" models and adherents of "empirical" models for fluid bed reactor performance. The fundamental approach employs the simple physical concepts of Section II such as bubble velocity, bed expansion, gas exchange, clouds, and wakes. There "first principles" of bubbles are combined into expressions such as Equation 10-20 for reactor performance. Dozens of such models with different combinations of assumptions have been developed. Empiricism may creep into such models in such aspects as the use of experimentally based correlations to predict bubble growth rate. In some cases bubble size becomes merely a curve fitting parameter in a reactor model rather than an independently determined property. The model has then become essentially empirical, although the form of the model expression has a theoretical basis.

The "empirical" models, on the other hand, regard the reactor as a black box having experimental characteristics of gas crossflow and backmixing which determine performance. Such models were used in the examples of Fluid Hydroforming and the Shell Chlorine Process in Section IV. These examples have demonstrated quite convincingly that tracer measurements of gas residence time and emulsion solids diffusivity can be used to give accurate predictions of fluid bed performance as a chemical reactor.

In theory the "fundamental" bubble models could be used to calculate the information gained empirically by the tracer tests, but deGroot (1967) and deVries et al. (1972) noted that for their experiments in large equipment, the available bubble models did not give predictions that were consistent with the data. Of particular interest are the scaleup effects on eddy diffusivity in Figure

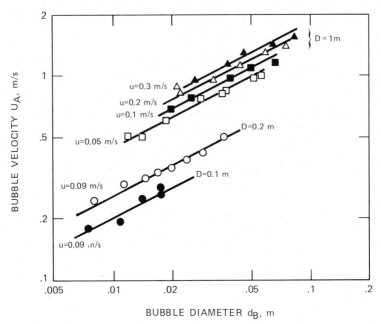

FIGURE 10-32 Measurements showing that bubbles of a given diameter rise faster in large beds than small ones.

10-17 and in the height of a gas transfer unit of Equation 10-53. Simple bubble theories of fluid bed performance do not adequately account for such effects.

The reason for the failure is indicated by Werther (1977). His extensive measurements on bubble flow patterns in fluid beds up to 1 m diameter showed not only bubble coalescence with height but also nonuniform horizontal concentrations of bubble flow. Preferred tracks were established with bubbles moving away from the walls and toward the center of the bed as they rose to the surface. Complementary solids circulation patterns were also present, which doubtlessly are responsible for the high solids mixing rates measured by other investigators. Werther (1978) quantified these effects by showing that bubbles of a given size rose faster in a large bed than in a small one. The data are shown in Figure 10-32. Using these experimental values for bubble velocity along with separately determined relations for bubble coalescence, mass transfer coefficient and nonideal bubble flow, Werther (1978) was able to predict reactor performance from a bubble based model.

While the Werther mechanistic approach is clearly the wave of the future, it presently requires more empirical data of a type that is much less easy to measure than the empirical crossflow/backmixing models. The purely empirical models have been calibrated over a somewhat greater range of bed diameters and operating conditions, and they accommodate backmixing effects which are not part of the Werther approach. Unfortunately there is not much empirical data which permits adjustment for gas or particle properties. It would seem then that perhaps the most responsible approach at present would be:

Obtain kinetic data from small slug flow fluid beds using the theoretical approach of Equations 10-19 and 10-20.

If the largest pilot plant is not in slug flow: (a) Estimate the height of a crossflow unit, H/N_C, from Equation 10-47 using conversion data and estimating solids diffusivity from Figure 10-17. (b) Using Equation 10-53, calculate the ratio of H/N_C for commercial bed height and diameter to H/N_C for the pilot plant. (c) Multiply the measured H/N_C from step (a) by this ratio to obtain H/N_C commercial. (d) Use Figure 10-17 and Section III-C to obtain commercial solids diffusivities and calculate reactor conversion from Equation 10-47.

If the largest pilot plant is in slug flow: (a) Estimate the largest bubble size which will be present according to Section III-A. (b) Estimate H/N_C for this bubble from Equation 10-18 using the value of H for the largest pilot plant. (c) From Equation 10-53 calculate H/N_C for a bed having a height of the largest pilot plant and a diameter of four times the value of D_B from step (a). (This is the smallest diameter bed which will not slug if sphere cap bubbles of diameter $4D_B$ are stable.) Likewise calculate H/N_C for commercial reactor dimensions. (d) Scale up the value obtained in step (b) by the ratio of H/N_C values calculated in step (c). (e) Use Figure 10-17 and Section III-C to estimate commercial solids diffusivity and calculate reactor conversion from Equation 10-47.

D. Uncertainties in Performance

Clearly the scaleup of fluid bed reactor performance is not yet a very exact science. The procedures given here can only give rough estimates. After the best estimate is made, sensitivity calculations should be made to evaluate the importance of accurate values of crossflow and backmixing and to study the cost involved in building a "safe" design. For instance, if N_D is greater than 3 or 4, conversion is sensitive to N_D only if very high conversions are of interest. If N_R is much greater than N_C or vice versa, an accurate estimate of the larger parameter is unimportant. If N_R is large, solids diffusivity will be unimportant.

Very often uncertainties in reactor performance can be accommodated by adding a safety factor to the height of the fluid bed. Providing for taller reactors at the design stage is usually cheap, since greater vessel "straight side" adds negligibly to vessel cost. If such conservatism ultimately proves unnecessary for obtaining design performance, it may still be a worthwhile preinvestment if the throughput of the reactor is later to be increased above original design. An alternative design safety factor would be to provide for easy corrective installation of baffles in the fluid bed in case the reactor performed below expectations. Baffles somewhat reduce bubble sizes and greatly reduce backmixing and the gulf stream circulation which contributes to high bubble velocity.

Above all the engineer cannot rely blindly on models or textbook equations in the scaleup of fluid beds. He must understand the models well enough to adapt them to his process, for each new process has a new set of important considerations, limiting conditions, and uncertainties.

NOMENCLATURE

A	Surface area of particle, m^2
C	Concentration
C_i	Inlet concentration of reactant
C_o	Outlet concentration of reactant
C_∞	Steady-state concentration from a tracer pulse
C_p	Gas heat capacity, J/kg K
C_{ps}	Solids heat capacity, J/kg K
d_i	Mean particle diameter of ith size fraction, m
d_o	Orifice diameter, m
d_p	Particle diameter, m
d_t	Diameter of heat transfer tube, m
d_{50}	Diameter of particle collected at 50 percent efficiency at dust loading L, m
d_{50}^0	Diameter of particle collected at 50 percent efficiency at dust loading L^0, m
D	Diameter of fluid bed or cyclone, m
D_B	Diameter of bubble, m
D_{BM}	Maximum bubble diameter, m
D_{BO}	Bubble diameter at distributor, m
D_C	Diameter of cloud surrounding bubble, m
g	Acceleration of gravity, m/s^2
h	Heat transfer coefficient, W/m^2 K
h_i	Height of cyclone inlet, m
H	Height of fluid bed or cyclone barrel, m
$H_{\max}$	Maximum height of slugging fluid bed, m
H_{mf}	Height of bed inventory at minimum fluidization conditions, m
I	Surface integral for diffusion from a slug
k	Rate constant for first order reaction in emulsion phase, s^{-1}
k_0	Attrition constant, m/s
k_b	Bed thermal conductivity, W/m K

NOMENCLATURE

- k_g Gas thermal conductivity, W/m K
- L Dust loading at cyclone inlet, kg/m³
- L^o Standard dust loading = 11.5×10^{-3} kg/m³
- L_s Length of slug, m
- M Slug shape factor = $4V_s/\pi D^3$
- n Number of grid holes
- n_i Number of particles in i the size fraction
- N_C Number of crossflow units for gas exchange between bubble and emulsion = $QH/V_B U_A$
- N_D Number of diffusion units for emulsion mixing = $UH^2/\delta_e H_{mf} \varepsilon$
- N_R Number of reaction units = kH_{mf}/U
- Nu_p Particle Nusselt number hd_p/k_g
- ΔP_B Pressure drop across fluid bed, Pa
- ΔP_G Pressure drop across grid or distributor plate, Pa
- Q Volumetric exchange rate of gas from a bubble, m³/s
- Q_D Exchange rate due to diffusion, m³/s
- Q_F Exchange rate due to gas throughflow, m³/s
- Re_p Particle Reynolds number, $d_p U \rho_g/\mu$
- t Time, s
- U Superficial gas velocity in bed; cyclone inlet velocity, m/s
- U_A Absolute rise velocity of a bubble, m/s
- U_B Rise velocity of a single bubble in a bed at minimum fluidizing conditions, m/s
- U_{mf} Superficial gas velocity at minimum fluidization, m/s
- v Gas flow rate, m³/s
- V Cyclone volume, m³
- V_B Bubble volume, m³
- V_j Velocity of dissipating entrainment jet, m/s
- V_o Orifice velocity, m/s
- V_T Single particle terminal velocity, m/s
- w Weight loss in attrition, kg
- w_i Width of cyclone inlet, m
- W Mass flux of entrained particles, kg/m² s
- W_i Weight of ith particle size fraction, kg
- X Distance, m

GREEK

- δ_e Eddy diffusivity of bed particles, m²/s
- δ_g Molecular diffusivity of gas, m²/s

ε Void fraction of bed
ε_{mf} Void fraction at minimum fluidization
θ Mean residence time, s
μ Gas viscosity, Pa s
ρ Expanded bed density, kg/m^3
ρ_g Gas density, kg/m^3
ρ_{mf} Bed density at minimum fluidization, kg/m^3
ρ_p Particle density, kg/m^3

REFERENCES

Anton, J. R., PhD Thesis, State University of Iowa, 1953.
Baird, M. H. I., and Davidson, J. F., *Chemical Eng. Sci.*, **17**, 87 (1962).
Bart, R., PhD Dissertation, Massachusetts Institute of Technology, 1950.
Barker, J. J., *Ind. Eng. Chem.*, **57** (5), 33 (1965).
Bradshaw, R. D., and Myers, J. E., *AIChE J*, **9**, 50 (1963).
Clift, R., Grace, J. R., and Weber, M. E., *Ind. Eng. Chem. Fundamentals*, **13**, 45 (1974).
Davidson, J. F., *Trans. Inst. Chem. Eng.*, **39**, 230 (1961).
Davidson, J. F., and Harrison, D., *Fluidised Particles*, The University Press, Cambridge, 1963.
Davidson, J. F., and Harrison, D., *Fluidization*, Academic Press, New York, 1971.
Davidson, J. F., Paul, R. C., Smith, J. S., and Duxbury, H. A., *Trans. Inst. Chem. Engrs.*, **37**, 323 (1959).
Davies, R. M., and Taylor, G. I., *Proc. Roy. Soc.*, **A200**, 375 (1950).
deGroot, J. H., Proc. Int. Symposium on Fluidization. Netherlands University Press, Amsterdam, p. 348, 1967.
deVries, R. J., vanSwaaij, W. P. M., Mantovani, C., and Heijkoop, A., Proc. 2nd International Symposium on Chemical Reaction Eng., Elsevier Publishing Corp., Amsterdam, pp. B9–59, 1972.
Donnadieu, G., *Rev., Inst. Franc du Petrole*, **16**, 1330 (1961).
Ferron, J. R., and Watson, C. C., *Chem. Eng. Prog. Symp. Series*, **62** (67), 51 (1966).
Fleurke, K. H., *The Chemical Engineer*, No. 216, CE 41 (1968).
Frantz, J. F., PhD Thesis, Louisiana State University, 1958.
Fritz, J. C., PhD Thesis, University of Wisconsin, 1956.
Frye, C. G., Lake, W. C., and Eckstrom, H. C., *AIChE J*, **4**, 403 (1958).
Harrison, D., and Leung, L. S., *Trans. Instn. Chem. Engrs.*, **39**, 409 (1961).
Harrison, D., and Leung, L. S., Symposium on the Interaction Between Fluids and Particles, Instn. Chem. Engrs., London, p. 127, 1962.
Heertjes, P. M., and McKibbins, S. W., *Chem. Eng. Sci.*, **5**, 161 (1956).
Holman, J. P., Moore, T. W., and Wong, V. M., *Ind. Eng. Chem.* (*Fund.*), **4**, 21 (1965).
Hovmand, S., and Davidson, J. R., Chapter 5 in *Fluidization*, J. F. Davidson and D. Harrison, Eds., Academic Press, New York, 1971.
Juveland, A. C., Dougherty, J. E., and Deinken, H. P., *Ind. Eng. Chem.* (*Fund.*), **5**, 439 (1966).
Kothari, A. K., M.S. Thesis, Illinois Institute of Technology, Chicago, 1967.
Krebs, R. W., p. 185 in *Fluidization*, D. F. Othmer, ed., Reinhold, New York, 1956.

Kunii, D. and Levenspiel, O., *Fluidization Engineering*, Wiley, New York, 1969 (Reprinted by Robert G. Krieger Publishing Co., Inc., Huntington, New York, 1977).

Leva, M., *Chem. Eng. Prog.*, **47**, 39 (1951).

Lewis, W. K., Gililand, E. R., and Girouard, H., *CEP Symp. Ser.*, **58**(38), 87 (1962).

Matsen, J. M., *AIChE Symp. Ser.*, **69**(128), 30 (1973).

Matsen, J. M., Hovmand, S., and Davidson, J. F., *Chem. Eng. Sci.*, **24**, 1743 (1969).

Matsen, J. M., and Fels, M., p. 658 in *Encyclopedia of Environmental Science and Engineering*, J. R. Pfafflin and E. N. Ziegler, eds., Gordon and Breach, New York, 1976.

May, W. G., *Chem. Engr. Prog.*, **55**(12), 49 (1959).

Merry, J. M. D., *AIChE J*, **21**, 507 (1975).

Mori, S., and Wen, C. Y., *AIChE J*, **21**, 109 (1975).

Quant, J. Th., van Dam, J., Engel, W. F., and Wattimean, F., *The Chemical Engineer*, No. 170, CE 224 (1963).

Richardson, J. F., and Ayers, P., *Trans. Inst. Chem. Engrs.*, **37**, 314 (1959).

Rowe, P. N., *Trans. Inst. Chem. Engrs.*, **39**, 43 (1961).

Rowe, P. N., Chapter 4 in *Fluidization*, J. F. Davidson and D. Harrison, eds., Academic Press, New York, 1971.

Rowe, P. N., *Chem, Engr. Sci.*, **31**, 285 (1976).

Rozenthal, E. O., Ph.D. Diss. Tekhnol, Inst. Food Industry, Moscow, 1955.

Shakhova, N. A., Ph.D. Thesis, Moscow Inst. of Chem. Machinery, Moscow, 1955.

Stewart, P. S. B., Ph.D. Dissertation, Cambridge University, 1965.

Sunkoori, N. R., and Kaparthi, R., *Chem. Eng. Sci.*, **12**, 166 (1960).

van Deemter, J. J., *Chem. Eng. Sci.*, **13**, 143 (1961).

van Deemter, J. J., p. 69 in *Fluidization*, J. R. Grace and J. M. Matsen, eds., Plenum Press, New York, 1980.

van Swaaij, W. P. M., and Zuiderweg, F. J., Proc. 2nd International Symposium on Chemical Reaction Engineer., Elsevier Publishing Corp., Amsterdam, pp. B9–25, 1972.

van Swaaij, W. P. M., and Zuiderweg, F. J., Proc. International Symposium on Fluidization and Its Applications. Ste Chimie Industrielle, Toulouse, 1973.

Volk, W., Johson, C. A., and Stotler, H. H., *Chem. Eng. Prog. Symp. Ser. No. 38*, **58**, 38 (1962).

Vreedenberg, H. A., *Chem. Eng. Sci.*, **9**, 52 (1958).

Vreedenberg, H. A., *Chem. Eng. Sci.*, **11**, 174 (1960).

Walton, J. S., Olson, R. L., and Levenspiel, O., *Ind. Eng. Chem.*, **44**, 1474 (1952).

Wamsley, W. W., and Johanson, L. N., *Chem. Eng. Prog.*, **50**, 347 (1954).

Werther, J., *Chem. Ing. Tech.*, **49**, 193 (1977).

Werther, J., *German Chem. Engr.*, **1**, 243 (1978).

Zenz, F. A., and Othmer, D. F., *Fluidization and Fluid-Particle Systems*, Rheinhold, New York, 1960.

Zenz, F. A., and Weil, N. A., *AIChE J*, **4**, 472 (1958).

11
LAMINAR FLOW PROCESSES

E. B. NAUMAN

I.	Major Issues in Scaleup	407
II.	Undisturbed Flows in Tubes	410
	A. The Equations of Change	410
	B. Isothermal Flows	412
	C. Laminar Flow Heat Exchangers	413
	D. Reacting Flows	415
III.	Motionless Mixers	416
	A. Blending Applications	417
	B. Heat Transfer Applications	418
	C. Reaction Applications	419
IV.	Moving Wall Devices	421
	A. Cavity Flows	421
	B. Extruders	424
	C. Stirred Tanks	426
V.	Overview	427
	Nomenclature	428
	References	429

I. MAJOR ISSUES IN SCALEUP

Consider the laminar flow of a constant viscosity fluid through a pipe having radius R_1 and length L_1. Suppose that the system is operating at volumetric flow rate Q_1, that the pressure drop across the system is ΔP_1, and that the mean residence time is $\bar{t}_1$:

$$\bar{t}_1 = \frac{V_1}{Q_1} = \frac{L_1}{\bar{u}_1} \tag{11-1}$$

where V_1 is the system volume and $\bar{u}_1$ is the mean axial velocity. It is desired to scale this system to a higher flow rate, Q_2. How should the new system be designed? There is no unique answer to this question. One might, for example attempt to push the greater flow rate through the same pipe; but this obviously changes many aspects of the system. If $Q_2 > Q_1$, then $\Delta P_2 > \Delta P_1$, $\bar{u}_2 > \bar{u}_1$, and $\bar{t}_2 < \bar{t}_1$.

A more common form of scaleup is to change the pipe dimensions but to maintain geometric similarity:

$$\frac{L_1}{R_1} = \frac{L_2}{R_2} \tag{11-2}$$

This form of scaleup allows $\Delta P_1 = \Delta P_2$ and $\bar{t}_1 = \bar{t}_2$. The volumetric flow rate increases as the total volume of the pipe and thus as the cube of the characteristic dimension:

$$\frac{Q_2}{Q_1} = \frac{R_2^3}{R_1^3} \tag{11-3}$$

In Chapter 8 it is shown that Equation 11-3 represents a most desirable form of scaleup since it keeps many important features of the system identical between large and small sizes. However, it is possible to maintain all features of the system at values which are independent of scale. In laminar flow, maintaining geometric similarity, constant pressure drop and constant residence time leaves the Reynolds number as a weakly increasing function of size:

$$\frac{(\text{Re})_2}{(\text{Re})_1} = \frac{R_2}{R_1} \tag{11-4}$$

Thus at sufficiently large sizes the flow will become turbulent, and the large system will behave very differently than the small system.

Since Reynolds number increases much more slowly with scale than volumetric flow rate, it is quite possible that the scaleup of a laminar system will remain laminar. However, it will not remain identical to the small system in aspects of chemical processing such as heat and mass transfer. These features of a laminar system will increase only as the square of the characteristic dimension so that a scaleup which depends on heat and mass transfer—and most do—must be done with care. The best approach is to use a mathematical model of the system and to scale according to the dictates of the model.

Fortunately, good models are often obtainable for laminar systems. This chapter outlines how such models can be developed and applied to scaleup calculations. Before doing so, however, we first give a qualitative example of the kinds of phenomena that can arise during laminar flow scaleup.

Suppose an exothermic liquid-phase chemical reaction is occurring in a capillary tube. Suppose further that the products of the reaction are more viscous than the reactants. The flow is laminar in the capillary tube, and we suppose it to remain laminar throughout the scaleup. Radial diffusion of heat is important in a capillary so that the contents of the reactor can be kept approximately isothermal at the wall temperature. The velocity profile will have the parabolic form of Poiseuille flow:

$$u_z(r) = 2\bar{u}(1 - r^2/R^2) \tag{11-5}$$

This means that fluid near the wall will move more slowly and will tend to react more completely.

Radial diffusion is easy to achieve in a capillary because of the short distance to be traversed; and this diffusion will tend to eliminate concentration gradients in the radial direction. Molecules will diffuse back and forth across the tube many times as they flow in the axial direction, sometimes being near the centerline and sometimes near the wall. Even though the velocity profile is parabolic, individual molecules will sample the different velocities in the profile many times and all will emerge with something near the average residence time, $\bar{t} = L/\bar{u}$.

If the aspect ratio, L/R, for the tube is reasonably large, say 100 or more, then axial diffusion will be negligible. The reaction will cause composition distributions to exist in the axial direction but not radially. The viscosity will increase gradually down the tube but will be uniform in the radial direction. The capillary tube will behave like a piston flow reactor.

When the tube radius is increased, the path length for diffusion will also increase but the time available will not change if constant $\bar{t}$ scaleup is used. This means that, for some sufficiently large radius, concentration gradients will arise due to greater reaction at the walls caused by the parabolic velocity profile. As a reactor, the larger tube will behave rather worse than the capillary tube. This statement is true even if the reaction has no effect on viscosity; it is dramatically true when the viscosity increases with increasing conversion as in a polymerization (Lynn and Huff, 1971). Material near the tube wall travels slowly, reacts further, becomes more viscous, and slows down still further. Meanwhile, material near the centerline must accelerate to maintain continuity of flow. This leads to elongated velocity profiles such as in Figure 11-1 and in extreme cases to hydrodynamic instabilities.

In large tubes, radial diffusion becomes negligible. Merrill and Hamrin (1970) state that it has a negligible effect on the reaction if

$$\frac{\mathscr{D}_A \bar{t}}{R^2} < 3 \times 10^{-3} \tag{11-6}$$

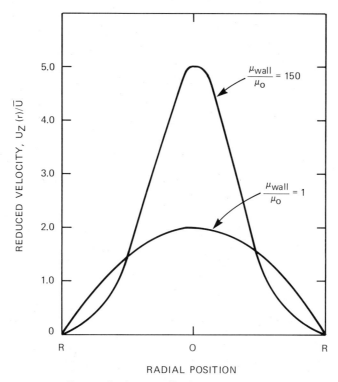

FIGURE 11-1 Elongated velocity profile due to high viscosities at tube wall.

where $\mathscr{D}_A$ is the molecular diffusivity of the reactant. Molecules initially at radial position $\xi = r/R$ will remain near that position throughout their stay in the reactor. The system will exhibit the full distribution of residence times corresponding to the parabolic velocity profile since the velocity averaging process is no longer operative. Reaction yields will generally be less than for smaller diameter tubes having the same $\bar{t}$. Achieving the same results on a large scale as on the small will require a larger aspect ratio, $L_2/R_2 > L_1/R_1$ and a higher pressure drop, $\Delta P_2 > \Delta P_1$. This is not the way one would normally like to scale a reactor.

This analysis has treated the system as though it remained isothermal during scaleup. This is satisfactory for a reasonable range of tube diameters since thermal diffusivities tend to be many orders of magnitude larger than molecular diffusivities. Eventually however, a tube size is reached where the heat of reaction can no longer be removed through the tube walls, as discussed by Valsamis and Biesenberger (1975). Hot spots develop with consequent changes in selectivity. Hot spots typically develop somewhere near the tube wall: near enough for the longer residence times to give a large exotherm, far enough away to be insulated from the cool wall. Figure 11-2 shows a typical temperature profile for a laminar flow reactor approaching thermal runaway.

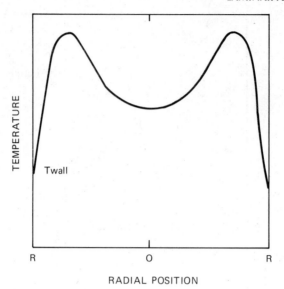

FIGURE 11-2 Temperature profile for laminar flow reactor approaching thermal runaway.

It may seem that a completely conservative approach to scaling up a laminar flow process is to design for no diffusion of mass or heat at all. This may give an uneconomic process. More typically, it will give an infeasible one since most chemical processes require some heat exchange or molecular level mixing. Even if it were feasible to neglect diffusional processes in the scaled up version, careful modeling of the pilot scale operation is necessary to account for the favorable effects of diffusion on the small scale so that they will not be expected on the large scale.

II. UNDISTURBED FLOWS IN TUBES

Some techniques for analyzing and scaling up forced laminar convection in long, straight, circular tubes will be outlined here. The methodology will be far from inclusive. The results are restricted to rather low Reynolds numbers and a host of secondary effects (which may become of primary importance in special cases) are ignored. Nevertheless, the results will be useful for a wide variety of practical scaleup problems. Also, they represent most of the situations for which the equations of change can be solved precisely.

A. The Equations of Change

Temperature, composition, and velocity profiles are governed by a set of partial differential equations which are known collectively as the equations of

UNDISTURBED FLOWS IN TUBES

change. For velocity, we assume cylindrical symmetry and consequently ignore u_θ. The dominant velocity component is the axial one, u_z, and is given by

$$0 = \frac{-dP}{dz} + \frac{1}{r}\frac{\partial}{\partial r}(r\tau_{rz}) \tag{11-7}$$

where, for a Newtonian fluid of variable viscosity,

$$\tau_{rz} = -\mu \frac{\partial u_z}{\partial r} \tag{11-8}$$

This formulation makes a number of approximations which consequently restricts its applicability to low Reynolds numbers, say Re < 100. One approximation is ignoring the radial component of velocity, u_r, but estimates of this component can be obtained from the continuity equation

$$\frac{1}{r}\frac{\partial}{\partial r}(\rho r u_r) + \frac{\partial}{\partial z}(\rho u_z) = 0 \tag{11-9}$$

Equations 11-7 through 11-9 allow calculation of the velocity profile and pressure drop for flow down a tube when the viscosity varies due to heat exchange or reaction. The temperature in the tube is given by the energy balance approximated as

$$C_v \frac{\partial(\rho u_r T)}{\partial r} + C_v \frac{\partial(\rho u_z T)}{\partial z} = \frac{\kappa}{r}\frac{\partial}{\partial r}\left(r\frac{\partial T}{\partial r}\right) - \Delta H \mathscr{R} + \mu\left(\frac{\partial u_z}{\partial r}\right)^2 \tag{11-10}$$

The first term in this equation represents radical convection and can be neglected when the axial velocity profile remains approximately constant. The last term in this equation represents heat generation by viscous dissipation. This is usually negligible for pipe flows but can be important in moving wall devices. The $\Delta H \mathscr{R}$ term represents heat of reaction and is coupled through the reaction rate to the equations of component continuity. For a single component A which has concentration C_A,

$$C_v \frac{\partial(u_r C_A)}{\partial r} + \frac{\partial(u_z C_A)}{\partial z} = \frac{\mathscr{D}_A}{r}\frac{\partial}{\partial r}\left(r\frac{\partial C_A}{\partial r}\right) + \mathscr{R}_A \tag{11-11}$$

A separate continuity equation should be written for each reactive component.

Equations 11-7 through 11-11 together define temperature, composition, and velocity profiles for low Reynolds number laminar flows in tubes. The simultaneous solution of these equations is now quite feasible and should be used as the basis for scaleup calculations.

B. Isothermal Flows

For a Newtonian fluid of constant viscosity, Equations 11-7 and 11-8 give

$$0 = \frac{dP}{dz} - \frac{\mu}{r}\left(\frac{d}{dr}r\frac{du_z}{dr}\right) \qquad (11\text{-}12)$$

The associated boundary conditions are zero slip at the wall,

$$u_z = 0 \quad \text{at} \quad r = R \qquad (11\text{-}13)$$

and radial symmetry

$$\frac{\partial u_z}{\partial r} = 0 \quad \text{at} \quad r = 0 \qquad (11\text{-}14)$$

Solution gives Equation 11-5 and the Poiseuille equation for pressure drop

$$\frac{\Delta P}{L} = \frac{P(z=0) - P(z=L)}{L} = \frac{8\mu Q}{\pi R^4} \qquad (11\text{-}15)$$

This result provides the basis for Equations 11-1 through 11-4. Isothermal laminar flow in long tubes scales very well. This is true even when the fluid is non-Newtonian.

For a power law (Ostwald–deWaele) fluid,

$$\tau_{rz} = -\zeta\left(\frac{\partial u_z}{\partial r}\right)^n \qquad (11\text{-}16)$$

where ζ and $n > 0$ are material constants. Substituting into Equation 11-7 and using the same boundary conditions as before gives

$$u_z(r) = \left(\frac{3n+1}{n+1}\right)\bar{u}\left[1 - \left(\frac{r}{R}\right)^{(n+1)/n}\right] \qquad (11\text{-}17)$$

This reduces to the Newtonian case, Equation 11-5, when $n = 1$. Corresponding to the Poiseuille equation for pressure drop we have

$$\frac{\Delta P}{L} = \frac{(6n+2)\zeta Q}{\pi n R^{(3n+1)/n}} \qquad (11\text{-}18)$$

which also reduces to the Newtonian case, Equation 11-15, when $n = 1$.

Scaleup with geometric similarity (Equation 11-12) and constant pressure drop gives

$$\frac{Q_2}{Q_1} = \left(\frac{R_2}{R_1}\right)^{(2n+1)/n} \qquad (\Delta P \text{ constant}) \qquad (11\text{-}19)$$

UNDISTURBED FLOWS IN TUBES

While some suspensions have $n > 1$, the far more common case is $n < 1$ which is exhibited by most polymer melts and solutions. Values of n in the range 0.2–0.5 are not uncommon. This results in volumetric flow rates increasing as the fourth through seventh power of the tube radius at constant pressure drop. This form of scaleup, while desirable for some purposes, gives a substantial decrease in mean residence time. A constant $\bar{t}$ scaleup satisfies Equation 11-3 rather than Equation 11-19 and results in substantial reduction in pressure drop as the scale increases:

$$\frac{\Delta P_2}{\Delta P_1} = \left(\frac{R_2}{R_1}\right)^{(n-1)/n} \quad (\bar{t} \text{ constant}) \tag{11-20}$$

Such a scaleup can be more desirable for non-Newtonian than that for Newtonian fluids. Larger size equipment can operate at lower pressures and thus remain mechanically feasible to construct.

C. Laminar Flow Heat Exchangers

Equation 11-10 can be applied to the heating of a Newtonian fluid in a tubular heat exchanger. Assuming constant physical properties, no heat of reaction and negligible viscous dissipation gives

$$2\bar{u}\left(1 - \frac{r^2}{R^2}\right)\frac{\partial T}{\partial z} = \alpha\left(\frac{\partial^2 T}{\partial r^2} + \frac{1}{r}\frac{\partial T}{\partial r}\right) \tag{11-21}$$

where $\alpha = \kappa/\rho C_v$. Typical boundary conditions are

$$T = T_{in} \quad \text{at} \quad z = 0 \tag{11-22}$$

$$\frac{\partial T}{\partial r} = 0 \quad \text{at} \quad r = 0 \tag{11-23}$$

$$T = T_{wall} \quad \text{at} \quad r = R \tag{11-24}$$

The situation specified here is known as the *Graetz problem*. An analytical solution is known but awkward to evaluate (Brown, 1960). Practical problems are readily solved numerically with even more complicated boundary conditions such as $T_{in} = T_{in}(r)$ and $T_{wall} = T_{wall}(z)$.

When expressed in dimensionless form, the Graetz problem becomes

$$2(1 - \xi^2)\frac{\partial \Omega}{\partial \lambda} = \left(\frac{\alpha \bar{t}}{R^2}\right)\left(\frac{\partial^2 \Omega}{\partial \xi^2} + \frac{1}{\xi}\frac{\partial \Omega}{\partial \xi}\right) \tag{11-25}$$

where $\xi = r/R$, $\lambda = z/L$, and $\Omega = (T - T_{wall})/(T_{in} - T_{wall})$. It is apparent from Equation 11-25 that the dimensionless temperature distribution depends only on the Graetz parameter, $\alpha \bar{t}/R^2$. Normally in scaleup we will have the same inlet and wall temperatures for the larger unit as for the small. Then the outlet temperature will also be the same if

$$\frac{\bar{t}_1}{R_1^2} = \frac{\bar{t}_2}{R_2^2} \tag{11-26}$$

where the thermal diffusivity, α, was assumed to be constant.

Scaleup at constant $\bar{t}$ is seen to require a longer length of the same diameter tube or multiple tubes in parallel. A single tube scaleup with geometric similarity gives

$$\frac{Q_2}{Q_1} = \frac{L_2}{L_1} = \frac{R_2}{R_1} \tag{11-27}$$

so that the throughput increases rather slowly with size. The same result is obtained for tubulent flow scaleups at constant Nusselt number. Typically, one scales in parallel when fluid temperature is the critical performance criterion. This conclusion also applies to non-Newtonian flows. The modified velocity profile, for example, Equation 11-17, affects heat transfer performance at a given tube diameter but does not alter the scaleup relations, Equations 11-26 and 11-27.

The heat transfer considerations thus far have neglected the temperature dependence of viscosity. Heating a viscous fluid will decrease its viscosity near the tube walls and thus flatten the velocity profile, improving heat transfer performance compared to the case of a constant viscosity system. Cooling will increase the wall viscosity, elongate the velocity profile, and decrease heat transfer performance. If either heating or cooling is continued far enough downstream the fluid temperature will asymptotically approach the wall temperature so that the velocity profile returns to its normal form.

Precise design of a laminar flow heat exchanger requires careful modeling of fluid properties followed by numerical solution of the simultaneous equations of motion and energy, Equations 11-7 through 11-10. In essence, this is designing from first principles and gives superior results compared to the older approach of using semiempirical correlations, as discussed by Kwant et al. (1973) and Popovska and Wilkinson (1977). Existing correlations are useful, of course, for preliminary estimates and to suggest the dimensionless groups appropriate for scaleup calculations.

Correlations can often be extended to non-Newtonian fluids by the simple expedient of adding a correction factor based on the power law index, n. For example, the well-known Leveque approximation becomes

$$Nu = 1.75 \left(\frac{3n + 1}{4n} \right)^{1/3} Gz^{1/3} \tag{11-28}$$

where $Gz = \pi R^2/\alpha \bar{l}$ is the Graetz number. Porter (1971) gives a comprehensive review of the literature on heat transfer to fluids in laminar flow.

D. Reacting Flows

We begin with the case of a tube diameter so small that radial temperature and composition gradients vanish. Molecular diffusion results in sufficient movement between streamlines so that all molecules emerge with nearly the same residence time. Piston flow prevails, and the component continuity equation becomes

$$\bar{u}\frac{dC_A}{dz} = \mathscr{R}_A \tag{11-29}$$

Temperature may be a function of axial position; but even so, Equation 11-29 can be integrated with ease. If a reactor is known to be operating in the piston flow region, pilot scale results can be scaled *down* with impunity. For a Newtonian fluid, one can scale down using Equations 11-2 and 11-3. Unfortunately, we usually want to scale up and this entails considerable risk for the reasons outlined in Section I.

At the other extreme, suppose the tube diameter is so large that the Merrill and Hamrin criterion, Equation 11-6, is satisfied for both molecular and thermal diffusivities. If the wall temperature is adjusted to give adiabatic operation, scaleup can be done with ease. Equations 11-2 and 11-3 are again used. The pressure drop will be constant if the fluid is Newtonian and will decrease at larger sizes if the fluid is pseudoplastic with $n < 1$. Design calculations can be based on

$$u_z(r)\frac{\partial C_A}{\partial z} = \mathscr{R}_A \tag{11-30}$$

for concentration and

$$u_z(r)\frac{\partial T}{\partial z} = \frac{-\Delta H \mathscr{R}}{\rho C_v} \tag{11-31}$$

for temperature.

Intermediate size tube diameters are inherently more difficult. The only sound approach is the simultaneous solution of Equations 11-7 through 11-11. This approach has become standard design practice since the work of Lynn and Huff (1971) and has been applied to systems with complex kinetics by Ghosh et al. (1975) and Wyman and Carter (1975). However, the design calculations reveal no feasible scaleup procedure. Single tube designs may have hydrodynamic instabilities as discussed by Lynn and Huff (1971) or thermal instabilities as reviewed by Valsamis and Biesenberger (1975). Even scaleup in parallel of multitube reactors may not be feasible due to maldistribution of flow between the tubes as shown by Joosten et al. (1981). The design engineer must then abandon the simple elegance of straight, open tubes and resort to more complicated strategies such as motionless mixers or mechanical agitation.

III. MOTIONLESS MIXERS

Motionless mixers can be defined as a set of stationary, flow diverting elements installed in a fluid stream. By dividing and recombining the flow, they promote radial mixing (i.e., mixing in a plane normal to the primary flow) and thus alleviate the unfavorable temperature and velocity profiles which would otherwise arise. At least seven commercial varieties are now available. Figures 11-3a and 11-3b show two of the simpler types. They, like most others, are designed as inserts into an otherwise empty tube.

Motionless mixers were originally conceived as a means for blending high viscosity fluids in laminar flow. Although they have been used for many other purposes, their design and operating principles can be easily understood by considering a typical blending application, the coloring of thermoplastics.

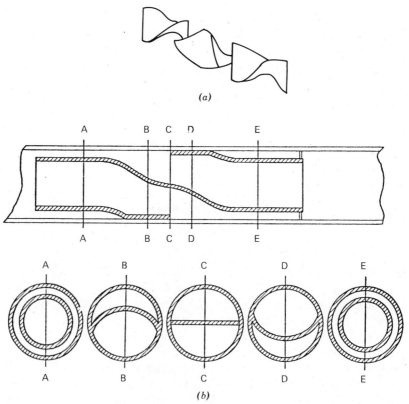

FIGURE 11-3 (a) Flow diverting elements of the Kenics type. (b) Motionless mixer for flow inversion.

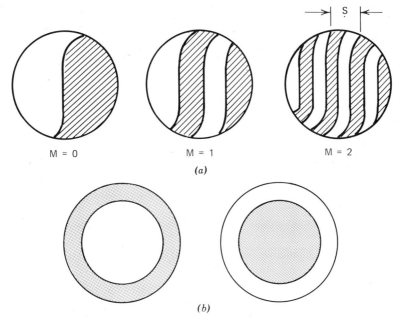

FIGURE 11-4 (a) Successive stages in flow division. (b) One stage of flow inversion.

A. Blending Applications

In coloring plastics, it is frequently advantagous to create a highly pigmented masterbatch and then to obtain a normally pigmented product by blending the masterbatch with unpigmented polymer. For simplicity suppose that equal volumes of masterbatch and unpigmented polymer are to be blended in a motionless mixer of the type shown in Figure 11.3a. The two fluid streams are introduced upstream of the mixer and flow side by side until the leading edge of the first mixing element is encountered. The leading edge divides the flow stream into two parts. The body of the mixer then reorients the parts by rotation in a helical channel. The two parts are recombined at the end of the first element, but the fluid stream is immediately redivided at a new location by the leading edge of the second mixing element.

The division/reorientation process is repeated M times in a motionless mixer containing M mixing elements. Figure 11-4a illustrates the cross section of the flowing polymer stream after $M = 0, 1, 2$ elements. The distance between successive colored regions is called the striation thickness, S, and in a mixer containing M elements,

$$\frac{S_M}{S_0} = 2^{-M} \qquad (11\text{-}32)$$

where S_0 is the initial striation thickness and is equal to the tube diameter for the coloring process under consideration.

If the coloring is done only for aesthetic reasons, blending is complete when the striation thickness has been reduced below the resolving power of the human eye. In scaleup we require S_M to be constant but suppose S_0 will vary linearly with size. This means that M must increase with increasing size but it increases very slowly

$$M_2 = M_1 + \frac{\ln R_2/R_1}{\ln 2} \qquad (11\text{-}33)$$

Typically, the length of a mixing element will have a fixed ratio to the tube size, say $L_E = 4R$. This means that the scaleup of a motionless mixer must violate geometric similarity with tube length increasing slightly faster than the tube diameter.

Example 11.1 Scaleup of Blender. Suppose a pilot plant mixer with $R = 1$ cm is operating at 25 kg/hr. A 16-element motionless mixer of the type shown in Figure 11-3a has given satisfactory results. Design a unit suitable for a production rate of 10000 kg/hr.

A reasonable design has $R = 8$ cm and uses 19 mixing elements. The length of the mixer will be 9.5 times longer than that of the pilot plant. The pressure drop for a Newtonian fluid will be 0.93 times that of the pilot plant while the mean residence time will be 1.52 times that of the pilot plant. An alternative design with $R = 7$ cm also uses 19 mixing elements. It has a pressure drop 1.38 times that of the pilot plant and a mean residence time which is 1.02 times that of the pilot plant. □

The powers of two which appear in Equation 11-32 are characteristic of the particular motionless mixer we have been considering. It is possible to design mixers which divide the fluid stream in three (or more) parts so that the striation thickness decreases as the third (or higher) power of M rather than as the second. Equation 11-33 should then be modified by replacing ln 2 with ln 3. Otherwise, the scaleup approach is identical.

A scaleup with complete geometric similarity (i.e., with $M_1 = M_2$) is possible if the increase in S_0 is avoided through the use of multiple injection ports.

B. Heat Transfer Applications

The flow patterns within motionless mixers destroy radial and tangential symmetry, making a first principles analysis difficult. Rather than using the equations of motion and energy as is done for empty tubes, motionless mixer design must rely on semiempirical correlations. Relatively few such correlations have been published although more extensive results are presumably available from vendors on a proprietary basis. However, the data are very dependent on fluid properties—particularly because of the temperature dependence of viscosity—and must be used with considerable caution.

For heat exchanger tubes containing motionless mixers of the type shown in Figure 11-3a, pressure drops from 5 to 100 times that in an empty, open type have been reported by Chen (1973) and Schott et al. (1975). The trade-off between using motionless mixers and using an empty tube of greater length depends on the pressure drop. When heat transfer is the only criterion of performance, motionless mixers rarely have an economic advantage. However, they may be a truly enabling technology when other operations are conducted simultaneously with heat transfer.

Where a motionless mixer is used in a pilot plant, scaleup can be more certain than a first principles design but will be uncertain in some respects. Scaleup in parallel is conservative except with severe cooling which can result in unequal flow distribution to the various tubes. In scaleup of a single tube, increasing M while maintaining $\bar{t}$ approximately constant should give nearly the same inside heat transfer coefficient for the larger unit as the smaller one. This approach comes close to maintaining geometric similarity and the ratio of surface area to volume decreases upon scaleup. Therefore, the amount of heat transferred per unit throughput of process fluid must decrease.

An increase in throughput with a constant $T_{out} - T_{in}$ can be accomplished by lengthening the tube while keeping its diameter constant. Although not confirmed experimentally, one might expect

$$\frac{Q_2}{Q_1} = \frac{L_2}{L_1} \qquad (11\text{-}34)$$

This same result holds when the flow through an empty tube is increased while keeping the diameter constant.

The discussion of motionless mixers has so far been restricted to the situation where mixing elements are distributed more or less continuously along the length of the tube. The motionless mixer is then said to be *distributed*. Another approach is to use *lumped* motionless mixers in which relatively short sections containing mixing elements are separated by long lengths of open tube. Analysis of systems with lumped motionless mixers is easier than for the distributed case. First principles calculations using the equations of change are possible as shown by Nauman (1979). Lumped motionless mixers of the flow division type, for example, Figure 11-3, give a modest but calculable gain in heat transfer performance.

Lumped motionless mixers of the flow inversion type (see Figures 11-3b and 11-4b) result in greater benefit. The trade-off with pressure drop appears to favor lumped motionless mixers over empty tubes and empty tubes over distributed motionless mixers.

C. Reaction Applications

Real benefits accrue from the use of motionless mixers in laminar flow reactors since this avoids the hydrodynamic instabilities associated with tubular polymerizers as studied by Lynn and Huff (1971). Moreover, by increasing radial mixing and wall heat transfer coefficients, motionless mixers postpone the appearance of thermal runaways. Overall, they enable laminar flow reactors to operate closer to piston flow even at large tube diameters. However, this is done at the expense of greater pressure drops and higher capital costs. The reaction system also is more difficult to design and optimize from first principles. However, first principles reactor designs based on solution of Equations 11-7 through 11-11 are feasible when the motionless mixers are lumped.

The scaleup of reaction systems using distributed motionless mixers remains an art if significant amounts of heat transfer accompany the reaction. Multitubular heat exchanger–reactors with distributed motionless mixers have been used commercially. However, no data are available regarding the distribution of flow between the various tubes. Existing applications are of the type where maldistribution of flow is unlikely to be a problem.

If the heat of reaction is low or if the reactor can be operated adiabatically, scaleup of a distributed motionless mixer is more certain. The recommended approach is to increase M according to Equation 11-33 while maintaining a constant $\bar{t}$. This forces some departures from strict geometric similarity. The length to diameter ratio will increase slowly with scale as will the pressure drop if the fluid is Newtonian.

Example 11.2 Scaleup of Reactor. Consider a 1000-fold scaleup from a pilot plant mixer operating at 10 lb/hr in a 128-in.-long, 2-in.-diameter tube which contains 32 mixing elements of the type shown in Figure 11-3a. Scaleup should be done at constant S_M and $\bar{t}$.

Constant residence time requires

$$\frac{L_2 R_2^2}{Q_2} = \frac{L_1 R_1^2}{Q_1} \tag{i}$$

Assume the length of a mixing element is two diameters so that $L_2 = 4R_2 M_2$. Substitute known values to obtain

$$M_2 R_2^3 = 32000 \tag{ii}$$

Equation 11-33 becomes

$$M_2 = 32 + \frac{\ln R_2}{\ln 2} \tag{iii}$$

Solve (ii) and (iii) simultaneously to obtain $R_2 = 9.7$ in. and $M_2 = 35.1$. If designed in exactly this fashion, the reactor would be 113.5 ft long and have a pressure drop 1.2 times that of the pilot unit. By comparison, a strictly geometric scaleup would give $R_2 = 10$ in., $L_2 = 106.7$ ft, and would contain 32 mixing elements. □

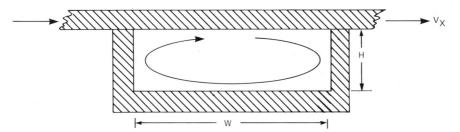

FIGURE 11-5 Wall driven flow in a rectangular cavity.

IV. MOVING WALL DEVICES

A. Cavity Flows

By moving wall device we mean any mechanically agitated system such as a stirred tank reactor or an extruder. A more technical definition has the zero-slip boundary condition applied at positions (the "walls") which are in relative motion. A simple example is flow in a rectangular cavity with a moving wall as shown in Figure 11-5. Suppose the process aim is heat transfer to the fluid in the cavity. This will certainly be enhanced by the motion of the upper wall.

Assuming $W/H \gg 1$, end effects in the cavity can be ignored, and the velocity profile will be a function of y only. Ignoring temperature dependence of viscosity

$$u_x(y) = V_x \left[3\left(\frac{y}{H}\right)^2 - 2\left(\frac{y}{H}\right) \right] \tag{11-35}$$

This profile, illustrated in Figure 11-6, consists of positive velocities for $y/H > 2/3$ and negative velocities for $y/H < 2/3$ which result from the pressure gradient in the channel. The energy balance written in rectangular

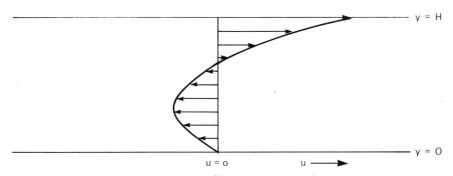

FIGURE 11-6 Velocity profile in cavity assuming $W/H \gg 1$.

coordinates becomes

$$\frac{\partial T}{\partial t} + u_x \frac{\partial T}{\partial x} = \alpha \frac{\partial^2 T}{\partial y^2} + \frac{\mu}{\rho C_v}\left(\frac{du}{dy^2}\right)^2 \qquad (11\text{-}36)$$

where u_x is given by Equation 11-35.

The unsteady nature of the heat transfer results from the assumption of no net flow in the cavity. The batch scaleup problem is one of determining how the batch time t varies with some characteristic dimension such as H. Typically, t will be defined as the time needed to achieve some desired (average) temperature for the contents of the cavity.

Equation 11-36 can be expressed using dimensionless variables as

$$\frac{\partial T}{\partial \theta} + (3\xi^2 - 2\xi)\frac{\partial T}{\partial \lambda} = \left(\frac{\alpha\tau}{H^2}\right)\frac{\partial^2 T}{\partial \xi^2} + \left(\frac{\mu}{\rho C_v \tau}\right)(6\xi - 2)^2 \qquad (11\text{-}37)$$

where $\lambda = x/H$, $\xi = y/H$, and $\theta = t/\tau$ where $\tau = H/V_x$. The temperature response of the system will depend only on the two dimensionless groups, $(\alpha\tau/H^2)$ and $(\mu/\rho C_v \tau)$. Scaleup at constant endpoint temperature will give

$$\theta_2 = \theta_1 \qquad (11\text{-}38)$$

provided

$$\left(\frac{\alpha\tau}{H^2}\right)_2 = \left(\frac{\alpha\tau}{H^2}\right)_1 \qquad (11\text{-}39)$$

and

$$\left(\frac{\mu}{\rho C_v \tau}\right)_2 = \left(\frac{\mu}{\rho C_v \tau}\right)_1 \qquad (11\text{-}40)$$

Scaleup at constant physical properties with Equations 11-39 and 11-40 results in $H_2 = H_1$ which is scaleup in parallel using many batches rather than a single large one. No other exact scaleup, by direct ratioing of groups or parameters, is possible when heat transfer through the walls and heat generation by viscous dissipation are simultaneously important.

Scaleup holding only Equation 11-40 constant is appropriate when heat generation by viscous dissipation predominates. Then Equation 11-40 gives $\tau_1 = \tau_2$ so that

$$t_2 = t_1 \quad \text{(viscous dissipation controlling)} \qquad (11\text{-}41)$$

and

$$\frac{(V_x)_2}{(V_x)_1} = \frac{H_2}{H_1} \qquad (11\text{-}42)$$

MOVING WALL DEVICES

Thus the wall speed increases with increasing size. With geometric similarity, batch size will increase as H^3 while batch time is constant. This is generally a very desirable form of scaleup, but it must be cautioned that diffusion of heat will become progressively less important as size increases. This results in increasing inhomogeneities in temperature within the cavity.

If viscous dissipation is negligible compared to heat transfer through the walls, only Equation 11-39 need be considered. This leads to

$$\frac{t_2}{t_1} = \frac{\tau_2}{\tau_1} = \frac{H_2^2}{H_1^2} \quad \text{(wall heat transfer controlling)} \quad (11\text{-}43)$$

and wall velocity decreases with scale

$$\frac{(V_x)_2}{(V_x)_1} = \left(\frac{H_2}{H_1}\right)^{-1} \quad (11\text{-}44)$$

The batch cycle time then increases as H_2. Productivity of the larger system, volume processed per unit time, thus increases with H rather than H^3.

Turning to applications other than heat transfer, scaleup based on Equation 11-40 will lead to a constant distribution of circulation times in the cavity, since the circulation time at any position, $\xi = y/H$, is proportional to $\tau = H/V_x$. Striation thickness will be identical when expressed using the dimensionless coordinate ξ but will in fact increase directly with H. Therefore, scaleup of blending at constant striation thickness leads to

$$\frac{t_2}{t_1} = \frac{H_2}{H_1} \quad (11\text{-}45)$$

while increasing the wall velocity as in Equation 11-42. Volume productivity of the batch blender will increase as H^2 for a scaleup with geometric similarity.

These examples have shown the productivity increasing as the first, second, or third power of the characteristic dimension for scaleup procedures which maintain geometric similarity. Typically, the cost of equipment for a geometrically similar scaleup will vary as $V^{0.6}$ to $V^{0.8}$ and thus as $H^{1.8}$ to $H^{2.4}$. Productivity must increase at least this rapidly for economy of scale to be realized. This means that the scaleup envisioned by Equations 11-43 and 11-44 is economically infeasible. Some element of complete similitude must be sacrificed.

The usual approach is to increase the linear velocity, V_x, according to Equation 11-42 and never to decrease it as suggested by Equation 11-44. With V_x increasing as H^1, the dimensionless group $\alpha\tau/H^2$ will decrease as H^{-2}, destroying thermal similitude. However, it may still be possible to maintain a nearly constant wall heat transfer coefficient during scaleup since the motion of the wall continuously exposes new surface. The time of exposure is proportional to τ and will be constant for scaleup procedure which increases V_x proportionally to H. This results in a constant heat transfer coefficient provided the approach of the fluid temperature to that of the moving wall is small so that we are always within an entrance region with respect to heat transfer rather than an asymptotic region.

Motion of the wall will have a decreasing benefit for heat transfer as the fluid temperature asymptotically approaches the wall temperature. Alternatively put, there must exist somewhere within the cavity a source of cold fluid to be brought into contact with the hot, moving wall. Scaleup with a constant heat transfer coefficient gives the total amount of heat transferred being proportional to area and thus to H^2. The amount of mass in the cavity will vary as H^3. Longer batch times will be required to transfer the same amount of heat per unit mass, in the asymptotic region of heat transfer, leading to Equation 11-45 as for blending.

This suggests that a batch, moving wall device can be scaled up as H^2 or H^3 provided the wall velocity is increased with H. Practical experience suggests this is generally true, and the same generalization will be seen to hold for moving wall devices operated in a continuous mode. However, heat and mass transfer examples exist which can only be scaled with H. Precise scaleup calculations naturally require a detailed process model.

B. Extruders

Single screw extruders are often used for food and polymer processing. Figures 11-7 and 11-8 illustrate the flow geometry. In the simplest applications the extruder simply acts as a conveyor for a viscous liquid without generation of significant back pressure. The relative motion of the extruder screw and barrel cause two velocity components, V_x and V_z. The cross-channel component, V_x, creates a circulating flow in the x–y plane much as that illustrated in Figures 11-5 and 11-6 and approximated by Equation 11-35. The down-channel component, V_z, superimposes a net velocity down the channel which is linear when there is no pressure generation in the downstream direction:

$$u_z = V_z \left(\frac{y}{H} \right) \qquad (11\text{-}46)$$

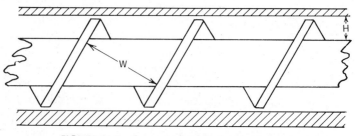

FIGURE 11-7 Geometry of a single screw extruder.

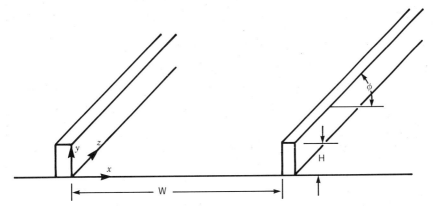

FIGURE 11-8 Unwrapped geometry of a single screw extruder.

Pressure generation in the z-direction results in a reverse flow—with some net throughput—in the z-y plane which is conceptually similar to that in the x-y plane (but see Tadmore and Gogos, 1979).

Siadat et al. (1979), Nauman (1977), and Lee et al. (1982) present detailed models for heat transfer and chemical reaction including consideration of nonisothermal and non-Newtonian cases. Single screw extruders are sometimes used to melt polymers as well as to convey and pump them. Analysis of this is reviewed by Tadmore and Klein (1970) and Lidor and Tadmore (1976); proprietary computer simulations are also available to help understand this mode of operation. Models for a variety of twin screw extruders have been given by Janssen (1978).

The following summary is intended to illustrate the general principles of extruder scaleup; the reader is referred to the specialized literature for specific applications.

In the absence of back pressure, the down-channel velocity component conveys material at a volumetric rate of

$$Q_D = WH\left(\frac{V_z}{2}\right) = WH\frac{DN\cos\phi}{2} \qquad (11\text{-}47)$$

where D is the diameter of the screw, N is its rotational velocity, and ϕ is the helix angle. For a geometrically similar scaleup, Q_D increases as $H^3 N$ so that scaleup as H^3 is possible if the screw RPM is held constant. Such scaleups are occasionally realized.

A twin screw extruder of the Werner and Pfleiderer variety may be scaled up nearly as H^3 when the primary objective is to melt and pump polymer pellets at a high rate. The extruder operates essentially adiabatically with the entire power output from the motor appearing as sensible heat in the extrudate. The motor power is dissipated in the polymer by mechanical friction prior to melting and by viscous dissipation after melting. Indeed, the energy input process is identical to that considered in the development of Equations 11-41 and 11-42. When there is a net flow in the z-direction, Equation 11-41 is equivalent to maintaining a constant residence time in the extruder (see Chapter 8) while Equation 11-42 expresses the condition that N be held constant.

A limit to scaling with H^3 at constant N is imposed when inhomogeneities in the extrudate become unacceptably large. In coloring operations, striations will ultimately become visible unless scaleup is conducted according to Equation 11-45. In film blowing, thermal inhomogeneities will give rise to variations in gauge thickness. It is theoretically possible to overcome blending problems by increasing the length to diameter ratio of the extruder upon scaleup. Equation 11-45 can be interpreted for a flow system as having L increase with H. A more common solution to such problems is to use a scaleup procedure in which throughput increases more slowly than as H^3. The throughput may be lowered either by reducing the screw speed or by installing external restrictions on throughput.

Typically, both approaches are used together. Variable speed drives and variable rate feeding devices are usually installed on extruders, and the best combination of operating conditions is found more or less empirically after plant startup. A useful rule of thumb is that throughput should increase as $H^{2.5}$ for many applications. Scaleup of an extruder as a polymer devolatilizer will typically be as H^2 with constant screw speed since this mass transfer operation obeys a surface renewal model except when foaming is important. For a more complete discussion of devolatilization, see Biesenberger and Sebastian (1983). Direct scaleup of extruders for critical heat transfer applications such as film blowing and cable extrusion will typically be as H^2 or even lower. There is a current trend to installing motionless mixers after the extruder to achieve a more economical scaleup.

C. Stirred Tanks

Stirred tank reactors sometimes operate in the laminar regime and follow the scaleup concepts outlined in this chapter. (See also Chapters 8 and 9.) Rather different agitator types are used for laminar flow than for turbulent flow. Baffles are uncommon. Anchors and helical ribbons are used with fluid viscositites above 100 poise. Single or multiple pitched blade turbines are generally preferred below 100 poise although the ratio of impeller diameter to tank diamter will generally be much higher than for turbulent flow applications. Values for the ratio of 0.5 to 0.75 are typical of polymerization reactors. Scaleup with constant power per unit volume is usually recommended although this is not always the conservative approach for the reasons outlined in Chapter 8. An extensive specialized literature is available regarding scaleup of laminar, agitated vessels including non-Newtonian and viscoelastic effects. Representative references to consider are Calderbank and Moo-Young (1959),

Norword and Metzner (1960), Moo-Young and Chan (1971), Ulbrecht (1974), and Novak and Rieger (1975).

Dynamic similarity in a stirred tank is achieved by operating at a constant Reynolds number, $\rho D^2 N/\mu$, but this form of scaleup is rare. The criterion of constant power per unit volume causes N to decrease much more slowly with scale;

$$\frac{N_2}{N_1} = \left(\frac{D_2}{D_1}\right)^{-2/3} \tag{11-48}$$

and for N_{Re} to increase with scale

$$\frac{(Re)_2}{(Re)_1} = \left(\frac{D_2}{D_1}\right)^{4/3} \tag{11-49}$$

so that a transition into turbulent flow may occur during scaleup. Barring such a transition, any heat or mass transfer limitation will become more severe upon scaleup. The argument of Equations 11-43 through 11-45 are qualitatively correct for a stirred tank in laminar flow. The productivity of the vessel will increase only as the first or second power of the diameter. Cooling by bulk boiling (autorefrigeration) and heating by direct injection of steam are exceptions that scale as D^3. Viscous dissipation also scales as D^3 when the criterion of constant power per unit volume is used. Reactions scale as D^3 unless there is a heat or mass transfer limitation.

V. OVERVIEW

We have attempted to give a general overview of scaleup techniques for laminar flow systems. The best of techniques are based on the simultaneous solution of the equations of motion, energy, and component continuity. Unlike turbulent flow, rigorous solutions are sometimes possible; and at first glance, it may appear that laminar systems are easier to scale. However, laminar systems frequently show large differences in composition or temperature which are eliminated in turbulent systems by eddy diffusion. In practice, this means that turbulent systems are easier to scale. For example, a turbulent reactor can be closely approximated by piston flow or perhaps by the axial dispersion model. Such simple models are rarely suitable for laminar flow. Instead, the equations which must be solved are typically more complex than their turbulent flow counterparts.

One usually finds that an exact scaleup—in the sense of maintaining complete similarity—is impossible. Compromises are necessary, and the scaleup becomes more of an art than a science. We have seen that complete similarity of velocity, temperature, and composition is rarely possible. One must choose to preserve those aspects of the process which are most germane to its economic or technical success. Often, this means scaling in parallel so that a tenfold increase in capacity requires a tenfold increase in capital cost. Avoiding such a costly form of scaleup is a goal for the process engineer which will remain valid for many years in the future. The goal becomes particularly elusive when effects such as viscous dissipation, non-Newtonian viscosities, or elasticity are important. The scaleup of processes for polymer fabrication, for example, blown film, has proven particularly difficult and has to proceed in small, incremental steps.

NOMENCLATURE

C_A	Concentration of molecular species A
C_v	Heat capacity
D	Diameter of stirred tank
$\mathscr{D}_A$	Diffusivity of molecular species A
H	Height of cavity
L	Length of extruder
M	Number of mixing elements
n	Power law index
N	Rotational velocity of screw or agitator
Re	Reynolds number
P	Pressure
Q	Volumetric flow rate
Q_D	Drag flow rate
r	Radial coordinate
R	Radius of tube
$\mathscr{R}$	Total reaction rate
$\mathscr{R}_A$	Reaction rate of species A
S	Striation thickness
t	Time
$\bar{t}$	Mean residence time
T	Temperature
$\bar{u}$	Mean fluid velocity
u_x, u_y, u_z	Velocity components in rectangular coordinates
u_r, u_z	Velocity components in cylindrical coordinates
V	System volume
V_x	Wall velocity component in x-direction
V_z	Wall velocity component in z-direction

W	Width of cavity
x, y, z	Rectangular coordinates

Greek

α	Thermal diffusivity
ΔH	Heat of reaction
ζ	Consistency of power law fluid
θ	Reduced time, t/τ
κ	Thermal conductivity
μ	Viscosity
ξ	Reduced coordinate, y/H or r/R
τ	Reduced coordinate, x/H
ρ	Density
τ	Time constant, H/V_x
τ_{rz}	Shear stress
ϕ	Helix angle in extruder
Ω	Reduced temperature

REFERENCES

Biesenberger, J. A., and Sabastian, D. *Polymer Reaction Engineering*, Wiley, New York, 1983.

Brown, G. M., *AIChE J.*, **6**, 179 (1960).

Calberbank, P. H., and Moo-Young, M., *Trans. Instn. Chem. Eng.*, **37**, 26 (1959).

Chen, S. J., *SPE ANTEC*, **19**, 258 (1973).

Ghosh, M., Foster, D. W., Lenezyk, J. P., and Forsyth, T. H., *AIChE J. Symp. Ser. No. 160*, **72**, 102 (1975).

Janssen, L. P. B. M., *Twin Screw Extrusion*, Elsevier, Amsterdam, 1978.

Joosten, G. E. H., Hoogstraten, H. W., and Ouwerkerk, C., *I & EC Proc. Des. Dev.*, **20**, 177 (1981).

Kwant, P. B., and Fierens, R. H. E., and van der Lees, A., *Chem. Eng. Sci.*, **28**, 1303 (1973).

Lee, L. Y. J., Ottino, J., Banz, W. E., and Mocosko, C. W., *Polymer Eng. Sci.*, **20**(13), 868 (1980).

Lidor, G., and Tadmor, Z., *Poly. Eng. Sci.*, **16**, 450 (1976).

Lynn, S., and Huff, J. E., *AIChE J.*, **17**, 475 (1971).

Merrill, L. S., Jr., and Hamrin, C. E., Jr., *AIChE J.*, **16**, 194 (1970).

Moo-Young, M., and Chan, K. W., *Can. J. Chem. Eng.*, **49**, 187 (1971).

Nauman, E. B., *AIChE J.*, **25**, 246 (1979).

Norwood, K. W., and Metzner, A. B., *AIChE J*, **6**, 436 (1960).

Novak, V., and Rieger, F., *Chem. Eng. J.*, **9**, 63 (1975).

Popovska, F., and Wilkinson, W. L., *Chem. Eng. Sci.*, **32**, 1155 (1977).

Porter, J. E., *Trans. Instn. Chem. Engrs.*, **49**, 1 (1971).

Schott, N. R., Weinstein, B., and LaBombard, D., *Chem. Eng. Prog.*, **71**(1), 54 (Jan. 1975).

Siadat, B., Malone M., and Middleman, S., *Poly. Eng. Sci.*, **19**, 787 (1979).

Tadmor, Z., and Klein, I., *Engineering Principles of Plasticating Extrusion*, van Nostrand-Reinhold, New York, 1970.

Tadmor, Z., and Gogos, C. G., *Principles of Polymer Processing*, Wiley, New York, 1979.

Ulbrecht, J., *The Chemical Eng.*, No. 286, 347 (June 1974).

Valsamis, L., and Biesenberger, J. A., *AIChE J.*, *Symp. Ser. No. 160*, **72**, 18 (1975).

Wyman, C. E., and Carter, L. F., *AIChE J. Symp. Ser. No. 160*, **72**, 1 (1975).

12

STAGEWISE MASS TRANSFER PROCESSES

J. R. FAIR

I.	Major Issues	432
II.	Fundamental Considerations	433
III.	Vapor–Liquid Systems—Distillation	435
	A. Major Issues	435
	B. Phase Equilibria	437
	1. Diffusion Coefficients	439
	2. Other Physical Property Data	439
	C. Prediction of Performance	439
	1. Stage Calculations	439
	2. Equipment Hydraulics—Tray Columns	442
	3. Mass Transfer Efficiency	450
	4. Liquid Mixing on Trays	456
	5. Entrainment from Trays	458
	D. Laboratory Scaleup	460
	1. Demonstration of Separability	460
	2. Prediction of Efficiency	461
	E. Idealizations and Assumptions	462
	F. Conclusions	462
IV.	Gas–Liquid Systems—Absorption and Stripping	463
	A. Major Issues	463
	B. Prediction of Performance	463
	1. Phase Equilibria	463
	2. Stage Calculations	465
	3. Equipment Hydraulics—Tray Columns	469

V. Liquid–Liquid Systems—Extraction	469
A. Major Issues	469
B. Extraction Notation	470
C. Phase Equilibrium	472
D. Prediction of Performance	474
1. Stage Calculations	474
2. Solvent Selection	479
3. Extraction Devices	481
4. Mass Transfer Relationships	489
5. Mass Transfer Efficiency of the Sieve Tray Extractor	490
E. Laboratory Studies for Scaleup	492
VI. Energy Considerations	493
Nomenclature	496
References	500

I. MAJOR ISSUES

Stagewise processing, as considered in this chapter, involves the intimate contacting of two or more immiscible or partially miscible phases in one or more discrete steps, or stages. The purpose of such contacting is to bring about transfer of mass between the phases; this implies diffusion under a concentration driving force within each phase, and it implies also transfer across one or more phase boundaries. More than one stage may be used in order to maximize the concentration differential, and flows between the stages are arranged to support this objective (Figure 12-1).

At each stage three important mechanisms are required:

Dispersion of one phase into another.
Transfer across the resulting phase interface.
Separation of the phases.

The dispersion is needed to provide the interfacial area for mass transfer, and the energy required for dispersion is supplied in various ways. The transfer

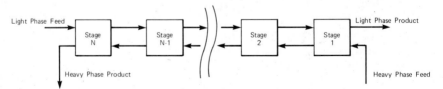

FIGURE 12-1 Simple countercurrent cascade of N stages, two contacting phases.

FUNDAMENTAL CONSIDERATIONS

across the interface is the primary objective of the process. The phase separation permits exit streams from the stage to flow to other stages or to exit from the entire system.

If the dispersion is fine enough, and if the dispersed phases are allowed to remain in contact long enough, a condition of *phase equilibrium* is ultimately reached. It is conventional to consider the equilibrium condition as the maximum attainable concentration change; stagewise processes are scaled up on the basis of their approach to equilibrium. The "perfect" situation is defined, and the "real world" situation is then considered on the basis of its fractional approach to perfection.

The practical limits to approaching equilibrium are apparent from the three mechanisms occurring during contacting of phases. The equipment and energy requirements for effecting intimate dispersion may be such that creation of extensive interfacial area is uneconomical. Keeping the phases in contact for long periods might seriously limit the practical capacity of the equipment. Separating the phases might require special devices if simple density differences do not promote rapid settling; in fact, the degree of dispersion and the ease of separation are often related, and one of them tends to limit the other. As a result of these factors, most stagewise mass transfer processes do not operate close to the equilibrium condition.

Practical equipment considerations limit the capacity and transfer efficiency of stage type contactors. This is likely to be more the case at the large scale than in the laboratory. It may be quite realistic, for example, to contact two liquid phases intensively in the small scale and thus to evaluate the equilibrium condition. The secret of scaleup is, then, to predict how close an approach to this condition is economically feasible in the plant prototype. Dealing with this secret is the thrust of this chapter.

II. FUNDAMENTAL CONSIDERATIONS

Consider as in Figure 12-2 a closed container in which a mixture of components i and j (a *binary* mixture) is observed to separate into two phases. Gibbs' phase rule requires that no more than two degrees of freedom, such as temperature and concentration, may be fixed to determine the system. If the contents of the container are mixed well and allowed to come to equilibrium, the phase compositions of one of the components will assume a given ratio:

$$K_i = \frac{y_i}{x_i} \qquad (12\text{-}1)$$

where

K_i = equilibrium ratio
y_i = mole fraction of i in the light phase
x_i = mole fraction of i in the heavy phase

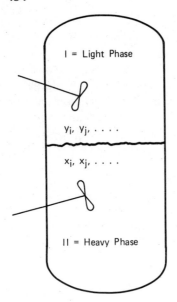

FIGURE 12-2 Equilibrium between phases I and II.

The *separation factor* results from a similar consideration of the other component:

$$\alpha_{ij} = \frac{K_i}{K_j} = \frac{y_i x_j}{x_i y_j} = \frac{y_i(1 - x_i)}{x_i(1 - y_i)} \qquad (12\text{-}2)$$

If a mixture containing more than two components (a *multicomponent* mixture) is placed in the container, Equation 12-1 still applies, but the last part on the right of Equation 12-2 does not apply. If a separation between, say, components i and k in the mixture is to occur, the separation factor $\alpha_{ik} = K_i/K_k = (y_i x_k)/(x_i y_k)$ cannot have a value of unity. One of the K values can have a value of unity, but their ratio cannot. It is always desirable to find conditions where the separation factor α is large, since fewer required stages for an overall separation will result. The conditions for such a maximized separation are constrained by the laws of thermodynamics, as will be shown later.

Because of their simplicity, binary mixtures are easiest to treat thermodynamically and experimentally, and the sources of equilibrium data for such mixtures are extensive. Multicomponent mixtures, on the other hand, can exhibit so many variations in concentration and degree of interaction of molecular species that they are covered less in the literature and represent unique situations when encountered for scaleup. Thus, the challenge of the multicomponent mixture situation should be recognized early in the scaleup process.

VAPOR–LIQUID SYSTEMS — DISTILLATION

The rate and degree of approach to equilibrium in a contactor are governed by the mass transfer process. A species must diffuse through one phase, overcome the resistance to transfer at the interface, and then diffuse through the other phase. The species must deal with other species moving in the same or opposite direction; there may be *unidirectional diffusion* or *counterdiffusion*. The diffusion within a phase may be by molecular means only, or it may be aided by turbulence, or *eddy diffusion*. Clearly, the fluid mechanics of the contacting situation can influence the rate of transport.

Transport within a phase can be represented by an integrated form of Fick's first law:

$$N_i = \frac{\mathscr{D}_i}{\Delta z} \Delta C_i \qquad (12\text{-}3)$$

where

N_i = mol/time transport of i, related to some interfacial area (a *molar flux*)
$\mathscr{D}_i$ = proportionality factor, known as the *diffusion coefficient*
ΔC_i = concentration difference of i in the phase, usually from a bulk condition to an interface condition
Δz = distance through which the diffusion occurs

Equation 12-3 represents a great simplification of basic mass transfer relationships, but is useful here for conceptual purposes. One might expect the net transport of the species to be enhanced by larger concentration gradients, shorter distances (due to intimate dispersion), and more extensive interfacial areas. The diffusion coefficient might be expected to be greater for smaller molecules, lower viscosity phases, higher temperatures, and so on. The conditions for enhancing mass transfer are likely to be greatly different between the laboratory, the pilot plant, and the commercial prototype. Understanding the differences is another challenge of the scaleup process.

In summary, equipment, operating conditions, properties of mixtures, flow effects, and economics might be expected to influence the scaleup process. The influences are functions of type of process, and are thus best presented on a process-type basis. In this chapter, vapor–liquid, gas–liquid, and liquid–liquid processes have been selected as both representative and important commercially.

III. VAPOR–LIQUID SYSTEMS — DISTILLATION

A. Major Issues

In the analysis of a distillation column, whether for new design or for process improvement/modification efforts, there is a logical sequence of steps that the designer must take. First, he must *define the system* that is to be separated. While this might be straightforward and obvious in many cases, it often is quite complex. The system definition must contain all components, including those in small concentrations that might cause problems with internal buildup or with unexpected contamination of products. Sometimes a complex mixture is to be separated, and it is necessary to work with some sort of a pseudocomponent approach. The textbook examples, such as the clean separation of a methanol/water mixture, do not happen very often in real life. So effort is needed at this first step in the sequence.

The next step is to *establish separation criteria*. What purities are required? What recoveries of the mixture components are to be achieved? Will today's purity requirement be satisfactory in tomorrow's environment? As all chemical engineers know, the establishment of product purity specifications strongly influences the cost of the distillation separation operation.

The third step is to *obtain reliable physical property data*. These data include the important solution thermodynamics parameters related to vapor–liquid equilibria (VLE). They also include such properties as phase densities, viscosities, surface tensions, and diffusion coefficients. It can be shown easily that the uncertainty of a design result is strongly dependent upon the accuracy of the physical property data which are used in the computations. Yet, the chore of obtaining the data is considered to be an onerous one by most chemical engineers, and the required diligence in handling the chore is all too often compromised.

The fourth step in the sequence is to *select a model for the determination of theoretical stages or transfer units*. Different models are applicable to different situations, and can range from rough graphical approaches to sophisticated and rigorous analytical approaches which require large computers for obtaining the solutions. The fifth step, then, is to *obtain the stages/transfer units*, the computations involved giving information on liquid and vapor flow rates in the column as well as concentration, temperature, and pressure profile data.

The sixth step is to *size the distillation column*. This includes the hydraulic analysis to establish operating ranges, pressure drop, and mass transfer efficiency. This, of course, is the ultimate objective of the scaleup process. The column dimensions so determined must represent an optimum combination of cost, reliability, and flexibility.

The six steps given above can apply to the bench or pilot scale as well as to the commercial scale. They may be considered *a posteriori*; for example, a laboratory distillation column may be used to demonstrate a separation, perhaps empirically ("What degree of separation is obtained with a column containing six feet of high-efficiency packing?"). However, before the resulting information can be used for scaleup, the steps must be addressed. Scaleup is, therefore, much more than simply relating small dimensions to large dimensions; an understanding of what happens in the laboratory is just as important as understanding what happens in the prototype.

B. Phase Equilibria

As mentioned earlier, equilibrium stage processes are based on the concept of physical equilibrium and require basic thermodynamic data for establishing the "ideal" stage separation. For distillation, vapor–liquid equilibria (VLE) are required, and background on the determination of such equilibria may be found in standard chemical engineering thermodynamics textbooks. Only a summary development will be included here. The separation factor (Equation 12-2) for distillation is the *relative volatility* which, for two components in a mixture is

$$\alpha_{ij} \equiv \frac{K_i}{K_j} = \frac{y_i x_j}{y_j x_i} \qquad (12\text{-}4)$$

The K value for component i in the mixture is

$$K_i = \frac{\gamma_i^L \Phi_i^L P_i^*}{\gamma_i^v \Phi_i^v P} \qquad (12\text{-}5)$$

where

γ_i^L, γ_i^v = the activity coefficients of i in the liquid and vapor
Φ_i^L, Φ_i^v = fugacity coefficients of i in the liquid and vapor
P_i^* = the vapor pressure of i
P = the total pressure

For moderate-to-low pressure systems, Φ values approach unity and the vapor mixture is ideal ($\gamma_i^v = 1.0$). Then,

$$K_i = \frac{y_i}{x_i} = \frac{\gamma_i^L P_i^*}{P} \qquad (12\text{-}6)$$

where the liquid-phase activity coefficient γ_i^L is the familiar "Raoult's law correction factor."

Equation 12-6 applies to many of the chemical systems that are distilled at moderate-to-low pressures. There have been many correlations published that permit the estimation of the liquid-phase activity coefficient, all being based on the Gibbs–Duhem thermodynamic relationship. A time-honored example is the model of Van Laar (1910):

$$\ln \gamma_i^L = A_{ij} \left[\frac{A_{ji} x_j}{A_{ij} x_i + A_{ji} x_j} \right]^2 \qquad (12\text{-}7)$$

$$\ln \gamma_j^L = A_{ji} \left[\frac{A_{ij} x_i}{A_{ji} x_j + A_{ij} x_i} \right]^2 \qquad (12\text{-}8)$$

The constants A_{ij} and A_{ji} are related to the infinite dilution activity coefficients:

$$\ln \gamma_i^{0,L} = A_{ij} \tag{12-9}$$

$$\ln \gamma_j^{0,L} = A_{ji} \tag{12-9a}$$

The Van Laar model is remarkably good for representing activity coefficient as a function of composition, as shown by Null (1970) and others. Tables of Van Laar constants for various binary systems are available from many sources such as Holmes and Van Winkle (1970). Improved models, not quite as simple to use, are those of Wilson (1964) and Renon and Prausnitz (1968). Extension to multicomponent systems is possible through the use of binary pair constants as proposed by Wilson (1964) and Chien and Null (1972).

For a binary system, the y–x equilibrium curve is represented by

$$y_i = \frac{\alpha_{ij} x_i}{1 + (\alpha_{ij} - 1) x_i} \tag{12-10}$$

The slope of this curve is

$$m = \frac{dy_i}{dx_i} = \frac{\alpha_{ij}}{\left[1 + (\alpha_{ij} - 1) x_i\right]^2} \tag{12-11}$$

(Note that as $x_i \to 0$, $m \to \alpha_{ij}$ and as $x_i \to 1.0$, $m \to 1/\alpha_{ij}$.) The slope given by Equation 12-11 enters into mass transfer calculations when there is a significant liquid-phase resistance. It can be approximated on a pseudobinary basis over sections of a y–x curve where it can be considered constant.

Sources of measured VLE are extensive, and a guide to the sources is the series of books by Hala and coworkers (1967, 1968) and Wichterle et al. (1976). There are also tabulations of the data themselves in the volumes by Hirata et al. (1975) and by Gmehling et al. (1979). One should use measured data whenever possible, after he has convinced himself that the data are reliable. When there is no source of experimental VLE, and when the requirements of the problem permit the inaccuracies of raw estimation, several methods are available:

For mixtures likely to behave ideally, simple vapor pressure ratios can be used to determine separation factors.

For regular solutions (in which there is no entropy of mixing) Scatchard–Hildebrand theory may be used. The approach is described in the book by Hildebrand, Prausnitz, and Scott (1970); a well-known method for predicting VLE in hydrocarbon systems, the Chao–Seader method (1961), is based on this theory.

VAPOR–LIQUID SYSTEMS — DISTILLATION

For certain combinations of homologs, the method of Pierotti and coworkers (1959) provides estimates of infinite solution activity coefficients. As noted in Equation 12-9, when these coefficients are known, the full concentration range can be filled in with equations such as that of Van Laar (Equation 12-8).

On the basis of molecular structures of the species in the mixture, the group contribution method UNIFAC developed by Fredenslund et al. (1975) provides activity coefficient values. The use of UNIFAC is summarized in the book by Fredenslund, Gmehling, and Rasmussen (1977).

1. Diffusion Coefficients

For prediction of mass transfer rates in distillation, values of the liquid and vapor diffusion coefficients ("diffusivities") are needed, and usually must be estimated. The recommendations of Reid, Prausnitz, and Sherwood (1977) should be followed when making estimates. For the liquid phase, the estimating methods provide coefficents for the dilute case only, and correction to the concentrated regions involves the use of VLE data; liquid-phase equilibria and diffusion rates are coupled.

2. Other Physical Property Data

Other properties important in the design and scaleup of distillation equipment include viscosities, densities, and surface tensions. Measurement and/or estimation of these properties is straightforward; however, caution must be applied when taking pure component properties into the context of mixtures that exist in distillation columns. The volume by Reid, Prausnitz, and Sherwood (1977) should be consulted for the recommended predictive methods.

C. Prediction of Performance

1. Stage Calculations

Individual contacting stages are combined to form the cascade generally known as a distillation column. Figure 12-3, taken from a paper by Wang and Wang (1981), shows such a cascade. The cascade has been made completely general permitting heat addition or removal at each stage, feed or withdrawal streams at each stage, reflux condensation in a separate heat exchanger, and vapor generation in a reboiler. The calculations for the stages required for a given separation may be handled in various ways. For a simple binary system, the well-known McCabe–Thiele or Ponchon–Savarit graphical methods may suffice. For multicomponent systems, especially those involving phase nonidealities, analytical solutions are required. Such solutions are best handled by computers, and a large body of published information on the computation techniques has been made available. Summaries of the techniques have been presented by Wang and Wang (1981), Fair and Bolles (1968), and others.

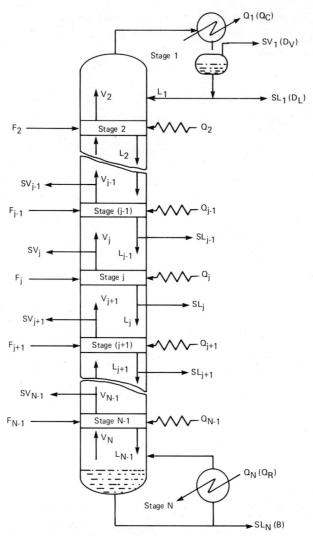

FIGURE 12-3 A generalized multistage separation column.

Table 12-1 shows areas of applicability of the several approaches to determination of theoretical stages.

In general, stage calculations are based on the limiting conditions of total reflux and infinite stages, as shown in Figure 12-4. An estimate of the required stages at total reflux may be obtained by use of Equation 12-12 developed by Fenske (1932).

$$N_M = \frac{\log\left[\left(\dfrac{x_i}{x_j}\right)_1 \left(\dfrac{x_j}{x_i}\right)_N\right]}{\log \alpha_{ij}} \tag{12-12}$$

VAPOR–LIQUID SYSTEMS — DISTILLATION

TABLE 12-1 Classification of Continuous-Distillation Problems by Methods of Calculation

System	VLE	Stages	Calculation	Remarks
Binary	Ideal	Few	Graphical	
Binary	Ideal	Many	Graphical/analytical	Computer useful
Binary	Nonideal	Few	Graphical	
Multicomponent	Ideal	Few	Analytical	Hand, if not repetitive
Binary	Nonideal	Many	Graphical/analytical	Computer useful
Multicomponent	Ideal	Many	Analytical	Computer essential
Multicomponent	Nonideal	Few	Analytical	Computer essential
Multicomponent	Nonideal	Many	Analytical	Computer essential

where N_M = the minimum number of stages (at total reflux) required for a separation of components i and j into distillate compositions of $x_{i,1}, x_{j,1}$ and residue compositions of $x_{i,N}, x_{j,N}$. Equation 12-12 is useful for approximating the effects of relative volatility and degree of separation (values of x), but it represents a condition not applicable to a column separating a flowing feed mixture.

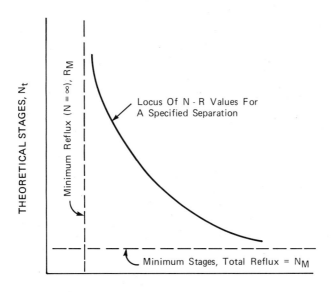

FIGURE 12-4 Stages — reflux curve for distillation.

The limiting condition of minimum reflux can be estimated by Equations 12-13 and 12-14 of Underwood (1948).

$$\frac{\sum \alpha_i x_{i,F}}{\alpha_i - \theta} = 1 - q \qquad (12\text{-}13)$$

$$\frac{\sum \alpha_i x_{i,D}}{\alpha_i - \theta} = R_M - 1 \qquad (12\text{-}14)$$

where q is the familiar term from McCabe–Thiele analysis and is the ratio of moles of saturated liquid in the feed to the total moles of feed. The use of Equations 12-13 and 12-14 involves solving Equation 12-13 for the root θ and then solving Equation 12-14 for the minimum reflux ratio R_M. The development and limitations of Equations 12-14 are given in Underwood (1948) and in many standard texts on distillation.

Through the use of the empirical Gilliland (1940) relationship, it is possible to estimate the required number of stages N at finite reflux ratio R:

$$\log\left(\frac{N - N_M}{N + 1}\right) = 0.75\left[1 - \log\left(\frac{R - R_M}{R + 1}\right)\right]^{0.5668} \qquad (12\text{-}15)$$

The form of Equation 12-15 is attributed to Eduljee (1975).

If a simple graphical approach, or an approximate analytical approach such as Equations 12-12 to 12-15 are not appropriate for the reliability required, then a rigorous stage-by-stage computation is called for. This comprises satisfying the so-called *M-E-S-H* equations (M = material balance, E = equilibrium, S = summation of phase mole fractions to unity, H = enthalpy balance) for every stage, recognizing that flows and compositions vary throughout the cascade of stages. Computer programs for handling the rigorous calculations are readily available, through computer service organizations as well as in the published literature.

2. Equipment Hydraulics — Tray Columns

Most stage-type distillation columns contain trays in which liquid flows laterally in contact with rising vapor. The best-known, nonproprietary device of this type is the crossflow sieve tray, shown in Figure 12-5, where the vapor is dispersed by means of small (3–6 mm) holes and the liquid is aerated to a froth or a spray. While the following discussion emphasizes sieve trays, it can be extended easily to other crossflow devices such as valve trays and bubble-cap trays.

VAPOR-LIQUID SYSTEMS — DISTILLATION

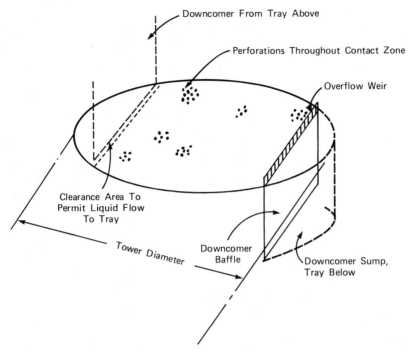

FIGURE 12-5 Diagram of crossflow sieve tray.

The dimensional variables to be considered in scaling up to a sieve tray column are:

Column diameter.
Spacing between trays.
Overflow weir length.
Weir height.
Downcomer dimensions.
Tray hole size and spacing.
Tray metal thickness.

When the tray spacing is large, higher linear velocities of vapor can be handled. When column diameter is large, greater throughputs are possible. When the overflow weir is high, there is greater liquid holdup on the tray, but also greater pressure drop in the flowing vapor. When the overflow weir length is short, large liquid flows can build excessive heads on the tray. When hole size is small, finer dispersion of vapor is possible, but there is greater danger of plugging. When holes are too closely spaced there is danger of liquid flowing down through the holes instead of across them to the overflow weir. It should be apparent that a judicious choice of the tray variables must be made for a design to be optimum.

The choice of variables is often dictated by the type of service. For high-pressure distillations, the liquid flows are usually large in comparison with the vapor flows, and the design is influenced by the ability of the trays and downcomers to accommodate the liquid flows. On the other hand, for distillations at high vacuum, the volumetric ratio of vapor to liquid is quite high and design is influenced by the ability of the trays to handle the vapor without excessive entrainment.

The design of tray columns follows a logical sequence. First, the *flow rates* of vapor and liquid are determined for the point of the column under consideration. This information is generally available from the stage calculations. Next, a *tentative diameter* of the column is established, based on the ability of an assumed tray spacing and hole geometry to handle the necessary vapor at a reasonably high approach to the limiting flow capacity. Then the *pressure drop* across the tray is calculated. Next, the dimensions of the downcomer are checked to ensure liquid handling capacity. Finally, the *tray efficiency* is estimated, based on mass transfer as well as fluid mechanical relationships.

Allowable flow rates of vapor are based on the tendency for liquid to be carried or blown off the tray at high velocities, leading to a condition of "flooding." The flood point is indicated on Figure 12-6, which represents a typical performance profile for a crossflow tray. The flood point has been studied extensively, and though difficult to measure precisely, it can be

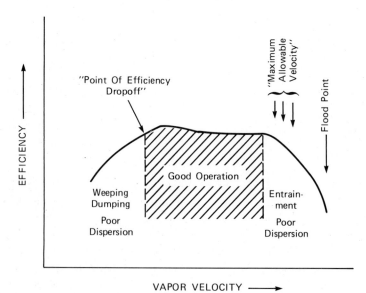

FIGURE 12-6 Performance profile for typical crossflow tray.

VAPOR-LIQUID SYSTEMS — DISTILLATION

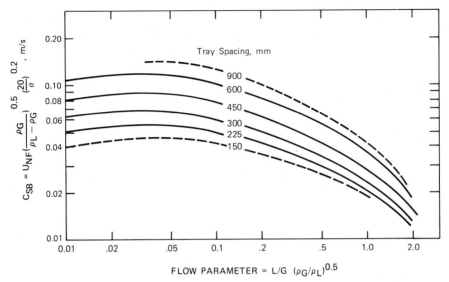

FIGURE 12-7 Flooding correlation for crossflow trays.

predicted with reasonable reliability and it serves as the upper bound for vapor capacity considerations. The vapor velocity at which flooding occurs has been correlated with mass flow rates, densities, and surface tension as shown in Figure 12-7. The capacity factor C_{SB} is

$$C_{SB} = U_{NF} \left(\frac{\rho_G}{\rho_L - \rho_G} \right)^{0.5} \left(\frac{20}{\sigma} \right)^{0.2} \quad (12\text{-}16)$$

The correlation, published originally by Fair (1961), has been verified by many later experiments and in general tends to be conservative. A predicted flood condition may not be reached at the operating value of U_{NF}. Importantly for scaleup, larger diameter towers tend to have greater capacity than smaller diameter towers. The value of U_{NF} in Equation 12-16 is found from

$$U_{NF} = \frac{\text{volumetric vapor rate at flood}}{\text{total tower area} - \text{area for downflow from tray}}$$

The denominator of this relationship is the "net area," or that which governs the velocity of approach to the tray.

The abscissa term of Figure 12-7 is a flow parameter, which represents a ratio of kinetic energies of liquid to vapor. When this ratio is low, roughly below a value of 0.1, a spray is formed on the tray due to the high kinetic energy of the vapor.

There are some important restrictions on the use of Figure 12-7. If there is a foaming tendency of the liquid, lower values of the flood limit result; it may be necessary to de-rate the column to 50 percent of its nonfoaming capacity. When the ratio of hole area to bubbling ("active") area is less than 0.10, the following corrections apply:

A_h/A_a	$U_{NF}/U_{NF,\text{chart}}$
0.10	1.0
0.08	0.9
0.06	0.8

Good design calls for a vapor loading of about 80 percent of flood. The design velocity (based on net area) is then calculated from

$$U_N = \left(\frac{\%\text{ flood}}{100}\right)(C_{SB} \text{ from Figure 12-7}) \tag{12-17}$$

After allowing for the cross-sectional area taken up by the downcomer taking liquid to the tray below, the tower cross-sectional area is calculated and the diameter obtained.

The pressure drop for vapor flow through a tray may be assumed equal to the simple sum of the drop through the holes and the residual drop through the froth or spray:

$$h_t = h_d + h_L \tag{12-18}$$

where the head terms are in heights of tray liquid. The drop through the holes ("dry drop") is calculated from a form of the orifice equation:

$$h_d = \frac{50.8}{C_v^2} \frac{\rho_G}{\rho_L} U_H^2 \tag{12-19}$$

where the heads are in millimeters of liquid (density ρ_L), and the velocity of vapor through the holes U_H is in m/s. The orifice coefficient is obtained from Figure 12-8 [Leibson and coworkers (1957)].

The residual drop h_L may be obtained from the equivalent clear liquid head at the overflow weir, corrected for aeration effects:

$$h_L = \beta(h_w + h_{ow}) \tag{12-20}$$

with the aeration factor β obtained from Equation 12-21 of Bolles and Fair (1982)

$$\beta = 0.19 \log L_w - 0.62 \log F_{vh} + 1.679 \tag{12-21}$$

where

L_w = equivalent clear liquid flow rate over the outlet weir, m³/s m
F_{VH} = flow factor through holes, $U_H \rho_G^{1/2}$, m/s (kg/m³)$^{1/2}$

VAPOR–LIQUID SYSTEMS — DISTILLATION

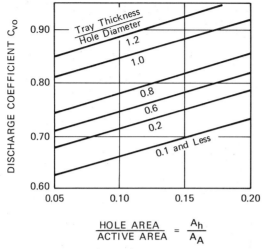

FIGURE 12-8 Discharge coefficients for sieve trays.

The weir crest h_{ow} in Equation 12-20 is calculated from the Francis weir equation:

$$h_{ow} = 664 L_W^{2/3} \qquad (12\text{-}22)$$

It should be noted that the use of equivalent clear liquid on the tray does not mean that such a type of liquid actually exists on the tray. Under normal commercial conditions, there is violent agitation on the tray, and the aerated mass is greatly expanded from the liquid volume alone. The head correlations take this difference into account empirically. The volume and characteristics of the two-phase mixture are important in the estimation of mass transfer efficiency, and will be summarized below.

The volume of the froth or spray on a tray may be taken as the product of the active (bubbling) area A_a and some effective froth height:

$$Q_f = \bar{Z}_f A_a \qquad (12\text{-}23)$$

At a point on the tray, the porosity is

$$\varepsilon = \frac{\bar{Z}_f - h_L}{\bar{Z}_f} = 1 - \frac{h_L}{\bar{Z}_f} \qquad (12\text{-}24)$$

Porosity varies vertically above the tray floor, as shown in Figure 12-9. Froth height also varies horizontally across the tray, hence the average term $\bar{Z}_f$ is used.

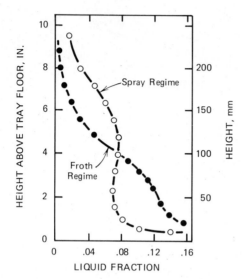

FIGURE 12-9 Vertical dispersion profiles, air–water on sieve trays.

Tray studies carried out several years ago, primarily with small bubble caps, as part of the AIChE-sponsored Distillation Efficiency Research Program (1958), included a number of visual observations of froth height. The observations were correlated by the equation

$$Z_f = 43.2 F_{va}^2 + 1.89 h_w - 40.6 \qquad (12\text{-}25)$$

where

Z_f = observed froth height, mm
F_{va} = flow factor through active area ($= U_a \rho_G^{1/2}$), m/s (kg/m³)$^{1/2}$
h_w = weir height, mm

At the flood point, Z_f was observed to be approximately equal to the tray spacing TS:

$$Z_{f,f} = 43.2 F_{va,f}^2 + 1.89 h_w - 40.6 \qquad (12\text{-}26)$$

This suggests a generalized approach to froth height estimation:

$$\frac{Z_f}{TS} = \frac{43.2 F_{va}^2 + 1.89 h_w - 40.6}{43.2 F_{va,f}^2 + 1.89 h_w - 40.6} \qquad (12\text{-}27)$$

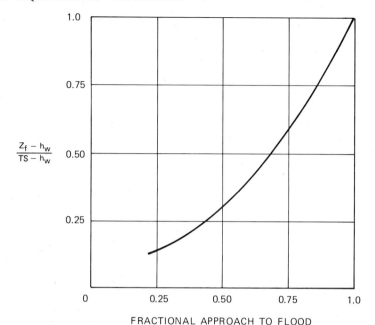

FIGURE 12-10 Chart for estimating height of froth, based on AIChE method.

For weir heights in the 25–75 mm range, Equation 12-27 becomes, approximately,

$$\frac{Z_f - h_w}{TS - h_w} = \left(\frac{F_{va}}{F_{va,f}}\right)^2 = \Gamma_f^2 \tag{12-28}$$

where Γ_f is the fractional approach to the flood point, based on vapor loading ($= U_n/U_{n,f}$). Equation 12-28 is shown in Figure 12-10 as an average curve representing 460- and 610-mm tray spacings, with 25- and 50-mm weirs.

The average *residence time for liquid* on a tray is equal to the volume of liquid on the tray, under steady flow conditions, divided by the rate of liquid flow to and from the tray:

$$\bar{t}_L = \frac{(1-\varepsilon)Z_f' A_a}{q} = \frac{h_L A_a}{q} \tag{12-29}$$

This residence time applies only to plug flow with unidirectional pattern, as discussed in the following sections. The distribution of residence times of liquid on a cross flow is a design parameter and the subject of a number of investigations. However, Equation 12-29 is of value in efficiency prediction, and can always be applied to a point on the tray.

Vapor is usually assumed to flow vertically upward on a tray, without recycle or stagnation. Even though the physical picture is different from this, the assumption appears valid for engineering design purposes. The holdup of vapor is the fraction, or porosity, or vapor in the froth. The *vapor residence time* is, then,

$$\bar{t}_G = \frac{\varepsilon Z'_f A_a}{Q_v} = \varepsilon \frac{Z'_f}{U_a} \qquad (12\text{-}30)$$

The evaluation and prediction of *interfacial area* on a crossflow bubble cap or sieve tray has been a challenge to chemical engineering researchers for many years. Only recently have real advances been made in meeting the challenge. Through the years there have been measurements of area by photographic or chemical methods, but the results have varied widely among researchers, and the techniques have at best been questionable.

Recent experimental work by Burgess and Calderbank (1975) and Calderbank and Pereira (1977) appears to represent a breakthrough in the measurement of interfacial area under actual sieve tray contacting conditions. The basis for the work is a four-point probe (electrical, for conducting systems; optical, for nonconducting systems) which when inserted at a point on a sieve tray can measure bubble size distribution and rate of bubble rise. Thus, data on size distribution are being collected and interfacial area is being computed directly from the size and rise rate measurements.

Fell [see Hai et al. (1977)] has installed the Calderbank type equipment in his laboratory in Australia, and has reported a few data. The results from the Calderbank and Fell laboratories are still under study, but a preliminary plot of the data is shown in Figure 12-11. While there is considerable scatter, it would appear that areas of the order of 300 m^2/m^3 froth and less are to be expected. Such values are considerably lower than those obtained earlier by photographic and chemical techniques.

The apparent leveling-off of the area with increasing throughput is noteworthy. This is thought to be due to bubble coalescence and the formation of very large bubbles as hole velocities become higher.

3. Mass Transfer Efficiency

The efficiency sought by the designer is the *overall column efficiency*:

$$E_{oc} = \frac{N_{t,c}}{N_{A,c}} = \frac{\text{number theoretical stages in column}}{\text{number actual stages in column}} \qquad (12\text{-}31)$$

where

$N_{t,c} = N_t - N_{rb} - N_{pc}$
N_t = calculated theoretical stages
N_{rb} = stages in reboiler
N_{pc} = stages in partial condenser

VAPOR–LIQUID SYSTEMS — DISTILLATION

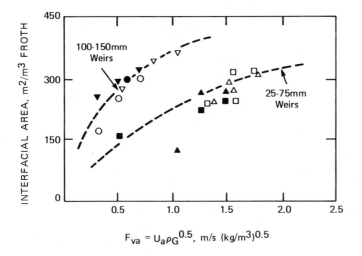

FIGURE 12-11 Interfacial area in sieve tray froths, obtained by direct bubble size–velocity measurements.

Symbol	System	Open Area,%	h_w,mm	d_H,mm
▼	Air-Water	4.8	150	6
○	Air-Water	10.8	150	9.5
●	Methanol	5.4	100	6
▲	Air-Water	5.0-5.9	25	6,12
△	Air-Water	10.7-11.0	25	6,12
■	Air-Water	5.0-5.9	75	6,12
□	Air-Water	10.7-11.0	75	6,12
▽	Benzene	5.4	100	6

If there is no partial condenser, $N_{pc} = 0$. If there is a kettle or once-through natural recirculation reboiler, $N_{rb} \sim 1.0$. If there is a thermosiphon reboiler, $N_{rb} \to 1.0$ at high recycle rates. It may be advisable to allow some extra trays in the column to account for uncertainties in design or for possible future changes in operating mode.

A value of E_{oc} may be obtained in several ways. There may be good operating experience with the same (or similar) system and the same type of contacting device; if the results of such experience are in the form of an evaluated efficiency, then the experience should be used. It is seldom, however, that directly applicable field experience is available. Another way to obtain a value of E_{oc} is to use carefully measured laboratory efficiencies, with an appropriate scale factor. This method will be discussed in some detail later. Alternatively, one can obtain E_{oc} through the use of mass transfer relationships.

The mass transfer method involves the estimation of several efficiencies in sequence. First, a *point efficiency* is determined. The value of this efficiency can vary across the tray, but is useful in that its value cannot exceed 100 percent. Next, a Murphree tray efficiency is estimated. This efficiency takes into account the concentration gradients on the tray. Next, the Murphree efficiency, which is calculated on a dry (nonentrainment) basis, is corrected for entrainment; this correction will be slight except when the vapor rate is in the vicinity of the flood point as shown in Figure 12-6. Finally, a value of E_{oc} is obtained from the Murphree efficiency by taking into account the variations in liquid to vapor ratio that occur across the entire column.

The transfer of material between vapor and liquid on a sieve tray occurs primarily in the froth (or spray) zone, and on a local basis may be postulated to occur as shown in Figure 12-12. For the cases shown, the liquid concentration changes horizontally as it passes the column of vapor, by an amount $\Delta x_{n,p} = (x_{n,p})_2 - (x_{n,p})_1$ and the vapor concentration changes by an amount $\Delta y = y_{n,p} - y_{n-1,p}$. The local, or *point efficiency* of the mass transfer process is defined by

$$E_{OG} = \left[\frac{y_n - y_{n-1}}{y_n^* - y_{n-1}}\right]_p \tag{12-32}$$

where the concentration y_n^* is that which would be reached if the exit vapor were in equilibrium with the exit liquid concentration $x_{n,p,2}$, that is, $y_n^* = K_{vL}(x_{n,p,2})$, K_{vL} being the vapor–liquid equilibrium ratio "K value" of Equation 12-5. In the application of Equation 12-32, the assumption is made that there are no vertical concentration gradients in the liquid phase. By definition, the point efficiency cannot exceed 1.0 (or 100 percent). An equivalent expression could be developed for the liquid-phase point efficiency E_{OL}, but practice is to standardize on vapor concentration unit expressions of efficiency. More will be said about this in connection with the tray efficiency.

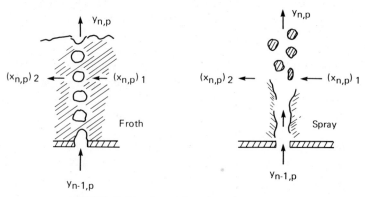

FIGURE 12-12 Contacting at a point on tray *n*.

VAPOR–LIQUID SYSTEMS—DISTILLATION

The expression E_{OG} takes into account the transfer through both phases. For vapor phase alone and liquid phase alone, the corresponding expressions are E_G and E_L, respectively. When there is no liquid-phase resistance to mass transfer, $E_{OG} = E_G$; similarly, when vapor-phase resistance is missing, $E_{OL} = E_L$.

The two-resistance model for mass transfer leads to the following relationship:

$$\frac{1}{N_{OG}} = \frac{1}{N_G} + \frac{\lambda}{N_L} \tag{12-33}$$

where

N_{OG} = number of overall transfer units, vapor concentration basis
N_G = number of vapor transfer units
N_L = number of liquid transfer units
λ = ratio of slopes of equilibrium and operating lines, that is,

$$\lambda = \frac{m}{L/V} = \frac{mV}{L}$$

Equation 12-11 elaborates the definition of m.

When N_{OG} is evaluated at a point on the tray, the basic expression for point efficiency is obtained:

$$E_{OG} = 1 - \exp(-N_{OG}) \tag{12-34}$$

For example, if 1.5 overall transfer units can be produced at a point on a tray,

$$E_{OG} = 1 - \frac{1}{e^{1.5}} = 0.777 \tag{12-35}$$

and the approach to equilibrium at a point will be 77.7 percent.

Liquid-phase transfer units are calculated from the relationship

$$N_L = k_L a_i \bar{t}_L \tag{12-36}$$

The interfacial area a_i and average liquid residence time were discussed earlier. While the contacting mechanisms are not understood well enough to enable highly reliable prediction of a_i and $\bar{t}_L$, the real difficulty is in predicting k_L, the liquid-phase mass transfer coefficient.

Two-film theory gives the following expression for k_L,

$$k_L = \frac{\mathscr{D}_L}{z_L} \tag{12-37}$$

with the film thickness z_L being a correlating parameter:

$$z_L = \frac{\mathscr{D}_L a_i \bar{t}_L}{N_L} \tag{12-38}$$

The penetration theory of Higbie (1935), which seems more representative of actual tray contacting mechanics, leads to

$$k_L = 2\sqrt{\frac{\mathscr{D}_L}{\pi \theta_L}} \tag{12-39}$$

with the exposure time θ_L being a correlating parameter:

$$\theta_L = \left(\frac{2 a_i \bar{t}_L}{N_L}\right)^2 \frac{\mathscr{D}_L}{\pi} = 1.27 \left(\frac{a_i \bar{t}_L}{N_L}\right)^2 \mathscr{D}_L \tag{12-40}$$

Higbie suggested that exposure time might be estimated from bubble diameter and rate of bubble rise (froth regime):

$$\theta_L = \frac{D_b}{U_{br}} \tag{12-41}$$

However, for the highly turbulent conditions on a sieve tray under load, Higbie's suggestion is completely unrealistic. Values of θ_L must be deduced from mass transfer data.

It should be noted that according to penetration theory, mass transfer rate is proportional to the square root of the diffusion coefficient; measurements of mass transfer rates show that indeed this is true. At the present time, the chief value of the penetration model is that it provides a basis for the influence of the diffusion coefficient.

Based on present knowledge, correlation on the basis of the volumetric coefficient $k_L a_i$ is more useful, and for design an empirical expression for sieve trays has been developed by the AIChE (1958) and Smith (1963):

$$k_L a_i = (0.40 F_{va} + 0.17)(3.875 \times 10^8 \mathscr{D}_L)^{1/2} \tag{12-42}$$

where

$F_{va} = U_a \rho_G^{1/2}$, m/s (kg/m³)$^{1/2}$
$\mathscr{D}_L$ = molecular diffusion coefficient in the liquid, m²/s

From Equations 12-40 and 12-42, an exposure time θ_L can be deduced:

$$\theta_L = \frac{10^{-8} a_i^2}{0.488 F_{va}^2 + 0.415 F_{va} + 0.0881} \tag{12-43}$$

VAPOR-LIQUID SYSTEMS — DISTILLATION

Therefore for typical values of $F_{va} = 1.0$ and $a_i = 175$ m^{-1}, Equation 12-43 gives an exposure time of 0.00031 s. For a typical bubble size of 2 mm and a rise velocity of 0.1 m/s, Equation 12-41 gives an exposure time of 0.002 s. The difference in values from Equations 12-41 and 12-43 is more than an order of magnitude. This underscores the unreliability of Equation 12-41.

Although the mechanisms of contacting are likely to be quite different in froth and spray contacting, our present knowledge is such that no distinction between them can be made in a design correlation. Studies that appear to span the transition range between froth and spray show no particular discontinuities or inflections, a fact that further confuses the impact of regime effects on tray design.

For additional background on liquid-phase mass transfer, the monograph by Sherwood, Pigford, and Wilke (1975) should be consulted.

Vapor-phase coefficients are calculated from the relationship

$$N_G = k_G a_i \bar{t}_G \tag{12-44}$$

From the two-film theory,

$$k_G = \frac{\mathscr{D}_G}{z_G} \tag{12-45}$$

Penetration theory, as originally developed, did not cover the vapor phase. Mechanistically, however, it can be applied as in the liquid phase,

$$k_G = 2\sqrt{\frac{\mathscr{D}_G}{\pi \theta_G}} \tag{12-46}$$

Studies of vapor-phase mass transfer on sieve trays have not led to design correlations for the volumetric mass transfer coefficients (similar to Equation 12-42), but the results from the AIChE (1958) studies of small bubble caps have led to a useful empirical correlation:

$$N_G = \frac{0.776 + 0.0046 h_w - 0.24 F_{va} + 0.0712 L_w}{(\text{Sc}_G)^{1/2}} \tag{12-47}$$

where

h_w = weir height, mm
L_w = liquid rate to tray, m^3/s m weir length
Sc_G = vapor-phase Schmidt number = $\left(\dfrac{\mu}{\rho \mathscr{D}}\right)_G$

Note that residence time $\bar{t}_G$ is not shown separately. If $\bar{t}_G = (\varepsilon Z_f)/U_a$ as from Equation 12-30, and Z_f is taken from Equation 12-25,

$$\bar{t}_G = \frac{\varepsilon \left(43.2 F_{va}^2 + 1.89 h_w - 40.6\right)}{U_a} \tag{12-48}$$

Finally, transfer units are combined to obtain efficiency. Liquid-phase transfer units are then calculated from Equations 12-36 and 12-42. Vapor-phase transfer units are calculated from Equation 12-47. The transfer units are then used in Equation 12-34 to obtain a point efficiency

$$E_{OG} = 1 - \exp(-N_{OG}) \tag{12-49}$$

where

$$N_{OG} = \frac{1}{1/N_G + \lambda/N_L}$$

4. Liquid Mixing on Trays

The point efficiency, or local efficiency, establishes approach to equilibrium at a point on the tray. If the tray froth or spray is completely mixed, such that there are no concentration gradients on the tray, then the local efficiency is the same as the overall tray efficiency, and is called the *Murphree tray efficiency* E_{mv}. The definition of Murphree efficiency is analogous to that of the point efficiency

$$E_{mv} = \frac{\bar{y}_n - \bar{y}_{n-1}}{y_n^* - \bar{y}_{n-1}} \tag{12-50}$$

where y_n^* is the average vapor composition in equilibrium with the exit liquid composition $x_{n,0}$. The bars over the vapor concentrations denote average values and imply complete mixing of the vapor. When there is complete mixing on the tray, then at all points $x = x_{n,0} = x_{n,p}$ and

$$E_{mv} = E_{OG} \quad \text{(complete mixing)} \tag{12-51}$$

If the liquid moves through the froth or spray in plug flow, and the vapor rising through becomes completely mixed before it enters the tray above,

$$E_{mv} = \frac{1}{\lambda}[\exp(\lambda E_{OG}) - 1] \tag{12-52}$$

Therefore, it is necessary to consider tray mixing effects in order to arrive at a value of E_{mv} for design. Entrainment effects must also be allowed. Figure 12-13 shows how the overall development of a design tray efficiency can be carried out.

Models used for representing liquid mixing on trays include:

Oliver and Watson (1956)	Liquid recirculation
Crozier (1956)	Froth recirculation
Johnson and Marangozsis (1958)	Liquid splashing
Gautreaux and O'Connell (1955)	Mixing pool
Foss, Gerster, and Pigford (1958)	Eddy diffusion

VAPOR–LIQUID SYSTEMS — DISTILLATION

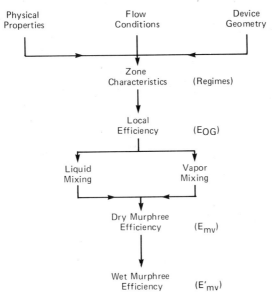

FIGURE 12-13 Chart showing various components of the estimation of Murphree tray efficiency.

A discussion of these models has been given by Sakata (1966). Only the mixing pool, and eddy diffusion models have been used by designers; our discussion here will be limited to them.

The mixing pool model was introduced by Gautreaux and O'Connell (1955) who visualized a crossflow tray as comprising a number of well-mixed stages, or pools, in series. Their general expression for mixing is

$$E_{mv} = \frac{1}{\lambda}\left(1 + \frac{\lambda E_{OG}}{n}\right)^n - 1 \qquad (12\text{-}53)$$

where n, the number of pools, can be estimated from Figure 12-14. When $n = 1$, there is only one stage on the tray, and Equation 12-53 reduces to $E_{mv} = E_{OG}$. When there is plug flow, $n \to \infty$, and Equation 12-52 is obtained. As an example, for a 2-m flow path, with $E_{OG} = 0.60$ and $\lambda = 1.0$, "average mixing" gives about six pools, and E_{mv} calculates to be 0.77. Thus, there is an enhancement of point efficiency of $0.77/0.60$ or 28 percent.

The eddy diffusion model postulates that there is counterflow mixing by eddy mechanisms to give a distribution of residence times on the tray. This model was employed in the AIChE distillation efficiency research work (1958) and has been used extensively to interpret concentration gradients in flowing systems. Experimental measurements have been made of eddy diffusion coefficients on sieve tray simulators, and the resulting correlation of Barker and Self (1962) is useful for design purposes.

$$\mathcal{D}_E = 6.68(10^{-3})U_a^{1.44} + 0.922(10^{-4})h_L - 0.00562 \qquad (12\text{-}54)$$

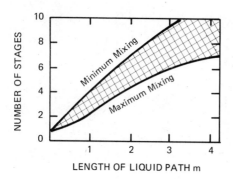

FIGURE 12-14 Chart for estimating number of mixing pools on a crossflow tray.

where

$\mathscr{D}_E$ = eddy diffusion coefficient, m²/s
U_a = vapor velocity through the active, or bubbling area of the tray, m/s
h_L = liquid head on tray, mm

The strong effect of the vapor velocity in Equation 12-54 should be noted. The velocity tends to be low in pressure fractionators, and the low value of $\mathscr{D}_E$ suggests an approach to plug flow. In vacuum fractionators, on the other hand, velocity tends to be high, and a closer approach to a well-mixed tray results.

The eddy diffusion coefficient is used to obtain a value of the dimensionless Peclet number,

$$\text{Pe} = \frac{W^2}{\mathscr{D}_E \bar{t}_L} \qquad (12\text{-}55)$$

where W is the length of travel across the tray. A high value of Pe indicates a close approach to plug flow; as Pe approaches zero, complete mixing occurs. Thus, a large diameter vacuum column would have some help from length of flow path in departing from complete mixing. The developed model for eddy diffusion includes an analytical expression relating E_{OG}, E_{mv}, Pe, and λ. A graphical representation is shown in Figure 12-15. For the example given for the pool model, the same enhancement of 28 percent with the eddy diffusion model would require a Peclet number of about 10.

5. Entrainment from Trays

At high vapor velocities, as shown in Figure 12-6, liquid entrainment has a deleterious effect on mass transfer efficiency. This is because the liquid recycle results in a departure from countercurrent flow conditions in the column. Conversion of a "dry" efficiency (as determined from Figure 12-14 or Figure 12-15, for example) to a "wet" efficiency involves an entrainment correction. A

VAPOR–LIQUID SYSTEMS — DISTILLATION

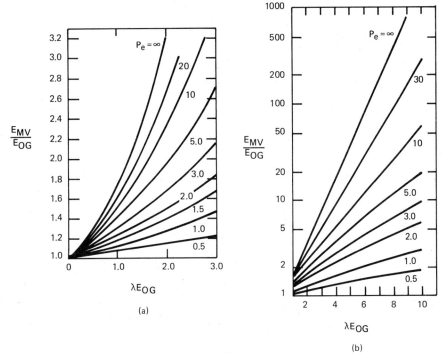

FIGURE 12-15 Charts for estimating effect of liquid mixing on tray efficiency.

simple but effective method for making this correction is based on a modification of the Colburn equation (1936) as developed by Fair and Matthews (1958):

$$\frac{E'_{mv}}{E_{mv}} = \frac{1}{1 + E_{mv}(\psi/(1-\psi))} \tag{12-56}$$

where

E'_{mv} = Murphree efficiency corrected for entrainment
ψ = Entrainment parameter, obtained from Figure 12-16

The abscissa group in Figure 12-16 is the flow parameter that is also used to predict flooding velocity in Figure 12-7.

After the wet Murphree efficiency is evaluated, there is one more step needed before the overall column efficiency is obtained from Equation 12-31. This step involves correcting for changes in slope of the operating and equilibrium lines throughout the column:

$$E_{OG} = \frac{\log[1 + E'_{mv}(\lambda - 1)]}{\log \lambda} \tag{12-57}$$

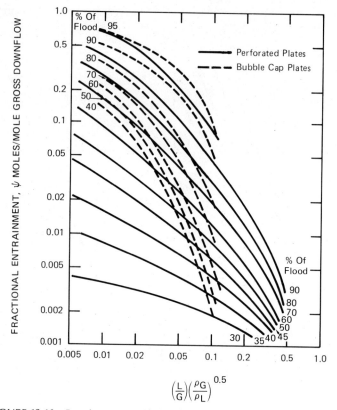

FIGURE 12-16 Entrainment correlation for sieve trays and bubble cap trays.

While the value of the wet efficiency E'_{mv} is used in Equation 12-57, allowance must be made for sections of the column that may not have enough entrainment to cause a correction of the dry efficiency.

D. Laboratory Scaleup

1. *Demonstration of Separability*

It is possible, and fairly common, to demonstrate in the laboratory or pilot plant that a mixture can be separated by distillation. There may be the presence of azeotropes, for example, that must be studied experimentally. Or there may be a suspicion of intermediate boilers that could tend to accumulate in central portions of a distillation column. Any of several laboratory devices can be used for this purpose, and there are high-efficiency packings that can be inserted in the laboratory column in varying heights to study reflux-stage effects. The techniques and equipment for laboratory distillations are described in the book edited by Perry and Weissberger (1965).

2. Prediction of Efficiency

Useful efficiency data can be gathered in the laboratory, but only if a small tray-type column is used. The greatest success in such scaleup work has been with the Oldershaw column (1941) which is available in glass or metal. This column contains sieve trays with downcomers and very small perforations. Typical sizes are 25-mm and 50-mm diameter, with tray spacings equal to diameter.

Recently, Fair and coworkers (1983) reported efficiency scaleup data for Oldershaw columns of glass and metal, and found that the small-scale efficiencies for the column were very close to the point efficiencies expected for the large columns. Figure 12-17 shows typical data; the large-scale tests were run in a 1.22-m column by Fractionation Research, Inc. The greatly different vapor velocities must be compensated for by normalizing with respect to a flood condition. This technique has been used successfully in a great many industrial design/scaleup projects.

The potential power of a scaleup method such as this bears further discussion. There are many cases of scaleup involving complex mixtures or highly nonideal mixtures that are difficult to evaluate with simple VLE data. Such mixtures can be processed with various numbers of trays (changing of tray number in the Oldershaw column is easily done) until the combination of trays and reflux is found that gives the desired separation. The commercial scale efficiency will be as great or greater than that of the small column. By this empirical approach, the problems of VLE evaluation and theoretical stage determination are bypassed.

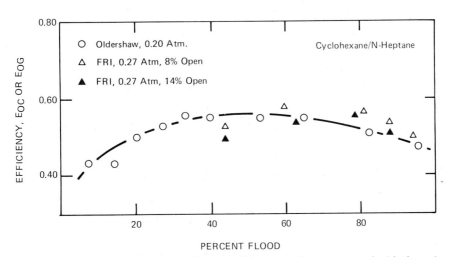

FIGURE 12-17 Efficiency of a 25-mm-diameter Oldershaw column compared with the point efficiency obtained in a 1.2-m-diameter column at Fraction Research, Inc.

E. Idealizations and Assumptions

Although the methodology for scaleup and design of distillation equipment is well known and straightforward, there are uncertainties in the process that must be recognized. Errors can arise in the equilibrium data from, for example, problems in the laboratory and a failure to determine thermodynamic consistency of the results. These errors can cause large changes in required stages, when close-boiling mixtures (with low values of the separation factor) are being considered. The effect of errors in VLE data can be estimated from the Fenske Equation 12-12. Another source of error in VLE data arises in the extension of binary data to multicomponent systems. It is a general problem that pure component properties do not relate predictably to properties of the same components in solution with other materials. Therefore, errors in physical properties of mixtures are ones to watch carefully.

Optimization of a distillation system design is handled in various ways. Experienced designers often do this by their "feel" and general experience. However, inroads of high energy costs have caused optima to shift. For example, it is now common to find an optimum reflux ratio set only 5 percent or so greater than the minimum reflux ratio; this is much lower than was found a decade ago. The most workable approach to optimizing a system is through the use of a process simulator such as FLOWTRAN, ASPEN, DESIGN, or PROCESS. These computer-aided design methods can generate economics data for an entire system and lend themselves to a case study approach to optimization. While these simulators have some internal optimization capability, not all of the important variables can be considered easily. An experienced process engineer with the resources of a system such as FLOWTRAN can quickly arrive at an optimum combination of plates, reflux, stream condition, tray spacing, entrainment ratio, and so on.

F. Conclusions

Distillation columns can often be designed without need for any laboratory work. The VLE and other properties of the system can be characterized using standard approaches and supporting theory. Stages can be calculated rigorously, if desired, through the use of computerized techniques. Equipment can be sized with reasonable reliability, based on a large body of performance data, especially for crossflow type tray columns. Perhaps the weakest link in the total process is the ad hoc prediction of mass transfer efficiency; here reliable models are still under development.

There are cases where some laboratory work is essential. These include the processing of systems without previous study and those which might be expected to behave quite nonideally, or which cannot be characterized in terms of pure components. Laboratory study of mass transfer efficiency is required when the system is expected to exhibit slow diffusion or other contributions to

GAS – LIQUID SYSTEMS — ABSORPTION AND STRIPPING

low efficiency. In such cases, the study must be made in a small tray-type unit, if meaningful scaleup data are to be obtained.

IV. GAS – LIQUID SYSTEMS — ABSORPTION AND STRIPPING

A. Major Issues

Gas–liquid operations such as absorption and stripping are similar to distillation. Many features of the equipment design are essentially the same. There are, however, some essential differences:

> The gas phase may consist largely of noncondensables, components present at temperatures above their critical points.
>
> Mass transfer is largely unidirectional, from gas to liquid or from liquid to gas; this affects the rates of diffusion since there is not present the "competing" species of the counterdiffusion process.
>
> There may be large heat effects due to heats of solution resulting in larger temperature gradients than would be encountered in distillation.

So long as these differences are taken into account, the methods for scaleup are the same as those for distillation: the system must be defined (and this may be difficult when scrubbing trace quantities from a gas stream), the degree of separation (absorption, stripping) must be defined, physical property data gathered, a model for determining stages selected, and the number of stages determined. Here only the differences from distillation will be emphasized.

A representative absorption–stripping system is shown in Figure 12-18, together with pertinent notation. It is convenient to number the stages oppositely between absorber and stripper; this permits use of identical equations for absorption and stripping efficiency determinations.

B. Prediction of Performance

1. Phase Equilibria

For gas–liquid systems, equilibria are often reported as gas solubilities. The basic considerations of equilibrium apply, however:

$$K_i = \frac{y_i}{x_i} = \frac{p_i}{Px_i} = \frac{H_i}{P} \qquad (12\text{-}58)$$

where H_i is the Henry's law coefficient and Dalton's law is assumed to apply. Many reported data in the literature are in terms of Henry's law coefficients.

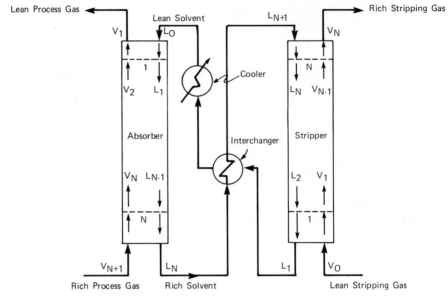

FIGURE 12-18 Absorption-stripping flow arrangement.

Other data are reported as Bunsen solubility coefficients, β_i:

$$\beta_i = \frac{\text{vol. of } i \text{ dissolved, based on STP}}{(\text{vol. of solvent})(\text{part. press. of } i, p_i)} \quad (12\text{-}59)$$

Also used is an Ostwalt coefficient L_i:

$$L_i = \frac{\text{vol. of } i \text{ dissolved, exptl. conditions}}{(\text{vol. of solvent})(\text{part. press. of } i, p_i)} \quad (12\text{-}60)$$

The correspondence between these terms is straightforward, but emphasis is given here so that confusion in reading the published data can be avoided.

There are many sources of published solubility data. These include the volumes by Seidell and Linke (1952, 1958, 1965), the compilation by Markham and Kobe (1941), the book edited by Dack (1975), and papers by Long and McDevit (1952) and Battino and coworkers (1966, 1973, 1977) published in *Chemical Reviews*. The "Solubility Data Series," sponsored by the International Union of Pure and Applied Chemistry, and edited by Kertes (1979) will comprise some 20 volumes when it is complete. By the end of 1981 seven volumes had appeared. Some of these references are bibliographical, others provide listings of solubilities.

Also published are many papers dealing with equilibria between gas–vapor mixtures and hydrocarbon liquids; these papers support studies of liquids recovery from natural gas streams and of other gas–liquid processing operations related to the fuels industry. The works by Hala and coworkers (1967, 1968), and Wichterle et al. (1976) provide some guidance in this area.

2. Stage Calculations

For absorption or stripping scaleup studies it is convenient computationally to divide them into "lean gas" and "concentrated gas" cases. The former cover instances where there is relatively little solute to be removed from the gas (absorption) or from the liquid (stripping). The dilute situation permits use of straight operating lines since there are no complications from heats of solution causing temperature gradients in the fluid phases. There are many practical cases where this lean situation prevails, the most common being the scrubbing of small amounts of contaminants from large volumes of gas.

Graphical methods are useful for lean gas cases, and stages can be stepped off easily, as shown in Figure 12-19. When there is more than one solute, separate equilibrium relationships are used, but the slopes of the operating lines (which represent the molar ratio of liquid to gas) are constant. A two-solute case is also shown in Figure 12-19. There is always a minimum solvent rate, analogous to the minimum reflux rate of distillation, which occurs when a "pinch" (infinite stages) occurs, and the location of this pinch is usually at the bottom of the column. Expressed in another way, the minimum solvent rate is that one which brings the exit solvent into equilibrium with the entering gas (absorption).

Analytical solutions of the lean gas case are straightforward, and involve the use of stripping and absorption factors.

$$\frac{V_m K_i}{L_m} = S_i = \text{stripping factor for } i \qquad (12\text{-}61)$$

$$\frac{L_m}{V_m K_i} = \frac{1}{S_i} = A_i = \text{absorption factor for } i \qquad (12\text{-}62)$$

In these expressions, V_m and L_m are molar flow rates of gas and liquid.

The Kremser equation may be used to compute the absorption recovery, or removal, of a solute i in a column of N equilibrium stages:

$$\frac{Y_{N+1} - Y_1}{Y_{N+1} - K_i X_0} = \frac{A^{N+1} - A}{A^{N+1} - 1} \qquad (12\text{-}63)$$

where

Y = moles i in gas/mole of solute-free gas
X_0 = moles of i in entering liquid/mole of solute-free gas
N = bottom theoretical tray
1 = top theoretical tray

Tray locations and numbering are shown in Figure 12-18.

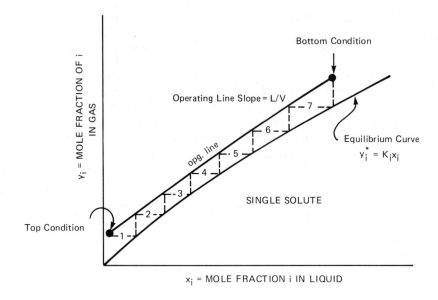
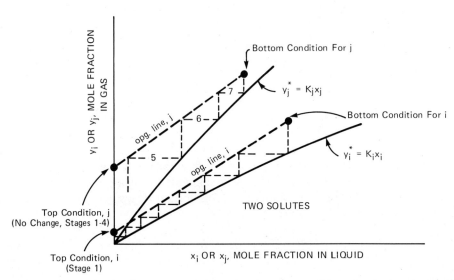

FIGURE 12-19 Graphical treatment of absorption, lean gas conditions, for single solute and two-solute cases.

GAS–LIQUID SYSTEMS–ABSORPTION AND STRIPPING

This expression allows for the presence of solute in the entering solvent. When there is no solute present in the solvent, $X_0 = 0$, and the left side of the expression is a simple fractional removal of the solute. The equivalent expression for stripping is

$$\frac{X_{N+1} - X_1}{X_{N+1} - Y_0/K_i} = \frac{S^{N+1} - S}{S^{N+1} - 1} \qquad (12\text{-}64)$$

where

$X =$ moles of i per mole of solute-free solvent
$Y_0 =$ moles of i in vapor per mole of solute-free solvent
$N =$ top theoretical tray
$1 =$ bottom theoretical tray

Again, if there is no solute in the stripping vapor, $Y_0 = 0$ and the left side represents a fractional stripping efficiency.

For multicomponent cases, especially when there is a rich gas to be processed (where the molar flows as well as temperatures can vary greatly throughout the column) it is usually necessary to utilize a rigorous model, exactly like the one used for distillation. The *M-E-S-H* equations can be used, and the only difficulty will be of convergence in those cases where both temperature gradients and temperature coefficients for solubilities are large. If a computer program is not available, the Edmister (1943) method may be used as a approximation. The Edmister method utilizes the Kremser form, but a modified nomenclature. An effective absorption factor A_e and an effective stripping factor S_e are used. If absorption stages are numbered from the top down as in Figure 12-18

$$E_{ai} = \frac{Y_{N+1} - Y_1}{Y_{N+1}} = \left[1 - \frac{L_0 X_0}{A' V_{N+1} Y_{N+1}}\right]\left[\frac{A_e^{N+1} - A_e}{A_e^{N+1} - 1}\right] \qquad (12\text{-}65)$$

where

$E_{ai} =$ fractional removal of solute i in the absorber
$Y_{N+1} =$ moles i entering/total moles of gas entering
$Y_1 =$ moles i leaving/total moles of gas entering
$L_0 =$ total moles of lean solvent entering
$X_0 =$ moles of i entering with solvent per total moles of lean solvent entering
$V_{N+1} =$ total moles of gas entering
$X_N =$ moles i leaving in rich solvent per total moles of lean solvent entering

$$A' = \frac{A_N(A_1+1)}{A_N+1} \tag{12-66}$$

$$A_e = \sqrt{A_N(A_1+1)+0.25} - 0.50 \tag{12-67}$$

$$A_N = \frac{L_N}{V_N K_N} \tag{12-68}$$

$$A_1 = \frac{L_1}{V_1/K_1} \tag{12-69}$$

$$V_N = V_{N+1}\left(\frac{V_1}{V_{N+1}}\right)^{1/N} \tag{12-70}$$

The Edmister method utilizes primarily terminal conditions and Equation 12-65 is applied to each solute in the entering gas. When there is no solute in the entering solvent, the first term on the right of Equation 12-65 becomes unity.

For stripping, with stages numbered oppositely, that is, from bottom to top, the analogous expression applies:

$$E_{si} = \frac{X_{N+1} - X_1}{X_{N+1}} = \left[1 - \frac{V_0 Y_0}{S' L_{N+1} X_{N+1}}\right]\left[\frac{S_e^{N+1} - S_e}{S_e^{N+1} - 1}\right] \tag{12-71}$$

where

E_{si} = fractional removal of solute i in the stripper
X_{N+1} = moles i entering/total moles liquid entering
X_i = moles i leaving/total moles liquid entering
L_{N+1} = total moles of liquid entering
V_0 = total moles of stripping gas
Y_0 = moles i entering with stripping gas per total moles of stripping gas entering
Y_N = moles i leaving in rich stripping gas per total moles of stripping gas entering

$$S' = \frac{S_N(S_1+1)}{S_N+1} \tag{12-72}$$

$$S_e = \sqrt{S_N(S_1+1)+0.25} - 0.50 \tag{12-73}$$

$$S_N = \frac{V_N K_N}{L_N} \tag{12-74}$$

$$S_1 = \frac{V_1 K_1}{L_1} \tag{12-75}$$

$$V_N = V_{N-1}\left(\frac{V_N}{V_0}\right)^{1/N} \tag{12-76}$$

Additional background on these and other methods for scaling up absorbers and strippers may be found in Sherwood, Pigford, and Wilke (1975).

3. Equipment Hydraulics—Tray Columns

The hydraulic calculation for absorption and stripping columns are exactly like those for distillation columns. One must watch for differences in the behavior of the liquid. Circulating liquid systems, such as absorber–stripper combinations, are prone to accumulate surface active impurities that can result in foaming problems. Tests for foaming under conditions that duplicate plant operations are advisable, but even then a small-scale system will not have the same accumulating tendency as the large system. If the scaled-up process turns out to have a foaming problem that was not anticipated, equipment capacities will be limited unless suitable antifoam agents are added to the system. A case study outlining the approaches and problems of handling foam in absorption–stripping systems has been published by Bolles (1967).

If foaming is known to occur, for example, from pilot plant studies, allowances can be made in the design of the columns. Usually this requires oversized downcomers (to allow time for foam collapse) and extra tray spacing (to contain the expanded volume of foam on the trays). It may also suggest use of packed columns as more suitable for handling the foam. Kohl and Riesenfeld (1979) is a valuable reference for design as well as operating information on absorbers and strippers; some of the well-known applications of the technology are described in detail.

Finally, a word of caution should be mentioned regarding heat effects in absorbers. Heats of solution may be such that the solvent temperature rises significantly in the column. When this happens, the solubility decreases and the absorber becomes ineffective. To handle temperature gradients intermediate cooling of the solvent may be required.

V. LIQUID – LIQUID SYSTEMS — EXTRACTION

A. Major Issues

Liquid–liquid stagewise mass transfer processes are typically extraction processes, at least those that concern the engineer with a scaleup project. Extraction has been practiced in the chemical and petroleum industries for many years, but there are far fewer extraction systems than there are distillation or absorption–stripping systems. Probably because of this relatively minor role, the technology for extraction scaleup and design is much less advanced than that for the other processes. However, for those separations where both extraction and distillation are feasible, extraction may have the potential for reduced energy requirements, and there is new interest in the unit operation today.

The major issues of scaleup in extraction are the same as those for distillation, except for some shifted emphasis. Initially, the *system must be defined*. Because of the natural areas of application, extraction systems are much harder to define precisely than distillation or absorption–stripping systems. Typically, a complex mixture is fed to the system, and a solvent is used to remove selectively components that affect the properties of the exit material. As an example, a widespread use of extraction is to separate an aromatics fraction from the product from a reforming operation. The extractor feed is a boiling range material, mostly C_6 to C_8 hydrocarbons. The function of the extractor is to produce a stream rich in aromatics that can be purified further by distillation. Initially purification by distillation is not feasible because of the very low separation factors involved. The total mixture to be handled is not precisely defined, and often is characterized as a simple pseudobinary such as heptane/xylene. *Separation criteria* are established as for any other separation, often in terms of product specifications instead of quantitative purities. *Physical property data* for extraction are centered on phase equilibrium, but interfacial tension, viscosity, and phase density difference play important roles in scaleup. Again, the equilibrium data are often reduced to a simple system, at least for characterization; for the example given above, the ternary system might be solvent–heptane–xylene.

Modeling of the separation in terms of theoretical stages is straightforward for ternary systems but much more complex for multicomponent systems. Rigorous models are available, however, for treating the multiple-stage, multicomponent case, and their basis could be represented in principle by the same diagram used for distillation as in Figure 12-3.

Determining the *dimensions of the extractor* is much less straightforward than the equivalent scaleup task in distillation. One reason for this is the preponderance of special, proprietary devices used in extraction, where the scaleup parameters, if at all available, are closely held by the proprietors of the equipment. Even so, the same principles as used in distillation are employed.

B. Extraction Notation

Extraction differs enough from distillation or absorption that a special discussion of its notation is in order. Figure 12-20 provides an introduction to this notation, with particular reference to the column-type extractor. First, it must be recognized that there will be a *light phase* and a *heavy phase*. Next, it must be recognized that a decision must be made as to which of the phases is to be dispersed. Following this, and for a simple countercurrent system, the basic immiscibility between solvent and feed must be recognized as enabling the separation. It is desirable to maintain consistent symbols for the following:

A = mass flow of nonextractant, variously called "carrier" or "raffinate"
B = mass flow of solvent
C = mass flow of solute, sometimes called the "distributed component"

LIQUID–LIQUID SYSTEMS—EXTRACTION

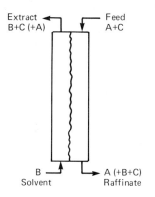

Light Solvent

A = "Carrier"
B = Solvent
C = Solute
    (Distributed Component)

If Solvent Dispersed, Main Interface At Top.

If Feed Dispersed, Main Interface At Bottom.

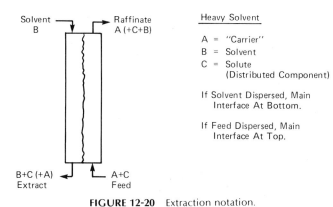

Heavy Solvent

A = "Carrier"
B = Solvent
C = Solute
    (Distributed Component)

If Solvent Dispersed, Main Interface At Bottom.

If Feed Dispersed, Main Interface At Top.

FIGURE 12-20 Extraction notation.

As shown in Figure 12-20, the feed stream becomes the *raffinate phase* and the solvent stream becomes the *extract phase*. The terminal flows from these phases are the raffinate and the extract. Note that the principal interface location depends upon which phase is dispersed.

A typical extraction system is shown in Figure 12-21. As contrasted with the simple systems of Figure 12-20, the feed stream is shown entering the extraction column toward the center, and the column is provided with reflux in the form of extractant that has been separated from the solvent. The stripper, which is a distillation column, is an important part of the extraction system, and it can use large amounts of energy if the proportion of solute in the feed is high and if a significant amount of extract reflux is needed.

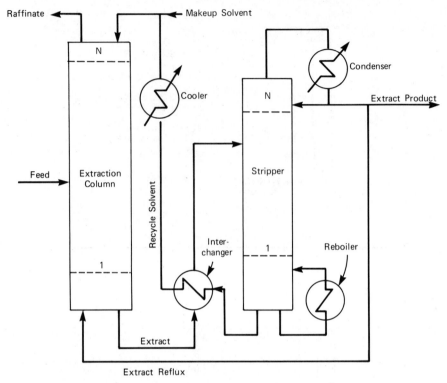

FIGURE 12-21 Extraction system, with extract reflux.

C. Phase Equilibrium

The simplest ternary system is the Type I system shown in Figure 12-22. For this system, the carrier and the solvent are essentially immiscible, while the carrier–solute and solvent–solute pairs are miscible. The diagram shows a single phase region and a two-phase region; for extraction to be carried out, compositions must be such as to fall into the two-phase region.

Phase equilibrium relations are indicated in Figure 12-22; the tie lines connect equilibrium phase compositions and thus provide a basis for selectivity:

$$\beta_{CA} = \frac{(X_C/X_A)_{\text{extract phase}}}{(X_C/X_A)_{\text{raffinate phase}}} \quad (12\text{-}77)$$

which will be recognized as the separation factor equivalent to relative volatility in distillation. Thus, $\beta_{CA} = K_C/K_A$, where the equilibrium ratio K is

$$K_C = \frac{X_{C,E}}{X_{C,R}} \quad (12\text{-}78)$$

LIQUID–LIQUID SYSTEMS—EXTRACTION

When liquid–liquid equilibria are determined in the presence of vapor (as is the usual case for laboratory determinations), the vapor is in equilibrium with both liquid phases. Then, Equation 12-6 in Section III-B on phase equilibria applies:

$$\frac{x_{i,E}}{x_{i,R}} = \frac{K_R}{K_E} = \frac{\gamma_{i,R}}{\gamma_{i,E}} \tag{12-79}$$

$$\beta_{ij} = \frac{\gamma_{i,R}\gamma_{j,E}}{\gamma_{i,E}\gamma_{j,R}} \tag{12-80}$$

Equations 12-79 and 12-80 utilize mole fractions for composition characterization, instead of weight fractions as commonly used in engineering work with extraction processes. Equation 12-6 is valid only for low pressures (fugacity coefficients equal to unity); for high-pressure equilibrium determinations, Equation 12-5 would be used. The important point to remember is that vapor–liquid equilibria (VLE) data can lead to liquid–liquid equilibria (LLE) data.

A final point regarding the Type I system: the *plait* point shown in Figure 12-22 is the intersection of the raffinate phase and extract phase equilibrium curves, and no separation can be made at that point. The analogy is with the azeotrope composition in distillation.

Figure 12-23 shows another type of ternary liquid–liquid system, one where there are immiscibilities between solvent and solute, and between solvent and carrier. The tie lines are indicated, and there is no plait point. With this type of system it is possible to obtain an extract that is essentially free of carrier, which is not possible with the Type I system shown in Figure 12-22. There are a few Type III systems, in which immiscibilities exit between all three pairs, but such systems are relatively rare in extraction system design. For all systems, temperature influences the locations of the phase envelopes, and a normally

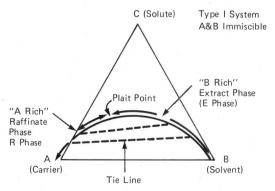

FIGURE 12-22 Phase diagram, Type I system.

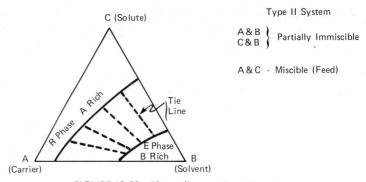

FIGURE 12-23 Phase diagram, Type II system.

immiscible system can become completely miscible if the temperature is raised sufficiently.

The equilateral triangle graphical representation is not the only way to deal with ternary LLE. The right triangle can be used, with the composition of one component being obtained by difference. There are also the equivalents of the y–x and H–y–x (Ponchon–Savarit) plotting approaches of distillation, as will be shown later. Finally, the graphical representation of four-component systems is possible, using three dimensions, but the usage of such a representation is quite limited.

The need for reliable liquid–liquid equilibrium data is clear. Such data can be measured with less difficulty than can vapor–liquid equilibria; the phases are brought to equilibrium in a suitable container and then allowed to separate *completely* before they are sampled for analysis.

Significant LLE data have been published, and should be consulted not only for possible use in scaleup but also for calibrating equipment for laboratory measurements. D'Ans-Lax (1967), Francis (1963) (1972), Himmelblau et al. (1959), Landolt-Bornstein (1950), Lewis (1953), Prausnitz et al. (1980), and Sorensen-Arlt (1979) either provide direct data or furnish guidance to the literature where the data can be found. Treybal (1973a) has listed representative values of the equilibrium ratio in Equation 12-78 for 278 systems. In general, the data reported have not been subjected to a thermodynamic consistency analysis, and allowance for errors should be made.

D. Prediction of Performance

1. Stage Calculations

It is convenient to treat extraction studies on an equilibrium stage basis, even if the equipment operates in a countercurrent mode (as in packed columns). For single stage extractions, a mixer–settler arrangement is used, with the stirred vessel designed to provide a close approach to equilibrium. For multiple-stage

LIQUID–LIQUID SYSTEMS—EXTRACTION

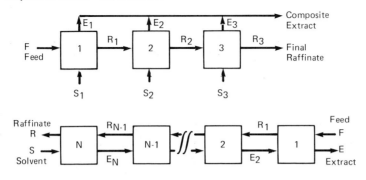

Net Flow To Left = R-S = F-E = Δ

FIGURE 12-24 Crosscurrent and countercurrent extraction.

extractions, both crosscurrent and countercurrent arrangements may be used, as indicated in Figure 12-24. The countercurrent system is more usual and is more efficient in its use of solvent. The stages are arranged as in distillation and it is often convenient to use distillation-type equipment such as tray columns and packed towers.

There is a minimum solvent rate that corresponds to the minimum reflux ratio in distillation. At this rate, an infinite number of stages would be required to make a given separation. Figure 12-25 illustrates the minimum rate for a simple Type I system, and also illustrates the graphical technique for stepping off equilibrium stages. For a given feed F and solvent S, values of the extract E and raffinate R can be obtained graphically, as shown. When extract and feed are in equilibrium (the line connecting them coincides with a tie line) then the resulting solvent rate is the minimum rate. If the solvent-to-feed ratio is

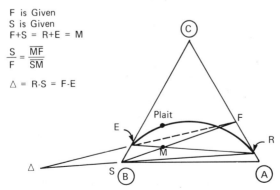

FIGURE 12-25 Minimum solvent rate, infinite stages.

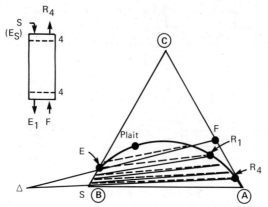

FIGURE 12-26 Higher-than-minimum solvent rate, four theoretical stages.

then increased (segment $\overline{MF}$ increases in proportion to segment $\overline{SM}$), stages can be stepped off, using the difference point as the pivot, as shown in Figure 12-26.

Since feed and extract pass each other on the bottom stage, equilibrium may limit the purity of the extract (on a solvent-free basis). To increase this purity, extract reflux may be used, as shown in Figure 12-27. The solvent stripper must be used in any case, in order that the solute may go on to its intended use and the solvent may be returned to the extractor. Returning a portion of the extract is equivalent to refluxing in distillation and it permits a much purer solute to be extracted. Figure 12-27 shows also how the analogy to heat in boilup is provided by the solvent in extraction.

Another graphical approach to stage determination embodies plotting the carrier–solute on a solvent-free basis. Figure 12-28 shows the equilibrium relationships for typical Type I and Type II systems. Note the analogy of the extract and raffinate phases to the vapor and liquid phases in distillation. A

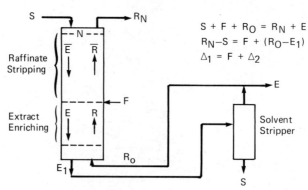

FIGURE 12-27 Use of extract reflux.

LIQUID–LIQUID SYSTEMS—EXTRACTION

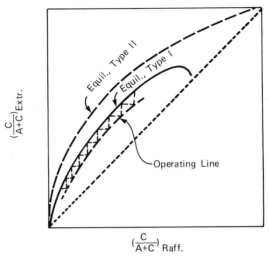

FIGURE 12-28 Solvent-free equilibrium diagram (with typical stages shown).

typical stage count for the Type I system is also shown in the figure. The solvent-free plot is especially useful for Type II systems, and has a complete analogy to the y–x McCabe–Thiele diagram for distillation. The effect of raffinate and extract reflux is seen readily, and the center area feed location is easily handled.

A summary of the stage flow relationships, for the center-fed extractor shown in Figure 12-21, is shown in Figure 12-29. The symbols and construction are generally taken from Treybal (1963). The corresponding McCabe–Thiele plot is shown in Figure 12-30.

Figure 12-31, which relates to Figures 12-29 and 12-30, shows still another graphical approach to extraction stage determination. The plot is known as a Janecke diagram, and its resemblance to the Ponchon type plot for distillation should be apparent. The horizontal scale is on a solvent-free basis, as in Figures 12-28 and 12-30, but it represents both extract and raffinate phases. The vertical scale is the mass ratio of solvent to the sum of the mass flow rates of the solute and the carrier. This type of plot underscores the analogy of heat in distillation to solvent in extraction. The location of the difference points Q and W, which have ordinate values of N_Q and N_W, respectively, is described in Figure 12-29. Equilibrium tie lines are dashed; for example, R_1 and E_1 are at equilibrium, which is simply a statement of the condition of equilibrium on Stage 1. The operating lines converge on the difference points, Q for the extract-enriching section and W for the raffinate-stripping section.

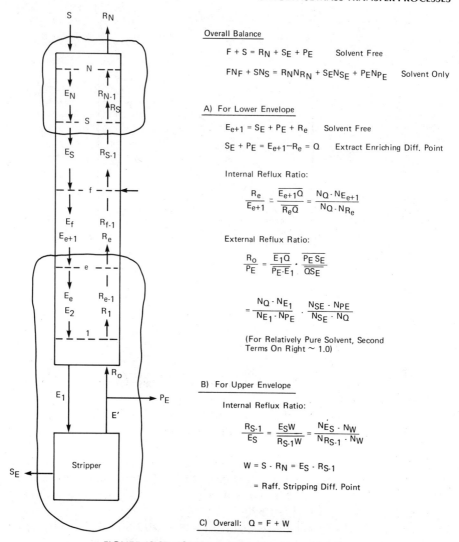

FIGURE 12-29 Countercurrent extraction with reflux.

For systems of more than three components, analytical approaches are required. It is possible, but not practical, to handle a four-component system on a three-dimensional plot. Rigorous-type models have been developed for handling multicomponent systems and require computers for their solution. In general, these models are proprietary, but a discussion of their approach has been published by Schiebel (1959). It appears that multicomponent models are not used very often, partly because of the lack of reliable multicomponent phase equilibria data and partly because the systems indicating their use are poorly defined. It is often satisfactory to use a pseudoternary system, with

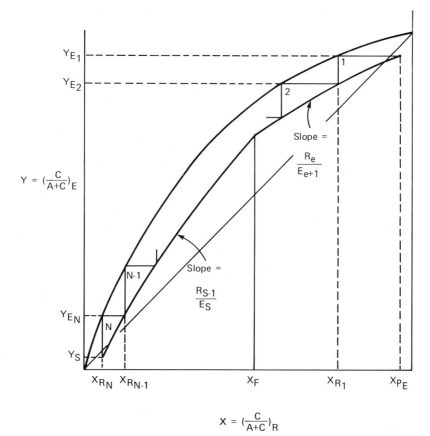

FIGURE 12-30 $Y-X$ diagram, Type II system, solvent-free basis.

pseudo components representing the properties of the design extract and raffinate streams.

2. Solvent Selection

The optimum solvent for a given separation is determined from a consideration of several criteria. Importantly, the "best" solvent for the laboratory or pilot plant development of a new process may not be feasible for the scaled-up plant. Some general criteria for solvent selection are:

Selectivity. A high value of the separation factor enables fewer stages to be used.

Equilibrium ratio. A high value of K_{CA} permits lower solvent/feed ratios.

Density. A high density difference between extract and raffinate phases permits more rapid phase separations and higher capacities in equipment.

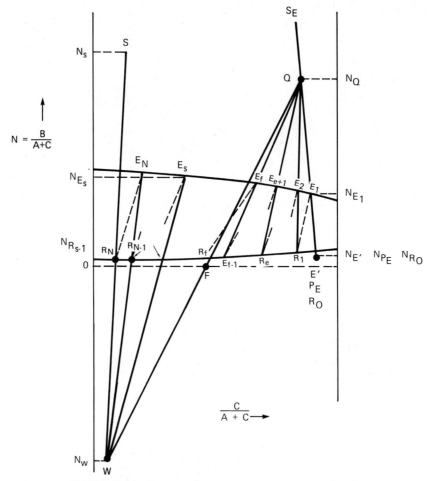

FIGURE 12-31 $N-Y-X$ diagram, Type II system (Janecke plot).

Insolubility of solvent. If the solvent is too soluble in the raffinate, significant solvent losses can occur.

Recoverability. It is desirable to make a clean separation of extractant and solvent in the stripper, without excessive energy requirements.

Interfacial tension. Low interfacial tension aids dispersion but hinders settling and phase separation.

Toxicity, flammability. These are important occupational health and safety considerations.

Cost. An excellent solvent, based on laboratory tests, may not be commercially available or may represent a very large initial cost for charging the system. Also, losses occur in operating systems and must be made up.

A discussion of these criteria may be found in Treybal (1980).

3. Extraction Devices

There are a great many different types of extraction devices in commercial use; many are proprietary and require the involvement of the proprietor in the scaleup process. A basic type of device is the mixer–settler system, useful if only a few theoretical stages are required but tending to be expensive if throughputs are high. A diagram of a mixer–settler system is shown in Figure 12-32. As can be seen, a considerable amount of piping and pumping is required for the three stages shown, but the contacting stages can be designed to approach 100 percent stage efficiency. Compact mixer–settler systems, with simple overflow–underflow arrangements, can be designed to minimize space and piping/pumping requirements.

Many extractors are of the tower type, and examples of this type are shown in Figure 12-33. The simplest of these, and the least efficient, is the spray extractor. The spray extractor comprises a vertical vessel with the only internal device being a distributor for the phase to be dispersed. As shown, the solvent is the heavy phase and is being dispersed through a perforated pipe distributor. The extract phase drops fall through the continuous phase (raffinate) and the solute diffuses from the continuous phase to the dispersed phase.

The spray extractor, while inexpensive, suffers from a low efficiency for two reasons: there is considerable backmixing in the continuous phase, thus lowering the available concentration driving force for diffusion, and the lack of reformation of drops penalizes the overall rate of mass transfer. (Experiments show that a significant amount of the total mass transfer occurs during drop formation, and devices that cause coalescence and reformation several times have a mass transfer rate advantage.) Measurements show that spray extractors do not normally produce more than two or three theoretical stages.

The packed extractor is a film type contactor which provides a reasonable approach to countercurrent flow of the phases but which suffers relatively low mass transfer rates. Like the spray extractor, it is relatively inexpensive but in typical installations does not produce more than three or four theoretical stages. It will be described in additional detail in Chapter 13.

The pulsed extractor became popular in the mid-1950s, largely through experience with small-scale units. The pulsing action is designed to create frequent renewals of the interfacial surface, thereby enhancing mass transfer rates. Perforated plates in the column minimize departures from effective countercurrent flow of the phases. When scaled up to large sizes, the pulse column was found to have lower efficiency because of the difficulty of propagating the pulses. To correct for this, the effect of pulsing was obtained by moving the plates in a reciprocating fashion. The pulse column is typified today by the Karr extractor (1959) which contains a series of perforated plates (without downcomers or upcomers) on one or more shafts, with the assembly being given a reciprocating movement.

Representative performance data for two reciprocating plate columns are given in Figure 12-34, taken from Karr and Lo (1976) and based on the o-xylene/acetic acid/water system. The important design variables appear to

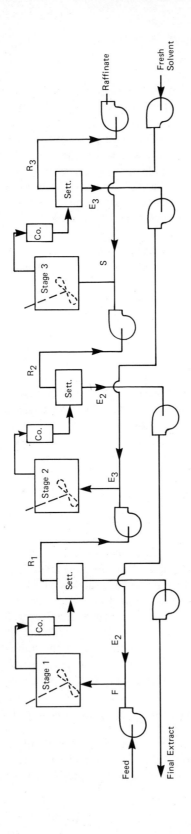

Extract = Heavy Phase; Raffinate = Light Phase
Co. = Coalescer   Sett. = Settler (Decanter)
FIGURE 12-32 Three-stage mixer–settler system.

LIQUID–LIQUID SYSTEMS—EXTRACTION

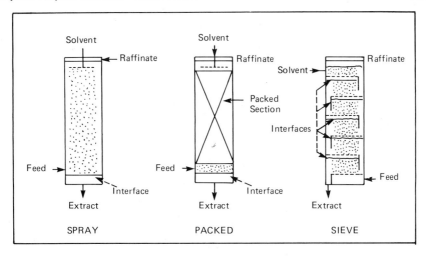

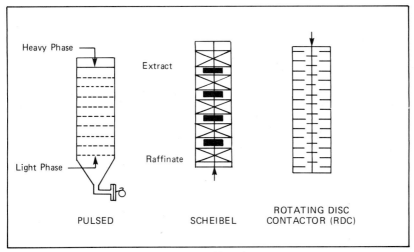

FIGURE 12-33 Representative column type extractors.

be length of stroke ("double amplitude") and reciprocating speed. The values of HETS (height equivalent to a theoretical stage) indicate stage efficiencies of the order of 5–10 percent; however, it is possible to use a very low tray spacing (1 in. in the example shown) and still maintain relatively high throughputs. A possible shortcoming of this type of device, other than its cost, is the fact that the reciprocating motion of the driver could give maintenance problems.

The Scheibel column is designed to simulate a series of mixer–settler extraction units, with self-contained mesh-type coalescers at each contacting stage. The dispersed phase holdup and mass transfer efficiency are controlled primarily by the speed of the agitators. Typical data for a Scheibel column

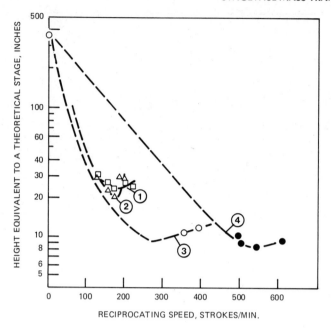

Symbol	Curve No.	Column Diameter, Inches	Phase Dispersed	Phase Extractant	Double Amplitude, Inches	Plate Spacing, Inches	Total Throughput, G.P.H./FT²
□	1	36	Water	Water	1	1	425
△	2	36	Water	Xylene	1	1	442
○	3	3	Water	Water	1	1	424
●	4	3	Water	Water	½	1	424

To convert inches to centimeters, multiply by 2.54. To convert strokes/minute to strokes/second, divide by 60. To convert gal/(hr-ft²) to m³/s.m² multiply by 1.131 (10⁻⁵).

FIGURE 12-34 Efficiency of the reciprocating plate extractor.

(1950) are shown in Figure 12-35, for the same system as represented in Figure 12-34, the acetic acid/o-xylene/water system. This system is considered a "difficult" system for mass transfer because of its relatively high interfacial tension; for "easy" systems the stage efficiency can exceed 100 percent. The reason why the apparent limitation of equilibrium can be exceeded is that in reality a Scheibel stage is two stages—one for agitation and one for coalescence. For the data of Figure 12-35, the minimum HETS is about 33 cm and is lower (i.e., efficiency higher) when the acetic acid (solute) is transferred from the aqueous phase to the hydrocarbon phase. The data in Figure 12-34 show

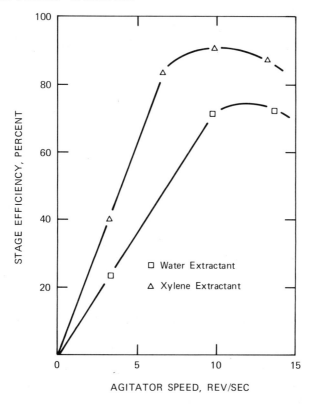

System: Acetic acid/o-xylene/water
Total Liquid Flow: 2.52 (10^{-4}) m/s (240 gal/hr).
Column Diameter: 30.5 cm. Total stage height: 30.5 cm.
FIGURE 12-35 Efficiency of the Scheibel column extractor.

the same effect of transfer direction on HETS. Flooding of the column would be reached at about 15 rps for the total liquid throughput (raffinate phase + extract phase) shown. Although moderately expensive, the Scheibel column gives very high contacting efficiency.

The rotating disk contactor (RDC) was introduced in the 1950s by the Shell companies [see Strand et al. (1962)] and has been used extensively in the petroleum industry for extractions involving hydrocarbon systems. Rotors on a central shaft create dispersion and movement of the phases, while stators provide the countercurrent staging. Like the Scheibel unit, the effectiveness of an RDC can be controlled to some extent by varying the speed of rotation of the disk dispersers.

The sieve tray extractor resembles a sieve tray distillation column. Downcomers (or upcomers) are provided to move the continuous phase downward or upward, depending on whether it is the heavy phase or the light phase. Tray perforations provide for drop formation at each stage, thus aiding the mass

TABLE 12-2 Classification of Extraction Devices

Force of Dispersion	Differential	Stagewise
Gravity only	Spray Packed	Sieve
Pulsation	Pulsed packed	Pulsed sieve Moving plate
Mechanical agitation	Rotating disk Oldshue/Rushton	Scheibel Mixer–settler
Centrifugal	Podbielniak De Laval	Westfalia

transfer process. The sieve tray device is nonproprietary, but because there is relatively little published information on its performance in large-scale equipment, engineering firms with such experiences tend to play the role of proprietor. More will be said later about the sieve tray extractor.

Extraction devices may be classified as shown in Table 12-2. The gravity type devices are largely nonproprietary and approximate methods are available for their scaleup and design. The centrifugal type devices are proprietary, and costly, but at modest throughputs can yield a relatively large number of stages. Table 12-3 shows approximate relative ratings of devices, as published by Todd (1962). Figure 12-36 from Todd shows approximate areas of application.

Table 12-4 shows approximate capacity and efficiency data for several types of column extractors, drawn in part from Laddha and Degaleesan (1978). The efficiency criterion is

$$E_L = (u_d + u_c)\left(\frac{1}{\text{HETS}}\right)$$

$$= \left\{\begin{array}{c}\text{combined phase}\\ \text{flow rate}\end{array}\right\}\left\{\begin{array}{c}\text{stages per}\\ \text{unit height}\end{array}\right\} \quad (12\text{-}81)$$

where

$$u_d = \frac{q_d}{A_t} = m^3/\text{hr m}^2 \quad (12\text{-}82)$$

$$u_c = \frac{q_c}{A_t} = m^3/\text{hr m}^2 \quad (12\text{-}83)$$

Note that E_L is not a fraction or a percentage.

A very thorough treatment of commercially available extraction equipment has been published by Lo (1979). Of particular interest in the context of the present work is the coverage by Lo of laboratory extraction devices.

TABLE 12-3 Ratings of Several Commercial Extractors

Contactor	Capital Cost	Operating and Maintenance Cost	Efficiency	Total Capacity	Flexi-bility	Volume Efficiency	Space		Ability to Handle Systems That Emulsify
							Vertical	Floor	
Spray	5	5	1	2	2	1	0	5	3
Baffle plate	4	5	2	4	2	3	1	5	3
Packed	4	5	2	2	2	2	1	5	3
RDC	3	4	4	3	5	4	3	5	3
Pulsed plate	3	3	4	3	4	4	3	5	1
Mixer–settler	2	2	3	4	3	3	5	1	0
Centrifugal	1	2	5	3	5	5	5	5	5

5 = high (desirable).
1 = low (undesirable).

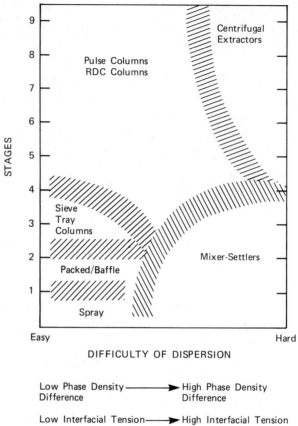

FIGURE 12-36 Areas of application of extraction devices.

TABLE 12-4 Approximate Efficiency and Capacity Characteristics of Extraction Columns

	$u_d + u_c$ m/hr	HETS, m	Stages/meter	E_L, hr^{-1}
Spray	15–75	3–6	0.3–0.15	3–7
Sieve	3–60	0.3–1.8	0.5–3.3	1–120
Packed	6–45	0.9–3	0.3–1.0	1–27
Karr	18–70	0.2–0.6	1.6–6.0	17–200
RDC	18–40	0.15–0.6	1.6–6.6	22–180
Scheibel	15–30	0.3–0.6	1.6–3.3	29–60

4. Mass Transfer Relationships

The two-resistance theory is applicable to extraction. The molar flux, based on interfacial area (for a given solute) is

$$N_s = k_R(C_R - C_{Ri}) = k_E(C_{Ei} - C_E) \tag{12-84}$$

$$= K_R(C_R - C_R^*) \tag{12-85}$$

$$= K_E(C_E^* - C_E) \tag{12-86}$$

where

k_R, k_E = mass transfer coefficients for the raffinate and extract phase
K_R = combined mass transfer coefficient, based on raffinate compositions
K_E = combined mass transfer coefficient, based on extract compositions
C_R, C_E = concentrations in raffinate and extract bulk phases
C_R^*, C_E^* = equilibrium concentrations referred to C_E and C_R

A representation for this transfer, based on diffusion from raffinate to extract, is shown in Figure 12-37. Subscripts for the continuous and dispersed phases could be substituted for the extract and raffinate subscripts in the figure and in Equations 12-84 to 12-86.

The mass transfer coefficients may be combined in the usual fashion:

$$\frac{1}{K_R} = \frac{1}{k_R} + \frac{1}{mk_E} \tag{12-87}$$

$$\frac{1}{K_E} = \frac{1}{k_E} + \frac{m}{k_R} \tag{12-88}$$

Here, $m \sim m_E \sim m_R$ (as in Figure 12-37). For the special case where the raffinate phase is the dispersed phase, Equations 12-87 and 12-88 lead to

$$\frac{1}{K_d} = \frac{1}{k_d} + \frac{1}{mk_c} \tag{12-89}$$

$$\frac{1}{K_c} = \frac{1}{k_c} + \frac{m}{k_d} \tag{12-90}$$

with subscripts d and c referring to the dispersed and continuous phases. The value of m is based on the slope of the equilibrium curve in which solute concentration in the dispersed phase is plotted as the ordinate scale.

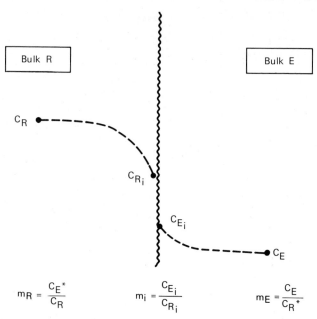

FIGURE 12-37 Example concentration gradients for solute transferring from raffinate phase to extract phase.

Evaluation of mass transfer coefficients is handled through models that relate to the contacting device used. Some knowledge of the hydraulic characteristics of the device is required, and this represents a severe limitation for most of the devices described above. Considerable study has been made of the crossflow sieve tray extractor, and the next section shows, by way of example, how the mass transfer efficiency of a sieve tray column might be predicted.

5. Mass Transfer Efficiency of the Sieve Tray Extractor

The sieve tray extractor is a popular type of nonproprietary device. As shown in Figure 12-33, it resembles a sieve tray distillation column, with light phase dispersed into a crossflowing continuous phase. The heavy phase moves from tray to tray via downcomers. Alternately, the heavy phase can be dispersed into a crossflowing light continuous phase. Then, upcomers serve as passages from tray to tray. The analogy to phase dispersion in distillation equipment is apparent.

The diagram of flows in a sieve tray column is shown in Figure 12-38. Presumed steps in the mass transfer process are:

Transfer during drop formation.
Transfer during drop rise (or fall).
Transfer during drop coalescence.

LIQUID–LIQUID SYSTEMS—EXTRACTION

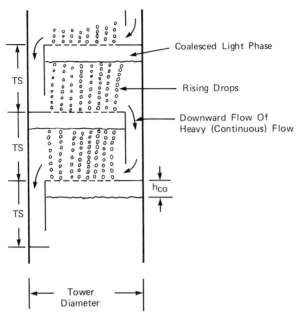

FIGURE 12-38 Flows in a sieve tray extraction column.

Of these, the last has been found to make relatively little contribution and thus emphasis in modeling is placed on the first two.

One model for predicting sieve tray column efficiency is described in detail in the book by Treybal (1980). It combines Murphree stage efficiency with an expression for total mass transfer between stages. The contribution for drop coalescence is considered to be 10 percent of that for formation. The general form of the efficiency expression is:

$$E_{md} = \frac{1.1 K_{fd} A_{fd} + K_{rd} A_{rd}}{q_D + 0.5 K_{rd} A_{rd} + 0.1 K_{fd} A_{fd}} \qquad (12\text{-}91)$$

where

E_{md} = overall tray efficiency (Murphree), based on dispersed phase concentrations

q_D = volumetric flow rate of dispersed phase

K_{fd} = overall mass transfer coefficient for drop formation, based on dispersed phase

K_{rd} = overall mass transfer coefficient for drop rise or fall, based on dispersed phase

A = interfacial area, based on equivalent overall mass transfer coefficient

Sources of the individual mass transfer rate expressions are given in the original reference (Treybal, 1980), and consideration must be given to two phenomena: whether the dispersed phase breakup is due to simple drop formation (low rates) or jetting (high rates), and whether the rising (or falling) drops can be characterized as stagnant, circulating, or oscillating. Assumptions of the Treybal model are:

Complete mixing of the continuous phase.

Plug flow of the dispersed phase.

Mean driving force for the drop rise (fall) region can be taken as the arithmetic average concentration difference.

Solute concentration is identical throughout the coalesced layer.

Other mechanistic models are those of Skelland and Conger (1973) and Pilhofer (1981).

For approximate work, an empirical method of Treybal (1963) is useful:

$$E_{oc} = \frac{5.65 H_t^{0.5} (u_d/u_c)^{0.42}}{\sigma} \qquad (12\text{-}92)$$

where

E_{oc} = overall column efficiency
H_t = tray spacing, meters
u_d, u_c = superficial velocities of the dispersed and continuous phases
σ = interfacial tension, mN/m (or dynes/cm)

The overall column efficiency is simply the total theoretical stages divided by the total number of actual stages. It is related to the Murphree efficiency as shown in Equation 12-57. In a study of a large amount of sieve tray column efficiency data, mostly for small-diameter columns, Rocha (1984) found that the empirical Treybal model, Equation 12-92, does a remarkably good job of predicting overall efficiencies.

E. Laboratory Studies for Scaleup

Extraction studies in the laboratory are often made to establish separability, and are not designed to provide scaleup information. The extractions may be carried out on a batch basis, or on a crosscurrent basis (as in Figure 12-24). The larger version of an extraction system will likely include countercurrent contacting, if several stages are involved. It is important to select a laboratory contactor that can provide scaleup information as well as demonstrate separability. This situation is analogous to the selection of an Oldershaw column for laboratory distillation studies, as discussed in Section III-D of this chapter.

ENERGY CONSIDERATIONS   493

Scaleup studies of several proprietary devices have been made. Figure 12-34 shows comparisons of Karr reciprocating plate columns with diameters of 7.6 and 91.4 cm (3 and 36 in.). By a reasonable application of mass transfer and fluid contacting fundamentals one might expect to scale up such a device successfully. It is usually possible to rent a laboratory-scale extraction column from the proprietor and on the basis of the latter's experience evolve a satisfactory set of scaleup criteria.

For the nonproprietary devices, such as the spray column and the sieve tray column, one must develop mathematical models based on small-scale information, and then use the models for scaleup studies. Such efforts are risky, and it is not uncommon for very healthy safety factors to be used, especially with respect to mass transfer efficiency. For such devices, the laboratory scale work is of relatively little use in scaleup, and thus one is better off to use a well-calibrated device, such as the Scheibel, at least to obtain a laboratory reading on required theoretical stages for a given separation.

VI. ENERGY CONSIDERATIONS

In this chapter three stagewise processes have been discussed in connection with the technology of scaleup and design. Usually it is apparent which of the processes should be selected for a given separation problem. Considerations behind the selection would include such factors as relative volatility, understanding of transfer mechanisms, cost of large-scale equipment, and difficulty of laboratory evaluation. These factors must enter into the final "bottom line" cost comparison of one process versus another. Clearly, this is chemical engineering practice as it has been known for decades.

An item entering into the cost analysis, one that is much more severe today than it was a decade ago, is the requirement for minimizing energy of separation. Thus, a few comments on the energy requirements of the methods are in order.

The conventional distillation column is quite inefficient thermodynamically. The heat added at the base often approximates the heat removed at the overhead, with the latter generally rejected to the atmosphere, through a cooling tower circuit or directly in an air-cooled condenser. If the distillation is carried out at subambient temperatures, the heat rejected overhead must be pumped (through refrigeration) before it can be rejected to the atmosphere.

Studies of energy reduction in distillation involve one or more of the following approaches:

Finding ways to utilize all or a portion of the heat in the column overhead vapor (instead of simply rejecting it to the atmosphere). Examples are exchanging the heat with a process stream, or pumping the heat to a level where it can serve as a useful heating medium or vaporizing a Rankine cycle fluid that in turn can drive a power turbine.

Decreasing the heat rejected by using a lower reflux ratio, or using intermediate reboilers and condensers.

Decreasing the amount of heat needed, by changing the separation sequence, decreasing heat leaks, using feed/product heat exchange, and so on.

Finding lower cost energy sources, such as surplus steam or a high-temperature liquid stream that needs to be cooled.

It is sometimes economical to install a vapor recompression ("heat pump") system if the overhead and bottom temperatures of the column are fairly close (close boiling separation). For new designs, it is usually desirable to use a low ratio of operating to minimum reflux; the cost of the required additional trays is offset by the savings in energy.

In absorption systems, the energy cost is related to the pressure loss for gas flow through the absorber, and the cost of heat to separate the solute from the solvent. The solute separation, carried out in the stripper, involves all the considerations given above for distillation.

For extraction, the major cost of energy is for the solvent stripper. The energy required for dispersing the phases (mechanical or pressure) is low in comparison. Thus, as for absorption, the energy analysis deals primarily with a distillation step.

Comparisons of energy requirements for several separation processes have been studied by Null (1980). While it is clearly difficult to make generalizations, he has presented guidelines that are of interest to process engineers. His comparisons between distillation and extraction are shown in Figures 12-39 and 12-40. The bases used for these figures represent extreme conditions.

In Figure 12-39, a completely nonvolatile solvent that does not contaminate the raffinate, and requires only a simple flash step for separation from the extract is assumed. If 60 percent of the feed, for example, is taken overhead in a distillation process ($D/F = 0.6$) and the required heating medium temperature for the distillation is 149°C, and if the distillation reflux ratio (R_D) is greater than 2.0, extraction should be considered as a viable alternate.

Figure 12-40 covers the opposite extreme where two solvent strippers are needed, one for the solvent–extract separation and one for the solvent–raffinate separation. If each of these strippers required a reflux ratio R_E of 2.0 at the same $D/F = 0.6$ for distillation, extraction would merit consideration if the distillation reflux ratio was greater than 4.0. For Figure 12-40, the temperature of the heating medium for solvent stripping is assumed to be the same as the temperature of the heating medium for distillation, an extreme situation, and not likely.

Any comparisons of distillation and extraction must take into account the state-of-the-art on equipment scaleup and design. One can be much more confident in distillation than in extraction, because of the much larger amount of development work done in distillation. There has been no extraction equivalent of Fractionation Research, Inc. in the area of large-scale equipment performance testing. However, for those cases clearly indicating a superiority

t_S = Required heating medium temperature for distillation column.

t_{SE} = Required heating medium temperature for extraction solvent stripper = $600°F, (316°C)$

R_D = Required reflux ratio for distillation.

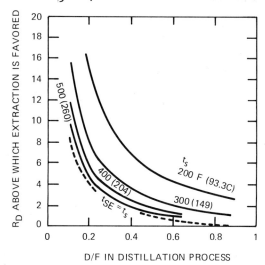

FIGURE 12-39 Extraction versus distillation selection, nonvolatile solvent.

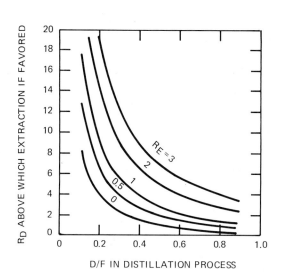

Heating medium temperatures for distillation and extraction solvent stripping are equal.

Reflux ratio for solvent-extract separation (R_E) equals reflux ratio for solvent-raffinate separation. Distillation reflux ratio = R_D

FIGURE 12-40 Extraction versus distillation selection, difficult solvent separation.

of extraction as a separation process, reasonable allowances for unknown scaleup factors may overcome the apparent economic penalties of proceeding with distillation as the selected method.

NOMENCLATURE

a_i	Interfacial area, m^2/m^3
A	Absorption factor
A', A_N	Edmister absorption factors
A_e	Effective absorption factor
A	Area, m^2
A_a	Active area on tray (bubbling area)
A_{fd}	Interfacial area for drop formation
A_h	Hole area on tray
A_{rd}	Interfacial area for drop rise
A_N	Net area on tray
A_t	Total tower cross sectional area
A_{ji}, A_{ij}	Van Laar constants
B	Bottoms flow, distillation, kg mol/s
C	Liquid phase concentration, kg mol/m^3
C^*	Equilibrium concentration
C_{SB}	Souders–Brown capacity coefficient, m/s
C_{vO}	Orifice discharge coefficient
d_H	Hole diameter, mm
d_p	particle (drop) diameter, mm
D	Distillate rate, kg mol/s
D_L	Distillate liquid rate
D_v	Distillate vapor rate
D_b	Bubble diameter, mm
$\mathscr{D}$	Molecular diffusion coefficient, m^2/s
$\mathscr{D}_E$	Eddy diffusion coefficient, m^2/s
E_{ai}	Fractional removal of i in absorber
E_L	Efficiency criterion, extraction, hr^{-1}
E_{md}	Murphree tray efficiency, dispersed phase composition basis, fractional
E_{mv}	Murphree tray efficiency, vapor composition basis, fractional
E'_{mv}	E_{mv} corrected for entrainment
E_{oc}	Overall column efficiency, fractional

NOMENCLATURE

E_{OG}	Local, or point, efficiency, overall resistance basis, fractional
E_{si}	Fractional removal of i in stripper
f	Fugacity, atmospheres
F	Feed rate, kg mol/s or kg/s
F_v	Vapor flow F factor, m/s (kg/m^3)$^{1/2}$
F_{vh}	F factor through holes, $U_h \rho_v^{1/2}$
F_{va}	F factor through active area, $U_a \rho_v^{1/2}$
G	Gas rate, kg/s
G_M	Gas rate, kg mol/s m^2
h	Head of liquid, mm
h_d	Head loss through dry tray
h_L	Head loss for vapor flow through aerated mass
h_{ow}	Weir crest, mm
h_w	Weir height, mm
H	Henry's law coefficient
H_t	Tray spacing, m
HETS	Height equivalent to a theoretical stage, m
k	Individual phase mass transfer coefficient, cm/s or m/s
K	Phase equilibrium ratio
K	Overall mass transfer coefficient, cm/s or m/s
K_c	Basis, continuous phase concentrations
K_d	Basis, dispersed phase concentrations
K_E	Basis, extract phase concentrations
K_{fd}	Basis, drop formation
K_R	Basis, raffinate phase concentrations
K_{rd}	Basis, drop rise
L	Liquid flow rate, kg mol/s
L_i	Ostwalt solubility coefficient
L_w	Liquid flow rate per length of outlet weir, m^3/s m
m	Slope of equilibrium line
n	Equilibrium stage designation
N	Number of equilibrium stages
N	Number of transfer units
N_{OC}	Overall transfer units, continuous phase composition basis
N_{OD}	Overall transfer units, dispersed phase composition basis
N_{OG}	Overall transfer units, gas composition basis
N_{OL}	Overall transfer units, liquid composition basis
N_A	Number of actual stages
N_i	Molar flux of component i, kg mol/s m^2 or g mol/s cm^2

N_M	Minimum number of theoretical stages
N_{pc}	Theoretical stages in partial condenser
N_Q	"N value" for extract enriching difference point Q, Figure 12-31
N_t	Theoretical stages
$N_{t,c}$	Theoretical stages in column
N_S	Molar flux of solute, kg mol/s m^2
N_W	"N value" for raffinate stripping difference point W, Figure 12-31
N_{rb}	Theoretical stages in reboiler
p	Partial pressure, kPa or atmospheres
p^*	Vapor pressure, kPa or atmospheres
P	Total pressure, kPa or atmospheres
q	Volumetric flow rate of liquid, m^3/s
Q	Heat input and removal rate, W
Q_c	Heat removal rate for condenser
Q_R	Heat addition rate for reboiler
Q	Extract enriching difference point, Figure 12-31
Q_f	Volume of froth, m^3
Q_v	Volumetric flow rate of vapor, m^3/s or m^3/hr
R	Raffinate phase
R	Reflux ratio
R_D	Reflux ratio for distillation
R_E	Reflux ratio for solvent stripping
R_M	Minimum reflux ratio
S	Solvent flow rate
S	Stripping factor
S_e	Effective stripping factor
S', S_N	Edmister stripping factors
Sc_G	Schmidt number for gas phase
SL	Side draw liquid flow rate, kg mol/s
SV	Side draw vapor flow rate, kg mol/s
t	Residence time, s
t_G	Residence time of gas or vapor in froth
t_L	Residence time of liquid in froth
t_s	Heating medium temperature for distillation, °C
t_{se}	Heating medium temperature for solvent stripping, °C
T	Absolute temperature, °K
TS	Tray spacing, mm
u_c	Superficial velocity of continuous phase, m/s or m/hr
u_d	Superficial velocity of dispersed phase, m/s or m/hr

u_{DC}	Liquid velocity in downcomer, m/s
U	Vapor or gas velocity, m/s
U_{br}	Velocity of bubble rise
U_h	Velocity through tray perforations
U_N	Velocity through tray net area
U_{NF}	Velocity through net area at flood
V	Vapor flow rate, kg mol/s
W	Length of liquid travel on tray, m
W	Raffinate stripping difference point, Figure 12-29
x	Mole fraction in liquid (or heavy phase)
X	Mass fraction or ratio
X_i	Moles of i in entering liquid per mole of solute-free gas
y	Mole fraction in vapor (or light phase)
y^*	Equilibrium mole fraction
Y_i	Moles of i in gas per mole of solute-free gas
z	Distance, m
Z_f	Height of froth, mm or m
Z_f'	Effective height of froth, mm or m
Z_L	Liquid film thickness, cm

Subscripts

A	Component A
C	Component C or continuous phase
D	Dispersed phase
D	Distillate
E	Extract phase
F	Feed
G	Gas or vapor
i	Interface
i	Component i
j	Component j
j	Stage j
L	Liquid
n	Tray n
N	Stage N
p	Point on tray
R	Raffinate phase
V	Vapor

Greek

α	Relative volatility or separation factor
β_{CA}	Separation factor in extraction
β_i	Bunsen solubility coefficient
β	Aeration factor
γ	Activity coefficient
γ^L	Liquid phase
$\gamma^{0,L}$	Liquid phase, infinite dilution
γ^v	Vapor phase
Γ_f	Fractional approach to flood
$\Delta\rho$	Phase density difference, kg/m^3
ε	Froth porosity, fractional
θ_L	Exposure time, liquid, s
θ_G	Exposure time, gas or vapor, s
θ	Root of Equations 12-13 and 12-14
λ	Ratio of operating line to equilibrium line
μ	Viscosity, mPa · s
ρ	Density, kg/m^3
σ	Surface or interfacial tension, mN/m (dynes/cm)
Φ^L	Fugacity coefficient, liquid phase
Φ^v	Fugacity coefficient, vapor phase
ψ	Fractional entrainment, mol/mol gross downflow

REFERENCES

American Institute of Chemical Engineers, *Bubble Tray Design Manual*, New York, 1958.

Barker, P. E., and Self, M. F. *Chem. Eng. Sci.*, **17**, 541 (1962).

Battino, R., and Clever, H. L., *Chem. Rev.*, **66**, 395 (1966).

Battino, R., and Wilhelm, E., *Chem. Rev.*, **73**, 1 (1973).

Battino, R., Wilhelm, E., and Wilcock, R. J., *Chem. Rev.*, **77**, 219 (1977).

Bolles, W. L., *Chem. Eng. Progr.*, **63**(9), 48 (1967).

Bolles, W. L., and Fair, J. R., "Distillation," in *Encyclopedia of Chemical Processing and Design*, Vol. 16, Dekker, New York, 1982.

Burgess, J. M., and Calderbank, P. H., *Chem. Eng. Sci.*, **30**, 743, 1107 (1975).

Calderbank, P. H., and Pereira, J., *Chem. Eng. Sci.*, **32**, 1427 (1977).

Chao, K. C., and Seader, J. D., *AIChE J.*, **7**, 598 (1961).

Chien, H. H., and Null, H. R., *AIChE J.*, **18**, 1177 (1972).

Colburn, A. P., *Ind. Eng. Chem.*, **28**, 526 (1936).

REFERENCES

Crozier, R. D., Ph.D. Thesis, University of Michigan, 1956.

Dack, M. R. J., ed., *Solutions and Solubilities*, Wiley, New York, 1975.

D'Ans-Lax, *Taschenbuch fur Chemiker und Physiker*, 3rd ed., Vol. I, Springer-Verlag, Berlin, 1967.

Edmister, W. C., *Ind. Eng. Chem.*, **35**, 837 (1943).

Eduljee, H. E., *Hydrocarbon Proc.*, **54**(9), 120 (1975).

Fair, J. R., and Matthews, R. L., *Petrol. Refiner*, **37**(4), 153 (1958).

Fair, J. R., *Petro/Chem. Eng.*, **33**(10), 45 (Sept. 1961).

Fair, J. R., and Bolles, W. L., *Chem. Eng.*, **75**(9), 156 (April 22, 1968).

Fair, J. R., Steinmeyer, D. E., Penney, W. P., and Brink, J. A., in *Chemical Engineers' Handbook*, C. H. Chilton and R. H. Perry, eds., 5th ed., Section 18, McGraw-Hill, New York, 1973.

Fair, J. R., Null, H. R., and Bolles, W. L., *Ind. Eng. Chem. Proc. Des. Devel.* **22**, 53 (1983).

Fenske, M. R., *Ind. Eng. Chem.*, **24**, 482 (1932).

Foss, A. S., Gerster, J. A., and Pigford, R. L., *AIChE J.*, **4**, 231 (1958).

Francis, A. W., *Liquid-Liquid Equilibriums*, Interscience, New York, 1963.

Francis, A. W., *Handbook for Components in Solvent Extraction*, Gordon and Breach, New York, 1972.

Fredenslund, A., Jones, R. L., and Prausnitz, J. M., *AIChE J.*, **21**, 1086 (1975).

Fredenslund, A., Gmehling, J., and Rasmussen, P., *Vapor–Liquid Equilibria Using UNIFAC*, Elsevier, Amsterdam, 1977.

Gautreaux, M. F., and O'Connell, H. E., *Chem. Eng. Progr.*, **52**, 232 (1955).

Gilliland, E. R., *Ind. Eng. Chem.*, **32**, 1220 (1940).

Gmehling, J., Onken, U., and Arlt, W., *Vapor-Liquid Equilibrium Collection* (Continuing Series), DECHEMA, Frankfurt, 1979–.

Hai, N. T., Burgess, J. M., Pinczewski, W. V., and Fell, C. J. D., "Mass Transfer in the Spray Regime on an Industrial Sieve Tray," Second Australian Conference on Heat and Mass Transfer, University of Sydney, Australia, 1977.

Hala, E., Pick, J., Fried, V., and Vilim, O., *Vapor-Liquid Equilibrium*, 2nd ed., Pergamon Press, Oxford, 1967.

Hala, E., Wichterle, I., Polak, J., and Boublik, T., *Vapor-Liquid Equilibrium at Normal Pressures*, Pergamon Press, Oxford, 1968.

Higbie, R., *Trans. AIChE*, **31**, 365 (1935).

Hildebrand, J. H., Prausnitz, J. M., and Scott, R. L., *Regular and Related Solutions*, Van Nostrand Reinhold, New York, 1970.

Himmelblau, D. M., Brady, B. L., and McKetta, J. J., *Survey of Solubility Diagrams for Ternary and Quaternary Liquid Systems*, Special Publ. No. 30, Bureau of Engineering Research, The University of Texas, Austin, Texas, 1959.

Hirata, M., Ohe, S., and Nagahama, K., *Computer Aided Data Book of Vapor–Liquid Equilibria*, Elsevier, Amsterdam, 1975.

Holmes, M. J., and Van Winkle, M. W., *Ind. Eng. Chem.*, **61**(1), 21 (1970).

Johnson, A. I., and Marangozsis, J., *Can. J. Chem. Eng.*, **36**, 161 (1958).

Karr, A. E., *AIChE J.*, **5**(4), 446 (1959).

Karr, A. E., Lo, T. C., *Chem. Eng. Progr.*, **72**(11), 68 (1976).

Kertes, A. S., ed., *Solubility Data Series*, Pergamon Press, Elmsford, NY, 1979–.

Kohl, A., and Riesenfeld, F., *Gas Purification*, 3rd ed., Gulf Publishing Co., Houston, 1979.

Laddha, G. S., and Degaleesan, T. E., *Transport Phenomena in Liquid Extraction*, Tata-McGraw-Hill, New Delhi, India, 1978.

Landolt-Bornstein, *Zahlenwerte und Funktionen aus Naturwissenshaft aus Physik*, *Chemie*, *Astronomie*, *Geophysik*, *und Technik*, 6th ed., Vol II (2b, 2c), Springer-Verlag, Berlin, 1950 + .

Leibson, I., Kelley, R. E., and Bullington, L. A., *Petrol. Ref.*, **36**(2), 127 (Feb. 1957).

Lewis, J. B., *Extraction: A Critical Review...*, H. M. Stationery Office, London, 1953.

Linke, W. F., and Seidell, A., *Solubilities of Inorganic and Metal-Organic Compounds*, Vol. II, American Chemical Society, Washington, 1965.

Lo, T. C., "Commercial Liquid-Liquid Extraction Equipment," Section 1.10 in *Handbook of Separation Techniques for Chemical Engineers*, P. A. Schweitzer, ed., McGraw-Hill, New York, 1979.

Long, F. A., and McDevit, W. F., *Chem. Rev.*, **51**, 119 (1952).

Markham, A. E., and Kobe, K. A., *Chem. Rev.*, **28**, 519 (1941).

Null, H. R., *Phase Equilibrium in Process Design*, Wiley-Interscience, New York, 1970. Reprint edition, R. E. Krieger, Huntington, NY, 1980.

Null, H. R., *Chem. Engr. Progr.*, **76**(8), 42 (1980).

Olander, D. R., *AIChE J.*, **12**, 1018 (1966).

Oldershaw, C. F., *Ind. Eng. Chem., Anal. Ed.*, **13**, 265 (1941).

Oliver, E. D., and Watson, C. C., *AIChE J.*, **2**, 18 (1956).

Perry, E. S., and Weissberger, A., eds., *Technique of Organic Chemistry, Vol. IV Distillation*, 2nd ed., Interscience, New York, 1965.

Pierotti, G. J., Deal, C. H., and Derr, E. L., *Ind. Eng. Chem.*, **51**, 85 (1959).

Pilhofer, T., *Chem. Eng. Commun.*, **11**, 241 (1981).

Pinczewski, W. V., and Fell, C. J. D., *Trans. Instn. Chem. Engrs.*, **52**, 294 (1974).

Prausnitz, J., Anderson, T., Grens, E., Eckert, C., Hsieh, R., and O'Connell, J., *Computer Calculations for Multicomponent Vapor-Liquid and Liquid-Liquid Equilibria*, Prentice-Hall, Englewood Cliffs, NJ, 1980.

Reid, R. C., Prausnitz, J. M., and Sherwood, T. K., *The Properties of Gases and Liquids*, 3rd ed., McGraw-Hill, New York, 1977.

Renon, H., and Prausnitz, J. M. *AIChE J.*, **14**, 135 (1968).

Rocha, A., Ph.D. Dissertation, The University of Texas (1984).

Sakata, M., *Chem. Eng. Progr.*, **62**(11), 98 (1966).

Scheibel, E. G., and Karr, A. E., *Ind. Eng. Chem.*, **42**, 1048 (1950).

Scheibel, E. G., *Petrol. Refiner.*, **38**(9), 227 (1959).

Seidell, A., *Solubilities of Inorganic and Organic Compounds*, Van Nostrand, New York, 1952.

Seidell, A., *Solubilities of Inorganic and Metal-Organic Compounds*, Van Nostrand, New York, 1958.

Sherwood, T. K., Pigford, R. L., and Wilke, C. R., *Mass Transfer*, McGraw-Hill, New York, 1975.

Skelland, A. H. P., and W. L. Conger, *Ind. Eng. Chem. Proc. Des. Devel.* **12**, 448 (1973).

Smith, B. D., *Design of Equilibrium Stage Processes*, McGraw-Hill, New York, 1963.

Sorensen, J. M., and Arlt, W., *Liquid-Liquid Equilibrium Data Collection*, Chemistry Data Series Volume V, Parts 1–3, Dechema, Frankfurt/Main, F. R. Germany, 1979.

Strand, C. P., Olney, R. B., and Ackerman, G. H., *AIChE J.*, **8**, 252 (1962).

Todd, D. B., *Chem. Eng.*, **69**(14), 156 (July 9, 1962).

Treybal, R. E., *Liquid Extraction*, 2nd ed., McGraw-Hill, New York, 1963.

Treybal, R. E., in *Chemical Engineers' Handbook*, R. H. Perry and C. H. Chilton, eds., 5th ed., pp. 15-7 to 15-12, McGraw-Hill, New York, 1973a.

REFERENCES

Treybal, R. E., in *Chemical Engineers' Handbook*, R. H. Perry and C. H. Chilton, eds., 5th ed., pp. 21-16 and 21-17, McGraw-Hill, New York, 1973b.

Treybal, R. E., *Mass Transfer Operations*, 3rd ed., McGraw-Hill, New York, 1980.

Van Laar, J. J., *Z. Physik. Chem.*, **72**, 723 (1910).

Wang, J. C., and Wang, Y. L. "A Review on the Modeling and Simulation of Multistaged Separation Processes," in *Foundations of Computer-Aided Chemical Process Design*, Vol. 2, American Institute of Chemical Engineers, New York, 1981.

Wichterle, I., Linek, J., and Hala, E., *Vapor-Liquid Equilibrium Data Bibliography*, Elsevier, Amsterdam, 1973. Supplement I (1976).

Wilson, G. M., *J. Am. Chem. Soc.*, **86**, 127 (1964).

13

CONTINUOUS MASS TRANSFER PROCESSES

J. R. FAIR

I. Major Issues	505
II. Fundamental Considerations	506
A. Phase Equilibrium and Intraphase Diffusion	506
B. Countercurrent Mass Transfer	506
C. Mechanics of Phase Contacting	513
III. Vapor–Liquid Systems—Distillation	516
A. Major Issues	516
B. Prediction of Performance	518
C. Liquid Distribution	518
D. Pressure Drop	519
E. Maximum Vapor Capacity	521
F. Mass Transfer Efficiency	522
G. Transfer Units and Theoretical Stages	526
H. Scaleup Procedure	527
I. General Comments	529
IV. Gas–Liquid Systems—Absorption and Stripping	530
A. Major Issues	530
B. Prediction of Performance	530
C. Mass Transfer Efficiency	531
D. Scaleup Procedure	533
E. General Comments	533

V. Liquid–Liquid Systems—Extraction	533
A. Major Issues	533
B. Nomenclature for Countercurrent Extraction	534
C. Prediction of Performance	534
D. Phase Distribution	535
E. Maximum Capacity (Flooding)	538
F. Mass Transfer Efficiency	538
G. Scaleup Procedure	544
H. General Comments	544
Nomenclature	545
References	547

I. MAJOR ISSUES

Countercurrent processing, as considered in this chapter, involves the intimate contacting of two or more immiscible or partially miscible phases, with the objective of transferring one or more components of the system across the boundary (or interface) between the phases in contact. Because of phase density differences or mechanical displacement, the contacting phases are caused to flow in opposite directions, commonly upward and downward. As contrasted to contacting in discrete steps, as covered in Chapter 12, the counterflowing phases here are in more or less continuous contact.

As stated, the purpose of the contacting is to bring about transfer of material between phases; this implies diffusion under a concentration driving force within each phase, and it implies also that there is movement across phase boundaries. If mass transfers during the total period of contact, then concentrations within the phases must change and there is the possibility that the total driving force for transfer can change throughout the total zone of contact. This zone can, if desired, be considered a differential volume or length, with differential changes in driving force concentrations occuring within that differential element. Hence, countercurrent mass transfer processes may be termed *differential mass transfer operations*. The total zone becomes the summation of the differential elements, obtained through conventional integration methods.

The general issues of equilibrium limitations, discussed in Chapter 12 for stagewise contacting, apply also to countercurrent systems. Practical equipment design considerations require that phase equilibrium not be approached too closely at any differential element. The closeness of approach is an economic consideration, readily quantified when information on equipment and operating costs is at hand. The purpose of this chapter is to provide concepts as well as techniques that enable the designer to develop optimum larger scale device specifications when scaling to a countercurrent contacting mode is favored.

II. FUNDAMENTAL CONSIDERATIONS

A. Phase Equilibrium and Interphase Diffusion

The concepts of phase equilibrium and differential mass transport have been discussed in Chapter 12. Equilibrium data must be available, either from laboratory experiments or from generalized estimation methods. Data on molecular diffusivity must also be available, usually in the form of diffusion coefficients measured or estimated. Affecting the extent and mobility of the interface will be the interfacial tension between contacting phases. Since the area for transport is an important term in the diffusion flux equations, interfacial tension data are normally required. Finally, the manner in which the phases are brought into contact plays an important role in the diffusion process, and this implies an understanding of the mechanics of fluid flow in the contacting device. Not all of these input data will always be available, at least for preliminary scaleup studies, and there are methods for estimating their effects; however, for final scaleup and design, reliable data should be obtained.

B. Countercurrent Mass Transfer

It is convenient to develop the diffusional mass transfer relationships in the context of a gas (or vapor) rising through a descending liquid. Variations from this situation, for example for liquid–liquid systems, will be developed when needed. At any point of liquid–gas contact, gradients of concentration for a diffusing species A[†] within the phases may be represented as in Figure 13-1. The molar flux of A is

$$N_A = k_{gA}(P_A - P_{Ai}) = k_{gA}P(y_A - y_{Ai}) \quad \text{gas} \quad (13\text{-}1)$$

$$N_A = k_{LA}(C_{Ai} - C_A) = k_{LA}\bar{\rho}_{M,L}(x_{Ai} - x_A) \quad \text{liquid} \quad (13\text{-}2)$$

In these expressions, N_A is the number of moles of A diffusing per unit time per unit interfacial area. It is often convenient to express this area per unit volume of the contactor.

Consider now a differential slice of a contactor having a cross-sectional area S (Figure 13-2). The slice has a height dZ and a volume $S\,dZ$. A material balance across the slice gives

$$x_A\,dL + L\,dx_A = -(y_A\,dV + V\,dy_A)$$

$$(\text{moles to liquid}) = (\text{moles from gas}) \quad (13\text{-}3)$$

It should be noted that Figure 13-1 and Equation 13-3 are based on flow from

---

[†] In this chapter diffusing species will be denoted as A, B, ..., instead of $i, j, \ldots$, to eliminate the confusion over the interface designation i.

FUNDAMENTAL CONSIDERATIONS

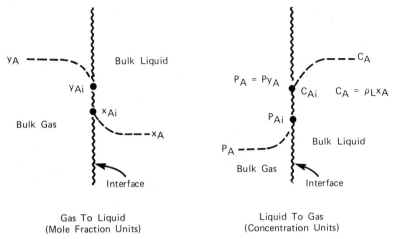

FIGURE 13-1 Transfer between bulk gas and bulk liquid.

gas to liquid, but with suitable changes of sign, they apply also to flow in the opposite direction.

Either side of Equation 13-3 represents the volume flux of A across the interface. Considering the right-hand side,

$$\frac{-(y_A dV + V dy_A)}{S dZ} = N_A a_v \qquad (13-4)$$

where a_v is the interfacial area per unit volume. Defining the molar gas velocity as $G_M = V/S$, Equation 13-4 can be transformed to

$$-y_A dG_M - G_M dy_A = N_A a_v dZ \qquad (13-5)$$

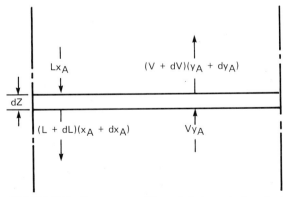

FIGURE 13-2 Flows across differential element of contactor.

For one component diffusing,

$$-dG_M = N_A a_v \, dZ \tag{13-6}$$

then

$$y_A N_A a_v \, dZ - N_A a_v \, dZ = G_M \, dy_A \tag{13-7}$$

and

$$dZ = \frac{-G_M \, dy_A}{N_A a_v (1 - y_A)} \tag{13-8}$$

Utilizing Equation 13-1,

$$dZ = \frac{-G_M \, dy_A}{k_{gA} a_v P(1 - y_A)(y_A - y_{Ai})} \tag{13-9}$$

Integrating over the entire tower height, from inlet composition of gas y_{A1} to outlet composition y_{A2}, and eliminating the negative sign by reversing the integration limits,

$$Z_T = \int_{y_{A2}}^{y_{A1}} \frac{G_M \, dy_A}{k_{gA} a_v P(1 - y_A)(y_A - y_{Ai})} \tag{13-10}$$

This is a basic expression, but its integration leads to some difficulties. If y_{Bm} is defined as a mean value of $1 - y_A$ over the contactor, and if both numerator and denominator of Equation 13-10 are multiplied by it,

$$Z_T = \int_{y_{A2}}^{y_{Ai}} \left(\frac{G_M}{k_{gA} a_v P y_{Bm}} \right) \left(\frac{y_{Bm}}{(1 - y_A)(y - y_{Ai})} \right) dy_A \tag{13-11}$$

The left term inside the integral is theoretically independent of concentration and pressure. It is defined as the *height of a gas–phase transfer unit*:

$$(HTU)_G = H_G = \frac{G_M}{k_{GA} a_v P y_{Bm}} \tag{13-12}$$

The total height of the contactor may be considered to be the product of the height of a transfer unit times the number of transfer units:

$$Z_T = H_G \cdot N_G \tag{13-13}$$

On the basis of Equation 13-13, and considering H_G to be constant (Equation 13-12), Equation 13-11 gives the definition of the *number of gas-phase transfer*

FUNDAMENTAL CONSIDERATIONS

units:

$$N_G = \int_{y_{A2}}^{y_{A1}} \frac{y_{Bm} \, dy_A}{(1 - y_A)(y_A - y_{Ai})} \tag{13-14}$$

Thus, the required height of the contactor Z_T can be calculated by means of Equation 13-13, with H_G taken from Equation 13-12 and N_G taken from Equation 13-14.

Equations 13-4 through 13-14 are based on the mass transfer in the gas phase only. A similar development for the liquid phase gives equivalent expressions:

$$H_L = \frac{L_M}{k_{LA} a_v \bar{\rho}_L x_{Bm}} \tag{13-15}$$

$$N_L = \int_{x_{A1}}^{x_{A2}} \frac{x_{Bm} \, dx}{(x_{Ai} - x_A)(1 - x_A)} \tag{13-16}$$

$$Z_T = H_L \cdot N_L \tag{13-17}$$

Accordingly, the height of the contactor may be calculated either on the basis of gas concentrations or liquid concentrations. The difficulty in using either phase separately is that interfacial concentrations (x_{Ai}, y_{Ai}) are needed, and when there is transfer resistance in both phases these concentrations will not normally be available. The values of H_G and H_L can be determined, since they are independent of concentration. The values of N_G and N_L may be combined on the basis of the two-film theory:

$$\frac{1}{N_{OG}} = \frac{1}{N_G} + \frac{\lambda}{N_L} \tag{13-18}$$

where

$$\lambda = \frac{m_e}{L_M/G_M} = \frac{\text{slope of equilibrium line}}{\text{slope of operating line}} \tag{13-19}$$

In Equation 13-18 the term N_{OG} is the *number of overall gas-phase transfer units*. The individual phase transfer units could be combined in terms of an overall liquid basis, but convention suggests that the gas-phase basis be adopted. The term N_{OG} is (cf. Equation 13-14)

$$N_{OG} = \int_{y_{A2}}^{y_{A1}} \frac{y_{Bm}^* \, dy_A}{(1 - y_A)(y_A - y_A^*)} \tag{13-20}$$

The heights of transfer units can also be combined:

$$H_{OG} = \frac{y_{Bm}}{y_{Bm}^*} H_G + \lambda \frac{x_{Bm}}{y_{Bm}^*} H_L \tag{13-21}$$

In Equations 13-20 and 13-21, y_{Bm}^*, y_{Bm}, and x_{Bm} are specially defined mole fractions of the nondiffusing species:

$$y_{Bm}^* = \frac{(1 - y_A) - (1 - y_A^*)}{\ln[(1 - y_A)/(1 - y_A^*)]} \qquad (13\text{-}22a)$$

$$y_{Bm} = \frac{(1 - y_A) - (1 - y_{Ai})}{\ln[(1 - y_A)/(1 - y_{Ai})]} \qquad (13\text{-}22b)$$

$$x_{Bm} = \frac{(1 - x_A) - (1 - x_{Ai})}{\ln[(1 - x_A)/(1 - x_{Ai})]} \qquad (13\text{-}22c)$$

Thus, y_{BM}^* deals with a mean value between the bulk gas composition and the composition of the gas that would be in equilibrium with the bulk liquid and, in a similar fashion, y_{Bm} and x_{Bm} are means of bulk and interface compositions.

The foregoing development is based on single component absorption or stripping, that is, unidirectional diffusion, and it applies to concentrated mixtures. For dilute mixtures (e.g., "lean gas" cases) the following terms

$$1 - y_A, \quad y_{Bm}, \quad y_{Bm}^*, \quad 1 - x_A, \quad \text{and} \quad x_{Bm}$$

approach unity. Accordingly, the use of Equations 13-10, 13-11, 13-12, 13-14, 13-15, 13-16, 13-20, and 13-21 is correspondingly simplified. There are many instances of practical importance where this simplification is justified. Mass transfer theory shows also that for equimolar counterdiffusion, as usually occurs in distillation, the same simplifications are proper, regardless of concentration level. Distillation applications will be discussed later in the chapter.

The slope of the equilibrium line m_e needed for the calculation of the term λ may be obtained from the y–x equilibrium relationship for simple absorption/stripping or binary distillation cases. For dilute absorption or stripping a simple $y_A^* = m_e x_A$ equilibrium relationship may suffice. For other cases the relationship may take the form

$$y_A^* = \frac{c_1 x_A}{1 + c_2 x_A} \qquad (13\text{-}23)$$

For binary distillations with a constant value of α_{AB}, the relative volatility, $c_1 = \alpha_{AB}$ and $c_2 = \alpha_{AB} - 1$. For multicomponent distillation systems, the slope m_e must be related to a pair of diffusing components, usually the key components to be separated, and the equilibrium ratio $K_A (= y_A^*/x_A)$ for the light key. Between compositions x_{A1} and x_{A2},

$$m_e = \frac{y_{A1}^* - y_{A2}^*}{x_{A1} - x_{A2}} = \frac{K_{A1} x_{A1} - K_{A2} x_{A2}}{x_{A1} - x_{A2}} = \overline{K}_A \qquad (13\text{-}24)$$

FUNDAMENTAL CONSIDERATIONS

where x_{A1} and x_{A2} are close enough for it to be considered that $K_{A1} \simeq K_{A2} = \overline{K}_A$. Since values of m_e and K_A can vary across the total concentration range, it may be necessary to divide the contactor into sections where average values of K, G_M, and L_M (and thus λ) can be considered constant.

For binary systems, if the equilibrium line and the operating line are both straight, it may be shown that

$$N_{OG} = N_t \frac{\ln \lambda (y^*_{Bm})}{\lambda - 1(y_{Bm})} \qquad (13\text{-}25)$$

where N_t is the number of equilibrium steps required for the same separation.

A given height of contactor can produce the concentration change that would be produced by a theoretical stage. This height is given the symbol HETP, *height equivalent to a theoretical plate* (HETP = Z_T/N_t), and use of Equation 13-25 leads to

$$H_{OG} = \text{HETP} \frac{\lambda - 1(y_{Bm})}{\ln \lambda (y^*_{Bm})} \qquad (13\text{-}26)$$

The λ in Equations 13-25 and 13-26 is defined by Equation 12-19. For cases in which the operating and equilibrium lines are both straight and parallel, and for distillation or lean gas absorption, $N_{OG} = N_t$ and H_{OG} = HETP.

The evaluation of H_{OG} is usually by means of individual values for the phases, H_G and H_L. However, the term may be obtained separately from

$$H_{OG} = \frac{G_M}{K_{OG,A} a_v P y_{Bm}} \qquad (13\text{-}27)$$

where K_{OG} is a combined mass transfer coefficient (in this case, for diffusing A), obtained from

$$\frac{1}{K_{OG,A}} = \frac{1}{k_{gA}} + m_e \frac{P}{\rho_{L,M}} \cdot \frac{1}{k_{LA}} \qquad (13\text{-}28)$$

or

$$\frac{1}{K_{OG,A}} = \frac{1}{k_{gA}} + \frac{H_A}{k_{LA}} \qquad (13\text{-}29)$$

where $\rho_{L,M}$ is the molar liquid density and H_A is the Henry's law constant for A in the liquid.

The overall coefficient K_{OG} is based on an overall concentration difference between the bulk gas composition and the concentration of gas that would be in equilibrium with the bulk liquid concentration. Thus, for diffusing species A,

512 CONTINUOUS MASS TRANSFER PROCESSES

Equation 13-1 is adapted as follows,

$$N_A = K_{OG}(p_A - p_A^*) = K_{OG}P(y_A - y_A^*) \qquad (13\text{-}30)$$

where

$p_A^* = H_A x_A$ and $y_A^* = K_A x_A$
H_A = Henry's law constant for A
K_A = equilibrium ratio for A
x_A = bulk liquid mole fraction of A

Since it is usually not possible to determine the interfacial area a_v, for a unit volume of contactor, the area term is usually carried with the mass transfer coefficient to give a volumetric coefficient. For species A, volumetric coefficients are $k_{gA}a_v$, $k_{LA}a_v$, and $K_{OG,A}a_v$. In combination,

$$\frac{1}{K_{OG,A}a_v} = \frac{1}{k_{gA}a_v} + \frac{H_A}{k_{LA}a_v} \qquad (13\text{-}31)$$

The units to be used for the mass transfer coefficients and heights of transfer units are important, and for the latter the simple dimension of length (meters or feet) is proper. For the mass transfer coefficient, the choice of concentration driving force leads to the coefficient units. Some examples are given below:

Liquid-Phase Transfer

$$C_A = \text{kg mol A}/\text{m}^3 \text{ liquid}$$

$$k_{LA} = \frac{\text{kg mol A}}{(\text{s})(\text{m}^2)(\text{kg mol A}/\text{m}^3)} = \frac{\text{m}}{\text{s}}$$

$$k_{LA}a_v = \frac{\text{m}}{\text{s}} \cdot \frac{\text{m}^2}{\text{m}^3}\text{contactor} = \frac{1}{\text{s}}$$

Gas-Phase Transfer

$$p_A = \text{atm partial pressure of A}$$

$$k_{gA} = \frac{\text{kg mol A}}{(\text{s})(\text{m}^2)(\text{atm A})}$$

$$k_{gA}a_v = \frac{\text{kg mol A}}{(\text{s})(\text{m}^2)(\text{atm A})} \cdot \frac{\text{m}^2}{\text{m}^3} = \frac{\text{kg mol A}}{(\text{s})(\text{m}^3)(\text{atm A})}$$

Gas-Phase Transfer

$$y_A = \text{mole fraction A}$$

$$k_{gA} = \frac{\text{kg mol A}}{(s)(m^2)(\text{mole fraction A})}$$

$$k_{gA}a_v = \frac{\text{kg mol A}}{(s)(m^2)(\text{mole fraction A})} \cdot \frac{m^2}{m^3}$$

$$= \frac{\text{kg mol A}}{(s)(m^3)(\text{mole fraction A})}$$

C. Mechanics of Phase Contacting

Countercurrent phase contacting is usually carried out in columns filled with packing elements ("packed columns") although in special cases tray type devices are used. The packed column (see Figure 13-3) contains a packed bed where the mass transfer takes place. The bed is formed by dumping packing elements into the column and onto a grid-type support device. The elements arrange themselves in a random fashion. Fluid flow is around and through these sections, and their efficacy for mass transfer depends upon their ability to keep the phases well dispersed such that the extended interfacial area and the turbulent transport enhance the flux of diffusing species. In contrast to dumped packings, there are "arranged packings" which comprise larger, specially formed sections that are placed in the column by hand.

The performance of a packed contactor depends on three primary considerations: packing shape and size, loadings of the flowing phases, and properties of the contacting fluids. So far as the packing material is concerned, there are many types on the market, and the properties of some of the more popular ones are shown in Table 13-1 (dumped packings) and Table 13-2 (arranged packings). Views of these materials are shown in Figure 13-4. There is one distinction between dumped packings that should be noted. Some of them (such as Raschig rings and Berl saddles) can orient themselves in a bed such that the fluids must flow around them whereas others (such as metal INTALOX® and PALL® rings) permit flow through themselves at any orientation. These latter types, with throughflow, tend to be more efficient for mass transfer while minimizing pressure drop due to form drag.

Tables 13-1 and 13-2 include data on small packing sizes (as well as large), since such sizes are commonly used in bench- and pilot-scale studies. For scaleup studies, the $\frac{1}{2}$-in. nominal packing size is the smallest recommended. In general, a column diameter-to-packing size ratio of at least eight is needed to ensure good heavy phase distribution; this indicates a minimum pilot scale column diameter of 4 in.

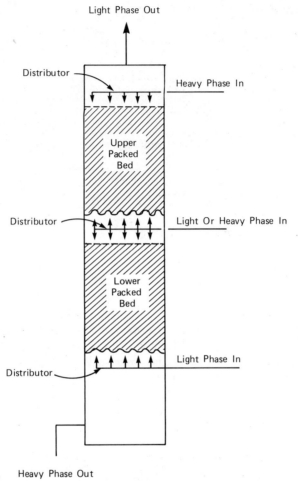

FIGURE 13-3 Packed contactor.

The rates of flow of the phases influence the mass transfer efficiency as well as the throughput capacity of a given packed column. Economics dictate the highest flow rates possible, within constraints of pressure drop and loss of countercurrent contacting at very high rates. Each packing type and size has an upper limit that may be estimated roughly on the basis of the sum of the two phase rates. A very high flow rate of the light phase can entrain upward a significant amount of the downflowing heavy phase, causing a loss of efficiency; conversely, a very high rate of heavy phase can entrain downward the light phase. Quantitative aspects of these phase flow limits will be given in connection with the appropriate separation method.

The general arrangement of a packed column is shown in Figure 13-3. The devices for distributing the phases are critical to the performance of the

TABLE 13-1 Characteristics of Dumped Packings

Packing Type	Nominal size, mm	Elements per m³	Bed Weight, kg/m³	Surface Area, m²/m³	ε Void Fraction	F_p Packing Factor, m⁻¹	$F_p \varepsilon^2$	Vendors[a]
INTALOX[x] saddles (ceramic)	13	730,000	720	625	0.78	660	402	Norton, Koch, Glitsch
	25	84,000	705	255	0.77	320	190	
	38	25,000	670	195	0.80	170	108	
	50	9,400	670	118	0.79	130	81	
	75	1,870	590	92	0.80	70	45	
INTALOX[x] saddles (metal)	25	168,400	350	n.a.[b]	0.97	135	127	Norton
	40	50,100	230	n.a.	0.97	82	77	
	50	14,700	181	n.a.	0.98	52	50	
	70	4,630	149	n.a.	0.98	43	41	
PALL[x] rings [Ballast rings, Flexirings] (metal)	16				0.92	230	195	Norton, Glitsch, Koch
	25	49,600	480	205	0.94	157	139	
	38	13,000	415	130	0.95	92	93	
	50	6,040	385	115	0.96	66	61	
	90	1,170	270	92	0.97	53	50	
Raschig rings (ceramic)	13	378,000	880	370	0.64	2000	819	Norton, Glitsch, Koch, others
	25	47,700	670	190	0.74	510	279	
	38	13,500	740	120	0.68	310	143	
	50	5,800	660	92	0.74	215	118	
	75	1,700	590	62	0.75	120	68	
Berl saddles (ceramic)	13	590,000	865	465	0.62	790	304	Koch, others
	25	77,000	720	250	0.68	360	166	
	38	22,800	640	150	0.71	215	108	
	50	8,800	625	105	0.72	150	78	
INTALOX[x] saddles (plastic)	25	55,800	76	206	0.91	105	87	Norton, Glitsch, Koch
	50	7,760	64	108	0.93	69	60	
	75	1,520	60	88	0.94	50	44	
PALL[x] rings (plastic)	16	213,700	116	341	0.87	310	235	
	25	50,150	88	207	0.90	170	138	
	50	6,360	72	100	0.92	82	69	

[a] Identification of vendors: Norton Company, Akron, Ohio; Glitsch, Inc., Dallas, Texas; Koch Engineering Co., Wichita, Kansas.
[b] n.a.—not available.

TABLE 13-2 Characteristics of Arranged Packings

Type		Surface Area, m^2/m^3	Void Fraction	Packing Factor, F_p, m^{-1}	Vendor
FLEXIPAC®	1	558	0.91	108	Koch
(metal)	2	246	0.93	72	
	3	134	0.96	52	
	4	69	0.98	30	
KOCH–SULZER®	—	490	> 0.90	66	Koch
MUNTERS®	12060	223	> 0.95	90	Munters
	19060	148	> 0.95	49	
	25060	98	> 0.95	33	

Other examples of arranged packings:

HYPERFIL®	Knitted mesh	Chem-Pro Equipment Corp., Hanover, New Jersey
GOODLOE®	Knitted mesh	Glitsch, Inc., Dallas, Texas
GLITSCH-GRID®	Stacked metal grids	Glitsch, Inc., Dallas, Texas

Note: Koch = Koch Engineering Co., Wichita, Kansas; Munters = Munters Corp., Fort Myers, Florida

packing. When there is one (or more) feed to a central section of the column, as shown in the figure, separate packing support and heavy phase redistribution devices are required. This arrangement is typical of distillation columns and of extraction columns in which there is extract reflux.

III. VAPOR – LIQUID SYSTEMS — DISTILLATION

A. Major Issues

The major issues for packed column distillation are exactly the same as for equilibrium stage distillation, as delineated in Chapter 12. The system must first be defined, with due regard to the presence of minor constituents of the mixture to be separated, and to the characterization of components that might not be easily identified. Next, the separation criteria must be established, since the design of the equipment is heavily dependent on the degree of separation required. In turn, reliable physical property information must be assembled in order that models for transfer units and for fluid mechanical factors might be implemented. Finally, the quantitative specifications of the equipment must be developed, including size of packing, dimensions of packed beds, type of distribution devices, and so on. As much care should be given to the design of the pilot- or bench-scale system as to the commercial scale system; this is a point often overlooked by research people in their propensity toward assembling laboratory equipment and then observing empirically how it performs.

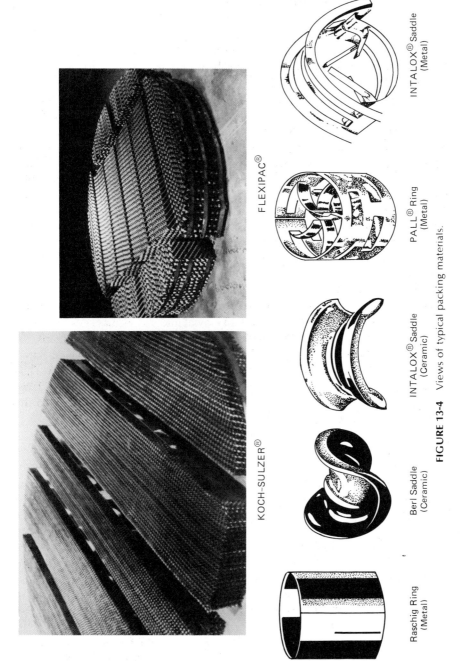

FIGURE 13-4 Views of typical packing materials.

B. Prediction of Performance

There are three major factors to consider in the evaluation of the performance of a new design or an existing fractionator. First, there is the maximum capacity of the column; this determines the minimum possible diameter of the bed. Second, there is the pressure loss in flow through the bed; this becomes quite critical in vacuum fractionations. Third, there is the capability of the bed to produce the required number of transfer units or theoretical stages. All of these factors are influenced by economic considerations, and general guidelines for them can be misleading when applied to specific cases. There are, for example, special packings of the arranged type that can enable difficult separations with a minimum of pressure drop (important especially in vacuum columns), but the initial cost of packing the column with these materials may be prohibitive except for the separation of high-valued products. As another example, the need for corrosion resistance in the separation of low-valued materials may limit packing selection to relatively inefficient materials that can be fabricated cheaply. Each case for scaleup and design should be evaluated separately.

The development and marketing of new packing elements is an active business, and in recent years a number of new elements have been placed on the market. Since there is not a readily available "Consumers Union" for testing these elements, reliance must be placed on the vendors' claims for performance. However, it is possible, on a limited basis, to have packings tested at Fractionation Research, Inc., Alhambra, California, U.S.A. The designer should check these claims carefully while at the same time recognizing that some of the newer packings are indeed superior in some respects to the well-documented traditional packings.

C. Liquid Distribution

Careful consideration must be given to the manner in which the liquid is distributed to the packing, at the top of the column as well as at the feed location (Figure 13-3). The three types of distributors commonly used are the trough type, the orifice/riser type, and the solid-cone spray nozzle. These are illustrated in Figure 13-5. The trough unit enables a broad range of liquid flow rates and is useful in research and multipurpose columns. The orifice unit confines the vapor–liquid contacting to the packed bed (there can be mass transfer in the distribution zone of the trough unit) but is more limited in liquid and vapor flows. The spray nozzle (or battery of nozzles) is not normally recommended because of its nonuniform pattern in larger sizes.

The packings that do not permit throughflow of liquid and vapor, such as Raschig rings and Berl saddles, can correct an initially poor distribution of liquid, and thus the selection of the distributor may not be overly critical. For the throughflow packings, such as PALL® rings, metal INTALOX® saddles,

VAPOR–LIQUID SYSTEMS — DISTILLATION

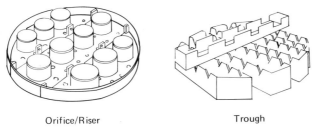

Orifice/Riser　　　　　　　　Trough

FIGURE 13-5 Types of liquid distributor for packed columns.

and the arranged elements, an initially poor distribution will not be corrected by the packed bed.

Guidance for liquid distributor design may be based on the number of liquid streams fed to the bed, per unit cross-sectional area. For tests of Pall rings in the 1.2-m distillation column at Fractionation Research, Inc. (FRI), satisfactory distribution was obtained by a trough unit feeding 40 streams/m² as discussed by Billet (1967). Tests at the BASF Company in Germany utilized an orifice unit with 96 streams/m² according to Billet (1967). Williamson (1951) found that for throughflow packings to maintain distribution requires 20 streams/m². Cornell et al. (1960) described a research-type orifice unit with about 120 streams/m². For general purposes, a minimum of 50 streams/m² is recommended, either from a trough type unit or from an orifice-riser unit.

D. Pressure Drop

The pressure drop of a vapor flowing upward through a packed bed, countercurrent to a downflowing liquid, is characterized graphically in Figure 13-6. At very low liquid rates, there is a significant cross-sectional area available to vapor flow, and pressure drop tends to be low. Increased liquid flow, in turn, decreases the area available for vapor flow, and pressure drop tends to be higher. Thus, the packed bed behaves as a plurality of irregularly shaped orifices whose area changes with changes in liquid flow rate. When the combined phase flow rates are high, there is holdup of liquid in the packing (the "loading zone") and with further increase of flows, flooding is reached.

Pressure drop in a packed distillation column may be estimated by the method of Eckert (1970) which is graphical in nature as shown in Figure 13-7. An analysis of this method has been made by Bolles and Fair (1979). The parametric lines on Figure 13-7 give readings of pressure drop in mm Hg/m of packed height. The abscissa scale is the dimensionless *flow parameter*;

$$FP = \frac{L}{G}\sqrt{\frac{\rho_g}{\rho_L}} \qquad (13\text{-}32)$$

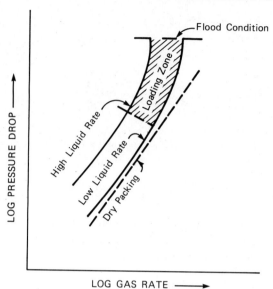

FIGURE 13-6 Pressure drop and loading characteristics of packed columns.

where L and G are mass flow rates. The ordinate scale is a form of *capacity parameter* originally introduced by Sherwood et al. (1938) and utilized for pressure drop purposes by Eckert (1970);

$$CP = \frac{G^2 F_p \Psi \mu_L^{0.2}}{\rho_g \rho_L g} \qquad (13\text{-}33)$$

where

G = gas velocity, kg/s m^2
F_p = packing factor (see Tables 13-1 and 13-2)
Ψ = ratio, density of water/density of liquid
μ_L = liquid viscosity, mPa s (centipoises)
ρ_g = gas density, kg/m^3
ρ = liquid density, kg/m^3
g = gravitational constant, 9.81 m/s^2

The use of Figure 13-7 is straightforward. Values of FP and CP are computed by Equations 13-32 and 13-33, and the pressure drop is read from the graph.

This generalized method is subject to correlational error. On the basis of a bank of distillation data, Bolles and Fair (1979) suggest a safety factor of 2.2 for 95 percent confidence in the predicted pressure drop, equivalent to doubling the value read from the graph of Figure 13-7. A better approach is to use measured data, often in the form of the example given in Figure 13-8.

VAPOR – LIQUID SYSTEMS — DISTILLATION

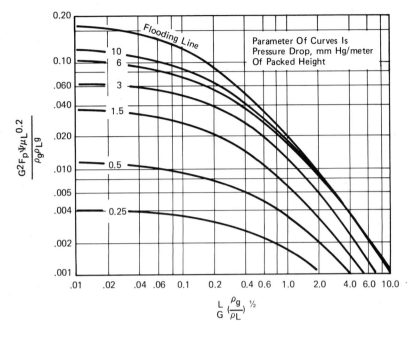

FIGURE 13-7 Generalized correlation for flooding and pressure drop in packings.

Unfortunately, there are no published generalized methods that are better than the one of Eckert.

E. Maximum Vapor Capacity

From a design standpoint, it is convenient to establish the maximum allowable vapor capacity of the packed column, usually termed "flood capacity." Although the actual flooding condition is difficult to measure precisely, it is a condition where a small increase in gas rate causes a large increase in pressure drop (Figure 13-6), indicating a rapid buildup of liquid inventory in the bed with consequent lowering of the rate of liquid flow from the bed. In theory, the height equivalent to a theroretical plate (HETP) goes to infinity at the flood point, because of loss of concentration driving force in the interphase transfer process.

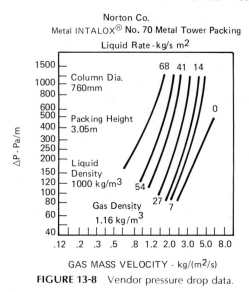

FIGURE 13-8 Vendor pressure drop data.

Experimental observations of the flood capacity of various packed columns have been correlated by Eckert (1970) as shown in Figure 13-7. The same correlation for pressure drop is used for flood capacity, and the same reliability limitations for pressure drop apply to flooding. Bolles and Fair (1979) found that a 1.32 safety factor should be applied to the flood velocity obtained from Figure 13-7, for 95 percent confidence that a premature flood would not occur. Thus, a flood vapor velocity obtained from Figure 13-7 would be divided by 1.32 in specifying the expected flood velocity. With this proviso, the use of the Eckert correlation is recommended.

F. Mass Transfer Efficiency

An estimate of the mass transfer efficiency of a packed column is necessary in order to specify the height of packing required for the separation. The packed height may be obtained from the simple relationship

$$Z_T = N_t \cdot \text{HETP} \tag{13-34}$$

where N_t is the number of theoretical stages required, and HETP is the "height equivalent to a theoretical plate," an empirical factor that must be based on general experience. Vendors of packing materials tend to use the HETP approach, and they provide HETP values to customers on the basis of general feedback from customers as well as from their own tests. Experimentally, it is easy to obtain values of HETP; one measures the separation, fits it to an appropriate model to obtain equivalent stages, and then divides the packed

VAPOR–LIQUID SYSTEMS — DISTILLATION

height used by the number of stages. Eckert (1970) suggests the following values of HETP for distillation systems with moderate values of surface tension, low viscosities, and good phase distribution:

(1) HETP = 0.7–0.8 m for 50-mm nominal size packings.
(2) HETP = 0.5–0.6 m for 37-mm nominal size packings.
(3) HETP = 0.4–0.5 m for 25-mm nominal size packings.

These values are for dumped packings and do not cover the throughflow type, except for PALL® rings. They also apply to the following pressure drop ranges (mm Hg/m):

(1) Raschig rings: 2.5 to 4.6.
(2) Ceramic INTALOX® saddles: 1.2 to 4.6.
(3) Metal PALL® rings: 0.6 to 4.6.

An alternate, and preferred, method for estimating the required height of packing is through the use of transfer units:

$$Z_T = H_{OG} \cdot N_{OG} \tag{13-35}$$

For distillation, Equation 13-27 reduces to

$$H_{OG} = \frac{G_M}{K_{OG,A} a_v P} \tag{13-36}$$

since for the equimolar counterdiffusion case, $y_{Bm} \approx 1.0$. Because there is normally mass transfer resistance in both phases, the value of H_{OG} is practically found from a modification of Equation 13-21, valid for distillation systems (i.e., with y_{Bm}, y_{Bm}^* and x_{Bm} having values of unity);

$$H_{OG} = H_G + \lambda H_L \tag{13-37}$$

where, for distillation,

$$H_G = \frac{G_M}{k_{GA} a_v P} \tag{13-38}$$

$$H_L = \frac{L_M}{k_{LA} a_v \rho_L} \tag{13-39}$$

Equations 13-38 and 13-39 follow from Equations 13-12 and 13-15, with appropriate simplifications.

Values for H_G and H_L, to be used in Equation 13-37, may be obtained from the general correlations of Bolles and Fair (1979, 1982). For ring type packings

(Raschig rings, PALL® rings, HY-PAK®),

$$H_G = \frac{0.017\Psi D_T^{1.24} Z_T^{0.33} Sc_G^{0.50}}{(Lf_1 f_2 f_3)^{0.6}} \quad (13\text{-}40)$$

For saddle type packings (Berl saddles, ceramic INTALOX® saddles),

$$H_G = \frac{0.029\Psi D_T^{1.11} Z_T^{0.33} Sc_G^{0.50}}{(Lf_1 f_2 f_3)^{0.5}} \quad (13\text{-}41)$$

where

Sc_G = gas-phase Schmidt number (dimensionless) = $\mu_g/(\rho_g \mathscr{D}_g)$
D_T = column diameter, m
Z_T = height of packed section, m
L = liquid mass rate, kg/s m²
$f_1 = (\mu_L/\mu_w)^{0.16}$, with $\mu_w = 1.0$ mPa s (1.0 cP)
$f_2 = (\rho_w/\rho_L)^{1.25}$, with $\rho_w = 1000$ kg/m³
$f_3 = (\sigma_w/\sigma_L)^{0.8}$, with $\sigma_w = 72.8$ N/m (72.8 dyn/cm)

The correlation parameter Ψ in Equations 13-40 and 13-41 may be obtained from Figure 13-9. A particular restriction should be noted: for column diameters larger than 0.6 m, the diameter correction for 0.6 m ($0.6^{1.11}$ or $0.6^{1.24}$) should be retained. In other words, no further correction is required for diameters larger than 0.6 m.

For both rings and saddle type packings, the liquid-phase correlation is

$$H_L = 0.258 \phi C_f Z_T^{0.15} Sc_L^{0.50} \quad (13\text{-}42)$$

where

Sc_L = liquid-phase Schmidt number (dimensionless) = $\mu_L(\rho_L \mathscr{D}_L)$
Z_T = height of packed section, m
ϕ = packing correlation parameter given in Figure 13-10
C_f = flooding correction factor given in Figure 13-11

C_f factor accounts in part for phase inversions that occur in the vicinity of the flood point.

An alternate method for predicting values of H_G and H_L has been published by Bravo and Fair (1982). It involves evaluation of coefficients k_{LA} and k_{GA} together with the interfacial area a_v from

$$a_v = 0.309 a_p \left\{ \frac{\sigma^{0.8}}{Z_T^{0.4}} \right\} \{Ca_L Re_G\}^{0.392} \quad (13\text{-}43)$$

VAPOR–LIQUID SYSTEMS — DISTILLATION

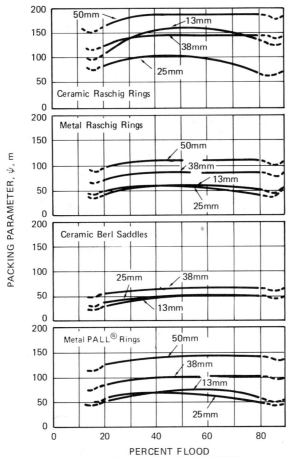

FIGURE 13-9 Correlation for gas-phase packing parameter ψ (Equations 13-40 and 13-41).

where

a_v = interfacial area, m^{-1}
a_p = packing surface area (Table 13-1), m^{-1}
σ = surface tension, N/m (dyn/cm)
Z_T = height of packed section, m
Ca_L = dimensionless capillary number for liquid = $L\mu_L/(\rho_L\sigma)$
Re_G = dimensionless Reynolds number for gas = $6G/(a_p\mu_G)$

Coefficients k_{LA} and k_{GA} are obtained from the Onda model discussed in Bolles and Fair (1979) and transfer unit heights H_G and H_L are obtained from Equations 13-38 and 13-39. In turn, H_{OG} is obtained from Equation 13-37.

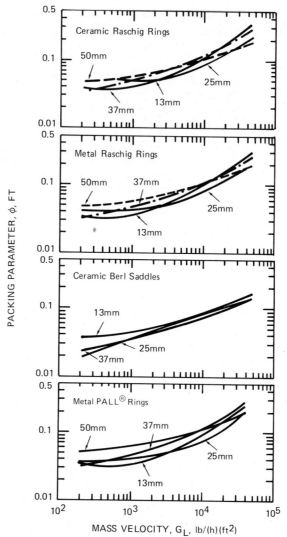

FIGURE 13-10 Correlation for liquid-phase packing parameter ϕ (Equation 13-42).

This method has the possibility of being applicable to new packings for which the specific surface area a_p is known. However, it has been tested only for the PALL® ring, among the throughflow packings.

G. Transfer Units and Theoretical Stages

For distillation (equimolar counterdiffusion) Equation 13-20 becomes

$$N_{\text{OG}} = \int_{y_{A2}}^{y_{A1}} \frac{dy}{y_A - y_A^*} \tag{13-44}$$

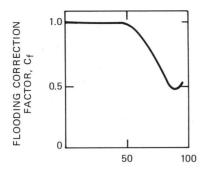

FIGURE 13-11 Flooding correlation factor for H_L.

where A denotes the diffusing species. For a simple binary distillation, where $y-x$ equilibrium data are available, evaluation of Equation 13-44 by graphical integration is straightforward. For multicomponent mixtures, where the value of N_G varies from component to component, it is usually expedient to use the distillation form of Equation 13-25:

$$N_{OG} = N_t \frac{\ln \lambda}{\lambda - 1} \tag{13-45}$$

If a value of HETP is needed, it may be obtained from a variation of Equation 13-26, specific for distillation:

$$\text{HETP} = H_{OG} \frac{\ln \lambda}{\lambda - 1} \tag{13-46}$$

As pointed out earlier, Equations 13-45 and 13-46 apply strictly when both operating and equilibrium lines are straight. When they are both straight and parallel, $N_{OG} = N_t$ and HETP $= H_{OG}$. When the lines are curved, the column should be divided into sections for analysis, with each section containing operating and equilibrium lines that can be considered straight.

Equations 13-45 and 13-46 indicate applicability of a binary form to multicomponent separations. It has been found that binary representations of multicomponent distillation mass transfer efficiencies represent reasonable approximations for engineering design purposes (Chan and Fair, 1984).

H. Scaleup Procedure

For distillations involving well-defined systems for which reliable physical property data (including vapor–liquid equilibria) are available, and for the more conventional dumped packing materials, the procedure for scaleup and design is straightforward. Required transfer units or theoretical stages are calculated. Tentative dimensions of the equipment are obtained from pressure drop, flooding, and empirical HETP procedures. The methods in Section F are then used to determine heights of transfer units or heights equivalent to a theoretical stage; in effect, this means checking the assumed bed height against the calculated bed height. Finally, correlation reliability parameters are applied to enable a high confidence in the final design.

The procedure is less straightforward if a packing material, for which generalized methods of design are not available, is selected. In this case, attempts may be made to fit existing models with the limited performance data available on the packing. These attempts should be supported by fundamental considerations regarding fluid flow and mass transfer, although the odd shapes of some of the newer packings make this difficult. In general, reliance must be placed on the vendor's research and experience with commercial installations.

When the system is not well defined, and when the available physical property information is approximate at best, the following procedure may be used:

(1) Run the system in a small column which has been calibrated for a similar system. The column may have plates, as for example the Oldershaw column (see Chapter 12), or it may have a special laboratory packing. The objective of this step is to ascertain the number of theoretical stages required for the given separation.

(2) If possible, as an alternate to step 1, run the system in a pilot scale column with at least 100 mm inside diameter and a packing nominal size of 13 mm

(3) Calculate the value of H_{OG} from the information obtained in step 2. (This will not be available in useful form from step 1.)

(4) Estimate values of H_G and H_L from Equations 13-40, 13-41, and 13-42 using Figures 13-9 and 13-10. From Equation 13-37 calculate the value of H_{OG}.

(5) If step 2 was used, compare the calculated value of H_{OG} with the measured value of H_{OG} (step 3). Make appropriate adjustments to H_G and H_L to bring agreement, that is, obtain adjusted values of the parameters in Figures 13-9 and 13-10. If step 3 was not used, then the estimated values of H_G and H_L must be used.

(6) For the same type of packing in the larger size needed for the commercial installation, estimate new values of the packing parameters (Figures 13-9 and 13-10), and for the larger value of the tower diameter and bed height, determine the final design. The height corrections in Equations 13-40 and 13-42 (and Equation 13-43, if the Bravo–Fair approach is used) involve the designer with a modest trial-and-error process.

VAPOR–LIQUID SYSTEMS — DISTILLATION

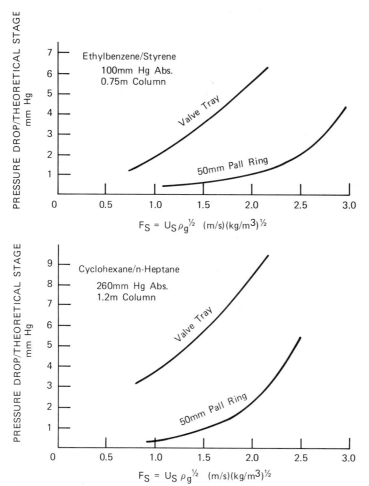

FIGURE 13-12 Pressure drop per theoretical stage for valve trays and metal PALL® rings, for vacuum distillation conditions.

I. General Comments

Packed distillation columns have become quite popular in recent years. This popularity stems primarily from the favorable pressure-drop mass transfer efficiency characteristics of the throughflow packings, either dumped or arranged. These characteristics assume signal importance when the distillation is conducted under vacuum, possibly also with the addition of diluent steam. A published comparison [Fair (1970)] between tray and packed columns is shown in Figure 13-12; the PALL® ring packing permits a significantly higher number of theoretical stages per unit of pressure drop. Even more dramatic differences would be obtained with the arranged KOCH–SULZER® and FLEXIPAC®, but these elements are significantly more expensive than the dumped PALL® ring packing.

Another unique application of packings is for systems with foaming tendencies. The packed bed serves as a built-in foam breaker, and this permits smaller column dimensions than would be needed for a tray column.

The traditional application of packed columns for corrosive service should not be overlooked. For this case, the corrosion-resistant ceramic packing materials often prove to be more economical than columns with high-alloy trays and downcomers.

IV. GAS – LIQUID SYSTEMS — ABSORPTION AND STRIPPING

A. Major Issues

The major issues for packed column absorption and stripping are no different from those for packed column distillation or, for that matter, from those for absorption and stripping in staged devices. Thus, reference should be made to Section III of this chapter and to Section IV of Chapter 12. A key difference between distillation and absorption or stripping is the unidirectional character of the mass transfer process. In distillation there is equimolar counterflow of the transferring species (or something close to it). In absorption, there is primarily unidirectional flow of solute (s) from the gas phase to the liquid phase, and the reverse holds for stripping. This difference is manifest in the mass transfer equations, as will be detailed later.

Thus, by way of summary and repetition the major issues for absorption–stripping are definition of system, establishment of separation criteria, assembly of reliable physical property data, determination of stages or transfer units, and final design of the physical system. As for packed column distillation, the type and size of the packing material plays a major role in the scaleup and design of packed absorbers and strippers.

B. Prediction of Performance

Methods and approaches to the sizing of absorbers and strippers do not differ from those for fractionators. The maximum flow capacity of the selected packing material must be determined (Section III-E of this chapter), and from this a tentative diameter of column is chosen (normally for about 50 to 60 percent of the maximum allowable gas velocity). The pressure drop (Section III-D) is estimated for the tentative total height of packed bed. Care is given to the choice of liquid distributor (Section III-C) in order to achieve the best possible mass transfer efficiency. Thus, the key point of departure for the design of absorbers/strippers, compared with fractionators, is mass transfer efficiency.

GAS–LIQUID SYSTEMS — ABSORPTION AND STRIPPING   531

C. Mass Transfer Efficiency

The required height of the packed bed may be estimated from either of the relationships:

$$Z_T = N_t \cdot \text{HETP} \tag{13-34}$$

$$Z_T = N_{OG} \cdot H_{OG} \tag{13-35}$$

Values of the height equivalent to a theoretical plate, HETP, may be obtained from packing vendors on the basis of background experience, or may be obtained from published listings (see Section III-F). In general, HETP values for absorbers and strippers will be somewhat larger than those for distillation, because of the normally lower concentrations of solute. As mentioned earlier, the transfer unit approach (Equation 13-35) is the preferred one, since it takes into account the individual phase resistances. Further, it is based on the mechanisms that actually take place in packed beds—countercurrent, or differential contacting.

For the determination of transfer units, Equation 13-20 is used:

$$N_{OG} = \int_{y_{A2}}^{y_{A1}} \frac{y_{Bm}^* \, dy_A}{(1 - y_A)(y_A - y_A^*)} \tag{13-20}$$

For a single diffusing species A this equation can be integrated in a stepwise fashion, and a graphical approach is often convenient. For such an approach, the term $y_{Bm}^*/[(1 - y_A)(y_A - y_A^*)]$ as ordinate is plotted against y_A over the range of y_{A1} (bottom of column) to y_{A2} (top of column). For the special case of *lean gas mixtures* (low concentration of species A), where the operating line is essentially straight, $y_{Bm}^* \simeq (1 - y_A)$ and Equation 13-20 reduces to

$$N_{OG} = \int_{y_{A2}}^{y_{A1}} \frac{dy_A}{y_A - y_A^*} \tag{13-47}$$

and integration is much more straightforward. A y–x plot for this condition is shown in Figure 13-13. If the driving force $y_A - y_A^*$ can be considered constant (e.g., a log mean average value), then

$$N_{OG} = \frac{y_{A2} - y_{A1}}{(y_A^* - y_A)_{\text{mean}}} \tag{13-48}$$

If *operating and equilibrium lines are straight* (lean gas case, K = constant), use of a mean driving force leads to

$$N_{OG} = \frac{1}{1 - \lambda} \ln\left[(1 - \lambda)\left(\frac{y_{A1} - Kx_2}{y_{A2} - Kx_2}\right) + \lambda\right] \tag{13-49}$$

where $\lambda = m_e(G_M/L_M)$ and, in this case, $K = m_e$.

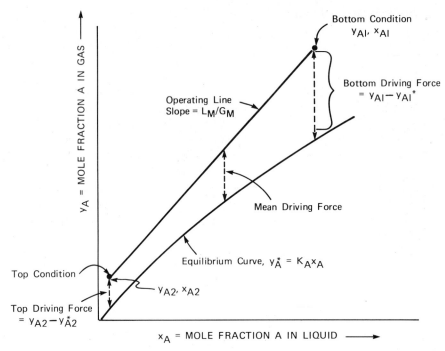

FIGURE 13-13 Absorption plot for lean gas case.

For a *concentrated gas mixture* (curved operating line) and a *straight equilibrium line* (K = constant), an approximate form of Equation 13-20 is

$$N_{OG} = \int_{y_{A2}}^{y_{A1}} \frac{dy_A}{y_A - y_A^*} + \frac{1}{2} \ln \frac{1 - y_{A2}}{1 - y_{A1}} \tag{13-50}$$

The second term on the right is known as the "Wiegand correction" (1940) and its use enables the simple integration of the first term on the right.

When both operating and equilibrium lines are significantly curved, Equation 13-20 should be handled exactly, on a stepwise basis. This is especially the case when there are large heat effects which influence the solute solubility (K value) separately from the concentration influence.

For the case of stripping, the equations for absorption apply, with limits and driving force reversed;

$$N_{OG} = \int_{y_{A1}}^{y_{A2}} \frac{y_{Bm}^* \, dy_A}{(1 - y_A)(y_A^* - y_A)} \tag{13-51}$$

and a plot of the y–x relationships shows the equilibrium curve to be above the operating line.

For absorption and stripping, heights of transfer units are calculated from Equation 13-12 (H_G), Equation 13-15 (H_L), and Equation 13-21 (H_{OG}). These

LIQUID – LIQUID SYSTEMS — EXTRACTION

equations simplify for the lean gas case, where $y_{Bm} \simeq 1$, $x_{Bm} \simeq 1$, $y_{Bm}/y_{Bm}^* \simeq 1$, and $x_{Bm}/y_{Bm}^* \simeq 1$. For other cases, the validity of these approximations should be checked.

The Bolles and Fair correlation for H_G and H_L applies to absorption/stripping as well as to distillation. Accordingly, Equations 13-40 to 13-42 and Figures 13-9 to 13-11 may be used for the estimation of heights of transfer units. The Bravo and Fair method (1982) has not been validated for the nondistillation cases. Finally, conversion between HETP and H_{OG}, and between N_{OG} and N_t, may be made by Equations 13-45 and 13-46, with recognition of the limitations of these equations.

D. Scaleup Procedure

The comments in Section III-H, dealing with the procedure for scaling up distillation columns, apply also to absorbers and strippers.

E. General Comments

The use of packings in commercial scale absorbers and strippers has been limited largely to cases where there is significant corrosion, and the traditional ring or saddle packings of ceramic materials have shown economic attractiveness. The premium cost of low-pressure drop, high-efficiency packings has usually not been justifiable, even for noncorrosive services. However, the use of packings in absorption and stripping service should not be ruled out, and there are situations where the throughflow elements can be attractive (e.g., flue gas scrubbing, foaming systems).

There are many absorption or stripping cases where there is a liquid-phase chemical reaction between the solute and the solvent. This affects the driving force, and in some cases the value of y_A^* is approximately zero. It also affects the liquid-phase mass transfer coefficient k_L, often enhancing it and thus reducing the value of H_{OG} when there is significant liquid-phase resistance in the nonchemical reaction comparative case. The methods of this chapter are extended to cover this mass transfer effect by Sherwood et al. (1975). The matter of absorption with chemical reaction (or chemical reaction influence by mass transfer) is treated in depth in Chapter 6 on fluid–fluid reactors. A number of commercial absorption and stripping processes involving reactive solvents are described in the book by Kohl and Riesenfeld (1979).

V. LIQUID – LIQUID SYSTEMS — EXTRACTION

A. Major Issues

Extraction devices of the countercurrent type are typified by the packed column and the spray column (see Chapter 12 for a general description of extraction devices). Some of the mechanically aided devices, such as the Karr column and the rotating disk contactor, operate in a mode that approaches counterflow. As in the case of gas–liquid systems, the geometry and mode of operation of the selected device affect the approach to analysis. In this section, models will be developed for the packed extractor, as the closest approach to a true countercurrent device, and extensions to other devices will then be noted.

The major issues in extractor scaleup and design do not differ from those discussed previously for other types of systems. The mixture to be separated must be defined, the degree of separation must be specified, mutual solubility and other property data must be available, and models must be available for the determination of the dimensions of the extractor. A difficulty arises in connection with extraction in that the raffinate phase, which may be compared with the gas phase in absorption, may be either the light phase or the heavy phase. Further, it may be either the dispersed phase or the continuous phase. This point is made to remind the reader of the need for care in keeping the extraction nomenclature straight.

B. Nomenclature for Countercurrent Extraction

In order to use analogies with gas–liquid systems, a slightly different nomenclature will be used in this chapter, compared with Chapter 12. Specific terms that will reappear are listed below.

Mass flow of carrier liquid $= A$
Mass flow of solvent $= B$
Mass flow of solute $= C$
Mole fraction of solute in raffinate phase $= x_c$
Mole fraction of solute in extract phase $= y_c$
Mass fraction of solute in raffinate phase $= x'_c$
Mass fraction of solute in extract phase $= y'_c$
Mass fraction of solute in raffinate phase, solvent-free basis, $= X$
$= [C/(A + C)]_R$
Mass fraction of solute in extract phase, solvent-free basis, $= Y$
$= [C/(A + C)]_E$
Selectivity $= \beta_{CA} = (y'_c/y'_A)/(x'_c/x'_A)$
Equilibrium ratio for solute $= K_C = y'^*_c/x'_c$; also $K_C = y^*_c/x_c$

C. Prediction of Performance

There are several major considerations in the evaluation of the performance of a new design or an existing extraction column. First, there is the selection of the device. In the case of packed extractors, this means the selection of the packing material. As in the case of gas–liquid contactors, the available

LIQUID–LIQUID SYSTEMS — EXTRACTION

experience will influence packing selection. Also the wettability of the packing material is an important consideration. Coupled with packing selection is the maximum flow capacity of the column; this takes into account the holdup of the dispersed phase as well as the reliability of the methods for predicting the flood point. Finally, there is the capability of the packed bed to produce the required number of transfer units or theoretical stages; among other things this depends on which of the phases is dispersed.

In general, the same packings used for gas–liquid and vapor–liquid systems can be used for liquid–liquid systems. Tables 13-1 and 13-2 of this chapter give pertinent dimensions of typical packings. The packing should be wetted by the continuous phase, otherwise the dispersed phase will tend to coalesce on the surface and decrease the available area for mass transfer. In this connection, ceramic and stoneware materials are generally preferentially wetted by aqueous phases; plastic and carbon are generally wetted by organic phases; metal materials can be wetted by either aqueous or organic phases, and some preliminary testing is needed. It is important that the packing be initially "conditioned" by the phase that preferentially wets it.

The choice of which phase to disperse will often depend on practical considerations, such as the hazard of the phase in larger quantity in the column (often the continuous phase) and the ratio of phases in the feed (it is often desirable to disperse the minority phase). A more technical consideration is which dispersion mode provides the higher rate of mass transfer. Many studies have shown that a given solute will transfer at an overall rate from dispersed phase to continuous phase that differs significantly from its overall rate from continuous phase to dispersed phase.

The mechanism of countercurrent flow in packed extractors is important when scaleup from laboratory to commercial is undertaken. If it is accepted that for the flowing phases there is a characteristic drop size, the extent of coalescence and breakup will be governed by the shape and size of the voids within the packed beds. A very small packing size will, for example, offer more obstruction to dispersed phased flow and thus provide more opportunity for coalescence and breakup than a large packing size. Thus, scaling from a small column to a large column, with associated increase in packing size, introduces a decrease in mass transfer efficiency.

A general diagram of a packed extraction column is shown in Figure 13-14. In the case shown, the light phase is dispersed, and the principal interface is at the top of the column. Alternately, the heavy phase could be dispersed; this would place the disperser in the bed at the top and the interface would be at the bottom. For the case of center-fed columns, packing support, phase redistribution, and feed entry would be combined.

D. Phase Distribution

As for gas–liquid and vapor–liquid systems, good distribution of the two liquid phases to the packed bed is important for good mass transfer. The more

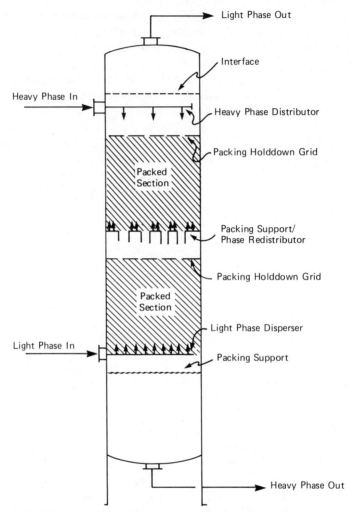

FIGURE 13-14 Flow diagram for a packed extraction column.

critical distribution problem is with the dispersed phase, where drops are formed by perforations that are normally in the size range of $\frac{1}{8}$-to $\frac{1}{4}$-in. diameter. Two commercially available distributors are shown in Figure 13-15. The device at the left is shown in position for dispersing the light phase; when inverted it disperses the heavy phase. It also serves as a bed support and redistributor (see Figure 13-14, center of column). The device at the right is a simple perforated pipe arrangement and can be inserted into the bed to distribute the dispersed phase, as indicated at the bottom of the bed in Figure 13-14. The devices in Figure 13-15 are suggestive of others that can be envisioned for phase distribution.

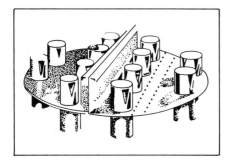
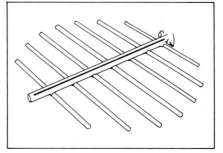

Orifice/Riser   Perforated Pipe

FIGURE 13-15 Two types of distributors for packed extraction columns.

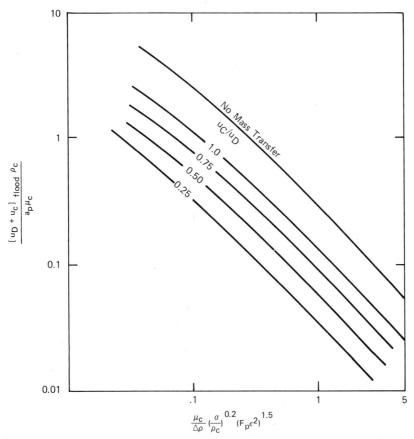

u_C = Continuous phase superficial velocity, m/s
u_D = Dispersed phase superficial velocity, m/s
ρ_C = Continuous phase density, kg/m^3
$\Delta\rho$ = Phase density difference, kg/m^3
F_p = Packing factor, m^{-1}
σ = Interfacial tension, mN/m (dynes/cm)
μ_C = Continuous phase viscosity, mPa s (centipoise)
a_p = Specific surface area of packing, m^{-1}
ϵ = Void fraction of packing

FIGURE 13-16 Generalized correlation for packed extractor flooding.

E. Maximum Capacity (Flooding)

A packed extractor can flood in much the same manner that a packed gas–liquid contactor floods. An excess flow rate of either phase can cause flooding, and correlations for extractors are often based on the sum of the two phase flow rates, as noted in Chapter 12.

A number of correlations of flooding in packed extraction columns have been published. None is considered particularly reliable for scaleup because of the small size of the columns for which flooding data have been measured. The most recent correlation is that of Eckert (1976) and Nemunaitis et al. (1971), shown in Figure 13-16. The correlation is an adaptation of an earlier one by Crawford and Wilke (1951) and points up the limitations in capacity caused by the presence of mass transfer. Thus, tests without a transferring solute show higher phase flow capacities than can be realized when a separation is being made. The graph of Figure 13-16 is entered from the abscissa scale, and for the evaluation of the abscissa parameter a combination of the packing factor F_p and the void fraction ε is needed. The appropriate terms are given in Table 13-1. Scatter in the basic data indicate that a safe approach to the flood point estimated should be no greater than 30 to 40 percent.

As might be expected, there is a "loading zone" for packed extractors, as in the case for packed absorbers or distillation columns (see Figure 13-6). In this zone there is an accumulation of dispersed phase, and the resulting coalescence has a detrimental effect on mass transfer rate. A low design approach to the estimated flood point is critical to good design, based on the present state of the art.

F. Mass Transfer Efficiency

The required height of the contacting zone may be estimated from either of the relationships

$$Z_T = N_t \cdot \text{HETS} \tag{13-52}$$

$$Z_T = N_{\text{OE}} \cdot H_{\text{OE}} = N_{\text{OR}} \cdot H_{\text{OR}} \tag{13-53}$$

where

HETS = height equivalent to a theoretical stage
N_t = number of theoretical stages required
N_{OE} = number of overall transfer units, based on extract phase concentrations
N_{OR} = number of overall transfer units, based on raffinate phase concentrations
H_{OE} = height of an overall extract phase transfer unit
H_{OR} = height of an overall raffinate phase transfer unit

LIQUID – LIQUID SYSTEMS — EXTRACTION

The analogy to gas–liquid mass transfer should be obvious. For absorption, the gas phase corresponds to the raffinate phase in extraction, at least in simple extraction systems. These equations do not deal with dispersed versus continuous phase designations; such is not necessary in the development of mass transfer expressions.

For the determination of transfer units, an analogy to Equation 13-20 is used:

$$N_{OR} = \int_{x_{c2}}^{x_{c1}} \frac{(1 - x_c)_m \, dx_c}{(1 - x_c)(x_c - x_c^*)} \tag{13-54}$$

where

x_c = mole fraction of solute C in the raffinate phase
x_c^* = mole fraction of C in the raffinate phase, that would be in equilibrium with the bulk concentration of C in the extract phase, $x_c^* = y_c/K_c$
$(1 - x_c)_m$ = logarithmic average of the nonsolute concentrations in the raffinate phase; for most instances,

$$(1 - x_c)_m \simeq \frac{(1 - x_c^*) + (1 - x_c)}{2} \tag{13-55}$$

For the case of a straight equilibrium line, the analog of Equation 13-50 is

$$N_{OR} = \int_{x_{c2}}^{x_{c1}} \frac{dx_c}{x_c - x_c^*} + \frac{1}{2} \ln \frac{1 - x_{c2}}{1 - x_{c1}} \tag{13-56}$$

If weight fractions are used,

$$N_{OR} = \int_{x'_{c2}}^{x'_{c1}} \frac{dx'_c}{x'_c - x'^*_c} + \frac{1}{2} \ln \frac{1 + rx'_{c2}}{1 + rx'_{c1}} \tag{13-57}$$

where

$$r = (M_{AB}/M_c)_R = \frac{(\text{mol. wt. nonsolute in raffinate})}{(\text{mol. wt. solute})} \tag{13-58}$$

The height of an overall transfer unit, based on raffinate phase concentrations is (in meters),

$$H_{OR} = \frac{R_M}{K_{OR,c} a_v (1 - x_c)_m} \tag{13-59}$$

where

R_M = rate of raffinate flow, kg mol/(s m² cross-sectional area of column)
$K_{OR,c}$ = overall mass transfer coefficient for solute C, kg mol/s m² mole fraction
a_v = interfacial area, m²

and other terms are as defined in Equations 13-54 and 13-55. The value of H_{OR} may also obtained from the individual phase terms

$$H_{OR} = H_R + \frac{R_M}{m_e E_M} H_E \qquad (13\text{-}60)$$

where

H_R and H_E = heights of transfer units for the raffinate and extract phases, m
E_M = rate of extract flow, kg mol/sec m²
m_e = slope of equilibrium line ($= dy_c^*/dx_c$)

Equations 13-54 through 13-60 may also be written for the extract phase, with appropriate substitution of y for x, and so on. This would be appropriate if the controlling resistance to mass transfer resided in the extract phase. However, as will be seen, the mass transfer technology for packed column scaleup and design is not far enough advanced to justify such refinements.

Unlike the case for gas–liquid and vapor–liquid systems, generalized correlations for use in Equations 13-52 and 13-53 are not available for packed columns. Representative data will be given in the next few paragraphs, in order that some perspective might be gained, but in general it will be necessary for scaleup experiments to be made. In the scaleup process, cautions as to changes in packing type or size, noted earlier, should be kept in mind.

Nemunaitis et al. (1971) reported efficiency data on a 0.46-m (18-in.) diameter column packed with 25-mm ceramic Raschig rings, 25-mm copper Pall rings, and 25-mm ceramic INTALOX® saddles. The system was kerosine–methylethylketone (MEK)–water, with the MEK as solute and the organic phase dispersed. Typical efficiency data are shown in Figure 13-17. For a packed height of 1.5 m the efficiency of the INTALOX® saddles and the PALL® rings was about the same, the PALL® rings being slightly more efficient. The Raschig rings were some 80 to 85 percent as efficient as the other packings. The aqueous phase was continuous, and the ceramic materials were appropriate for this situation. Apparently the metal PALL® rings were adequately wetted by the continuous phase.

Of more significance in the Nemunaitis work was the confirmation that much of the total mass transfer occurs during initial drop formation at the disperser. Figure 13-18 shows the column height effect, as reported by Eckert

LIQUID – LIQUID SYSTEMS — EXTRACTION

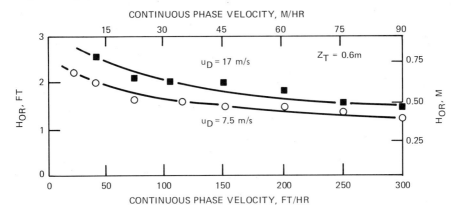
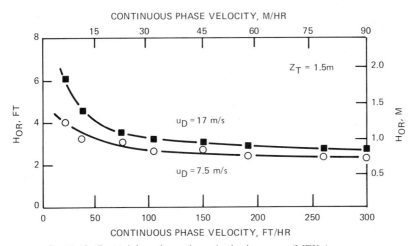

FIGURE 13-17 Heights of transfer units for kerosene/MEK/water system.

(1976), with comparisons with an empty column, operating as a simple spray column. At this scale of operation, it would appear that the packing characteristics play a relatively minor role in the mass transfer process. Nemunaitis et al. (1971) recommend a minimum packing size of 13 mm for scaleup purposes; below this size there is interference with drop flow as mentioned earlier.

Smith and Beckmann (1958) studied packed column mass transfer with binary systems of partial miscibility, in order to arrive at transfer unit values for the individual phases. The systems were MEK/water and methyl isobutyl carbinol (MIBC)/water. A 0.10-m (4-in.) column was used, packed to 0.91 m

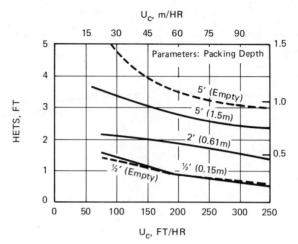

FIGURE 13-18 Effect of packed height on packing efficiency, kerosene/MEK/water system.

height with 13-mm ceramic Raschig rings. The data were correlated as follows:

$$H_C = 0.25\left(\frac{u_C}{u_D}\right)^{0.63}, \text{m} \quad \text{(MEK/water)} \quad (13\text{-}61)$$

$$H_D = 0.34 \text{ m} \quad (13\text{-}62)$$

$$H_C = 0.27\left(\frac{u_C}{u_D}\right)^{0.78}, \text{m} \quad \text{(MIBC/water)} \quad (13\text{-}63)$$

$$H_D = 0.44 \text{ m} \quad (13\text{-}64)$$

In these expressions the subscript C refers to the water (continuous) phase and subscript D to the organic (dispersed) phase. Either phase served as the raffinate phase, depending upon direction of mass transfer. Equations 13-61 and 13-62 give a reasonable approximation to the results of Nemunaitis et al. (1971) when packed height and packing size are taken into account.

Another study of packed columns, large enough to be considered useful for scaleup, was made by Leibson and Beckmann (1953), using the system toluene–diethylamine (solute)–water. Data for several sizes of ceramic Raschig rings, and for a 0.15-m (6-inch) column diameter, are shown in Figure 13-19. These data are for transfer of solute from the water phase, and thus the subscript for the raffinate phase may be used. The organic phase was the dispersed phase, and an estimate by Treybal (1963) gives $m_e = 0.74$.

It is possible to improve the mass transfer efficiency of a packed column by pulsing it with a suitable reciprocating pump, and a number of studies of pulsed packed and tray columns have been reported. Typical of these is the work by Brandt et al. (1978), shown in Figure 13-20. The value of HETS is

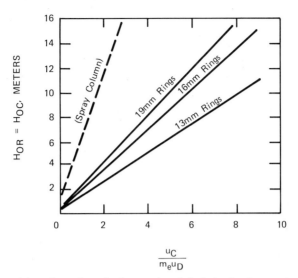

FIGURE 13-19 Heights of transfer units for transfer of diethylamine from water to toluene in a 6-in. column packed with ceramic Raschig rings.

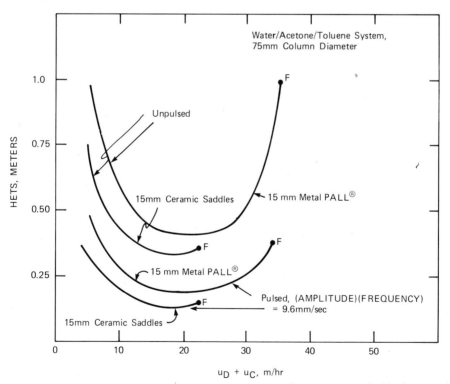

FIGURE 13-20 Effect of pulsing on mass transfer efficiency of a packed bed.

543

approximately cut in half, for the conditions studied. For a more complete report on the effects of pulse conditions on mass transfer and flooding in packed extractors, the paper by Bender et al. (1981) should be consulted. A related technique for improving the efficiency of packed extractors is controlled cycling, in which the raffinate and extract flows are alternated in an on–off manner. Laboratory data for such a mode of operation have been reported by Szabo et al. (1964), but there have been no reports on large columns with controlled cycling.

Removal of the packing, but with maintenance of the same liquid levels, gives the so-called spray column, in which the dispersed drops rise (or fall) through the continuous phase. While many studies of spray columns have been reported, the larger amount of axial mixing in the continuous phase makes them quite inefficient at large scales, and they are generally not used commercially.

G. Scaleup Procedure

A procedure similar to that given in Section III-H should be used for packed liquid–liquid extraction columns. The required theoretical stages or transfer units should be determined, either in a calibrated laboratory column or by calculation, if the system is well defined and amenable to analytical treatment. For the laboratory column, the York–Scheibel column, described in Chapter 12, can give a close approximation to the theoretical stage count, since it can be operated very close to 100 percent efficiency. Alternately, laboratory mixer–settler units can be used, but under the same phase ratio as planned for the commercial unit.

The efficiency of a laboratory packed column can then be determined, and the minimum size for reliable scaleup is 100-mm (4-in.) diameter, in order that a minimum packing size of 13 mm can be used and still maintain the desired 8:1 column diameter: packing nominal size ratio.

Because of the relatively poor state of the art in packed column scaleup and design, conservative approaches are always recommended. In this chapter, relatively little has been said about the effects of axial mixing of the phases; under high loading conditions, this can detract from efficiency, and the modest approach to flood suggested earlier (30 to 40 percent approach) should be used.

H. General Comments

At this point it should be clear to the reader that the technology for design of large countercurrent extractors is not nearly as well developed as for the equivalent devices in gas–liquid and vapor–liquid service. This is generally because extraction is not as well explored, and hence not as much used, as absorption, stripping, and distillation as a commercial separation process. Chapter 12 includes comments on the potential energy conservation advantages of extraction over distillation, in certain applications. One would

NOMENCLATURE

a_p	Specific surface area of packing, m²/m³
a_v	Interfacial area, m²/m³
c_1, c_2	Constants in Equation 13-23
C_A	Concentration of species A in liquid, kg mol/m³
C_{Ai}	Concentration at the interface
C_f	Flooding correction factor for packings (Figure 13-11)
CP	Capacity parameter (Equation 13-33)
D_T	Tower diameter, m
$\mathscr{D}_g$	Diffusion coefficient, gas phase, m²/s
$\mathscr{D}_L$	Diffusion coefficient, liquid phase, m²/s
E_M	Extract flow rate, kg mol/s m²
F_p	Packing factor, m⁻¹
FP	Flow parameter (Equation 13-32)
g	Gravitational constant, 9.81 m/s²
G_M	Gas or vapor rate, kg mol/s m²
H	Height of a transfer unit, m
H_E	Height of a transfer unit, extract phase
H_G	Height of a transfer unit, gas phase
H_L	Height of a transfer unit, liquid phase
H_{OE}	Height of an overall transfer unit, extract basis
H_{OG}	Height of an overall transfer unit, gas basis
H_{OR}	Height of an overall transfer unit, raffinate basis
H_R	Height of a transfer unit, raffinate phase
HETP	Height equivalent to a theoretical plate, m
HETS	Height equivalent to a theoretical stage, m
H_A	Henry's law coefficient for species A, atm/mole fraction
k_{gA}	Mass transfer coefficient for species A in gas phase, kg mol/s m² atm
k_{LA}	Mass transfer coefficient for species A in liquid phase, m/s
K_A	Equilibrium ratio for species A $(= y_A^*/x_A)$
K_c	Equilibrium ratio for solute C, liquid–liquid system, $(= y_c^*/x_c)$
L	Liquid rate, kg mol/s
L'	Liquid rate, kg/s

L_M	Liquid rate, kg mol/s m^2
m_e	Slope of equilibrium line
M	Molecular weight
M_{AB}	Molecular weight of nonsolute species
M_c	Molecular weight of solute C
N_A	Molar flux for species A, kg mol/s m^2
N	Number of transfer units
N_G	Number of gas phase transfer units
N_L	Number of liquid phase transfer units
N_{OE}	Number of overall transfer units, extract basis
N_{OG}	Number of overall transfer units, gas basis
N_{OR}	Number of overall transfer units, raffinate basis
N_t	Number of theoretical plates or stages
p_A	Partial pressure of species A, atm
p_{Ai}	Partial pressure at interface
P	Total pressure, atm
r	Ratio for converting to mass basis (Equation 13-58)
R_M	Raffinate rate, kg mol/s m^2
S	Cross-sectional area, m^2
Sc	Dimensionless Schmidt number ($= \mu/\rho \mathscr{D}$)
Sc$_G$	Schmidt number for gas
Sc$_L$	Schmidt number for liquid
u_C	Superficial velocity of continuous phase, m/s or m/h
u_D	Superficial velocity of dispersed phase, m/s or m/h
V	Vapor or gas rate, kg mol/s
x	Mole fraction in liquid or in raffinate
x_A	Mole fraction of species A in liquid
x_{Ai}	Mole fraction of A in liquid, at interface
x_A^*	Mole fraction of A in liquid, at equilibrium
x_{BM}	Mean value of mole fraction of nondiffusing species (Equation 13-22)
x_c	Mole fraction of solute C in raffinate
x_{ci}	Mole fraction of C in raffinate, at interface
x_c^*	Mole fraction of C in raffinate, at equilibrium
x'	Mass fraction in raffinate
x_c'	Mass fraction of solute C in raffinate
y	Mole fraction in vapor or in extract
y_A	Mole fraction of species A in vapor
y_{Ai}	Mole fraction of A in vapor, at interface

y_A^*	Mole fraction of A in vapor, at equilibrium
y_{Bm}, y_{Bm}^*	Mean values of mole fraction of nondiffusing species (Equation 13-22)
y_c	Mole fraction of solute C in extract
y_{ci}	Mole fraction of C in extract, at interface
y_c^*	Mole fraction of C in extract, at equilibrium
y'	Mass fraction in extract
y_c'	Mass fraction of solute C in extract
Z	Height, m
Z_T	Total height of packed zone, m

Subscripts

A	Diffusing species
B	Nondiffusing species
C	Dispersed phase
E	Extract phase
G or g	Gas phase
L	Liquid phase
m	Mean
M	Molar basis
R	Raffinate phase

Greek

α_{AB}	Relative volatility between species A and species B
β_{AC}	Separation factor in extraction, species A to species B
$\Delta\rho$	Phase density difference, kg/m³
λ	Ratio of the slopes of the equilibrium line and operating lines
μ	Viscosity, mPa s (or centipoises)
ρ	Density, kg/m³ ($\bar{\rho}$ = average density)
ρ_M	Molar density, kg mol/m³
σ	Surface or interfacial tension, mN/m (or dyn/cm)
Ψ	Density ratio, water to liquid being processed

REFERENCES

Bender, E., Berger, P., Leuckel, W., and Wolf, D., *Int. Chem. Eng.*, **21**(1) 29 (1981).

Billet, R., *Chem. Eng. Progr.*, **63**(9) 53 (1967).

Bolles, W. L. and Fair, J. R., *I. Chem. E. Symp. Ser.*, (London) No. 56, Sect. 3.3, 35 (1979).

Bolles, W. L., and Fair J. R., *Chem. Eng.*, **89**(14) 109 (July 12, 1982).

Brandt, H. W., Reissinger, K. -H., and Schroeter, J., *Chem. -Ing. -Tech.*, **50**, 345 (1978).

Bravo, J., and Fair, J. R., *Ind. Eng. Chem., Proc. Design Devel*, **21**, 162 (1982).

Chan, H., and Fair, J. R., *Ind. Eng. Chem., Proc. Design Devel*, **23**, 820 (1983).

Cornell, D., Knapp, W. G., Close, H. J., and Fair, J. R., *Chem. Eng. Progr.*, **56**(8), 48 (1960).

Crawford, J. W., and Wilke, C. R., *Chem. Eng. Progr.*, **47**(8), 423 (1951).

Eckert, J. S., *Chem. Eng. Progr.*, **66**(3), 39 (1970).

Eckert, J. S., *Hydrocarbon Processing*, **55**(3), 117 (1976).

Fair, J. R., *Chem. Eng. Progr.*, **66**(3), 45 (1970).

Kohl, A., and Riesenfeld, F., *Gas Purification*, 3rd ed., Gulf Publ. Co., Houston, 1979.

Leibson, I., and Beckmann, R. B., *Chem. Eng. Progr.*, **49**(8), 405 (1953).

Nemunaitis, R. R., Eckert, J. S., Foote, E. H., and Rollison, L. R., *Chem. Eng. Progr.*, **67**(11), 60 (1971).

Sherwood, T. K., Pigford, R. L., and Wilke, C. R., *Mass Transfer*, McGraw-Hill, New York, 1975.

Sherwood, T. K., Shipley, G. H., and Holloway, F. A. L., *Ind. Eng. Chem.*, **30**, 765 (1938).

Smith, G. C., and Beckmann, R. B., *AIChE J.*, **4**, 180 (1958).

Szabo, T. T., Lloyd, W. A., Cannon, M. R., and Speaker, S. S., *Chem. Eng. Progr.*, **60**(1), 66 (1964).

Treybal, R. E., *Liquid Extraction*, 2nd ed., McGraw-Hill, New York, 1963.

Wiegand, J. H., *Trans. AIChE*, **36**, 679 (1940).

Williamson, G. J., *Trans. Instn. Chem. Engrs.*, **29**, 215 (1951).

14

SOLID – LIQUID SEPARATION PROCESSES

L. SVAROVSKY

I.	Major Issues in Solid–Liquid Separation Processes		550
	A. Efficiency of Solids and Liquid Separation		551
	B. Cake Washing		553
	C. Dewatering		554
	D. Solid–Solid Separation		554
	E. Pretreatment of Suspensions		555
II.	Fundamental Considerations		556
	A. Description of the Particle(s)		556
		1. Selection of Relevant Characteristic Particle Size	556
		2. Description of Size Distribution and Mean Size	556
		3. Particle Shape and Density	557
		4. Other Particle Properties	558
	B. Coagulation and Flocculation		558
		1. Perikinetic Flocculation	558
		2. Orthokinetic Flocculation	558
	C. Interaction of Particles and Fluids		559
		1. Gravity Settling at Low Concentrations	559
		2. Hindered Settling at High Concentrations	560
		3. Centrifugal Sedimentation at Low Concentrations	562
		4. Centrifugal Sedimentation at High Concentrations	562

		D. Flow Through Packed Beds	563
		1. Limitations of the Carman–Kozeny Equation	563
		2. The Basic Cake Filtration Equation—Darcy's Law	564
		3. Separation in Deep Beds	565

III. Solid–Liquid Separation Equipment   565
    A. Gravity Settling Tanks and Thickeners   565
    B. Centrifuges and Hydrocyclones   566
    C. Flotation   568
    D. Surface Filters—Cake Filtration   569
    E. Deep Bed Filters   571

IV. Small-Scale Studies for Equipment Design and Selection   572
    A. Gravity Settling Tanks and Thickeners   572
    B. Centrifuges and Hydrocyclones   573
    C. Flotation   574
    D. Cake Filters and Dewatering Screens   574
    E. Deep Bed Filters   575

V. Scaleup Techniques   576
    A. Gravity Settling Tanks and Thickeners   576
    B. Hydrocyclones   578
        1. Dimensional Analysis at Low Solids Concentrations   578
        2. Dimensional Analysis at High Solids Concentrations   581
        3. Grade Efficiency Curves   583
    C. Sedimenting Centrifuges   583
    D. Flotation   586
    E. Cake Filters and Dewatering Screens   587
    F. Deep Bed Filters   589

VI. Uncertainties   590
    A. State of Flocculation and Pretreatment   590
    B. End Effects in Equipment   590
    C. Deviations from Stokes' Law in Particle Settling   590
    D. Compressibility of Cakes   590
    E. Reentrainment   591

Nomenclature   591
References   593

I. MAJOR ISSUES IN SOLID – LIQUID SEPARATION PROCESSES

As the name suggests, solid–liquid separation is the separation of two phases, solid and liquid, from a suspension. The processes available for carrying out this operation are often referred to as "mechanical separation processes" because the separation is accomplished by purely physical means. This does not preclude chemical or thermal pretreatment which is increasingly used to enhance the separation that follows. A summary of the main classes of

MAJOR ISSUES IN SOLID–LIQUID SEPARATION PROCESSES

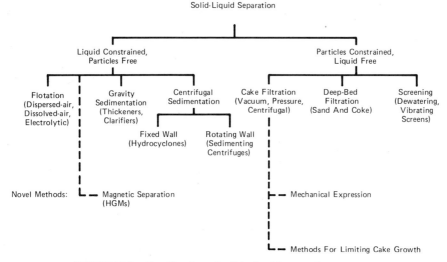

FIGURE 14-1 Classification of solid–liquid separation processes.

equipment available for solid–liquid separation is given in Figure 14-1; some novel methods undergoing rapid development are also shown in the figure.

A. Efficiency of Solids and Liquid Separation

A solid–liquid separation is never complete. As can be seen from the schematic diagram of a separator in Figure 14-2, there may be some solids leaving in the liquid (overflow) stream and some of the liquid will inevitably leave with the bulk of the solids through the underflow. In order to be able to compare the performance of different separators there has to be some way to measure this imperfection of separation.

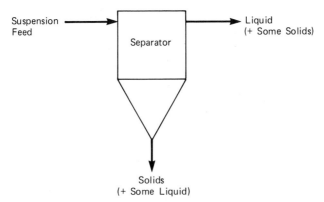

FIGURE 14-2 A schematic diagram of a separator.

The most common way to assess the degree or efficiency of separation is by considering the solids and the liquid separately in two independent factors. One is the mass fraction of the solids recovered, often called the separation efficiency or solids retention (this does not take the liquid into account), and the other is the moisture content or concentration of the recovered solids, which is a measure of imperfections in the separation of the liquid. The relative importance of the two factors varies from application to application. Both factors represent major scaleup issues.

The efficiency of solids separation is best expressed as a grade efficiency curve, shown in Figure 14-3, since performance of most separators is highly size dependent. This concept of grade efficiency is particularly useful for hydrocyclones, centrifuges, or settling tanks. The curve is constant for a particular set of operating conditions and relatively independent of the size distribution of the feed solids. With filters, the grade efficiency curve is strongly influenced by the quantity of the solids accumulated on or inside the filter medium but its position and shape for a clean medium is useful in filter rating.

Details about the properties and the use of grade efficiency curves can be found in Chapter 3 of Svarovsky (1981). If a single number is needed to replace the grade efficiency curve, the size corresponding to a 50 percent grade efficiency (x_{50}) on the curve is often used as the "cut" size. This is often done with sedimenting centrifuges and hydrocyclones. This concept of "cut size" is extremely useful in scaleup because it allows comparison of different machines

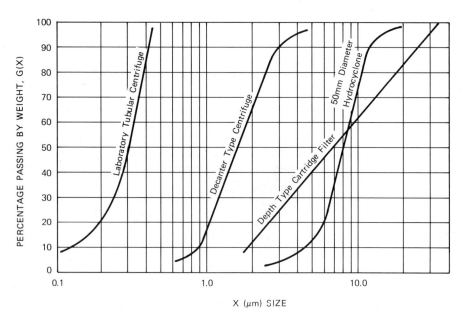

FIGURE 14-3 Grade efficiency curves of various equipment.

MAJOR ISSUES IN SOLID-LIQUID SEPARATION PROCESSES

on the basis of a number which is largely independent of the feed solids content. Scaleup of hydrocyclones or sedimenting centrifuges based on cut size is discussed in Section V.

The second criterion of separation performance is how much "misplaced" liquid, that is, the liquid separating to the underflow, has to be accepted with the separated solids. In filtration this is usually expressed as mass ratio either of moisture to dry cake or in any similar way. With hydrocyclones, sedimenting centrifuges, and settling tanks, where much more water leaves with the solids than in filtration, mass or volume concentration of solids in the underflow is used. Whichever is the way to describe the affect, the dryness of the separated solids is an important and often neglected issue in scaleup.

Sometimes there is a need for a combined, single criterion of efficiency in which separation of the solids and the liquid are considered simultaneously. Several different criteria exist; these are fully reviewed in Svarovsky (1979).

Separation efficiencies are obviously the basis of scaleup if the large-scale equipment must produce the same performance as was obtained on a small scale. However, many scaleup procedures are based solely on bench scale tests. Then the procedure must also include equipment selection. Sizing of the surface filters, for example, may be based on filter leaf tests. However, the large-scale filter must be chosen from a wide variety of commercial surface filters available today.

While equipment selection to give the desired performance characteristics under the proposed operating conditions is critical, we will concern ourselves almost exclusively with scaleup for specific types of equipment assuming the choice has already been made.

B. Cake Washing

Washing is used to replace the mother liquor in the solids stream, usually in a cake form, with a wash liquid. The growing importance of washing is determined by the demands for increasing purity of most products on the market combined with the increasingly poorer quality of available raw materials. Since cake washing is often multistaged, it may represent a dominant fraction of the total installation cost. The quality of washing is characterized by the washing curve shown in Figure 14-4 where the ratio, x/x_0, of the instantaneous to the initial mother liquor concentration in the cake is plotted against the quantity of washing of liquid used. This is usually expressed as the wash ratio T, that is, the number of void volumes used. The step function in Figure 14-4 represents the ideal optimum washing obtained with plug displacement while the curve is typical for practical washing operations. Knowledge of the washing curve is critical to scaleup of solids washing.

Several models exist to describe and predict washing performance from the minimum of experimental measurements. Washing can be cocurrent or countercurrent. It often is done in several stages and it can be enhanced by cake

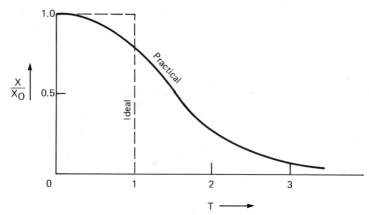

FIGURE 14-4 Typical washing curve.

compression as discussed in Chapter 15 of Svarovsky (1981). In some cases of high specific resistance of the cake and/or cake cracking, washing by reslurrying of the cake may be advantageous.

C. Dewatering

Dewatering is directed at reducing the moisture content of cakes. It is achieved either by mechanical compression, by air displacement under vacuum, by pressure, or by drainage in a gravitation or centrifugal system. In dewatering by mechanical compression, the major issue in scaleup is the compressibility of the cake, usually expressed by constants in an empirical exponential equation relating cake voidage to applied pressure.

In dewatering by air (or gas) displacement the important issue is the threshold pressure which has to be exceeded in order that gas may enter the filter cake. In addition, one must determine the irreducible saturation level which sets the lowest moisture content achievable by gas displacement and finally kinetic dewatering characteristics. Similarly, in dewatering by gravity or centrifugal action the irreducible saturation and kinetic dewatering characteristics must be known.

D. Solid – Solid Separation

The particle-size dependent nature of some solid–liquid separation equipment such as hydrocyclones or sedimenting centrifuges can be used for solid–solid separation (classification of the solids) in which either the coarse or the fine particles are removed from the product (degritting and desliming, respectively). Even when solids classification is not required it may be desirable before the solid–liquid separation stage so that the material in each different size range may be treated by the type of equipment best suited to it. The major issue in solid–solid separation is the steepness of the grade efficiency curve because this determines the amount of misplaced material (fines in the coarse product and coarse particles in the fine product). Scaleup of such solid–solid separation machines is obviously based on a cut size concept as discussed in Section I-A.

E. Pretreatment of Suspensions

Conditioning or pretreatment of the feed suspension is an important aspect of solid–liquid separation. The purpose of pretreatment is to alter some important property of the suspension so as to improve the performance of a separator that follows. A suspension can be pretreated by:

Addition of inert filter aids, as body feed or precoat, or both, in order to improve filterability of the resulting cake in filtration. Pretreat materials include diatomaceus earth, expanded perilitic rock, cellulose, nonactivated carbon, ashes, and ground chalk. Using filter aids as precoats on a relatively open filter medium reduces penetration of fine solids into or through the filter medium. They also facilitate easier cleaning of the filter medium. As a body feed, the filter aid is continuously added to the feed suspension. Since the proportion of the filter aid needed is generally around 1 : 1, this is only feasible when relatively low solids concentrations are being filtered.

Flocculation or coagulation can be carried out using mineral electrolytes or polyelectrolytes (synthetic organic polymers). This increases the effective particle size improving both sedimentation rate and the permeability of filter cakes in equipment such as surface and depth filters, thickeners, or sedimenting centrifuges. The basic principle involved in chemical pretreatment is neutralization of surface charges on particles, reducing interparticle repulsion.

Mineral coagulants such as alum or lime produce particles up to 1 mm in size. Synthetic polyelectrolytes (flocculation agents) flocculate colloidal particles into giant, often interconnected flocs. Synthetic polyelectrolytes have undergone fast development in the past decade and this led to a remarkable improvement in the use of many types of separation equipment. Since these agents are relatively expensive using the correct dosage is critical and has to be carefully optimized. Overdosages are not merely uneconomic but they may inhibit the flocculation process, causing operating problems such as blinding of filter media or mud balling and underdrain constriction in sand filters.

As surface charges are also affected by pH, control of pH is essential with synthetic polyelectrolytes. While reduction or elimination of the repulsive barrier is a necessary prerequisite of successful flocculation, the perikinetic flocculation that brings particles together through Brownian motion is often assisted by orthokinetic flocculation creating particle collisions through the fluid motion. Scaleup of "orthokinetic" flocculators, which are most often some form of paddle devices, is based on the product of mean velocity gradient and time (for a constant volume concentration of the flocculating particles). The concepts are discussed further in Section II-B.

Entirely physical pretreatment processes such as crystallization, freezing, temperature adjustment, thermal treatment, and aging have been used. Each pretreatment method should be considered jointly with the type of equipment being considered; it should be included in the basic scaleup tests.

II. FUNDAMENTAL CONSIDERATIONS

A. Description of the Particle(s)

1. Selection of Relevant Characteristic Particle Size

An irregular particle can be described by a number of sizes; Svarovsky (1981) has grouped the definitions as follows:

Equivalent sphere diameters.
Equivalent circle diameters.
Statistical diameters.

An equivalent sphere diameter is the diameter of a sphere that would have the same property as the particle itself (e.g., the same volume, the same settling velocity, etc.). Equivalent circle diameters are the diameters of a circle having the same property as the projected outline of the particles. Statistical diameters are obtained when a linear dimension is measured (by microscopy) parallel to a fixed direction.

There are a wide variety of methods for particle size measurement; each measuring different sizes. When selecting a method, it is best to take one that measures that size which is most relevant to the property or the process that is to be controlled. For example, in hydrocyclones, sedimenting centrifuges, or gravity settling tanks, it is most relevant to use one of the sedimentation methods which measure the Stokes diameter. The Stokes diameter is the diameter of a sphere of the same density as the particle itself and, the particles, assuming Stokes' law, would fall in the liquid at the same velocity. In filtration, on the other hand, it is the surface–volume diameter (or diameter of a sphere having the same surface-to-volume ratio as the particle) which is most relevant to the separation mechanisms.

2. Description of Size Distribution and Mean Size

Very few if any particulate systems are monosized. Most show a distribution of sizes; depending on what properties are measured the distribution can be by number, surface, or mass. While a conversion from one size characterization to another is theoretically possible, this requires assuming a constant shape factor through the distribution. This is not true in many cases. Conversions are therefore to be avoided whenever possible by choosing a measurement method which measures the desired distribution directly. Except for a few specialized applications, the most relevant distribution for solid–liquid separation is usually the mass distribution.

If a population of particles is to be represented by a single number there are many different measures of central tendency or "mean" sizes as discussed in Chapter 2 of Svarovsky (1981). As to which is to be chosen to represent the population, once again this depends on the application and namely what property is of importance and should be represented. Details and examples of presentation of size distributions and evaluation of different mean sizes can also be found in Svarovsky.

3. Particle Shape and Density

Particle shape affects many of the secondary properties relevant to solid–liquid separation such as the particle–fluid interaction and packing. Shape can be characterized by factors defined quite differently; as with definitions of particle size, the choice of a shape factor is made in items of relevance to the application in question. The most frequently used shape factors are:

Sphericity—ratio of the surface area of a sphere having the same volume as the particle to actual particle area; the reciprocal is known as the coefficient of rugosity or angularity.

Circularity—ratio of the perimeter of a circle having the same area as the projected area of the particle to the actual particle perimeter.

Surface shape—coefficient of proportionality relating the surface area of the particle to the square of its measured diameter.

Volume shape—coefficient of proportionality relating the volume of the particle to the cube of its measured diameter.

Surface–volume shape—ratio of surface to volume shape coefficients.

Other shape factors can be defined as ratios of two different measurements of particle size as obtained from comparison of particle size distributions. The shape factor is then the multiplier which brings the results into agreement. For example, shape factors relating Stokes diameter to the sieve diameter or to the equivalent volume diameter as measured by Coulter Counter can be defined.

Particle density is of great importance in separation by gravity or centrifugal sedimentation. Those processes work if there is a finite density difference between the solid density of the particles, ρ_s, and the liquid density, ρ_L. The relevant solids density in such cases is the density measured by wet pycnometry. Particle density influences solid–liquid separation processes such as deep bed filtration or centrifugal filtration through its effects on particle inertia and settling behavior.

4. Other Particle Properties

Other particle properties can also influence the separation processes. Particle rigidity and strength affects the compressibility of cakes. However, little is known quantitatively about the nature of the relationships. The angle of friction between the particle and the filter medium is relevant for applications such as crossflow filtration where the particle medium friction controls cake discharge. Limited studies have been made on tests for failure properties of wet cakes similar to tests now standard in testing dry powders.

B. Coagulation and Flocculation

Coagulation and flocculation are used to increase the effective particle size improving thereby the settling rates in sedimentation process or the permeability and drainage of filter cakes. Although the terms coagulation and flocculation are often used interchangeably, the processes are different. Coagulation is the agglomeration of primary particles into particles up to 1 mm in size. Flocculation involves not only the agglomeration of particles but also their interconnection into giant loose flocs up to 1 cm in size. Flocculating agents create favorable conditions for floc formation through neutralization of surface changes and the reduction of interparticle repulsion. Flocculation occurs through particle–particle collisions. Depending on the mechanism involved in the collisions, the process is divided into perikinetic and orthokinetic flocculation.

1. Perikinetic Flocculation

Perikinetic flocculation is the first stage of floc formation induced by the Brownian motion. This is a second order process quickly diminishing with time and largely completed in a few seconds. The higher the initial concentration of the solids the faster is the rate of flocculation.

Theories of colloid stability attribute flocculation to a change in the balance between the van der Waals attractive forces and the repulsive electric double layer forces at the liquid–solid interface. Bridging flocculation in which polymer molecules are adsorbed on more than one particle is due to charge effects, van der Waals forces, or hydrogen bonding. The rate of flocculation can be determined from the Smoluchowski rate law which states that the rate is proportional to the square of the particle concentration, inversely proportional to the fluid viscosity, and independent of particle size.

2. Orthokinetic Flocculation

Orthokinetic flocculation is induced by liquid motion which produces shear within the suspension. Orthokinetic flocculation leads to particle size growth being a function of both shear rate and particle concentration. Large-scale one-pass water clarifiers use orthokinetic flocculators prior to introducing the suspension into a settling tank [Purchas (1977)].

Flocculation can also result from particle–particle collision caused by differential rates of settlement. This effect can be quite pronounced in large size plants where large particles in "fast fall" will capture small particles settling more slowly.

Flocculation can also result from the shrinkage and densification of loose and bulky flocs through application of fluctuating mechanical forces [Ide and Katoka (1979)]. This force can exude liquid from the floc. Pelletlike flocs can be produced by slow stirring of blanket zones in sludge blanket clarifiers by rotating paddles. As a result, higher overflow rates than are reached with conventional blanket clarifiers can be attained.

C. Interaction of Particles and Fluids

1. Gravity Settling at Low Concentrations

At concentrations below 0.5 volume percent particles are so far apart that they do not influence each other. Therefore their settling behavior is as if they were alone in an infinite expanse of the liquid. Depending on the value of the particle Reynolds number, Re_p, three different regimes of particle settling have been identified:

The laminar or Stokes regime.
The transition regime.
The fully turbulent or Newton region.

The upper limit of the Stokes region is usually taken as $Re_p = 0.2$ which corresponds to a 60 μm particle of density 2650 kg m^{-3} settling in water. Since in solid–liquid separations only very fine particles (which are most difficult to separate) are of concern, only the Stokes regime of particle settling need be considered. The terminal settling velocity U_g under gravity of a spherical particle of diameter x is determined from Stokes' law as follows:

$$U_g = \frac{x^2 \Delta \rho\, g}{18\mu} \qquad (14\text{-}1)$$

where

$\Delta \rho =$ the solid–liquid density difference
$\mu =$ liquid viscosity
$g =$ gravity acceleration

Equation 14-1 can be applied to nonspherical particles provided the size is obtained from sedimentation measurements as an equivalent Stokes' diameter.

There is also a bottom limit of particle size—usually about 1 μm—below which Stokes' law cannot be assumed to apply in gravity sedimentation. Brownian diffusion is significant below 1 μm resulting in settling rates lower than predicted by Stokes' law.

2. Hindered Settling at High Concentrations

As the concentration of the suspension increases particles get closer together and interfere with each other's motion. When the particles are not distributed uniformly the overall effect is a net increase in settling velocity since the return flow resulting from volume displacement will predominate in particle sparse regions. This is the now well-known cluster formation which is only significant in nearly monosized suspensions.

Widely dispersed suspensions clusters do not survive long enough to affect the settling behavior. Since the return flow is nearly uniformly distributed, the settling rate steadily declines with increasing concentration. This is referred to as hindered settling and can be quantitatively approached by:

A Stokes' law correction introducing a multiplying factor.

Adopting "effective" fluid properties for the suspension, different from those of the pure fluid.

A determination of bed expansion from a modified Carman–Kozeny equation.

These approaches yield essentially identical results:

$$\frac{U_p}{U_g} = \varepsilon^2 f(\varepsilon) \qquad (14\text{-}2)$$

where

U_p = hindered settling velocity of a particle
U_g = terminal settling velocity of a single particle as calculated from Stokes' law (Equation 14-1)
ε = volume fraction of the fluid (voidage)
$f(\varepsilon)$ = "voidage function"

The voidage function will have different forms depending on the approach adopted. However, the differences between the available expressions for $f(\varepsilon)$ are not great and are frequently within experimental accuracy. The most important forms are as follows:

Carman–Kozeny Equation

$$f(\varepsilon) = \frac{\varepsilon}{10(1-\varepsilon)} \qquad (14\text{-}3)$$

Brinkman's Theory Applied to Einstein's Viscosity Equation

$$f(\varepsilon) = \varepsilon^{2.5} \tag{14-4}$$

Richardson and Zaki Equation

$$f(\varepsilon) = \varepsilon^{2.65} \tag{14-5}$$

For irregular or nonrigid particles (e.g., flocs) the exponent in Equation 14-4 and 14-5 can be considerably larger than for spheres. Equations 14-3, 14-4, and 14-5 apply only to systems such as coarse mineral suspensions where flocculation is absent. Suspensions of fine particles, because of the very high specific surface of the particles, often flocculate, showing a different behavior. With increasing concentration such suspensions can develop an interface which becomes sharper as the concentration increases. The slurry is then said to be in the "zone settling region."

Particles below the interface, if the size range is not more than 6:1, settle "en masse." That is, the particles all settle at the same velocity irrespective of their size. There are two possible reasons for this: either the flocs become of roughly similar size and settle at the same velocity or the flocs are locked into a loose plastic structure and sweep down as a "web." However, the settling rates of the interface (and of the solids below it) for many practical suspensions can still be described by Equation 14-5. The value of U_g must be determined by extrapolation of an experimental log-linear plot of U_g vs ε to $\varepsilon = 1$. The value of this intercept can be used to make an indirect measurement of the floc size. The slope of the plot determines the value of the exponent in Equation 14-5.

The concentration at which zone settling can first be observed depends very much on the material and its state of flocculation. No guidance can be given on the value of the concentration at which this may occur. Addition of flocculation (or dispersing) agents can drastically change the concentration. Only experimental studies can yield its value.

At higher concentrations still the flocs can become significantly supported mechanically from underneath as well as hydraulically. The suspension is then said to be in compression or compression settling. Solids in compression continue to consolidate; the rate of consolidation depends not only on the concentration but also on the structure of the solids. This in turn depends on the pressure and flow conditions. The problem is closely related to cake filtration and expression. For intermediate concentrations between those of zone settling and fully established uniform compression, channeling can occur particularly in slowly raked large-scale thickeners. Under those conditions a coarser structure of pores becomes interconnected in the forms of channels.

Most studies on the consolidation process of solids in compression use the basic model of a porous medium with point contacts which gives a general mass and momentum balance. This must be supplemented by a model describing filtration and deformation properties. Probably the best model to date is that of Kos (1979) which uses two parameters to define the characteristic behavior of suspensions. This model can be potentially applied to the processes of sedimentation, thickening, cake filtration, and expression.

3. Centrifugal Sedimentation at Low Concentrations

The terminal settling velocity of a particle under centrifugal acceleration can be calculated from Stokes' law by substitution of Rw^2 into Equation 14-1 for g (where w is angular speed and R the radius of particle position). However, there is an important difference: particle motion is no longer at constant velocity since as particles move radially outward they are continuously accelerated and Rw^2 increases. However, just as in gravity settling, particle inertia can be neglected. The particles are assumed to move at their respective terminal velocities.

The upper size limit for applicability of Stokes' law to centrifugal sedimentation is reduced because of the acceleration involved. If separation factor G is defined as a number of g's,

$$G = \frac{Rw^2}{g} \tag{14-6}$$

then the top size limit of Stokes' law, X_c, for a centrifuge can be calculated from the limiting size under gravity, X_g, as follows

$$X_c = \frac{X_g}{\sqrt[3]{G}} \tag{14-7}$$

The distance covered by a particle under gravity acceleration in a given time is calculated simply as $U_g \times t$. Under centrifugal acceleration the terminal velocity expression

$$U_c = \frac{dR}{dt} = \frac{x^2 \Delta\rho\, Rw^2}{18\mu} \tag{14-8}$$

must be integrated from R_0 (the initial radius) to R (radius after time, t)

$$\frac{R(t)}{R_0} = \exp\left(\frac{x^2 \Delta\rho\, w^2 t}{18}\right) \tag{14-9}$$

The settling time of a given particle size, x, from R_0 to R can then be calculated from Equation 14-9.

4. Centrifugal Sedimentation at High Concentrations

Centrifugal sedimentation at higher concentrations is not as hindered as in the case of gravity sedimentation since the higher concentrations exist within the

FUNDAMENTAL CONSIDERATIONS

settling suspension only for a fraction of the total settling time rather than for the total time as under gravity. This is due to the continuous dilution effect; that is, as particles settle radially they spread out and the concentration at a given radius falls steadily with time. However, the settling is still hindered for some period of time and this must be taken into account since it increases the settling times. Baron and Wajc (1979) have concluded that settling in the centrifugal field leads to much the same behavior as in gravity settling. A visible interface between the supernatant liquid and the settling suspension does form and the concentration below the interface is the same down to the sediment level. However, the concentration is not constant but decreases with time because of the dilution effect, until the interface ceases to be visible. Total settling times are naturally much longer than those obtained at lower concentrations. Baron and Wajc's analysis uses the experimental settling flocs curve obtained under gravity. In principle the analysis permits scaleup of sedimenting centrifuges from gravity experiments. Further development in this area is anticipated.

D. Flow Through Packed Beds

1. Limitations of the Carman–Kozeny Equation

Flow through packed beds under laminar conditions can be described by the well-known Carman–Kozeny equation,

$$\frac{Q}{A} = \frac{\Delta p}{\mu L} \frac{\varepsilon^3}{C(1-\varepsilon)^2 S_0^2} \qquad (14\text{-}10)$$

where

- Q = volumetric flowrate
- A = face area of the bed
- L = depth of the bed
- Δp = applied pressure drop
- ε = voidage of the bed (porosity)
- S_0 = volume specific surface of the bed
- μ = liquid viscosity
- C = constant

The constant C depends on particle size, shape, and porosity; in low porosity ranges its value is about 5. Unfortunately while Equation 14-10 is a reasonable approximation for incompressible cakes over narrow porosity ranges, it cannot be used for compressible cakes. However, it does show the high sensitivity of pressure drop to cake porosity and to the specific surface of the solids.

2. The Basic Cake Filtration Equation — Darcy's Law

Darcy's equation combines the constants in Equation 14-10 into one factor, K, the permeability of the bed.

$$K = \frac{\varepsilon^3}{C(1-\varepsilon)^2 S_0^2} \tag{14-11}$$

K is a constant only for incompressible solids. For compressible cakes the value of K depends on the applied pressure, liquid velocity, and solids concentration. Therefore, use of the Darcy's equation in cake filtration testing and scaleup presents major problems even when there is a linear relationship between the liquid velocity and pressure drop.

For scaleup studies, the Ruth form of Darcy's law is preferred,

$$\frac{Q}{A} = \frac{\Delta p}{\mu R} \tag{14-11a}$$

where R is known as the bed resistance. Bed resistance in cake filtration consists of the medium resistance in series with the resistance of the deposited cake (assuming no penetration of solids into the filtration medium). A general filtration equation can then be written as

$$\frac{Q}{A} = \frac{\Delta p}{\alpha \mu c (V/A) + \mu R} \tag{14-12}$$

where

α = specific cake resistance
μ = liquid viscosity
c = solids concentration in the feed
V = filtrate volume collected since the commencement of filtration
R = medium resistance

Equation 14-12 provides a basis for analysis of cake filtration data. Feed liquid flow rate, Q, and filtrate volume, V, can be related as follows.

$$\frac{dV}{dt} = Q \tag{14-13}$$

Equation 14-13 assumes that the volume of the solids and the liquid retained in the cake is negligible. This is reasonable for low concentrations; however significant errors can result at high cake solids concentrations and high cake moisture contents. The usual approach for allowing for the nonapplicability of

Equation 14-13 is by using a "corrected" value for the concentration c in Equation 14-12.

3. Separation in Deep Beds

In deep bed filtration, particles pass into the medium and are collected within the bed by several different collecting mechanisms: gravity settling, Brownian diffusion, inertial deposition and interception, and piezophoresis. These mechanisms constitute the so-called transport step of bringing particles onto the collecting surface of the material making up the bed. The attachment of particles to the bed results from the electrical double layer and molecular forces existing at the interfaces between the particles and the grains of the bed.

Mathematical models exist to describe the effect of these forces on particle trajectories in filter pores. They also describe the process of gradual clogging with the associated pressure drop and filtrate quality. However, the complexity of the process requires the use of several empirical coefficients.

The simple models characterize particle removal as a change in concentration over the depth of the filter bed, L:

$$-\frac{dc}{dL} = \lambda c \qquad (14\text{-}14)$$

The parameter λ is a measure of efficiency and depends on the approach velocity, the particle size of the feed solids, the grain size of the media, the liquid viscosity, and the amount of solids already deposited in the bed. The parameter λ can be expressed as a function of filtration time and bed depth. Equation 14-14 can also be written with an additional term on the right-hand side to account for the stripping of the solids deposited during filtration [Purchas (1977)].

III. SOLID – LIQUID SEPARATION EQUIPMENT

A general classification of solid–liquid separation equipment is given in Figure 14-1. Solid–liquid separation equipment is described briefly in this section. Purchas (1977) provides an extensive discussion of commercial equipment.

A. Gravity Settling Tanks and Thickeners

The most commonly used thickener type is the circular basin. The flocculant-treated feed stream enters a central feedwell which disperses it gently into the thickener. Conventionally feed enters the top. The suspension is allowed to find the height in the basin where its density matches the density of the suspension already inside and spreads out at that level. Other thickeners introduce the feed through the side or through the bottom and disperse it into the compression layer; this eliminates the zone settling layer.

The liquid overflow is collected in a trough around the periphery of the basin. Conventional thickeners are constructed of steel (in sizes up to 25 m in diameter) or concrete (in sizes up to 170 m in diameter) with the floor sloped toward the underflow discharge in the center. Raking mechanisms slowly turning around the central column promote solids consolidation in the compression zone and aid discharge. Smaller basins may have a covered top.

Thickener types vary with application. Deep cone-based compression tanks are used for coal washery slurries, so that the highly flocculated sediment can be compacted by the weight of the material. The tray thickener is a series of circular thickeners stacked around a common central shaft, each having its own raking mechanism. The lamella-type thickener saves space by using flat inclined plates in the settling tank to promote solids contacting and settling along and down the plates. Corrugated and other plate configurations have also been used.

Clarifiers are built with emphasis on the clarity of the overflow, with feed concentrations being typically lower than for most thickeners. Gravity clarifiers can resemble circular thickeners but more often are rectangular basins with the feed at one end and overflow at the other. Settled solids are pushed to a discharge trench by paddles or blades on a chain mechanism. Flocculants are commonly added prior to the clarifier. In the water industry [Purchas (1977)] the common clarifiers are the "one pass" clarifiers with either a rectangular or circular basin and with paddle-type flocculators; the sludge blanket clarifiers, where the feed flows through a blanket of flocculated solids; and the solids recirculation clarifiers, where settled solids are fed back to mix with the incoming raw water. Newer developments include lamella clarification in "rapid settling clarifiers" or ballasted sand clarifiers.

B. Centrifuges and Hydrocyclones

Equipment available for centrifugation is divided into fixed-wall devices such as hydrocyclones and rotating-wall devices such as sedimenting centrifuges. The difference between the two is in the type of flow used in the machines. In hydrocyclones, the flow pattern approaches that of a free vortex, with high velocity gradients that cause shear, and may break agglomerates or flocs. This is not desirable for separation, but is useful for classification. Hydrocyclones are used extensively for separation and classification because of their reliability and low cost. In separation applications hydrocyclones are used primarily for thickening.

In sedimenting centrifuges, a forced vortex is developed that creates very little shear, making this equipment compatible with coagulation and flocculation. Centrifuges may also be used for solids classification. The suspension is fed and rotated at high speed in an imperforate bowl. Liquid is removed

SOLID–LIQUID SEPARATION EQUIPMENT

through a skimming tube or over a weir. The solids may remain in the bowl (batch processing) or be continuously or intermittently removed from it.

Industrial units can be distinguished by the design of the bowl and the mechanism of solids discharge as follows:

Tubular.
Multichamber.
Imperforate basket.
Scroll type.
Disk centrifuges.

The disk centrifuges can be further subdivided into solids-retaining, solids-ejecting, and nozzle types.

Tubular, multichamber, and solids retaining disk machines are suitable only as clarifiers since the bowls have to be cleaned manually. Imperforate basket and the solids ejecting disk machines are suitable for moderate feed concentrations having intermittent solids discharge during which the feed may have to be briefly interrupted. Only the nozzle-disk-type and the scroll-type centrifuges are truly continuous both in operation and solids discharge. The latter is well suited for particularly high solids concentrations of up to 50 percent by volume in the feed.

A conventional hydrocyclone consists of a cylindrical section joined to a conical section. Feed is injected tangentially into the upper part of the cylindrical section developing a strong swirling motion within the cyclone. Liquid containing the fine particle fraction is discharged out the top through a centrally located cylindrical tube called the overflow pipe or vortex finder. The remaining liquid, containing the coarse fraction, discharges through the underflow orifice at the cone tip.

Manufacturers tend to deviate from an "optimum" design so that the same cyclone can cover a range of flow rates and cut sizes. These versatile units have an extendable cylindrical section, a variable size vortex finder, and different conical-section angles. Most commercial units are equipped with a variable underflow orifice through replaceable push-in nozzles or by hydraulically, pneumatically, or mechanically operated continually variable orifices.

Industrial hydrocyclones vary in size from 20 mm to 1.5 m diameter and can be operated either as single units or in parallel multicyclone arrangements. Such arrangements are achieved by piping several units separately in parallel; by nesting them in a spiderlike arrangement round a central distributor column with common feed, overflow, and underflow; or by building several small units into cylindrical or square boxes.

Small-diameter units have correspondingly small underflow orifices. These are susceptible to blockage. The multicyclone arrangements recently developed in England by Mozley incorporate a manual or automatic device designed for periodic clearing of the underflow orifices with a pointed rod.

C. Flotation

Flotation is a gravity separation process based on the attachment of air or gas bubbles to solid (or liquid) particles. These are then carried to the liquid surface where they accumulate as float and are skimmed off. The process consists of two stages: the production of suitably small bubbles, and their attachment to the particles. Depending on the method of bubble production, flotation is classified as dissolved air, electrolytic, or dispersed air.

Dispersed air flotation is basically a solid–solid separation process and has been used for many years in mineral processing to concentrate base metals. Large bubbles of about 1 mm are used in dispersed air flotation. These are produced by agitation combined with air injection (froth flotation) or by bubbling air through porous media (foam flotation). Only relatively coarse solids can be floated this way, so solid feeds must be ground down only to 300 μm. Selectivity is based on the relative wettability of solid surfaces. Only minerals having a specific affinity for air bubbles rise to the surface, others are wetted by water and sink.

Wettability and frothing are controlled by three classes of chemical reagents: frothers, collectors, and modifiers. Separation is carried out in cells typically from 1.1 to 1.8 m^2 in flow area, where the feed is mixed and injected with air. Using many cells in series provides sufficient time for floating all of the valuable mineral.

Dissolved air and electrolytic flotation have made considerable inroads into solid–liquid separation applications since they can be applied to fine particle suspensions. They offer a viable alternative to gravity sedimentation, operating at much higher separation rates, and using smaller, more compact equipment of lower capital but higher operating costs.

Dissolved air flotation produces fine bubbles of less than 100 μm diameter. The process is based on the higher solubility of air in water as pressure increases. Part or all of the feed is saturated with air, and then the pressure is reduced so that fine air bubbles appear and are available for flotation. There are three ways of accomplishing this:

Saturation at atmospheric pressure and flotation under vacuum.

Saturation under static head with upward flow resulting in bubble formation (microflotation).

Saturation at pressures higher than atmospheric (200–700 kPa) and then flotation under atmospheric conditions.

The latter approach is most common.

Usually only a portion of the effluent is saturated with air and recycled at rates of 25–50 percent, or even 100 percent. Recycling minimizes clogging problems and flow variations in the influent, and lowers the danger of floc breakup. The process is carried out in circular or rectangular cells. The feed, often treated with flocculating agents, enters the inlet mixing chamber, as does the recycle nearly saturated with air. The amount of air required is determined by a mass ratio of air to dry solids of 0.015 to 0.030. Solids accumulate in the flotation zone and are skimmed off by the sludge removal mechanism. The effluent exits at the other end of the unit, with a portion being recycled through a dissolution tank (saturator).

In electrolytic flotation (electroflotation), hydrogen and oxygen gas bubbles are generated by electrolysis. Instead of the saturator, a rectifier system supplying 5–20 V (DC) at a current of approximately 100 A/m^2 of electrode is required. The potential difference required to maintain the necessary current density for bubble generation depends on the electrical conductivity of the feed slurry. This unit is very similar to the dissolved air unit. In fact, the same unit can be used for both processes if electrodes and external recycle saturation system are provided. As with dissolved air units, a flocculation stage normally precedes the flotation stage.

The main development in electrolytic flotation has been in electrode design. Many materials have been used, from carbon and graphite to titanium and platinum. Lead dioxide coated titanium anodes have probably the longest useful life. If the unit is made of steel, all internal surfaces must be coated.

D. Surface Filters — Cake Filtration

In cake filtration, particle deposition takes place on the face of a thin medium via a screening mechanism. As soon as a layer of cake appears on the filter face, deposition shifts to the cake itself, and the medium acts only as a support. As the cake grows, resistance to flow increases. At a constant pressure drop, this results in a gradual fall in feed flow rate.

Conventional cake filtration devices rely on the formation of an undisturbed cake, which requires that both the particles and the suspending liquid approach the filter medium at a right angle. There are several new devices and methods where the cake growth is limited by mechanical or hydraulic means in order to maintain high flow rates. Mechanical squeezing of the cake can be combined with conventional cake filtration.

There are basically three types of filters differing in the required magnitude of the pressure drop:

Vacuum.
Pressure.
Centrifugal.

Some belt filter presses use gravity drainage in a preliminary stage; this is probably the only situation where a static gravity head is used to generate the driving force for cake filtration.

In pressure filters the driving force is the liquid pressure generated by pumping or through the application of gas pressure on the surface of the feed slurry in a vessel. Alternatively, pressure can be generated by a membrane pressing down the suspension or on the cake in the so-called variable chamber filters. Pressure filter equipment includes:

Conventional

Plate and frame presses.
Rotary drum pressure filter.
BHS Fest filter.
Leaf type pressure vessel filters (horizontal, vertical or rotating leaves).
Vertical recessed plate—endless traveling cloth automatic press.

Variable Chamber (with Cake Squeeze)

Membrane plate presses.
Plate press filter.
Cylindrical membrane presses (with or without the initial conventional pressure filtration).
Belt filter presses, combining gravity drainage of highly flocculated slurries with mechanical squeezing of cake between two running belts.

Pressure filters can also combine pressure filtration with a method of limiting cake growth. Examples are: the side filter centrifuge, the dynamic filters with rotating elements, and crossflow filtration in porous tubes or leaf-type elements.

In vacuum filters the driving force for filtration results from the application of a suction on the filtrate side of the medium. The most common batch operated vacuum filters are the vacuum leaf filter and the vacuum Nutsche filter. Many other filters can be regarded as modifications of the latter: the double-tipping fan filter and the horizontal rotating pan filter (with a scroll discharge or a tilting pan version). The horizontal endless cloth vacuum filter, which resembles a belt conveyor in appearance is an offspring of the pan filter. The top strand of the cloth is used for filtration, cake washing, and drying. The bottom return strand is for tracking and washing of the cloth. These filters are often classified by the method used to support the filter medium: for example, the rubber belt, the reciprocating tray, and the indexing cloth types. Savings in floor area and installed costs can be made by using rotary vacuum filters of the drum or disk type. However, the important advantage of using a horizontal filtering surface is then lost. Horizontal belt and rotary vacuum filters are continuous.

In centrifugal filters, the driving force for filtration is the centrifugal action on the fluid. Essentially they consist of a rotating basket fitted with a filter medium. A general classification contrasts fixed-bed centrifuges, such as the perforated basket centrifuge or the peeler centrifuge, with moving bed centrifuges, such as the conical screen centrifuge or the pusher centrifuge.

While virtually any type of screen can be used for dewatering, some are designed specially for this purpose. A good example is the sieve bend which incorporates some of the principles of an inertial separator along with those of a screen. Essentially, a unit consists of flat surfaced, wedge bars mounted with their axes horizontal but arranged one above the other so that they form a surface with horizontal slits in it. The uppermost bars are located directly above one another but subsequent ones are shifted laterally and rotated slightly until the upper surface of the latter ones form an angle of about 45° with the horizontal. The result is a concave surface.

Feed material is introduced on the uppermost wedge bar surface and withdrawn as an undersize product between the bars and an oversize, dewatered product from the lower surface. The feed slurry enters vertically downward and tangentially over the width of the upper surface. It flows down the concave surface moving always at right angles to the openings between the wedge bars. A thin layer of slurry on the underside is deflected due to the drag on it as it passes over the bars. This portion flows out between the bars while the oversize continues over the surface.

The unit is unique in that the thickness of the layer deflected determines the particulate size separated and not the openings between the bars as is the case in ordinary screening. Usually the separation size is near one-half the spacing between the bars, although as the cut size is decreased this comes closer to the opening size. The fact that the particles passing through the grid are always of lesser size than the openings gives the sieve bend good nonclogging attributes. At the present, the working range of sieve bends is from about 8 mesh to 50 μm diameter particles.

Cartridge filters use an easily replaceable cartridge made of paper, cloth, or various membranes having pore sizes down to 0.2 μm. The suspension is simply pumped, sucked, or gravity-fed through the filter. Their capacity is limited. Dewatering screens and cartridge filters are only used for special application.

E. Deep Bed Filters

As the name suggests, deep bed or depth filters separate solids inside a deep bed of granular material. The bed has to be periodically cleaned by backwash. If frequent cleaning is to be avoided, only very low feed solids concentrations can be handled in deep bed filtration.

The most common arrangement for a deep bed filter is the downflow gravity design. Where a single medium is used and the backwash is sufficient to fluidize the medium, size segregation can result in the finest material residing in the upper layers of the filter. This will hinder penetration of feed solids into the medium and prevent effective use of the bed.

Segregation can be avoided by backwashing with air and water simultaneously so as to minimize bed expansion. The use of dual or multimedia filters also overcomes this problem. The key is selection of coarse material of low density for the top layer, and progressively finer materials of increasing density for the lower layers.

The upflow filter is not commonly used. This configuration avoids the use of different density media, since the size segregation that occurs during filtration helps the solids-holding capacity of the bed. Backwashing is generally performed using unfiltered water and air simultaneously. A retaining grid prevents bed carryover and allows higher filtration rates. Compared to downflow filters, upflow filters provide less protection against unfiltered water passing through to the effluent side, mainly because the backwash is in the upward direction.

The constraints of batch operation can be avoided by selecting a continuous filter, which continuously backwashes a portion of the medium. Among these units are moving-bed filters, Hydromation in-depth filters, radial-flow filters, and traveling backwash filters. All of these have been described by EPA (1975). However, onstream reliability of continuous depth filters has not been firmly established. Various designs of pressure type depth filters are also available.

IV. SMALL-SCALE STUDIES FOR EQUIPMENT DESIGN AND SELECTION

Design of equipment for solid–liquid separation can be done only in a few instances from fundamental principles and primary particle properties such as size and its distribution, density, shape, and so on. Scaleup techniques for most types of equipment require specialized empirical constants or "secondary" properties of the particulate system such as the specific cake resistance or the critical settling flux (see Section V). The state of flocculation in particular has a profound effect on all critical properties and, in testing for these properties, choice and optimization of flocculants have to be considered at the same time.

A. Gravity Settling Tanks and Thickeners

For the simplest case of particulate clarification when no flocculation takes place during settling (either the flocculation process is completed before entering the settling tank or the suspension is entirely nonflocculent) the basic test is the so-called "short tube procedure." This consists of settling in and decantation from a large measuring cylinder in order to evaluate the settling rate, that is, the specific overflow rate which would produce a satisfactory clarity of the overflow.

The "long tube procedure" is designed for systems where flocculation (or deflocculation) takes place during settling and thus the settling tank's performance depends not only on the specific overflow rate but also on the residence time in the tank. Tests are conducted in a vertical tube that is as long as the expected depth of the clarifier, under the assumption that a vertical element of suspension that has been clarified, will maintain its shape as it moves across the tank.

When the overflow clarity is independent of overflow rate and depends only on retention time (as is the case for high solids removal from a flocculating suspension), the required time can be determined from a simple test of residual solid concentrations in the supernatant as a function of retention time, under condition of mild shear. This is sometimes called the second order test procedure because the flocculation process follows a second order rate reaction.

The design of the sludge blanket clarifiers is based on the "jar test" and a simple measurement of the blanket expansion and settling rate as discussed by Purchas (1977). Different versions of the jar test exist. However, it essentially consists of a bank of stirred beakers used as flocculators in which the flocculant addition is optimized to produce the maximum floc settling rate. Visual floc size evaluation is usually also included.

The critical settling flux essential for evaluation of zone settling-layer area demand of a gravity thickener is measured either using the Coe and Clevenger (1916) method or the simpler Talmage and Fitch (1955) procedure. The former consists of a series of settling tests in a measuring cylinder where the initial settling rates of a visible interface within the settling suspension are measured for different initial solids concentrations. The Talmage and Fitch procedure simplifies the tests because it requires only one test at any concentration providing it is in the zone settling regime.

Both methods should give an identical critical settling flux and therefore identical pool areas, but this is not so in practice. Usually the Coe and Clevenger method leads to underdesign of thickener area while the Talmage and Fitch procedure leads to overdesign. With highly flocculent slurries the area demand of the compression layer may exceed that of the zone settling layer and the compression zone also has a depth demand. Kos (1979) has described a "multiple-batch upflow" test for compression zone evaluation. This is by no means standard yet, however, and its reliability remains to be determined.

B. Centrifuges and Hydrocyclones

As flocculation is largely ineffective in hydrocyclones, these can be designed from known particle properties and required mass recoveries using the scaleup relationships desired in Section V-B. The relationships are reasonably reliable for the well-documented and tested "optimum" designs and at low solids concentrations. Nonstandard designs can be tested easily on a small scale providing that some basic installation rules are observed. High feed solids concentrations present problems since the scaleup of hindered settling in a hydrocyclone is not yet well established.

While some basic settling tests can be carried out in a bottle-type laboratory centrifuge, the scaleup of sedimenting centrifuges must be based on a small scale test in laboratory machines of the same type as the commercial machines to be used. The purpose of the tests is to establish a flow rate for the suspension under conditions giving satisfactory overflow clarity. The results can then be scaled up using the Sigma theory reviewed in Section V-C. Flocculating agents should be incorporated in the tests since unlike hydrocyclones, the shear effects in a sedimentary centrifuge are much weaker and the flocs are not broken. Once again, little is known about the effects of hindered settling in sedimenting centrifuges and this area is still under development.

C. Flotation

Tests required for scaleup of dissolved air flotation can be carried out at three different levels:

Batch, bench scale tests.
Batch flotation column tests.
Continuous tests.

Batch tests are useful for optimization of coagulant type and dosage of pH. However, the only meaningful scaleup data can be obtained from continuous tests with a scaled down laboratory flotation unit. In this unit, both clarification and thickening can be observed and evaluated simultaneously. Such experiments can be designed to yield the constants K_1 through K_7 discussed in Section V-D, or they can be used to establish the optimum specific overflow rate. Many manufacturers of dissolved flotation equipment have such laboratory units, often mobile, and can carry out such tests on a routine basis. A full description of the tests with worked examples is given in Purchas (1977).

The situation is similar for electrolytic flotation. Batch tests are suitable only for evaluation of gas rate production and the conditions for bubble attachment. Data for design of plant can be obtained only in continuous experiments. This can be done using test equipment for dissolved air flotation that has been suitably adapted for electrolysis.

D. Cake Filters and Dewatering Screens

A wide variety of test apparatus is used to obtain the secondary cake properties needed for scaleup of surface filters. Some measure of filterability can be obtained using the "capillary suction time (CST) apparatus" as discussed in Purchas (1977). CST measures the rate of saturation on a sheet of absorbent paper exposed to the suspension. There is some correlation between CST and specific cake resistance; however, this is not very reliable. The value of CST measurement is mainly in quick evaluations of flocculant addition.

Direct measures of the specific resistance, α, and medium resistance, R, can be obtained with a simple Buchner funnel. However, more representative results are obtained with specially designed filter leaves, with or without a shim, which can also provide information about moisture contents, cake washing characteristics, and the suitability of precoats. Since the tests are carried out by application of vacuum on the filtrate side, they are limited to pressure drops of less than 1 bar. This is sometimes not sufficient to provide enough information about compressibility of the cake (constants α_0 and n in Equation 14-49). A pressure apparatus is then necessary. Laboratory pressure filter cells or compression permeability cells can be used at elevated pressures although permeability cells have been criticized for using mechanical pressure on the cake to simulate hydraulic pressure. Wherever mechanical squeezing is under consideration, the cake compressibility constant for Equation 14-53 can be obtained with either membrane or piston laboratory presses.

The most reliable data for scaleup are obtained from experiments on small or pilot scale versions of actual filtration apparatus. Such equipment is readily available on the market in the form of small rotary drum vacuum filters, plant-and-frame presses, and so on. The scaleup of filtering centrifuges is different, however, because the driving force for filtration is not only due for the pressure drop created by the centrifugal head but also due to centrifugal forces acting on the liquid flowing through the cake. Zeitsch [in Svarovsky (1977)] recognized and described a test procedure with a specially designed filter breaker to measure the "intrinsic permeability" of the cake. Purchas (1977) has described a similar test.

The DSM sieve bend scales up quite well if the Reynolds number is above 250. Laboratory tests with a geometrically similar machine can be used to establish the effective cut size, the moisture content of the product, and the underflow to throughput ratio.

E. Deep Bed Filters

A laboratory filterability test [Ives in Svarovsky (1981)] can be used for rapid assessment of suspension characteristics and pretreatment procedures. Studies on a model deep filter smaller in diameter than the full-scale filter, but using the same suspension in the same fluid and flowing at the same rate through the same depth of the same porous medium, can be scaled up directly.

Ives [in Purchas (1977)] points out that scale of the fluid action and particle separation in a deep bed filter is limited by the boundaries of the pores in the medium, not by the size of the filter container. The only constraint is that the wall-to-wall distance should be at least 50 times the largest grain size so that wall effects can be neglected. Model deep filters should be fitted with pressure manometers so that the pressure distribution along the height of the bed can be measured. These tests also permit study of filter washing and medium expansion during washing. Model deep bed filters are available for rental from some manufacturers.

A preliminary assessment of suitability of a filter medium can be made from the basic primary properties of the medium such as particle size, shape, density, durability, settling velocity, and so on.

V. SCALEUP TECHNIQUES

A. Gravity Settling Tanks and Thickeners

Gravity settling tanks for clarification without coagulation or flocculation (removing small amounts of solids) are identical in principle as with laminar settling chambers for cleaning gases (Svarovsky, 1980). The grade efficiency curve $G(x)$ is

$$G(x) = \frac{U_g A}{Q} \tag{14-15}$$

where

U_g = terminal settling velocity of a particle of size x
Q = feed flow rate equal to overflow rate
A = plan area of the tank

Equation 14-15 is derived assuming laminar flow in the tank and no end effects.

Tank area is the only design parameter affecting the theoretical separational performance irrespective of the shape or depth of the pool. Equation 14-15 can be rewritten in terms of the more conventional dimensionless groups as

$$\text{Stk Fr} = G(x) \tag{14-16}$$

If Stokes' law is assumed to set the particle settling velocity, the Stokes number is defined as

$$\text{Stk} = \frac{x^2 \Delta p}{18\mu} \cdot \frac{Q}{AH} \tag{14-17}$$

and the Froude number

$$\text{Fr} = \frac{HgA^2}{Q^2} \tag{14-18}$$

H being a characteristic dimension of the tank, for example, the height. However, height cancels out in Equation 14-16.

Equation 14-15 provides a basis for scaleup in that for the same performance the flow rate is proportional to area. Measurement of grade efficiency curves in settling tanks is rather difficult, particularly on a large scale since the feed solids must remain constant over the large residence times involved. The specific overflow rate (or "overflow volume flux"), Q/A, giving satisfactory overflow clarity from simple settling tests is, therefore, usually measured (see Section IV-A).

SCALEUP TECHNIQUES

When clarification is not "particulate" and flocculation takes place in the settling tank, overflow clarity depends not only on the overflow rate but also on the retention time. Under such conditions data are obtained on the long tube procedure discussed in Section IV-A. Where the effect of the retention time dominates, the overflow rate can be ignored. For coagulation clarifiers used predominantly in the water industry, scaleup is usually based on the overflow rate determined from jar tests. These tests should be designed to select the best flocculant and to determine the settling rates giving a clear supernatant liquid.

Thickeners are basically gravity settling tanks which in addition to producing a clear overflow are also designed to have a thick underflow with minimal water content. Feed into a thickener is generally more concentrated than with clarifiers and quite often exhibits hindered settling behavior as discussed in Section II-C-2.

There are basically three layers in an operating thickener: the topmost "clarification" layer, the "zone settling" layer, and the "compression" layer at the bottom. Each layer requires an area for its function; ideally the thickener should be designed on the largest area. Generally, the function of the clarification layer is to prevent particles that have escaped from the zone settling layer or from the feed, from leaving with the overflow. This duty is often less important than the thickening duty; thickener area is usually chosen on the basis of the zone settling or compression layer requirements.

Conventional design and scaleup of thickeners operated on mineral and metallurgical slurries is based on the area demand of the zone settling layer. This assumes that the compression zone imposes only a solids retention (and hence depth) demand, but no independent demand on area. The method assumes that the solids on their way downward from the feed layer to the underflow continuously increase in concentration from the feed to the underflow concentration (as determined from time retention tests). Total solids flux (mass flow rate of solids per unit area) which the different layers in the thickener are capable of accommodating then go through a minimum between the feed zone and the underflow. This minimum solids flux, G_c, determines the minimum thickener design area. A thickener of this or greater area should not overflow solids through backing up of the zone-settling layer in the thickener. Moreover, backup should be so small as to introduce no demand on the depth allocated to the zone-settling layer of the thickener.

The area, A, of the thickener is calculated from the critical solids flux, G_c, and the feed flow rate, Q, and concentration, c_f, using the simple mass balance in Equation 14-19.

$$G_c A = Q c_f \tag{14-19}$$

This mass balance assumes complete separation of all of the feed solids.

Design and scaleup of the thickeners centers around the determination of the critical solids flux, G_c. This value cannot be estimated from the primary properties of the particulate system due to the unpredictable effect of flocculation; it must be obtained experimentally. Meaningful G_c data can be obtained for large-scale or pilot-scale thickeners. However, with few exceptions, this is impractical due to the scale and cost of such experiments.

If the settling velocity of the solids is assumed to be a function of concentration only [$u = u(c)$], then this function should be unique for a given suspension and identical for batch settling or a continuous operation. This is the basic philosophy of the conventional Coe and Clevenger and Talmage and Fitch tests which differ only in the manner G_c is determined from the experimental tests. The Talmage and Fitch method requires simpler tests as discussed in Section IV-A.

If the function $u = u(c)$ is known, then the critical flux, G_c, corresponds to the minimum on the total flux G curve, as given by Equation 14-20

$$G = \frac{u}{(1/c) - (1/c_u)} \qquad (14\text{-}20)$$

where the underflow concentration c_u is determined from retention time tests. The thickness of the compression zone can also be determined from the retention time tests on the assumption that the solids concentration reached in the compression layer in a batch test after a given time will be the same as in the compression layer of a continuous thickener.

B. Hydrocyclones

The hydrocyclone literature is full of studies of the effects of different design parameters such as the cone angle, vortex finder diameter, cyclone length, and so on, on pressure drop and efficiency. This information is contradicting, confusing, and quite often useless. What is needed is a choice of standard designs of geometrically similar liquid hydrocyclones, like those existing for gas cyclones [Svarovsky (1980)] fully documented with test data on operating characteristics. Only then could a meaningful scaleup be carried out and result in a reliable design without further extensive testing.

The suggested scaleup of hydrocyclones is based on currently available theories of particle separation rewritten into dimensionless forms. This enables a qualified judgment to be made as to whether a hydrocyclone is viable for a process and, if so, to make a preliminary design. However, further testing, preferably on a commercial scale will be required.

1. Dimensional Analysis at Low Solids Concentrations

At low solids concentrations (less than 1 percent by volume) the flow pattern in the cyclone is not affected by the presence of particles in the flow and particle–particle interaction is also negligible. Consequently, since few particles are in the underflow, the underflow to throughput ratio can be kept low and it also can be assumed to have no effect on the cut size, x_{50}. Dimensional analysis for this case [Svarovsky (1977)] results in three dimensionless groups which relate six variables; cut size, x_{50}, cyclone diameter, D, liquid viscosity, μ, and density, ρ, density difference between the solids and the liquid, $\Delta\rho$, and suspension flowrate, Q.

The relationship can be written as

$$\text{Stk}_{50} = f\left(\text{Re}, \frac{\Delta\rho}{\rho}\right) \quad (14\text{-}21)$$

where

$$\text{Re} = \frac{vD\rho}{\mu} \quad (14\text{-}22)$$

v is the characteristic velocity calculated from the cross section of the cyclone body, that is,

$$v = \frac{4Q}{\pi D^2} \quad (14\text{-}23)$$

Stk_{50} is the Stokes number defined as

$$\text{Stk}_{50} = \frac{x_{50}^2 \Delta\rho v}{18\mu D} \quad (14\text{-}24)$$

The pressure-drop flow-rate relationship for a hydrocyclone is the same as that for any other flow device,

$$\text{Eu} = f(\text{Re}) \quad (14\text{-}25)$$

where the Euler number, Eu, is the well-known pressure loss factor defined as

$$\text{Eu} = \frac{\Delta p}{\rho v 2/2} \quad (14\text{-}26)$$

Some studies base their analysis on different reference velocities and characteristic dimensions. Inlet velocity and inlet diameter have been frequently used, leading to "inlet" Reynolds number, Re_i, Stokes number, Stk_i, and Euler number, Eu_i. Other Reynolds numbers have also been defined.

For geometrically similar hydrocyclones the choice of characteristic velocity and dimension is arbitrary, as long as it is specified. The basis adopted in this chapter directly involves the most important cyclone dimension, its diameter D. Therefore, it results in the simplest and mot direct scaleup.

Most published studies, whether theoretical or experimental, show that the dimensionless group $\Delta\rho/\rho$ has no effect on Stk_{50} and thus both Stk_{50} and Eu depend only on Re. This makes possible scaleup of geometrically similar cyclones from basic bench type tests on smaller units.

Rietema (1961) has derived from residence time theory and verified experimentally a characteristic cyclone number depending only on geometric proportions of the cyclone. Optimum sets of cyclone proportions are obtained by minimizing the characteristic cyclone number. It can be shown from his results that

$$\text{Stk}_{50}\text{Eu} = 0.0611 \tag{14-27}$$

for

$$\frac{L}{D} = 50, \quad \frac{l}{D} = 0.4, \quad \frac{b}{D} = 0.28, \quad \frac{e}{D} = 0.34$$

where

L = overall length of the cyclone
l = length of the vortex finder (outlet pipe projecting into the cyclone)
b = inlet diameter
e = diameter of vortex finder

These results are applicable to "long cone" cyclones with the included angle of the conical section not exceeding 30°.

The pressure-drop flow-rate relationship for Rietema hydrocyclones was approximated by Gerrard and Liddle (1975) as

$$\text{Eu} = 24.38\text{Re}^{0.3748} \tag{14-28}$$

The "normal cyclone configuration" developed by Bradley (1965) has the relative dimensions

$$\frac{L}{D} = 6.43, \quad \frac{l}{D} = \frac{1}{3}, \quad \frac{b}{D} = \frac{1}{7}, \quad \frac{e}{D} = \frac{1}{5}$$

Since the underflow diameter is $D/15$ and length of cylindrical section is $D/2$, this is equivalent to an included angle for the conical section of 9°. The equilibrium orbit theory developed by Bradley gives very simple results for negligible underflow to throughput ratio.

$$\text{Stk}_{50} = 1.535 \times 10^{-5} \tag{14-29}$$

$$\text{Eu} = 7240 \tag{14-30}$$

This assumes that the exponent n in the free vortex flow formula is 0.8 and the entrance velocity loss $\beta = 0.45$. Since both constants and particularly β depend on Re, Equations 14-29 and 14-30 apply only for Re of 5570 in

Bradley's experiments. Mitzmager and Mizrahi (1964) have shown that

$$\beta \sim Re^{0.1615}$$

Since n is approximately constant, the following more general correlations can be obtained for normal cyclone configurations

$$Eu = 446.5 Re^{0.323} \qquad (14\text{-}31)$$

and

$$Stk_{50} Eu = 0.1111 \qquad (14\text{-}32)$$

Equations 14-31 and 14-32 are similar to those for Rietema's cyclone (Equations 14-27 and 14-28). Hydrocyclones operated at low feed concentrations can be described in the same manner as indicated here. Knowledge of the two basic equations

$$Eu = f(Re)$$

$$Stk_{50} Eu = \text{const.} \qquad (14\text{-}33)$$

allows calculations of any unknown variables if the other variables are known or assumed.

The common design problem is establishing the diameter, D, and the number of cyclones, N, to be operated in parallel to treat a given flow rate, Q, at a pressure drop Δp while achieving a specific cut size. This can be done by writing down equations for the hydrocyclone type chosen, substituting Q/N for flow rate and solving for the two unknowns, N and D. This, of course, assumes Δp is set independently. If the pressure drop is an unknown variable, then there are an infinite number of solutions. Such a case lends itself to optimization. Gerrard and Liddle (1975) have shown how operating costs and fixed charges can be optimized.

2. Dimensional Analysis at High Solids Concentrations

At feed solids concentrations higher than 1 percent by volume there are complex influences on both the cut size and the grade efficiency curve. The increased solids concentration results in changes in flow patterns as well as hindered settling from particle–particle interaction. Changes in flow patterns result in changes in the pressure loss factor, Eu. This paradoxically decreases with increasing concentration due to the reduction in centrifugal head. Both of these effects lead to an increase in cut size.

There is also an indirect effect on the volumetric underflow to throughput ratio, R_f. As the volume of the solids going to the underflow increases, the underflow orifice has to be opened up to accommodate the increase. R_f is then no longer negligible. The beneficial effect of R_f can be partly taken into account by using a reduced grade efficiency $G'(x)$ defined as

$$G'(x) = \frac{G(x) - R_f}{1 - R_f} \qquad (14\text{-}34)$$

where $G(x)$ is the unmodified grade efficiency.

$G'(x)$ is usually an S-shaped curve passing through the origin. The size corresponding to 50 percent is defined as "reduced" cut size x'_{50}. The effect of increasing R_f goes beyond the benefits of the dead flux taken into account in Equation 14-34. The value of R_f, determined primarily by the size of the underflow orifice, directly affects x_{50}.

Dimensional analysis for higher concentrations therefore has to include two additional dimensionless groups C (fraction by volume) and R_f:

$$\text{Stk}'_{50} = f\left(\text{Re}, \frac{\Delta\rho}{\rho}, C, R_f\right) \qquad (14\text{-}35)$$

Based on the few isolated empirical correlations available in the literature Svarovsky (1977) proposed the following expression for x'_{50},

$$\text{Stk}'_{50} = K_1(1 - R_f)^{1/2} \exp(K_2 c) \qquad (14\text{-}36)$$

Equation 14-36 has been shown to hold well for concentration above 8 percent by volume. The empirical constants K_1 and K_2 depend on cyclone configuration as well as on the test solids. For limestone in an AKW cyclone K_1 and K_2 were found to be 9.056×10^{-5} and 6.461, respectively. Plitt (1977) has shown $K_2 = 6.3$ for silica test solids.

Control of the underflow orifice is essential to providing the required discharge capacity. If the underflow concentration reaches 40 to 50 percent by volume, the cut size becomes primarily a function of the capacity of the underflow orifice. The requirements of high total efficiency (mass recovery) and a thick underflow cannot be maximized at the same time. A compromise must be found depending on their relative importance for the overall process.

If a high mass recovery of solids is required, the underflow orifice should be relatively open in order to keep the underflow concentration low, say below 12 percent by volume for limestone according to Svarovsky and Bavishi (1977). If however, a thick underflow is required, the capacity of the underflow orifice can be reduced to give up to about 50 percent by volume (depending on the solids) but at a reduction of mass recovery.

Semiempirical dimensionless correlations presented in this section apply only to cyclones operating with a free discharge from both underflow and overflow. Conditions of back pressure or syphoning can significantly change the operation of a hydrocyclone and invalidate the known correlation. Syphoning in the overflow in particular is very detrimental to cyclone performance mainly due to the resulting flow instabilities. It is therefore important to break the syphon by venting the overflow pipe before taking the flow down below the vortex finder level.

3. Grade Efficiency Curves

The shape of the reduced grade efficiency curve, $G'(x)$, is not significantly affected by cyclone configuration. $G'(x)$ can be expressed as a function of dimensionless size, x/x_{50}. Several empirical equations have been proposed to describe the curve: the Rosin–Rammler–Bennett type equation proposed by Bradley (1965), the simpler Rosin–Rammler equation by Plitt (1977), an exponential equation by Draper et al. (1969), and a cumulative log-normal distribution by Gibson (1979). At low concentrations, all the equations give essentially the same shape of $G'(x)$ and there is little basis to choose among them. At high concentrations, however, the slope of the curves (which determines the sharpness of cut) is reduced. Moreover, the slopes become, as shown by Draper et al. (1969), a function of the solids properties such as particle density and shape.

Predictions of the expected grade efficiency curve for an actual cyclone can be made by determining the cut size from the existing correlations and deducing $G'(x)$ from one of the empirical formulae referred to above. The reduced grade efficiency curve, $G'(x)$, so obtained can then be converted into a prediction of the actual grade efficiency curve, $G(x)$, by means of an estimated value of R_f using Equation 14-34.

C. Sedimenting Centrifuges

Scaleup of sedimenting centrifuges is, not surprisingly, most frequently based upon solids recovery or clarity of the overflow. However, there are situations when the limiting factor may be solids loading, underflow concentration, or scrolling torque. Scaleup of sedimenting centrifuges based on solids recovery has been dominated for the past three decades by the sigma (Σ) concept [Ambler (1952)]. This is a simple relationship between the operating variables related to the particles and the fluid expressed as the terminal settling velocity, U_g, of the cut size, x_{50}, under gravity conditions; the size, speed, and geometric proportions of the centrifuge expressed in the sigma factor; and the suspension flow rate, Q, as follows

$$2U_g \Sigma = Q \qquad (14\text{-}37)$$

with U_g obtained from Stokes' law as

$$U_g = \frac{x_{50}^2 \Delta \rho g}{18\mu} \qquad (14\text{-}38)$$

where

$\Delta\rho$ = density difference between the solids and the liquid
μ = liquid viscosity
g = gravity acceleration

Σ has a dimension of area and it theoretically represents the area of a gravity settling tank capable of the same separational performance as the centrifuge in question.

The analogy with a gravity settling tank is a strong one since the grade efficiency of a laminar settling tank of area A is Equation 14-15

$$G(x) = \frac{U_g A}{Q} \tag{14-15}$$

For $G(x_{50}) = \frac{1}{2}$,

$$\frac{1}{2} = \frac{U_g A}{Q} \tag{14-39}$$

This is identical to Equation 14-37 if A is replaced by Σ. In terms of dimensionless groups used for scaleup of hydrocyclones, Equation 14-37 can be written as

$$\text{Stk}_{50} \text{Fr} = \tfrac{1}{2} \tag{14-40}$$

where the Stokes number is

$$\text{Stk}_{50} = \frac{x_{50}^2 \Delta\rho}{18\mu} \frac{Q}{\Sigma D} \tag{14-41}$$

and the Froude number is

$$\text{Fr} = \frac{D g \Sigma^2}{Q^2} \tag{14-42}$$

D is a characteristic dimension of the centrifuge, for example, its diameter. Comparisons with gravity settling are unreal, however, since this ignores the effects of Brownian diffusion, convection currents, and flocculation. Therefore, it is better to refer to Σ as an index of the centrifuge size. For each of the five centrifuge types discussed in Section III-B there is a basic equation for calculation of Σ from centrifuge dimensions and speed. This equation is, in its general form,

$$\Sigma = \frac{\omega^2}{g} \pi K \tag{14-43}$$

where ω is the angular speed of the centrifuge and K is a geometric factor pertaining to the particular centrifuge. K values for typical centrifuges are given in Svarovsky (1981). Gravity acceleration is not a factor in determining sigma. Gravity acceleration g in Equation 14-38 cancels with the g in Equation 14-43 when the two equations are substituted in the basic statement of the sigma given in Equation 14-37.

Calculated sigma factors from simplifying assumptions ignore end effects and particle reentrainment. More reliable values based on experiments can be established. Extensive work on comparisons over the past two decades have shown that while experimental results differ from calculated values when different types of centrifuges are being considered, scaleup between centrifuges of the same type is fairly reliable. Scaleup is based on a simple application of Equation 14-41,

$$\frac{Q_1}{\Sigma_1} = \frac{Q_2}{\Sigma_2} \tag{14-44}$$

keeping the value of x_{50} constant.

Attempts have been made to extend the application of Equation 14-44 to "cross-type" scaleup between different centrifuge configurations through an efficiency factor μ_i,

$$\frac{Q_1}{\mu_1 \Sigma_1} = \frac{Q_2}{\mu_2 \Sigma_2} \tag{14-45}$$

where μ_i are relative efficiencies of different types. This has not been a productive approach. Cross-type scaleup is not reliable unless more factors other than just the sigma values are considered. There have been many attempts to modify the Σ theory since its conception. Important improvements have been directed at the introduction of more realistic flow patterns in specific types of centrifuges, with account taken of end effects and the introduction of particle shape factors.

Although experimental values of Σ can be obtained from the test data through use of the grade efficiency curve, one may prefer to measure the fraction of solids unsedimented $(1 - E_T)$ where

$$E_T = \frac{x_c(1 - x_0)}{x_0(1 - x_c)}$$

x_0 = initial mother liquor concentration

x_c = concentration in centrate from centrifuge

$1 - E_T$ is plotted against the ratio of the measured flow rate to calculated Σ value. The Σ value can be varied by changing the speed of rotation. The curve often is a straight line on log-probability paper. While a function of the size distribution of the feed, it can be used to find the ratio of Q/Σ for an acceptable efficiency with a given feed material. Extrapolation of the data beyond the measured portion of the graph should be made with caution.

The major shortcoming of the sigma concept is that cut size, which is basic to the sigma concept, is insufficient as a criterion of separation efficiency. Different total efficiencies can be obtained for a given cut size, if the size distribution of the feed particles differs. The only way to describe fully the performance of a sedimenting centrifuge is by a grade efficiency curve. Knowledge of this curve allows accurate and reliable predictions of total efficiencies with different feed solids. However, the projections are subject to the operating characteristics of the centrifuge, the state of dispersion of solids, and other variables remaining constant.

Rather than trying to modify the sigma concept to make it more flexible and complete, one should employ the grade efficiency concept. This, of course, requires a large number of tests along with some careful calculations. Studies are in progress to fill this gap. Presently the best information on extensions of the sigma concept is due to Gibson (1979) who carried out large-scale experiments on disk type and scroll type centrifuges. One of his many important findings was that the reduced grade efficiency curves of these two types of centrifuges nearly fitted a cumulative log-normal function. The sharpness of cut characterized by the geometric standard deviation σ_g was surprisingly poor even with the nozzle type disk centrifuge: Gibson (1979) obtained a σ_g of 2.35 for 1.6-mm nozzles and 2.61 with 2-mm nozzles in an Alfa–Laval QS 412 centrifuge. This may have been due to the effect of flocculation of the china clay used in tests. Analysis of other data published by Allen and Baudet (1979), taken in an Alfa–Laval solids ejecting type laboratory disk centrifuge LAPX 202 also with china clay, gives a somewhat better value of $\sigma_g = 2.14$.

Gibson (1979) has also published some extremely valuable graphical and analytical correlations relating the operating cut size of the two types of centrifuges with suspension flow rate and feed solids concentration. In principle, these permit mathematical modeling of large scale separation plants.

D. Flotation

The basic design parameter for dissolved air flotation is the specific overflow rate, v_L, similarly to settling tanks, However, here v_L represents the downward velocity of the liquid and its value depends on the air–solids mass ratio, a_s, the specific mass of air adsorbed on the particles. Experiments on algal wastewater and activated sludge by Purchas (1977) show that an empirical relation between v_L and a_s existed in the form

$$v_L + u_g = K_1 a_s^{K_2} \qquad (14\text{-}46)$$

where u_g is the settling velocity of an average particle at $a_s = 0$ and K_1 and K_2 are constants for a particular suspension. However, addition of surface active agents such as detergents or coagulants is sometimes necessary. Their

SCALEUP TECHNIQUES

presence can alter physical properties so they must be included in the tests. The second important design parameter is the float solids concentration c_f. This has been found to be independent of a_s since dewatering of the float material is principally due to drainage of the interstitial water in the float material above the water level.

The empirical relationship between c_f, the depth of the float above water level, H_w, and solids flux, G_s, (mass flow rate of solids introduced per unit area of flotation unit) has been found to be as follows:

$$C_f = K_3 H_w^{K_4} G_s^{-K_5} \qquad (14\text{-}47)$$

Experimentally H_w has been related to the depth of the float below the water level H_B and a_s by

$$\frac{H_B}{H_w} = \frac{K_6}{a_s^{K_7}} \qquad (14\text{-}48)$$

Unfortunately, constants K_3, K_4, K_5, K_6, and K_7 are specific to a given solid–liquid system.

In a manner similar to gravity thickeners, both the clarification duty and thickening duty of a flotation unit make independent area demands. The procedure recommended by Purchas (1977) is to select an area needed for satisfactory thickening first and then check the clarification demand. The area requirements would then be increased if necessary.

E. Cake Filters and Dewatering Screens

Scaleup of conventional cake filtration is based on Equation 14-12. Solutions of this equation are given by Svarovsky (1981) for many kinds of operations such as constant pressure, constant rate, or the variable pressure variable rate. The basic problem encountered in scaleup of cake filters is establishing effective values of the medium resistance and the specific cake resistance.

The medium resistance, R, often varies with time. This behavior results when some of the solids penetrate the medium as it compresses under applied pressure. For convenience, the resistance associated with the piping and the feed and outlet ports is also sometimes included in R.

The specific cake resistance, α, is the most troublesome parameter; α is needed to calculate the resistance to flow from the quantity of cake deposited on the filter. In practice, α depends on the approach velocity of the suspension, the degree of flow consolidation that the cake undergoes with time, the feed–solids concentration, and, most importantly, the applied pressure drop, Δp. α is not constant since Δp changes in most practical cakes, due to the compressibility of the cake.

The specific cake resistance, α, often decreases with velocity and feed concentration. Rushton et al. (1979) showed that α may go through a maximum when it is plotted against solids concentration. The strongest effect of α is that of pressure. This is often expressed as

$$\alpha = \alpha_0 (\Delta p_c)^n \tag{14-49}$$

where α_0 is the cake resistance at the unit applied pressure drop, Δp_c is pressure drop across the cake, and n is an empirical exponent. Since each layer in a cake is subject to a different pressure drop, the only simple way to deal with the problem is to define an average value, α_{av}.

A rather widely used definition of the average is

$$\alpha_{av} = \Delta p_c \Big/ \int_0^{\Delta p_c} \frac{d(\Delta p_c)}{\alpha} \tag{14-50}$$

which when used with Equation 14-49, gives

$$\alpha_{av} = (1 - n)\alpha_0 (\Delta p_c)^n \tag{14-51}$$

This can be substituted for α in Equation 14-12.

For higher feed concentrations, the volumes of the feed slurry and the filtrate differ significantly. A correction based on the effective concentration, $c_{corrected}$, must be applied. The solids concentration in the feed appears in Equation 14-12 to express the growth of the cake over time. There is an obvious inconsistency in this approach, since the concentration c applies to the feed while the volume V refers to the filtrate. In order to reconcile the two the value of c used in Equation 14-12 must be corrected for high solids concentrations by

$$c_{corrected} = \frac{1}{1/c - 1/\rho_s - (m - 1)/\rho} \tag{14-52}$$

where

m = mass ratio of wet to dry filter cake
ρ_s = solids density
ρ = liquid density

The filter size calculated from Equation 14-12 is almost always larger than necessary and therefore on the safe side. Studies are still in progress to quantify and explain the complex effects of cake resistance. Filter design procedures will likely have to change accordingly. One thing is certain: liquid-filtration theory will never be simple.

Whenever mechanical squeezing is applied in addition to conventional filtration, the additional dewatering of a compressible cake can be worked out from the reduction in cake porosity. The basic empirical relationship usually used for the analysis is similar to Equation 14-49.

$$\varepsilon = \varepsilon_0 (\Delta p)^{-\lambda} \qquad (14\text{-}53)$$

where ε_0 is porosity at unit pressure drop and λ is an empirical exponent. Scaleup of filtration centrifuges must consider in addition to the pressure drop from the centrifugal head, the centrifugal "mass" forces on the liquid flowing through the cake. Analysis of the filtration cycle is complex and the reader is referred to Svarovsky (1981) and Purchas (1977). The critical operating parameters for a dewatering screen of a given design are the slot width, s, and the suspension feed velocity on top of the sieve bend, v. If a Reynolds number is defined as

$$\text{Re} = \frac{vs}{\gamma} \qquad (14\text{-}54)$$

then the performance of the screen will become independent of Re above some value. For a DSM sieve bend this is at a Reynolds number greater than 250.

The effective cut size, x_{50}, is then about half the slot width, s, and the underflow-to-throughput ratio is constant as is the moisture content of the separated solids. The scaleup procedure from small-scale gravity feed tests is quite simple. Since the minimum slot size available is 50 μm which gives a cut size of 25 μm (for a gravity feed velocity of 3 m/s), the feed velocity has to be increased up to 12 m/s by pump pressure feeding for fine particles of less than 100 μm in size.

F. Deep Bed Filters

Most of the available mathematical models for deep bed filtration characterize the removal of particles as a change in concentration as in Equation 14-14,

$$-\frac{dc}{dL} = \lambda c \qquad (14\text{-}14)$$

as discussed in Section II-D-3. λ, a measure of efficiency, depends on approach velocity, particle size of the feed solids, grain size of the media, liquid velocity, and the quantity of solids already deposited in the bed. Mean values of λ are generally used.

Considerable empirical data on values of λ are available in the literature. Equation 14-14 can also be written with an additional term on the right-hand side to account for stripping of the solids deposited during filtration. Data from laboratory, pilot scale, or (preferably) full-scale tests on the feed suspension are essential for developing reliable design information and optimizing pretreatment methods.

VI. UNCERTAINTIES

A. State of Flocculation and Pretreatment

By far the greatest uncertainty in scaleup of solid–liquid separation is the state of flocculation and the events that will influence flocculation. There is still a considerable lack of knowledge about aspects of flocculation mechanism uncertainty about the action of many flocculants and also about the physical properties of flocs. Scaleup of flocculation equipment is consequently very doubtful and such equipment is designed more by experience and guess work than by calculation. Much the same applies to other methods of pretreatment such as aging, heat treatment, or freezing followed by thawing.

B. End Effects in Equipment

These represent a major area of uncertainty in scaleup, particularly for gravity and centrifugal clarifiers, hydrocyclones, thickeners, and flotation units. Turbulence and nonuniform flow distribution, for example, short circuiting and surface flows, and so on, are the major causes. Nonideality can be allowed for means of efficiency factors such as area detention efficiencies, as discussed by Purchas (1977). Unfortunately, such efficiency factors themselves depend on equipment size and their values can be established only from experience with similar units.

C. Deviations from Stokes' Law in Particle Settling

Scaleup of many dynamic separators is largely based on Stokes' law. Deviations observed in practice present major difficulties. Limitations of Stokes' law have been discussed in Section II-C. Flocculation and concentration effects are the most troublesome in scaleup of sedimenting centrifuges, hydrocyclones, and gravity clarifiers. Convection currents that develop in almost any prolonged settling situations can also be a major problem.

D. Compressibility of Cakes

In cake filtration the greatest uncertainty is in the flow properties of the cakes. Most practical cakes are compressible and this plays havoc with filtration theory. Specific cake resistance is very sensitive to porosity and this in turn varies through the cake depending on pressure, approach velocity, feed concentration, and even on time as flow consolidation takes place. The general rule is that as many as possible of the above conditions must be kept constant in scaleup tests. This is very difficult to do. Cake cracking, particularly in washing applications, is a major area of difficulty.

E. Reentrainment

Reentrainment refers to the pickup of particles already separated back into the flow. This occurs in deep beds, gravity clarifiers, sedimenting centrifuges, and hydrocyclones. The magnitude of reentrainment often depends on the scale of the equipment. Although reentrainment impacts on the effective cut size it does not scale up to the same degree as the major separation mechanisms. Equipment and surface finish and other aspects of workmanship often affect reentrainment.

NOMENCLATURE

a_s	Air-to-solids mass ratio
A	Face area of bed or of a thickener
b	Cyclone inlet diameter
c, C	Solids concentration
c_f	Feed solids concentration
c_0	Overflow concentration
c_u	Underflow concentration
D	Cyclone diameter
e	Vortex finder diameter
E_T	$\dfrac{x_c(1 - x_0)}{x_0(1 - x_c)}$, Total efficiency or recovery
Eu	Euler number
Fr	Froude number
g	Gravity acceleration
G	Separation factor
$G(x)$	Grade efficiency
$G'(x)$	Reduced grade efficiency
G_c	Critical flux
G_s	Solids flux
H	Height
H_b	Depth of float below liquid level
H_w	Height of float above liquid level
K	Constant, permeability of bed
l	Length of vortex finder
L	Depth of bed, cyclone length
n	Exponent
N	Number of cyclones in parallel

Q	Volumetric flowrate
R	Radius, medium resistance
R_f	Underflow-to-throughput ratio
Re	Reynolds number
Re_p	Particle Reynolds number
S	Distance between bars in sievebends
S_0	Volume specific surface of bed
Stk	Stokes number
Stk_{50}	Stokes number corresponding to x_{50}
Stk'_{50}	Reduced Stokes number
T	Wash ratio, number of void volumes
u	Particle velocity
U_g	Particle terminal settling velocity
U_p	Hindered settling velocity of a particle
v	Liquid velocity
v_1	Specific overflow rate in flotation
V	Volume of filtrate
w	Angular speed
x	Particle size, mother liquor concentration
x_c	Concentration in centrate from centrifuge
x_0	Initial mother liquor concentration
x_{50}	Cut size
α	Specific cake resistance
α_0	Specific cake resistance at unit applied
α_{av}	Average specific cake resistance
β	Entrance velocity loss—hydrocyclones
Δp	Pressure drop
$\Delta \rho$	Density difference between solids and liquid
ε	Voidage, porosity
ε_0	Porosity at unit pressure drop
λ	Efficiency parameter
μ	Liquid viscosity
γ	Kinematic viscosity
ω	Angular speed
ρ	Liquid density
ρ_s	Solids density
σ_g	Geometric standard deviation
Σ	Capacity factor

REFERENCES

Allen, T., and Baudet, M. G., *Powder Technology*, **18**, 131–138 (1977).

Ambler, C. M., *Ind. Eng. Chem.*, **53** (6), 430 (1961).

Baron, G. and Wajc, S., "Hindered Settling in Centrifuges," Synopse 686, *Chem. Eng. Tech.*, No. 4, 333 (1979).

Bradley, D., *The Hydrocyclone*, Pergamon Press, Oxford, 1965.

Coe, H. S., and Clevenger, G. H., *Trans. Am. Inst. Mining Met. Engs.*, **55**, 356 (1916).

Draper, N., Dredge, K. H., and Lynch, A. J., Paper 22, 9th Commonwealth IMM Congress (1969).

Elsken, J. C., and Ehinger, G. A., "New High Capacity Stationary Screens," *Chem. Eng. Prog.*, **59** (1), 76–80 (1963).

EPA, "Process Design Manual for Suspended Solids Removal," U.S. Environmental Protection Agency Technology Transfer, 825/1-75-883a., Washington, DC., (Jan. 1975).

Gerrard, A. M., and Liddle, C. J., "The Optimal Selection of Multiple Hydrocyclone Systems," *The Chemical Engineer*, pp. 295–296 (May 1975).

Gibson, K., "Large Scale Tests on Sedimenting Centrifuges and Hydrocyclones for Mathematical Modelling of Efficiency," *Proc. of the Symposium on Solid-Liquid Separation Practice*, Yorkshire Branch of the I. Chem. E., Leeds, England, pp. 1–10 (March 27–29, 1979).

Ide, T., and Katoka, K., "A Technical Innovation in Sludge Blanket Clarifiers," The Second World Filtration Congress 1979, pub. by Filtration Society, Olympia, London, pp. 377–385 (Sept. 18–20, 1979).

Kos, P., *Theory of Gravity Thickening of Flocculent Suspensions and a New Method of Thickener Sizing*, The Second World Filtration Congress 1979, pub. by Filtration Society, Olympia, London, pp. 595–603 (Sept. 18–20, 1979).

Mitzmager, A., and Mizrahi, B. S., *Trans. Inst. Chem. Eng.*, **42**, T152–T157 (1964).

Plitt, L. R., "The Analysis of Solid-Liquid Separations in Classifiers," *The Canadian Mining and Metallurgical (CIM) Bulletin*, pp. 42–47 (April 1977).

Purchas, D. B., ed., *Solid-Liquid Separation Equipment Scaleup*, Uplands Press., London, 1977.

Rietema, K., *Cyclones in Industry*, Elsevier, Amsterdam, 1961.

Rushton, A., Hosseini, M., and Hassan, I., "The Effects of Velocity and Concentration on Filter Cake Resistance," *Proceedings of the Symposium on Solid-Liquid Separation Practice*, Leeds, U.K., Yorkshire Branch of the I. Chem. E., pp. 78–91, (March 27–29, 1979).

Svarovsky, L., and Bavishi, A., Proc. European Congress on "Transfer Processes in Particle Systems," Nuremberg, Dechema, Frankfurt-am-Main (March 28–30, 1977).

Svarovsky, L., "Efficiency of Separation Processes," in *Progress in Filtration and Separation*, R. J. Waleman, ed., Elsevier, Amsterdam, 1979.

Svarovsky, L., *Solid-Gas Separation*, Elsevier, Amsterdam, 1980.

Svarovsky, L., ed., *Solid-Liquid Separation*, 2nd Edition Butterworths, London, 1981.

Talmage, W. P., and Fitch, E. B., *I & EC*, **47**, 38 (1955).

15

THE ENVIRONMENTAL CHALLENGES OF SCALEUP

P. B. LEDERMAN

I.	Evaluation of Major Environmental Issues	595
	A. The Environmental Challenge	595
	B. The Three Environmental Sinks	596
	C. Regulation	596
	1. Development/Regulatory Interactions	596
	2. Agency Responsibility	599
	3. Support of Environmental Permit Applications	600
	4. Federal Statutes and Regulations	600
	a. Air Emissions	601
	b. Water Discharges	604
	c. Solid/Hazardous Discharges to the Land	607
II.	Control and Minimization of Waste	612
	A. Hazardous Waste from Pilot Units	612
	B. The Chemical Industry as a Regulated Industry	612
	C. Meeting Environmental Regulations	612
	D. Minimizing Environmental Concerns	613
III.	Environmental Considerations and Scaleup	614
	A. Development of Environmental Data	615
	B. Environmental Impacts—A Part of Process Evaluation	616
	C. Disposal Alternatives	616

D.	Challenges to New Products	618
E.	Right to Know	618
IV.	The Environmental Challenge	619
References		619

I. EVALUATION OF MAJOR ENVIRONMENTAL ISSUES

A. The Environmental Challenge

It has been almost 100 years since Ibsen wrote *The Enemy of the People* in which environmental considerations were a central theme. Yet it has been only in the last two decades that environmental considerations and challenges became a primary concern of and impacted significantly on the business and technical community. Particularly in the United States, land, air, and water were considered "free." As long as processes did not create local nuisances, such as objectionable odors, environmental considerations were not part of design and manufacturing philosophy. Designers did consider the safety of operating personnel from the point of view of catastrophes, but in most cases did little, or nothing, to design into the systems safeguards from chronic as well as acute exposures to environmental insults.

Some basic steps to minimize pollution were taken in certain cases. These included the elimination of floating oil in rivers using API separators. In particular cases the discharge of toxic materials into waterways, which also served as sources of drinking water, were eliminated. This was usually done only in those cases where an industrial source could be identified as the source of a drinking water treatment plant problem. Phenol discharge from refinery waste streams was such a chemical.

Solid waste was disposed of either in or on the land, and sludges were ponded. Land was considered an essentially free commodity. The out-of-sight, out-of-mind syndrome was prevalent throughout society and the cost of environmentally acceptable disposal was not considered in design or economic evaluations.

Highly visible air pollution was given attention in the 1950s and early 1960s. The cleanup of downtown Pittsburgh by switching from coal to gas in the 1950s is particularly noteworthy. Los Angeles was a leader in the 1950s and early 1960s in identifying the causes of smog and establishing scenarios to minimize releases which cause smog. In most cases this did not affect industrial sources, with the exception of power generating facilities, in the mid-1960s; development facilities were only affected in the 1970s.

Beginning with Earth Day 1970 the casual approach to pollution has changed drastically. The process designer, the chemist, and the process development engineer must consider environmental aspects of any process. In many cases these will become paramount in developmental efforts. Public pressure, as well as statutes and regulations, require that we minimize waste discharges and design for zero environmental insult.

This new design philosophy poses added constraints and requires a somewhat different development and design approach to any process. The costs involved in meeting environmental, occupational health and safety, and consumer product safety requirements are so significant that they can, and often do, become prime considerations in whether to proceed with process development and scaleup.

B. The Three Environmental Sinks

There are three environmental sinks: the air, the water, and the land. In the early years of environmental regulation it was quite common that only one or two media sinks were of concern. For chemical plants, discharges to flowing waters were of primary concern because this was the regulated sink. Ocean disposal was an important sink, but was of secondary concern compared to rivers. For the power industry, discharges to the air were of primary concern. Always the land served as an available final disposal source; until 1976 this was at best a semiregulated sink. It is clear that removing a pollutant from one medium only causes it to be discharged to another medium, in a different form in many cases. For example, removing sulfur oxides for stack gases created large amounts of calcium sulfite/sulfate sludge which had to be disposed of on the land. As long as land was cheap it was easier to pond these sludges than to find a use for them or minimize their impact on the land. It is evident, as shown in Figure 15-1 that pollutant removal from one sink affects pollutant discharge to another. It is only now, more than ten years after the passage of the first environmental statutes, that the last "free" sink has been closed. The real challenge today, in scaleup, is to minimize total pollutant volume. The discharge of any pollutant must be accomplished in an environmentally acceptable manner. The process of addressing the environmental challenges and regulations in the scaleup process is complex.

C. Regulation

1. Development / Regulatory Interactions

The process of arriving at the environmental regulations specific to a process is a long one. The development engineer may have to deal with regulations of two to five agencies, depending on the industry. The most common of these will be the Environmental Protection Agency (EPA) and the Occupational Safety and Health Administration (OSHA). However, the Food and Drug Administration (FDA), the Nuclear Regulatory Agency (NRC), the Department of Transportation (DOT), and/or others may have some jurisdiction, as shown in Figure 15-1. It is advisable to check with the people in an industrial organization who

EVALUATION OF MAJOR ENVIRONMENTAL ISSUES

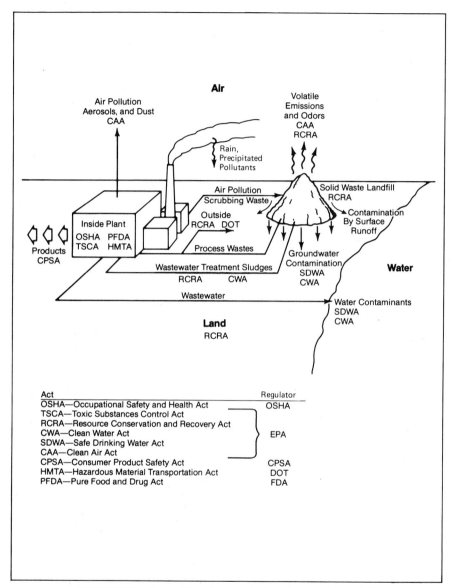

FIGURE 15-1   The multimedia impact of environmental regulation.

are responsible for governmental regulations and regulatory affairs. They are familiar with regulations and regulatory agencies which have jurisdiction.

It is important that one begin groundwork with federal and state agencies at the earliest possible moment. Be sure to do so only with the advice of the governmental regulatory people in your company. A major concern in any development should be confidentiality of information, particularly with the

598   THE ENVIRONMENTAL CHALLENGES OF SCALEUP

effect of the Freedom of Information Act. Yet it is important to determine at an early stage the possible impediments that may exist from an environmental regulatory point of view, in the development of a new process or the bringing on of a new product. Discussions should be on an informal basis and rulings should not be requested. However, one should determine whether your product will be regulated under FDA procedures, or will be the concern of the

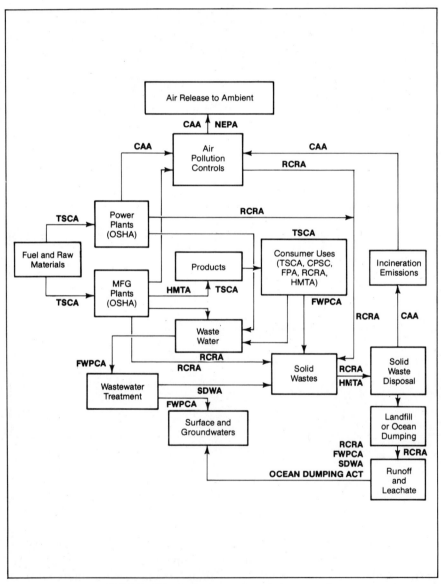

FIGURE 15-2 Environmental regulation scenario.

TABLE 15-1 Environmental Responsibilities of Federal Agencies

Agency	Responsibility
Environmental Protection Agency (EPA)	Sets and enforces regulations on discharges to the environment: air, water, land. Regulates use of pesticides, herbicides, insecticides. Regulates release, manufacture, and use of toxic substances.
Department of Transportation (DOT)	Controls transport of hazardous materials.
Occupational Safety and Health Administration (OSHA)	Regulates work place safety and environmental conditions.
Food and Drug Administration (FDA)	Regulates use of substances in food, drug, and cosmetic applications.
Department of Interior (DOI)	Regulates surface mining rehabilitation.
Office of Surface Mining (OSM) Mine Safety Administration (MESA)	Establishes and enforces mine safety standards.
Nuclear Regulatory Commission (NRC)	Regulates nuclear discharges.
Department of Energy (DOE)	Energy/environmental research and development; nuclear waste disposal.

Consumer Product Safety Administration, or will fall under toxic substances regulations. Again, early discussions and early identification are important to the successful completion of the scaleup to a full scale plant.

2. Agency Responsibility

The mandates of various federal agencies are relatively clear. In some cases state agencies have taken over the role of the federal agency. In other cases, particularly in the area of nuclear and toxic substances there are some overlaps. Table 15-1 summarizes basic responsibilities of the federal agencies. Regulating environments within a plant generally falls to OSHA, while regulating the environment outside of a plant and discharges from a plant fall to EPA. If a substance is a pesticide or is toxic, EPA is the permitting agency, but if it is used in a food or drug application, FDA takes primacy. In nuclear matters, the NRC generally has primacy, but disposal of nuclear wastes is the responsibility of the DOE, with EPA having oversight responsibility. In a mining situation, the Department of Interior's Mine Safety Administration plays an important part in developmental efforts, while restoration of the land will be regulated by rules written by the Department of Interior, but approved and enforced by EPA.

3. Support of Environmental Permit Applications

It is important in any developmental scale of operation to begin early the process of enlisting the support of the body politic. New facilities, as illustrated by the siting of new power plants, require five to ten years from conception to startup. Much of this time will be taken up by answering environmental considerations. The first step in doing this is writing an environmental impact statement (EIS). This will require a knowledge of the land to be used, the type of plant, the discharges that can be anticipated, the land use patterns, geology, and the socioeconomic climate. The EIS, which often can be several feet thick, will be reviewed by local and federal agencies, as well as the environmental community. It will serve as a basis for discussions and possible litigation as to whether the plant should be permitted.

It is often useful to meet on an informal basis with local political and environmental groups, that is, the local power structure, to gain their understanding and support, either active or by silence, of the particular project under consideration. This type of support can easily save millions of dollars and years of time. If during these informal discussions, opposition develops, every effort should be made to answer those concerns. It will often be less expensive to make concessions that do not affect the process technology per se than to "stand pat."

It is important that, during the process of developing the environmental impact statement and having informal meetings with concerned citizen groups, concerns of key federal and state statutes and officials be addressed. The earlier this can be done, the smoother will be the developmental process.

4. Federal Statutes and Regulations

Environmental regulation has not been a major concern in the scaleup process until fairly recently. However, environmental legislation, even though unenforced and having no effect on the process industries, covered pollution of harbor waters as early as 1890. Today, there are federal, and usually state, laws covering all aspects of environmental concern. Those of greatest concern in process development and scaleup are summarized on Table 15-2 and their interaction is depicted in Figure 15-2. There are a number of additional laws, such as the Hazardous Materials Transport Act and the Safe Drinking Water Act, which affect the scaleup process indirectly by establishing standards which supplement those established under the primary laws. With the passage of RCRA in 1976 and the regulation of solid, hazardous waste in 1980, there is no unregulated sink. This imposes new challenges to the developer to minimize waste generation on all media.

Although a legislative history and detailed discussion of the laws is beyond the scope of this discussion, it may be found in Lederman (1983). However,

EVALUATION OF MAJOR ENVIRONMENTAL ISSUES

TABLE 15-2 Environmental Regulatory and Related Acts

Act	Agency
Federal Water Pollution Control Acts (FWPCA)	EPA
Clean Air Acts (CAA)	EPA
Resource Conservation and Recovery Act (RCRA)	EPA
Federal Pesticide Act (FIFRA)	EPA
Toxic Substances Act (TSCA)	EPA
Safe Drinking Water Act (SDWA)	EPA
Marine Protection, Reservoirs, and Sanctuaries Act (Ocean Dumping)	EPA
Hazardous Materials Transport Act	DOT
Food, Drug, and Cosmetics Act	FDA
Consumer Product Safety Act	CPSC
Energy Supply and Environmental Coordination Act (ESECA)	DOE
Deepwater Port Act	DOT, EPA
Surface Mining Control and Reclamation Act	DOI
Occupational Safety and Health Act	OSHA

brief analyses of the regulations seem appropriate. It is these regulations which set the tone of the challenge the process developer must meet.

a. AIR EMISSIONS

The Clean Air Act provides for control of emissions into the atmosphere from all sources. Much of its impact has been on the automobile and power industries, as these are two major sources of pollutants. However, the legislation is clear in that it provides for setting criteria pollutant emission standards for all new sources. Hazardous pollutants may also be regulated and will be discussed further. Much of the regulation of atmospheric pollution has been given over to the states, who have developed State Implementation Plans to meet ambient air quality standards given in Table 15-3. These air quality standards have been established as required by the Clean Air Act with the public health and welfare in mind. The primary standards, established for all six criteria pollutants, establish levels which, if exceeded, affect public health. The secondary standards, sometimes more stringent, were established to protect property. These standards serve as a basis for establishing point-source discharge criteria.

In a region where those standards are not met, it will be difficult, if not impossible, to build new plants. Thus, the Clean Air Act and the resulting body of regulations can significantly affect location of a new facility (even a pilot facility) that emits any criteria pollutants: SO_2, particulates, NO_x, hydrocarbon, oxidant. It may also mitigate against making process changes.

TABLE 15-3 National Ambient Air Quality Standards (g/m^3), Typical Criteria Air Pollutant

Pollutant	Primary[a]	Secondary[b]
Sulfur dioxide		
1 yr (avg)	80	—
24 hr (annual max)	365	—
3 hr (avg)	—	1,300
Particulates		
1 yr (geometric mean)	75	60
24 hr (annual mean)	260	150
Nitrogen oxides		
1 yr (annual mean)	100	100
Hydrocarbons		
(0600–0900)	160	160
Carbon monoxide		
(1 hr)	40,000	40,000
(8 hr)	10,000	10,000
Oxidants		
(1 hr)	260	260
Lead (Quarterly)	1.5	1.5

[a] Public health is basis.
[b] Public welfare is basis.

The Clear Air Act requires EPA to set emission performance standards for new plants for all criteria pollutants and other emissions, as may be appropriate. Existing air pollution sources need not meet New Source Performance Standards. However, in order to meet Air Quality Criteria for the criteria pollutants (Table 15-3), the states must set standards for existing stationary sources. These are sometimes less stringent than New Source Performance Standards, but more often than not follow them. When an existing source is modified in a way that alters the process capacity significantly, increases the emissions, or is reconstructed at a cost equal to 50 percent of a new facility cost, the New Source Performance Standards apply.

The EPA is expected to issue New Source Performance Standards for most of the process industries. These standards are achievable by the best technology currently available. This is both a technology and economics based standard. The technology must be demonstrated to be technically and economically feasible, at least at one commercial site. Alternate control technologies that provide the same or lower emissions will be acceptable as long as the emission level is met. The fact that application of emission control technology in a given situation is uneconomic may not provide an escape for industry from meeting the standard. The affected industrial segments are listed in Table 15-4.

EVALUATION OF MAJOR ENVIRONMENTAL ISSUES

TABLE 15-4 Process Industry New Source Performance Standards

Representative Categories Under Development	Representative Categories To Be Developed

Organic chemicals
   Storage, transfer
   Manufacture
Potash production
Nitrates production
Polymer/resin plants
Textile processing
Urea
Wood pulping
Coal gasification
Petroleum refining
	Synthetic fibers
	Ammonia
	Borax/boric acid
	Detergents
	Uranium refining

The process industries for which New Source Performance Standards are to be issued are varied. The list suggests the industries that may be affected by regulations in the future unless there are changes in the law or more likely the regulatory environment. Where standards are to be developed, regulations will not be issued for probably ten years. The standard-setting process includes fact finding, drafts, internal review, public review and comment, and revision prior to final issuance. This usually takes two to three years. Standards are reviewed every five years.

Only those few materials listed in Table 15-5 have been identified as hazardous air pollutants. However, the question of hydrocarbon and hazardous pollutant control, with respect to both levels and control methodology, will be addressed in the future in much greater depth than before. Twenty additional hydrocarbons have already been proposed for listing as hazardous air pollu-

TABLE 15-5 Typical Hazardous Pollutants Regulated Under Clean Air Act

Asbestos
Beryllium
Mercury
Vinyl chloride
Lead[a]
Benzene[b]
Arsenic[b]

[a] Now a criteria pollutant.
[b] No standard set.

tants. Health effect studies and control methodology alternative studies, which are essential background to standard setting, are just getting under way.

The proper balance between environmental control and cost, so important to sound development, will continue to be a focal point of discussion. While more cost/benefit data will become available, this will be insufficient to abate emotionally charged discussion.

b. WATER DISCHARGES

The Clean Water Acts of 1972 (amended in 1977) have had a much greater influence on the process industry and development of new processes than the Clean Air Act. The legislation and resulting regulations had three basic aims:

Eliminate pollutant discharges into navigable waters by 1985.
Ensure that the waters shall be suitable for swimming and fishing by 1983.
Prohibit the discharge of toxic pollutants.

In contrast to the air legislation, the water legislation depends almost entirely on effluent limitations. Water quality standards were required in 1965. However, they play a secondary, but nevertheless important, role, particularly where drinking water is of concern. Initially, water quality and effluent guidelines were established for traditional municipal sanitary waste pollutants. Only in 1979 were criteria established for the specific chemical species listed in Table 15-6. The additional criteria given in Table 15-7 were expected based on primary drinking water standards. The timing of their issuance is highly uncertain.

The legislation is comprehensive and only parts of Titles III and IV of the Clean Water Acts are of interest to the process industry. Of particular interest in these titles are the sections covering effluent guidelines, pretreatment stan-

TABLE 15-6 Typical Suggested Water Quality Standards[a]

Pollutant	Maximum
Arsenic and compounds	130 mg/L
Beryllium	1.46 mg/L
Cadmium	3.92 mg/L
CCl_4	1400 mg/L
Chlordane	0.36 mg/L
Chloroform	1200 mg/L
2,4-Dichlorophenol	110 mg/L
Lead	1.39 mg/L

[a] *Federal Register*, 15 March 1979.

TABLE 15-7 Maximum Contaminant Levels for Safe Drinking Water Act

Contaminant	Level mg/L
Arsenic	0.05
Barium	1
Cadmium	0.010
Chromium	0.05
Lead	0.05
Mercury	0.002
Nitrate (as N)	10
Selenium	0.01
Silver	0.05
Endrin	0.0002
Lindane	0.004
Methoxychlor	0.1
Toxaphene	0.005
2,4-D	0.1
2,4,5-TP	0.01

dards, toxic pollutants, oil and hazardous materials spill liability, and the establishment of the NPDES permit system.

During the period from 1974 to 1977 the emphasis was on setting limits and controlling conventional pollutants such as BOD, COD, and total dissolved solids. Under the 1972 act, EPA attempted to establish a toxic substances list. The first proposed list included aldrin, dieldrin, benzidine and its salts, cadmium and its compounds, cyanide, DDT, endrin, mercury, PCBs, and toxaphene. This list was attacked in the courts. No list has been published because the data on which limitations could be based are very scarce. The strict time limits and hearing requirements of the law made establishment of standards almost impossible. As a result of court action, PCBs were withdrawn from the market. The use of other materials was also sharply curtailed as a result of other court decisions. These were, however, individual actions and do not provide a broad basis for toxic pollutant control.

The emphasis on toxic or hazardous pollutant control was initiated in 1976 as a result of a suit, *NRDC* vs. *Train*, which resulted in an out-of-court consent decree called the Flannery Agreement. This requires EPA to promulgate best available technology (BAT) effluent standards in 21 industries for 65 toxic substances, as listed in Table 15-8. The 1977 amendments adopted this approach.

New industrial plant discharges are regulated under Section 306 of the Clean Water Act which deals with national standards of performance. This section requires EPA to define BATs for new water discharge sources for 36

TABLE 15-8 Flannery Agreement Toxic Pollutants

Acenaphthene
Acrolein
Acrylonitrile
Aldrin/dieldrin
Antimony and compounds
Arsenic and compounds
Asbestos
Benzene
Benzidine
Beryllium and compounds
Cadmium and compounds
Carbon tetrachloride
Chlordane (technical mixture and metabolites)
Chlorinated benzenes (other than dichlorobenzenes)
Chlorinated ethanes (including 1,2-dichloroethane, 1,1,1-trichloroethane, and hexachloroethane)
Chloroalkyl ethers (chloromethyl, chloroethyl, and mixed ethers)
Chlorinated naphthalene
Chlorinated phenols (other than those listed elsewhere; includes trichlorophenols and chlorinated cresols)
Chloroform
2-chlorophenol
Chromium and compounds
Copper and compounds
Cyanides
DDT and metabolites
Dichlorobenzenes (1,2- 1,3-, and 1,4-dichlorobenzenes)
Dichlorobenzidine
Dichloroethylenes (1,1- and 1,2-dichloroethylene)
2,4-dichlorophenol
Dichloropropane and dichloropropene
2,4-dimethylphenol
Dinitrotoluene
Diphenylhydrazine
Endosulfan and metabolites
Endrin and metabolites
Ethylbenzene
Fluoranthene
Haloethers (other than those listed elsewhere; includes chlorophenylphenyl ethers, bromophenylphenyl ether, bis-(dichloroisopropyl) ether, bis-(chloroethoxy) methane, and polychlorinated diphenyl ethers)
Halomethanes (other than those listed elsewhere; includes methylene chloride, methylchloride, methylbromide, bromoform, dichlorobromomethane, trichlorofluoromethane, dichlorodifluoromethane)
Heptachlor and metabolites
Hexachlorobutadiene
Hexachlorocyclohexane (all isomers)
Haxachlorocyclopentadiene
Isophorone
Lead and compounds
Mercury and compounds
Naphthalene
Nickel and compounds
Nitrobenzene
Nitrophenols (including 2,4-dinitrophenol and dinitrocresol)
Nitrosamines
Pentachlorophenol
Phenol
Phthalate esters
Polychlorinated biphenyls (PCBs)
Polynuclear aromatic hydrocarbons (including benzanthracenes, benzopyrenes, benzofluoranthene, chrysenes, dibenzanthracenes, and indenopyrenes)

Selenium and compounds
Silver and compounds
2,3,7,8-tetrachlorodibenzo-p-dioxin (TCDD)
Tetrachloroethylene
Thallium and compounds
Toluene
Toxaphene
Trichloroethylene
Vinyl chloride
Zinc and compounds

EVALUATION OF MAJOR ENVIRONMENTAL ISSUES

industry categories. The standards issued under this section are called New Source Performance Standards. They are generally less stringent than conventional (301) and toxic (307) BAT requirements, but more stringent than best practical technology (BPT) requirements. The standards are based on application of the best available demonstrated control technology (BADT), operating practice, or alternative effecting zero discharge of designated pollutants.

c. Solid / Hazardous Discharges to the Land

The last sink to be regulated was the land. The Resource Conservation and Recovery Act (RCRA) of 1976 (amended in 1984) included two key sections dealing with encouragement of recycling, reuse, and control of the disposal of hazardous wastes. The latter has received and continues to receive the major emphasis, with a large body of regulations which require special cradle-to-grave care, record keeping, and disposal and management of hazardous wastes. Regulated wastes, while ostensibly solid, have been defined to include liquids, semiliquids, and contained gases. They may be hazardous because they are ignitable, reactive, corrosive, toxic, or acutely hazardous. Some wastes are listed specifically; however, many are not listed, but will be considered hazardous because they meet specific criteria. The initial lists of wastes or materials which may be in waste streams include over 300 streams and substances. This list already covers many halogenated organics and heavy metals, and will undoubtedly continue to grow.

The potential exposure under this act is much greater than either the Clean Air or Clean Water Acts. Careful consideration for proper disposal must be given during the early development stages. Absolute amounts, not concentration, are the guiding factors, with amounts as low as 1 kg being covered.

There may be cases where disposal costs are so high that alternate process routes must be developed. It has become obvious that hazardous wastes should be segregated, where possible, or detoxified prior to leaving the process. These regulations are still being developed and must be tested before an in-depth analysis of their effects can be made. Ultimately, their effects will be significant in any developing technology. Of particular interest is 40 CRF 261, Appendix VIII summarized in Table 15-9, listing chemicals that serve as a basis for classifying a waste as hazardous.

TABLE 15-9 Chemical Constituents Which Serve as a Basis for Hazardous Designation[a]

Acetaldehyde	Acrolein
(Acetato)phenylmercury	Acrylamide
Acetonitrile	Acrylonitrile
3-(α-Acetonylbenzyl)-4-hydroxy-coumarin and salts	Aflatoxins
	Aldrin
2-Acetylaminofluorene	Allyl alcohol
Acetyl chloride	Aluminum phosphide
1-Acetyl-2-thiourea	4-Aminobiphenyl
	(*Continued on next page*)

TABLE 15-9 (Continued)

6-Amino-1,1a,2,8,8a,8b-hexahydro-8-(hydroxymethyl)-8a-methoxy-5-methylcarbamate azirino(2′,3′,3,4) pyrrolo(1,2-a)indole-4,7-dione (ester) (Mitomycin C)
5-(Aminomethyl)-3-isoxazolol
4-Aminopyridine
Amitrole
Antimony and compounds, N.O.S.[b]
Aramite
Arsenic and compounds, N.O.S.
Arsenic acid
Arsenic pentoxide
Arsenic trioxide
Auramine
Azaserine
Barium and compounds, N.O.S.
Barium cyanide
Benz(c)acridine
Benz(a)anthracene
Benzene
Benzenearsonic acid
Benzenethiol
Benzidine
Benzo(a)anthracene
Benzo(b)fluoranthene
Benzo(j)fluoranthene
Benzo(a)pyrene
Benzotrichloride
Benzyl chloride
Beryllium and compounds, N.O.S.
Bis(2-chloroethoxy)methane
Bis(2-chloroisopropyl) ether
Bis(chloroethyl) ether
N,N-Bis(2-chloroethyl)-2-naphthylamine
Bis(2-chloroisopropyl) ether
Bis(chloromethyl) ether
Bis(2-ethylhexyl) phthalate
Bromoacetone
Bromomethane
4-Bromophenyl phenyl ether
Brucine
2-Butanone peroxide
Butyl benzyl phthalate
2-sec-Butyl-4,6-dinitrophenol (DNBP)
Cadmium and compounds, N.O.S.

Calcium chromate
Calcium cyanide
Carbon disulfide
Chlorambucil
Chlordane (alpha and gamma isomers)
Chlorinated benzenes, N.O.S.
Chlorinated ethane, N.O.S.
Chlorinated naphthalene, N.O.S.
Chlorinated phenol, N.O.S.
Chloroacetaldehyde
Chloroalkyl ethers
p-Chloroaniline
Chlorobenzene
Chlorobenzilate
1-(p-Chlorobenzoyl)-5-methoxy-2-methylindole-3-acetic acid
p-Chloro-m-cresol
1-Chloro-2,3-epoxybutane
2-Chloroethyl vinyl ether
Chloroform
Chloromethane
Chloromethyl methyl ether
2-Chloronaphthalene
2-Chlorophenol
1-(o-Chlorophenyl)thiourea
3-Chloropropionitrile
alpha-Chlorotoluene
Chlorotoluene, N.O.S.
Chromium and compounds, N.O.S.
Chrysene
Citrus red No. 2
Copper cyanide
Creosote
Crotonaldehyde
Cyanides (soluble salts and complexes), N.O.S.
Cyanogen
Cyanogen bromide
Cyanogen chloride
Cycasin
2-Cyclohexyl-4,6-dinitrophenol
Cyclophosphamide
Daunomycin
DDD
DDE
DDT

Diallate
Dibenz(a,h)acridine
Dibenz(a,j)acridine
Dibenz(a,h)anthracene(Dibenzo(a,h) anthracene)
7H-Dibenzo(c,g)carbazole
Dibenzo(a,e)pyrene
Dibenzo(a,h)pyrene
Dibenzo(a,i)pyrene
1,2-Dibromo-3-chloropropane
1,2-Dibromoethane
Dibromomethane
Di-*n*-butyl phthalate
Dichlorobenzene, N.O.S.
3,3'-Dichlorobenzidine
1,1-Dichloroethane
1,2-Dichloroethane
trans-1,2-Dichloroethane
Dichloroethylene, N.O.S.
1,1-Dichloroethylene
Dichloromethane
2,4-Dichlorophenol
2,6-Dichlorophenol
2,4-Dichlorophenoxyacetic acid (2,4-D)
Dichloropropane
Dichlorophenylarsine
1,2-Dichloropropane
Dichloropropanol, N.O.S.
Dichloropropene, N.O.S.
1,3-Dichloropropene
Dieldrin
Diepoxybutane
Diethylarsine
O,*O*-Diethyl-*S*-(2-ethylthio)ethyl ester of phosphorothioic acid
1,2-Diethylhydrazine
O,*O*-Diethyl-*S*-methylester phosphorodithioic acid
O,*O*-Diethylphosphoric acid, *O*-*p*-nitrophenyl ester
Diethyl phthalate
O,*O*-Diethyl-*O*-(2-pyrazinyl)phosphorothioate
Diethylstilbestrol
Dihydrosafrole
3,4-Dihydroxy-α-(methylamino)methyl benzyl alcohol

Di-isopropylfluorophosphate (DFP)
Dimethoate
3,3'-Dimethoxybenzidine
p-Dimethylaminoazobenzene
7,12-Dimethylbenz(a)anthracene
3,3'-Dimethylbenzidine
Dimethylcarbamoyl chloride
1,1-Dimethylhydrazine
1,2-Dimethylhydrazine
3,3-Dimethyl-1-(methylthio)-2-butanone-*O*-[(methylamino)carbonyl]oxime
Dimethylnitrosoamine
α,α-Dimethylphenethylamine
2,4-Dimethylphenol
Dimethyl phthalate
Dimethyl sulfate
Dinitrobenzene, N.O.S.
4,6-Dinitro-*o*-cresol and salts
2,4-Dinitrophenol
2,4-Dinitrotoluene
2,6-Dinitrotoluene Di-*n*-octyl phthalate
1,4-Dioxane
1,2-Diphenylhydrazine
Di-*n*-propylnitrosamine
Disulfoton
2,4-Dithiobiuret
Endosulfan
Endrin and metabolites
Epichlorohydrin
Ethyl cyanide
Ethylene diamine
Ethylenebisdithiocarbamate (EBDC)
Ethyleneimine
Ethylene oxide
Ethylenethiourea
Ethyl methanesulfonate
Fluoranthene
Fluorine
2-Fluoroacetamide
Fluoroacetic acid, sodium salt
Formaldehyde
Glycidylaldehyde
Halomethane, N.O.S.
Heptachlor
Heptachlor epoxide (α, β, and γ isomers)

(*Continued on next page*)

TABLE 15-9 (Continued)

Hexachlorobenzene
Hexachlorobutadiene
Hexachlorocyclohexane (all isomers)
Hexachlorocyclopentadiene
Hexachloroethane
1,2,3,4,10,10-Hexachloro-1,4,4a,5,8,8a-hexahydro-1,4 : 5,8-endo,endo-dimethanonaphthalene
Hexachlorophene
Hexachloropropene
Hexaethyl tetraphosphate
Hydrazine
Hydrocyanic acid
Hydrogen sulfide
Indeno(1,2,3-c,d)pyrene
Iodomethane
Isocyanic acid, methyl ester
Isosafrole
Kepone
Lasiocarpine
Lead and compounds, N.O.S.
Lead acetate
Lead Phosphate
Lead subacetate
Maleic anhydride
Malononitrile
Melphalan
Mercury and compounds, N.O.S.
Methapyrilene
Methomyl
2-Methylaziridine
3-Methylcholanthrene
4-4'-Methylene-bis-(2-chloroaniline)
Methyl ethyl ketone (MEK)
Methyl hydrazine
2-Methyllactonitrile
Methyl methacrylate
Methyl methanesulfonate
2-Methyl-2-(methylthio)propionaldehyde-o-(methylcarbonyl) oxime
N-Methyl-N'-nitro-N-nitrosoguanidine
Methyl parathion
Methylthiouracil
Mustard gas
Naphthalene
1,4-Naphthoquinone
1-Naphthylamine
2-Naphthylamine
1-Naphthyl-2-thiourea
Nickel and compounds, N.O.S.
Nickel carbonyl
Nickel cyanide
Nicotine and salts
Nitric oxide
p-Nitroaniline
Nitrobenzene
Nitrogen dioxide
Nitrogen mustard and hydrochloride salt
Nitrogen mustard N-oxide and hydrochloride salt
Nitrogen peroxide
Nitrogen tetroxide
Nitroglycerine
4-Nitrophenol
4-Nitroquinoline-1-oxide
Nitrosamine, N.O.S.
N-Nitrosodi-N-butylamine
N-Nitrosodiethanolamine
N-Nitrosodiethylamine
N-Nitrosodimethylamine
N-Nitrosodiphenylamine
N-Nitroso-N-propylamine
N-Nitroso-N-ethylurea
N-Nitrosomethylethylamine
N-Nitroso-N-methylurea
N-Nitroso-N-methylurethane
N-Nitrosomethylvinylamine
N-Nitrosomorpholine
N-Nitrosonornicotine
N-Nitrosopiperidine
N-Nitrosopyrrolidine
N-Nitrososarcosine
5-Nitro-o-toluidine
Octamethylpyrophosphoramide
Oleyl alcohol condensed with 2 moles ethylene oxide
Osmium tetroxide
7-Oxabicyclo(2.2.1)heptane-2,3-dicarboxylic acid
Parathion
Pentachlorobenzene
Pentachloroethane
Pentachloronitrobenzene (PCNB)
Pentacholorophenol
Phenacetin
Phenol

Phenyl dichloroarsine
Phenylmercury acetate
N-Phenylthiourea
Phosgene
Phosphine
Phosphorothioic acid, O,O-dimethyl ester, O-ester with N,N-dimethyl benzene sulfonamide
Phthalic acid esters, N.O.S.
Phthalic anhydride
Polychlorinated biphenyl, N.O.S.
Potassium cyanide
Potassium silver cyanide
Pronamide
1,2-Propanediol
1,3-Propane sultone
Propionitrile
Propylthiouracil
2-Propyn-1-ol
Pyridien
Reserpine
Saccharin
Safrole
Selenious acid
Selenium and compounds, N.O.S.
Selenium sulfide
Selenourea
Silver and compounds, N.O.S.
Silver cyanide
Sodium cyanide
Streptozotocin
Strontium sulfide
Strychnine and salts
1,2,4,5-Tetrachlorobenzene
2,3,7,8-Tetrachlorodibenzo-p-dioxin (TCDD)
Tetrachloroethane, N.O.S.
1,1,1,2-Tetrachloroethane
1,1,2,2-Tetrachloroethane
Tetrachloroethene (Tetrachloroethylene)
Tetrachloromethane
2,3,4,6-Tetrachlorophenol
Tetraethyldithiopyrophosphate
Tetraethyl lead
Tetraethylpyrophosphate

Thallium and compounds, N.O.S.
Thallic oxide
Thallium (I) acetate
Thallium (I) carbonate
Thallium (I) chloride
Thallium (I) nitrate
Thallium selenite
Thallium (I) sulfate
Thioacetamide
Thiosemicarbazide
Thiourea
Thiuram
Toluene
Toluene diamine
o-Toluidine hydrochloride
Tolylene diisocyanate
Toxaphene
Tribromomethane
1,2,4-Trichlorobenzene
1,1,1-Trichloroethane
1,1,2-Trichloroethane
Trichloroethene (Trichloroethylene)
Trichloromethanethiol
2,4,5-Trichlorophenol
2,4,6-Trichlorophenol
2,4,5-Trichlorophenoxyacetic acid (2,4,5-T)
2,4,5-Trichlorophenoxypropionic acid (2,4,5-TP) (Silvex)
Trichloropropane, N.O.S.
1,2,3-Trichloropropane
O,O,O-Triethyl phosphorothioate
Trinitrobenzene
Tris(1-azridinyl)phosine sulfide
Tris(2,3-dibromopropyl) phosphate
Trypan blue
Uracil mustard
Urethane
Vanadic acid, ammonium salt
Vanadium pentoxide (dust)
Vinyl chloride
Vinylidene chloride
Zinc cyanide
Zinc phosphide

[a] 40 CFR 261.1, *Federal Register*, **45**(98), 33132, (19 May 1980).
[b] N.O.S. indicates general class; not otherwise specified.

II. CONTROL AND MINIMIZATION OF WASTE

A. Hazardous Waste from Pilot Units

Disposal of any waste stream from a pilot or laboratory unit must also be carefully considered. The disposal must be covered by the same chain-of-custody as that from a full-scale facility. Each site must have a generator number, manifest its waste, have it hauled away by an approved hauler to an approved disposal site, and maintain files for three years. The waste generator is responsible for proper packing, labeling, and analysis of the waste, and for ensuring that it gets to the designated disposal facility. Liability does not cease with disposal at the designated site, although this is not totally clear at this time. If the waste is to be stored, treated, or disposed of on-site, a permit will be required; the site must then meet stiff criteria as a storage or disposal facility.

B. The Chemical Industry as a Regulated Industry

The Toxic Substances Control Act (TSCA) makes the chemical industry a product-regulated industry, much the same as the food, drug, and pesticide industries. TSCA is long and highly detailed. The congressional intent underlying the regulatory action is stated in Section 2 of the act:

> ... that human beings and the environment are being exposed to a large number of chemical substances and mixtures, that there are some whose manufacture, processing, distribution, use, or disposal may present an unreasonable risk, and that the effective regulation of interstate commerce also necessitates the regulation of intrastate commerce in such chemical substances and mixtures.

The act requires manufacturers to develop adequate data with respect to the effect of chemical substances and mixtures on health and the environment. This act is of particular concern to developers because of its major impact on new chemicals (those not covered in a list of currently produced chemical substances). The new chemicals are subject to premarket notification prior to manufacture or use, which requires extensive test data on health and environmental effects. The extent of the data required for one substance is so significant that testing costs in the range of 100,000 to 1 million dollars are possible. This investment may be a major factor in deciding at an early stage whether to carry on development and scale up a process.

C. Meeting Environmental Regulations

It should be obvious that the environmental legislation, not even considering the occupational health legislation, is all pervasive. This brief discussion of the federal statutes and regulations has only scratched the surface. It is important, perhaps critical, that prior to proceeding with the development project, extensive discussions are held with environmental and health experts to determine potential risks involved as the development proceeds. In-depth work at an early stage can minimize the potential for derailing a project for legal and regulatory rather than process technical rationale. It cannot be expected that development scientists and engineers will be experts in the environmental arena. Because knowledge of the law and regulations and the limitations that are imposed is important in making design and operating decisions, early consultation with experts is important.

D. Minimizing Environmental Concerns

There are many areas where toxic materials are used and environmentally unacceptable waste products are generated. Key areas of concern include:

Trace impurities in feed.
Side reaction products.
By-products.
Separation inefficiencies.
Product contaminants.
Effect of recycle.
Fugitive emissions.

Feed stocks, unlike laboratory chemicals, contain traces of impurities. At times these impurities are harmless and pass through without affecting reactions or creating unwanted side reactions. At other times they are hazardous or are precursors of hazardous wastes. In these latter cases the developer must consider whether or not a pure feed stream will eliminate a hazardous waste. The costs of obtaining the pure feed stream must be compared to the cost of handling the hazardous waste.

Ferric chloride, obtained as a waste product from titanium dioxide manufacture has been used, for example, as a coagulant in sewage treatment plants. More often than not the waste ferric chloride contains trace, ppm, levels of mercury. This mercury ends up in the treatment works' sludge. The treatment works could then be considered a generator of hazardous waste. If this is the only acutely hazardous material contained in the sludge there would be an economic decision for the treatment operator regarding the use of this inexpensive by-product ferric chloride as compared to a pure form not containing mercury. Without the mercury present, the sludge would probably not be considered hazardous and the disposal cost would be a factor of 10 less.

Typical impurities in process feed streams include aromatics in organic feeds, arsenic and heavy metals in metallurgical feed stocks, and trace quantities of sulfur and chloride in a variety of feeds. Therefore, it is important in scaleup to evaluate feed stock impurities in terms of their potential environmental impact.

Side reactions, particularly in chlorinations and brominations, can result in by-products that create serious environmental challenges. One of the best examples of this is the trace quantities of dioxins in chlorinated products, such as in Agent Orange, as well as in other chlorinated hydrocarbons. The typical route for inadvertent dioxin generation is pyrolytic dimerization of chlorophenols at temperatures above 170°C. It is important to control reaction conditions to minimize the side reactions that result in toxic by-product formation. By-products often fall into the same category as discussed above—they may be valuable or they may become hazardous waste.

Separation inefficiencies in a design can lead to a product and/or waste becoming hazardous. Most organic reactions are followed by purification steps and one of the bottom streams is usually a tar. This tar can contain polynuclear aromatics and chlorinated hydrocarbons, as well as acids and sulfonated organics, if the separation is improperly designed or operated. It would be difficult to prove that this particular still-bottoms product is not hazardous. Therefore, it is important to find means of disposing of this in an environmentally acceptable manner. However, other separations inefficiencies can lead to some of these materials being found in otherwise innocuous side streams.

Recycling to achieve minimum pollutant load is to be encouraged. However, care must be taken to ensure that side reaction products do not build up to a concentration where they become significant contributors to the pollution load in a bleed stream. For example, it is important to purge wastes regularly to avoid dioxin buildup in chlorination of phenols.

To date fugitive emissions have not been managed by environmental protection regulations. However, OSHA has regulated them and it is evident that this will be of considerable concern in the future. A typical example of this is fugitive emission in steel mills resulting in a high dust load. Another is fugitive emissions in hydrocarbon processing plants; these are already of concern in California and will be of more concern as time proceeds. It is important that design of facilities ensure that there are no leaks. This is done in the production of hydrazine. The typical route to the production of this fuel has been the condensation of dimethylnitrosamine, which has been identified as a carcinogen. A plant using double and triple seals and purging all spaces to a double afterburner incinerator was built to ensure that no nitrosamines would leak into the atmosphere. The same sort of precautions are being taken in the production of vinyl chloride to ensure zero leaks.

III. ENVIRONMENTAL CONSIDERATIONS AND SCALEUP

In the final analysis, a careful check of potential environmental factors prior to full scaleup must be made. Based on this, a determination should be made whether or not process alternatives should be developed.

It is obvious that environmental and health considerations must play a key role in the development of new processes and new products. When then should these environmental considerations be factored into the decision process?

ENVIRONMENTAL CONSIDERATIONS AND SCALEUP

Because of the high risks involved, it is important to make decisions, even if based on only preliminary or extrapolated data, at the earliest time. Costs for handling environmental problems such as testing and waste disposal must be factored into every evaluation step. Environmental and health considerations should be among the factors in the decision matrix at the start of laboratory work, at key laboratory program evaluation points as information becomes available, when deciding to bring the project to bench scale, when deciding to bring the project to full scale, and when deciding to commercialize it.

If testing is required, time will be an important factor. Often data sufficient for approval of a new process or product will take several years to develop. Thus, health and environmental testing must begin at the earliest possible time. A number of products ready for commercial development have recently been taken off the market when the developers decided, at the last minue, not to go through the premarketing notification, testing, and evaluation that was required by EPA.

A prime example of the refusal to go through the required procedure was the removal of a phthalate ester from the PVC plasticizer market after the development cycle had been completed. A great deal of money was spent needlessly in developing a process, only to abandon the project because of the costs of meeting environmental concerns. An early discussion with regulators on an informal confidential basis may be in order, to determine whether the particular product and/or process will pose problems and will require extensive testing. It should be noted, also, that lack of data or inconclusive data will draw out the approval procedures, thus tying up a great deal of invested capital unproductively.

A. Development of Environmental Data

It is often difficult at the early stages of development to obtain experimentally the data necessary to determine what the environmental problems may be. There is usually insufficient product to evaluate the presence of toxicants or hazardous materials. The effect of actual versus laboratory-pure feed stocks cannot be determined; nor can the effects of full scale equipment, process conditions, and resulting by-products be established. What then can be done to obtain data to establish design parameters? Several approaches which have borne fruit in the past and which should be considered at the early stages of development include:

Thermodynamic and kinetic evaluation for potential by-products.
Comparison with other reactions.
Consult a panel of experts experienced in the field.

All of these methodologies have been used in the past to identify potential hazardous impurities as well as to develop feed stocks for processes where sufficient testing had not been carried out. While not every impurity can be established either qualitatively or quantitatively, it is often possible to have a

90 percent success rate on the qualitative evaluation of potentially hazardous materials. This will serve to develop analytical scenarios, as well as economic impact scenarios and critical design check lists for the developing process. As data become available this data base can be refined and the economic impacts on the process development honed. This pollutant parameter analysis should be carried out on a continuing basis as the process goes through its various development stages.

B. Environmental Impacts — A Part of Process Evaluation

It is important throughout the development to define, as well as possible, the following parameters:

Potential feed stocks, their composition, and trace impurities.
Waste streams and their composition.
Utility of waste streams.
Disposal alternatives and costs.
Effect of reaction conditions on yield and toxic product contamination.
Catalyst, solvent, and other material utilization and discharge.

These parameters affect the process economics to a great extent. Environmental discharge minimization must be balanced with economic considerations.

Throughout the scaleup and development process it is important to develop alternatives. Use of alternative feed stocks and their economic impact is one of the keys because of potential environmental problems caused by feed stock impurities.

It is important to establish whether it is feasible to handle waste streams separately, to minimize the degree to which hazardous waste streams contaminate nonhazardous waste streams. Further, if waste streams cannot be successfully disposed of or used beneficially, it is important to determine alternatives for concentrating waste streams and segregating hazardous portions for separate treatment and disposal. In process treatment, neutralization or destruction of waste is an important alternative to off-site or on-site disposal. Development of scenarios for potential cost-saving and/or process-saving technologies should be evaluated along with the technical process scenarios.

C. Disposal Alternatives

We cannot deal at length with waste disposal alternatives. Other references deal with these and should be consulted as scenarios are developed. However, a brief discussion of various scenarios will be useful at this point.

An obvious, although not often feasible, approach to hazardous waste management is to not release the hazardous waste outside of its own process.

This may entail detoxification, neutralization, or recycle to extinction. Sometimes an alternate processing scheme can avoid hazardous intermediates and, thus, a potentially hazardous product impurity or waste.

Where wastes are produced and found to be hazardous the first question that must be answered is whether to handle those wastes in a single purpose or a multipurpose installation. In the past, wastes generally have been collected and an end-of-pipe, multipurpose waste handling and disposal system has been utilized. With increasing disposal costs, as high as $500 to $2,000 per cubic yard for solid hazardous waste, multipurpose waste disposal facilities that handle both hazardous and nonhazardous substances may not be economically advantageous. In addition, as opposed to waste streams that are handled outside the process or off-site, current regulations encourage in-process treatment by not imposing heavy regulatory demands on those types of streams.

Along with the use of single purpose waste treatment, concentration of the waste streams should be considered. As waste treatment and disposal costs are usually volume sensitive, a more concentrated stream of hazardous waste will be less costly to dispose of than a dilute, high-volume stream. Pretreatment to remove or detoxify hazardous materials may be the most economical solution for liquid wastes. The economics of concentration (by evaporation, filtration, distillation, or crystallization) as compared to the cost of disposing of large volumes of dilute hazardous waste must be carefully examined. Past economics are not a good guide, as disposal costs are no longer at the "zero" level that they have been in the past.

Destruction, detoxification, and/or neutralization should also be considered. These, along with other types of treatments, may be economically viable. In some cases they may even be an economic necessity because no one will take the waste as generated. Destruction of chlorinated or unchlorinated organic material in high-temperature or long-residence incinerators will probably become standard practice within the next few years. Detoxification of extremely hazardous substances, such as dioxin, prior to disposal or reuse may be the only alternative available. Neutralization of acids and bases has been practiced extensively in the past and will probably be practiced to a greater extent in the future. Where possible, chemical treatments such as sulfide precipitation of heavy metals, with or without recovery, may become economically attractive.

Ultimate disposal will either be by destruction or in landfills. Unfortunately, landfills suitable for disposal of hazardous or toxic substances will be at a premium. Land for disposal of even nonindustrial and nonhazardous industrial waste will become more scarce. One can anticipate the economics of scarcity taking hold. Additionally, liability of both the generator and the landfill owner are increasing considerably. While landfills often remain the ultimate disposal site of choice, they will become less attractive and harder to find.

At this writing, underground injection is still a viable ultimate disposal technology in some parts of the country. It is expected that, within a short time, new underground injection wells will be banned and many of the existing wells will be closed down because of their potential impact on drinking water aquifiers. Here again, scarcity will take hold and only in particular situations will underground injections be attractive.

The future will bring new technologies for waste management. Most of these will be geared to reduction and reuse of the waste. Some wastes may be stored for eventual reuse. As a general guide, it should be noted that the most economically viable process will be one that either does not generate waste or minimizes waste generation.

D. Challenges to New Products

The discussion to this point has been primarily concerned with disposal of hazardous substances and/or waste. However, this is not the only challenge to the developer. TSCA and regulations emanating from it provide for consumer protection for new products. This will, of course, complete the cradle-to-grave concept in the production and utilization of many materials of manufacture. At present, we are only seeing the tip of the iceberg in terms of new product testing and premarket notification. EPA and the states are becoming active in this area. In the past few years, Pennsylvania has issued a recall order, similar to those issued for years in the automobile industry, on a polyurethane–formaldehyde foam on which there were a number of consumer complaints. We have not reached the point where batch-by-batch certification of such items as polyvinylchloride are required, but that is possible. Testing of such products as phthalic anhydride for environmental and health effects has just been ordered by the EPA under the TSCA. Early identification of potential risks is important in the development of new products, new processes, and new uses for old products.

E. Right to Know

A new issue facing those in development and scaleup efforts is Workplace Hazard Communication or "Right-to-Know." As a result primarily of the hazardous waste situation, a great deal of activity in the environmental and working community has been expended on developing "Right-to-Know" legislation. This generally consists of a set of regulations which require that management communicate with workers the hazards which they might face resulting from exposure to chemicals. There are federal, state, and local regulations which can apply. At the moment it is unclear which regulation or regulations will preempt the other. The number of chemicals listed varies from several hundred to several thousand. In some cases the regulations require that in addition to worker training, the hazards be communicated to local emergency response organizations and, in the case of New Jersey, the local community. There are generally provisions for protection of "trade secrets" but these are both cumbersome and may not fully protect the development process. The communications required from a technical viewpoint include the development of material safety data sheets which include toxicological as well as environmental and physical information. Regulations also require training and in some cases identification of the chemicals contained in vessels and lines. Implementation of these regulations is still at an early stage but they will certainly impact those developing and scaling up processes to an extent at least as great as the other environmental statutes.

IV. THE ENVIRONMENTAL CHALLENGE

What does this all mean to the process designer and developer? What are the minimum requirements to be met? Minimum requirements include:

Identification of all potentially toxic or hazardous substances.
Pathways for all substances in the process, whether they are major or trace constituents.
Potential environmental and health effects of all reactions and products.
Potential for accidental catastrophic incidents, such as spills.
Early evaluation of potential environmental standards for emissions to the atmosphere, water, and land.

It is obvious by now that in process development a whole new thought process is required. Not only must we think about the technical challenges of scaleup but also additional challenges, translated into technical challenges, of environmental constraints and restraints. This places the burden on the developer to identify pitfalls that may arise. Without early identification many projects will be carried through scaleup only to end up on the junk heap because of environmental and health considerations. This new dimension provides a real challenge for the process developer—one that can and must be met for us to continue to enjoy the benefits of a technological society.

REFERENCES

DeRenzo, D. J., ed., *Unit Operations for Treatment of Hazardous Industrial Wastes*, 1978.

Lederman, P. B., *Environmental Regulations and the Process Industry*, Encyclopedia of Process Technology, J. McKetta and W. Cunningham eds., Dekker, New York, 1983.

Minor, P. S., *The Industry/EPA Confrontation*, McGraw-Hill, New York, 1976.

National Water Quality Commission Report to Congress, U.S. Government Printing Office, 1976.

Sax, N. I., *Dangerous Properties of Industrial Materials*, Van Nostrand, New York, 1979.

PL 90-148, Air Quality Act of 1967, as amended by PL 91-604, 1970; and PL 95-95, Clean Air Act Amendments of 1977 (16 November 1977).

PL 92-500, Federal Water Pollution Control Act of 1972 (18 October 1972), amended by PL 95-217 (28 December 1977).

PL 94-580, Resource Conservation and Recovery Act of 1976 (21 October 1976), as amended by PL 96-463, Used Oil Recycle Act of 1980 (15 October 1980) and PL 96-482 (21 October 1980).

16

EVALUATING MATERIALS OF CONSTRUCTION IN PILOT PLANT CORROSION TESTS

P. E. KRYSTOW

I.	Major Issues in Conducting Small-Scale Corrosion Tests	621
II.	Principles of Corrosion Testing in Small-Scale Units	623
	A. Aims and Purposes	623
	B. Considerations in Developing Corrosion Tests	623
	C. Materials Selection Criteria	624
	1. Factors Affecting Materials Selection	624
	2. Commonly Used Materials of Construction	625
	3. Nature of Corrodents	628
	D. Guidelines for Anticipating and Preventing Corrosion	629
	1. Uniform Corrosion	630
	2. Pitting	631
	3. Galvanic Corrosion	632
	4. Crevice (Concentration Cell) Corrosion	632
	5. Intergranular Corrosion	634
	6. Stress Corrosion	634
	7. Dezincification	635
	8. Corrosion Fatigue	635
	9. Erosion/Corrosion	636
	10. Hydrogen Damage	636
III.	Conducting Corrosion Tests in Pilot Plants	638
	A. Selection of Materials To Be Tested	638

B.	Test Specimen Selection	639
C.	Surface Preparation, Measurement, and Weighing	643
D.	Exposure Techniques and Test Duration	643
E.	Examination and Cleaning of Specimens After Exposure	644
F.	Evaluation of Nonmetallic Materials	645
IV.	Special Laboratory Tests to Supplement Pilot Plant Data	647
	A. Available Laboratory Tests	647
	B. Rapid Corrosion Testing Techniques	649
V.	Uncertainties in Pilot Plant Corrosion Testing	652
References		653

Premature, unexpected failure of construction materials is a serious concern in commercial units. Failures that cause unscheduled shutdowns usually lead to costs many times greater than the actual cost of the equipment involved. Material failures can range from a catastrophic event resulting in an explosion and complete destruction to a mild corrosion situation leading to excessive metal contamination with resulting unacceptable product quality.

Therefore, it is important at the very outset of the process demonstration studies to determine the appropriate construction material requirements for the commercial unit. In many cases, where the environment is unique, it becomes mandatory to conduct appropriate corrosion tests in the pilot plant or at the very least, in laboratory tests. While the key purposes of scaleup studies are gathering of process information and developing optimum operating conditions and data to permit commercial operation, it is also important to conduct appropriate corrosion tests whenever possible in such studies.

Through an organized corrosion testing program at the process demonstration stage, it is often possible to predict and avoid material inadequacies in the commercial unit and at the same time establish optimum, economic commercial construction. Not all processes need to be tested—only those that present unique process conditions where the corrosion, mechanical, and metallurgical data are lacking.

I. MAJOR ISSUES IN CONDUCTING SMALL-SCALE CORROSION TESTS

There are three major issues in setting up a small-scale corrosion test program that must be considered at the very outset. First and foremost, it is necessary to determine the need for corrosion testing; that is, whether there is a potential for corrosion problems to exist in the new process and if so, whether they are unique. There is certainly no need to conduct tests if the environments are not corrosive or corrosion data are already available. Some guidelines will be provided in this chapter to assist in determining both potential corrosion problems and the possible uniqueness of the process environment.

The second, and extremely important, issue concerns the validity of corrosion data obtained in the small-scale tests as contrasted to what is actually experienced in a commercial unit. Simply stated, the data obtained in the pilot plant and laboratory tests can be valid and give predictable performance for a commercial unit; but only if variables significantly affecting corrosion exist in both small-scale and commercial units, are recognized, and are compensated for by appropriate additional testing.

In many cases, there may be sufficient differences between the two units that even with supplemental tests (laboratory or other) it may not be possible to determine the exact corrosion rates to be expected in the commercial unit. Nevertheless, it is almost always possible, based on corrosion data obtained in a pilot unit and supplemental laboratory tests, to select materials that can be expected to provide acceptable corrosion protection and equipment life. This is so even though the exact corrosion rate of these materials cannot be predicted accurately. In other words, it is possible through tests to select materials which can be expected to provide reasonable service for a minimum of five years.

In situations where uncertainty exists about the corrosion performance of the pilot unit and the commercial unit, corrosion data must be collected during the initial running of the commercial unit. In this manner, a more accurate corrosion prediction can be made and replacement of materials or maintenance in the commercial unit can be scheduled at an appropriate time. Often when confirming tests are run in the large unit, a material that is expected to provide five year's service based on the small-scale tests, provides a substantially longer life. This is because corrosive conditions in small-scale tests tend to be more severe than those in a commercial unit operation.

The key to obtaining valid data for extrapolation to the commercial unit is the recognition of the variables involved and the carrying out of additional corrosion tests as required to compensate for some of these variables. Section II, "Principles of Corrosion Testing in Small Scale Units," describes the important factors in establishing the corrosion testing program.

The third issue concerns the desirability for carrying out corrosion tests in the pilot units rather than through laboratory corrosion tests. The smaller the test apparatus, the more difficult it is to extrapolate data accurately to a larger unit. Laboratory corrosion tests should be conducted only to cover those special conditions in the commercial unit which do not exist in the pilot unit operations but which may have a significant effect on materials performance. Since pilot unit process conditions are often varied significantly in the course of process studies, it is also worthwhile to perform corrosion testing after all process data are collected. A special run (minimum two weeks) in which corrosion test samples are exposed at the process conditions that are planned for the commercial unit is particularly important if there are unique process conditions that affect corrosion.

II. PRINCIPLES OF CORROSION TESTING IN SMALL-SCALE UNITS

A. Aims and Purposes

The prime purpose of corrosion testing is to gather data that will permit the design and construction of an economical commercial unit that is free of significant material failures and will not result in excessive metal contamination of product. Often the pilot unit and laboratory equipment is made up of special corrosion resistant materials that are not typical of the construction materials used for commercial plants. This is often done to assure trouble-free operation in the small-scale units and also to avoid concern of product contamination since a wide variety of process conditions will be studied. Therefore, it is important to test the suitability of the more economical, conventional construction materials that would be considered for commercial units since it is often too costly to use the special materials employed in the small scale units.

B. Considerations in Developing Corrosion Tests

In Table 16-1 are summarized the important process related considerations for establishing a small-scale corrosion test program. If the pilot unit conditions are not similar to those contemplated for the commercial unit, then it is necessary to devise other tests (often of a laboratory nature) that will supplement and more adequately reflect the scaled-up conditions. For example, the velocity, temperature, heat transfer effects, residence time, and feed impurities may all be sufficiently different in the small-scale unit from the commercial unit as to require special evaluation studies. Such situations usually exist in the case of the smaller pilot units; particularly those that operate batchwise and

TABLE 16-1 Process Considerations in Developing Corrosion Test Programs

Are process environments unique; is published corrosion data adequate?
Are differences of pilot unit from commercial plant significant?
Will there be an impact on corrosion from the operating conditions?
How similar or dissimilar are
   velocity, temperature, pressure
   residence time, heat transfer, pressure drop in equipment
   concentration and impurities in raw materials
   volume of liquor in area of metal exposed
What is the impact of operating variables on corrosion, that is, feed changes, upsets, shutdown conditions, and so on?

TABLE 16-2 Metallurgical Considerations in Developing Corrosion Test Programs

---
Selection of economic materials to be tested for commercial units.

Types of materials for testing, for example, plate, tubing, castings, valves, plastics, protective coatings, ceramics.

Fabrication requirements, for example, weldability, machinability, and so on. Susceptibility to the special forms of corrosion; are special laboratory tests required?

Evaluation of required mechanical properties, for example, allowable stress, notch toughness, stress rupture, and so on.

What are possible corrosion control measures, for example, corrosion inhibitors, neutralizers, and so on? How should they be tested?

---

are used to demonstrate portions of a commercial process that will operate continuously.

In addition to the evaluation of the process variables affecting the performance of construction materials, it is necessary to evaluate the metallurgical factors as highlighted in Table 16-2. Studies that establish the corrosiveness of the environment, possible materials including cost, required fabrication techniques, and mechanical properties, as well as the determination of whether unique corrosion and metallurgical phenomena exist are important considerations in any corrosion test program. Whenever possible the same construction materials should be used in the pilot unit as those being considered for the large unit.

C. Materials Selection Criteria

Once it is established that the proposed process environments have some degree of uniqueness and a literature search fails to reveal appropriate corrosion data, it will be necessary to determine the most suitable construction materials for the test program. Numerous texts are available providing useful information both on corrosion of construction materials in chemical environments and materials selection criteria. The text *Corrosion Engineering* by Fontana and Greene (1967, 1978) is an excellent reference for quickly obtaining supplementary information on the available corrosion data for chemical environments.

1. Factors Affecting Materials Selection

Corrosion resistance and cost are the two important factors governing the selection of construction material. However, other factors such as strength, fabricability, weldability, resistance to brittle failure, availability, and appearance are also vital. For example, mechanical strength becomes an important consideration if the process involves high pressure or elevated temperatures. Fabricability must be evaluated if special forming and shaping characteristics are required. Weldability can be a special problem with the more corrosion resistant materials such as titanium or the high chromium alloys. Resistance to brittle failure is especially important if the temperatures of operation go below $-29°C$ ($-20°F$). Availability may become a problem when special highly corrosion resistant materials are necessary since delivery of equipment made of special materials may require one or two years. Finally, appearance is important in the food or pharmaceutical industry which must have metals that maintain a lustrous finish and do not rust. It is essential to consult with the materials manufacturers for the latest available information when the alternate factors besides corrosion resistance must be considered.

2. Commonly Used Materials of Construction

In Table 16-3 are listed the metals that are most often used in commercial units. These materials provide acceptable corrosion resistance for most environments at reasonable costs. Carbon steel, of course, is the most desirable of the construction materials since it is the most economical, readily welded, and available in all forms. When process temperatures rise above about 454°C (850°F), then the low chromium–molybdenum alloys are favored since they provide improved oxidation resistance and have higher elevated-temperature strength. Also, the chromium–molybdenum alloys are used when hydrogen is present since they provide additional resistance to hydrogen attack. For added corrosion resistance, particularly to organic acids, phenols, oxidizing environments, and so on, the stainless steels and sometimes aluminum alloys are used. Austenitic stainless steels (e.g., 300 series), aluminum, and so on, are employed where a rust-free surface is desired or where product contamination is a concern. The heat resistant steels provide superior high temperature strength

TABLE 16-3 Commonly Used Metals

Carbon steel and cast iron
Carbon–molybdenum alloy steel
C–$\frac{1}{2}$ Mo
Chromium–molybdenum alloy steels
$1\frac{1}{4}$ Cr–$\frac{1}{2}$ Mo
$2\frac{1}{4}$ Cr–1 Mo
5 Cr–1 Mo
Type 405, 410 stainless steel
Type 304, 316, 321, 347, 309, 310 stainless steel
HH, HK, HP heat resisting steels
Monel
Brass and bronze
Aluminum

TABLE 16-4 Special Corrosion Resistant Construction Materials

Chromium–molybdenum (9 Cr, $\frac{1}{2}$ Mo)	Nickel steels ($3\frac{1}{2}$ Ni, etc.)
Nickel	Lead
Inconels	Cupro–nickels ($\frac{70}{30}$, $\frac{90}{10}$, etc.)
Incoloys	High performance metals (Ta, Zr, Cb, etc.)
Hastelloys	Titanium
Duplex alloys	Noble metals (Au, Pt, Ag)
Silicon cast iron	Proprietary alloys as manufactured by Cabot, Carpenter, Sandvik, Jessop, Uddeholm, VDM, and other metal suppliers.
Ferritic stainless steels	

even at temperatures in excess of 816°C (1500°F). Nonferrous alloys such as Monel, brass, bronze, and so on, are used primarily for aqueous environments including sea water and reducing environments, and so on.

In Table 16-4 are listed the specialty metals and alloys that are used when improved corrosion resistance is required. These alloys are more costly but can serve where the commonly used metals prove to be unacceptable due to corrosion. However, use of these alloys requires careful evaluation of the alternate factors such as weldability, availability, fabricability, and so on. Those aspects need to be reviewed with the materials manufacturers for specific recommendations.

Nonmetallic materials are an important and growing class of special construction materials that can be considered if the commonly used metals and alloys prove inadequate. Nonmetallics generally offer superior corrosion resistance at substantially lower cost than the special metallic materials listed in Table 16-4. However, they, particularly plastics, are generally limited to lower temperatures and pressures. Furthermore, they should be avoided when handling corrosive flammable fluids since in the event of a fire they deteriorate rapidly.

Special nonmetallics such as glass linings, carbon, graphite, acid resistant brick, refractories, and so on, offer unique corrosion resistance to many chemical environments. Therefore, they find significant applications, particularly at higher temperatures. However, equipment constructed with these special nonmetallic materials are subject to higher maintenance. Moreover, they require careful installation and operation.

In Tables 16-5, 16-6, 16-7, and 16-8 are listed the important thermoplastic, elastomeric, reinforced thermoset plastic, and special nonmetallic materials which are used in commercial units. Unfortunately, it is beyond the scope of this chapter to review the chemical resistance properties of each of these construction materials. However, Fontana and Greene (1967, 1978), the NACE Corrosion Data Survey [Hamner (1974, 1975)], and 1979 *Yearbook of Los Angeles Rubber* (1979) are useful references for obtaining quick information on the suitability of the various metallic and nonmetallic materials in different

TABLE 16-5 Thermoplastics — Available in Solid and/or Lined Construction[a]

Fluorocarbons, Teflon[b] TFE (260°C or 500°F), Teflon[b] FEP (400°F), Tefzel[b] (300°F)
Chlorinated polyether, Penton[b] (121°C or 250°F)
Polyvinylidene fluoride, Kynar[b] (121°C or 250°F)
Polyvinylidene chloride, Saran[b] (200°F)
Polypropylene, PP (99°C or 210°F)
Polyethylene, PE (38 to 82°C or 100 to 180°F)
Polyvinyl chloride, PVC (71°C or 160°F)
Chlorinated polyvinyl chloride, CPVC, reinforced (99°C or 210°F)

[a] Maximum temperature given in parentheses.
[b] Tradename.

TABLE 16-6 Elastomers — Available in Lined Construction[a]

Soft natural rubber (66°C or 150°F)
Hard natural rubber, ebonite (82°C or 180°F)
Hypalon[b] (93°C or 200°F)
Neoprene (93°C or 200°F)
Butyl rubber (93°C or 200°F)
Hycar[b] (121°C or 250°F)
Viton[b] (204°C or 400°F)

[a] Maximum temperature given in parentheses.
[b] Tradename.

TABLE 16-7 Reinforced Thermoset Plastics — Available in Solid and/or Lined Construction[a]

Polyester glass fiber reinforced (93°C or 200°F)
Epoxy, asbestos reinforced (149°C or 300°F)
Epoxy, glass fiber reinforced (149°C or 300°F)
Phenolic asbestos reinforced (149°C or 300°F)
Furan, asbestos reinforced (149°C or 300°F)
Furan, glass fiber reinforced (93°C or 200°F)
Furan, carbon reinforced (93°C or 200°F)

[a] Maximum temperature given in parentheses.

TABLE 16-8 Special Nonmetallic Materials (Available in Solid and/or Lined Construction)

Cement and concrete—Used for handling neutral solutions, particularly salt water.

Reinforced cement (Gunite)—Lining for small drums in neutral solutions.

Acid brick—Linings handling strong hot acids, particularly H_2SO_4.

Carbon brick—Linings handling strong hot acids where acid brick inadequate.

Impervious graphite (Karbate^a)—Resin impregnated, good for heat exchangers in strong acids at temperatures to 170°C or 338°F.

Glass and glass linings—Useful for reactors and piping, particularly when hot HCl is handled and metals are expensive or inadequate.

Wood—cypress, redwood, oak, pine—Cumbersome but useful for cooling towers and atmospheric pressure tanks handling acids, etc. (pH 2–11).

Castable and refractories—high-temperature refractories (Alumina, Mullite, etc.)—Generally used as insulation to avoid use of costly heat resistant alloys, i.e., furnace casings and processes involving elevated temperatures.

^{a} Tradename.

---

chemical environments. Also, nearly all manufacturers publish corrosion resistance charts for their special manufactured materials that provide useful data on materials performance.

For convenience, in Tables 16-5, 16-6, and 16-7 the maximum temperatures are listed for a number of nonmetallic materials. These temperatures are only applicable to low-pressure applications with nonaggressive fluids such as water. The temperatures are significantly reduced at higher pressures and with increased aggressiveness of the process fluid. For special nonmetallic materials (Table 16-8) maximum temperature limits are difficult to provide and are highly dependent on the chemical environment and therefore are not included.

3. Nature of Corrodents

There are two basic types of corrodents present in process streams that must be considered to determine the appropriate material(s) of construction. The first are those added during processing such as the examples shown in Table 16-9. While the list is by no means complete, it illustrates the nature of corrodents that are added during processing. Since these are known compounds, they are readily recognized and can be coped with by proper materials selection.

A few of the more typical subtle corrodents are listed in Table 16-10. Here, the corrodents are formed by decomposition or by reaction of chemicals (or impurities) during the processing operation. When decomposition or chemical reactions are occuring to yield corrosive gases or acids, the corrosion rates can become unpredictable; corrosion test studies are then required.

TABLE 16-9 Typical Corrodents Added During Processing

Sulfuric acid
Caustic
Ammonia
Calcium chloride
Hydrofluoric acid
Aluminum chloride
Phosphoric acid
Hydrogen

TABLE 16-10 Typical Corrodents Formed During Processing Operations

Hydrochloric acid
H_2S or SO_2
Carbonic acid or CO_2
Naphthenic acids
Organic sulfur compounds
Formic acid
Ammonia

D. Guidelines for Anticipating and Preventing Corrosion

To proceed with corrosion testing in a pilot plant, one must have a knowledge of the forms of corrosion and the methods used to prevent corrosion. Selection of corrosion resistant materials of construction is the most appropriate method of preventing corrosion, but not the only method. While our emphasis is on material selection, in Table 16-11 is a list of the important methods for preventing corrosion.

Materials selection is very contingent on the nature of the corrosion encountered. Therefore, we will review the alternative forms of corrosion listed in Table 16-12; a more extensive discussion will be found in Fontana and Greene (1967, 1978).

TABLE 16-11 Methods for Preventing Corrosion

Change environment
Change design
Materials selection
Anodic and cathodic protection
Coatings and/or linings
Corrosion allowance
Inhibitors
Neutralizers

TABLE 16-12 Forms of Corrosion

Uniform or general attack
Pitting
Galvanic corrosion—two metals corrosion
Crevice corrosion
Intergranular corrosion
Stress corrosion
Dezincification—selective leaching
Corrosion fatigue
Erosion/corrosion
Hydrogen damage

1. Uniform Corrosion

Uniform corrosion is illustrated in Figure 16-1. It is characterized by a reaction that proceeds over the entire metal surface or over a large area of a surface; it is the most common type of attack.

The "no corrosion" illustration in Figure 16-1 is typical of metals exposed to nonelectrolytes such as hydrocarbons. The "metal loss" situation occurs when the metal is exposed to a process fluid which forms soluble corrosion products, for example, aluminum exposed to caustic solutions or carbon steel in dilute nitric acid. "Corrosion scale" is found when an insoluble corrosion product forms such as is experienced when dilute sulfuric acid contacts lead.

The rate of uniform corrosion can be readily measured by determining mils (0.001 in.) penetration per year (MPY) or millimeters per annum (mm/a). This is the only form of corrosion for which an overall corrosion rate (MPY or mm/a) measurement is significant. Other forms of corrosion are generally localized; therefore, an overall corrosion rate is not a satisfactory indicator of corrosion.

The formula for calculating the corrosion rate is given in Table 16-13. Table 16-14 has been prepared as a guide for using MPY (mm/a) values to determine corrosion allowance specifications. A rule of thumb guide for corrosion allowance specifications is also given in Table 16-14.

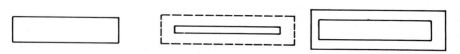

No Corrosion   Metal Loss   Corrosion Scale

FIGURE 16-1 Forms of uniform corrosion.

TABLE 16-13 Calculation of Corrosion Rate

Corrosion rate

$$\text{Mils/year (MPY)} = \frac{534W}{DAT}$$

Inches/year (IPY) = 0.001 × MPY
Millimeters/year (mm/a) = 0.0254 × MPY

where

W = weight loss, mg
D = density of metal in g/cm^3
A = area, in.^{2}
T = time, hr

TABLE 16-14 Corrosion Allowances

Corrosion Rate, MPY (mm/a)	Corrosion Condition	Corresponding Corrosion Allowance Specification, in. (mm)
0–3 (0–0.08)	Unattacked	$\frac{1}{16}$ in. (1.5 mm)
4–12 (0.09–0.30)	Slight attack	$\frac{1}{8}$ in. (3.0 mm)
13–16 (0.31–0.41)	Mild attack	$\frac{3}{16}$ in. (4.5 mm)
17–25 (0.41–0.63)	Heavy attack	$\frac{1}{4}$ in. (6.0 mm)

2. Pitting

Pitting is a form of localized corrosion in which rapid penetration takes place at several random small areas on a metallic surface. The appearance of pitting is unpredictable; its occurrence depends on the alloy and factors of exposure.

Pitting commonly occurs at metal surface imperfections. The imperfections lead to a break in the continuity of the metal surface; pitting results when there is contact with an electrolyte. An electrolytic cell is formed, the anode of which is the active metal at the imperfection and the cathode is the normal metal surface. The potential difference of this cell results in current flow with attendant rapid corrosion of the metal.

The break in the continuity of the metal surface is the result of some heterogeneity such as a rough spot, a nonmetallic inclusion, a scratch, or an indention. Pitting may also develop under deposits that prevent access of oxygen to the surface of the metal or cause differences in solution composition or concentration over the metal surface. It may also be found where breaks occur in a normally protective film on the metal surface, for example, oxides, insoluble corrosion products, and so on.

Certain environments accelerate pitting. In general, pitting, particularly of stainless steel, is most likely to occur when chlorides and other halogens are present in an oxidizing environment.

Naturally, information on the overall corrosion rate (MPY or mm/a) is of little value when pitting occurs. Since this attack is local the corrosion will be concentrated at several small specific areas. The rate of the pit penetration into the metal surface is more significant. When such information is available, it may be used as a guide for selecting construction materials. However, any construction materials subject to severe pitting should not be specified even though the overall corrosion rate may be acceptable.

3. Galvanic Corrosion

Galvanic corrosion occurs when two dissimilar metals contact each other forming a galvanic cell upon exposure to a corrosive environment. The galvanic current causes an increase in the corrosion rate of the anodic metal over that which would occur if there were no contact with a dissimilar metal. The relative areas of the anodic and cathodic metals have a marked effect on the severity of the damage. This is particularly important when the anodic area is small compared with the cathodic area. Often the corrosion is located near the point of contact and can appear as deep grooves or pitting in the anodic metal.

When use of coupled dissimilar metals is unavoidable, galvanic corrosion can often be reduced by:

Use of electrically insulated joints.

Avoiding use of a small anodic area in contact with a large cathodic area.

Making both metals cathodic through externally applied current.

Providing a more anodic third metal, to be preferentially corroded and replaced periodically.

With a knowledge of a galvanic behavior of metals and alloys, it is possible to set up a series that will indicate the tendencies of metals and alloys to form galvanic cells and to predict the probable direction of the galvanic effects. The series in Table 16-15 shows the relative potential of the various metals in seawater. Other environments result in different potentials and changes in the order can occur.

4. Crevice (Concentration Cell) Corrosion

Crevice (concentration cell) corrosion occurs when different concentrations of a corrosive solution contact different areas of the same metal. Therefore, conditions such as cracks, crevices, scale, or surface deposits, which permit stagnant solution to contact the metal, promote crevice corrosion. When different concentrations of solutions contact a metal, an electropotential difference is established and current flows from the area exposed to one concentration to the area exposed to the other concentration.

TABLE 16-15 Galvanic Series of Metals and Alloys for Seawater

Corroded end (anodic or least noble)

Magnesium
Magnesium alloys

Zinc

Aluminum 2S

Cadmium

Aluminum 17ST

Steel or iron
Cast iron

Chromium–iron (active)

Ni resist

18-8 chromium–nickel–iron (active)
18-8-3 chromium–nickel–molybdenum–iron (active)

Lead–tin solders
Lead
Tin

Nickel (active)
Inconel (active)
Hastelloy C (active)

Brasses
Copper
Bronzes
Copper–nickel alloys
Monel

Silver solder

Nickel (passive)
Inconel (passive)

Chromium–iron (passive)
18-8 chromium–nickel–iron (passive)
18-8-3 chromium–nickel–molybdenum–iron (passive)
Hastelloy C (passive)

Silver

Graphite
Gold
Platinum

Protected end (cathodic, or most noble)

Crevice corrosion can be minimized by avoiding accumulations of deposits on metal surfaces, sharp corners, loose threaded or gasket joints, and other conditions that would favor establishing stagnant areas of solution. Some metals are more subject to crevice corrosion than others; Monel, for example, is more resistant than stainless steel to crevice attack.

5. Intergranular Corrosion

Intergranular corrosion is an attack confined to the grain boundaries of the metal. This occurs when the grain interfaces (boundaries) become more reactive than the grain itself. Often its severity is not evident on visual examination; however, intergranular corrosion can completely destroy the mechanical properties of the metal to the depth of its penetration.

Austenitic stainless steels such as Type 304, 18 Chromium, 8 Nickel are susceptible to severe intergranular attack by many corrosives when not properly heat treated or if exposed to temperature ranges of about 340 to 760°C (650 to 1400°F). Susceptibility to intergranular corrosion for Type 304 results from precipitation of chromium carbides at the grain boundaries either during fabrication or during use. Fortunately, the condition is eliminated during fabrication by reannealing and quenching from temperatures above 1065°C (1950°F) before use.

Where reannealing after fabrication is impractical or the service temperatures are within the carbide-producing temperature range, extra low carbon alloys (0.03 percent C maximum) or alloys with stabilizing agents such as columbium or titanium must be used. This minimizes the tendency for chromium carbide precipitation.

Almost all austenitic stainless steel process equipment can be subject to this type of corrosion. Particularly affected are furnace tubes, towers, drums, and piping that are fabricated of welded alloy steel Type 304.

6. Stress Corrosion

Stress corrosion occurs under the combined action of residual or externally applied tensile stresses and corrosive agents. Residual stresses can remain after forming, welding, or heat treatment; sometimes they can also result from thermal gradients or from metallographic structural changes that cause a change in volume. This differs from stress corrosion fatigue where the corrosion product film is repeatedly cracked by reversal of stress or repeated stressing from vibrations or repeated expansions and contractions. Stress corrosion is extremely insidious since it occurs without warning and is characterized by little if any metal loss. Austenitic stainless steels exposed to trace amounts of chlorides are particularly susceptible to stress corrosion.

Stress corrosion makes itself evident through crack formation with little loss of ductility. Attack usually occurs along localized paths; pits or crevices serve to concentrate the stress, destroying protective films and keeping the anodic material exposed to the corrosive environment. Where the metal is subjected to high external or residual stresses, corrosion will progress at an accelerating rate as the stress increases.

Stress corrosion is avoided by controlling relevant metallurgical factors by proper selection or alloys and by minimizing environmental factors promoting cracking. The Materials Technology Institute (1979) has developed guidelines for controlling stress corrosion.

Caustic embrittlement is a form of stress corrosion that only occurs when ferrous alloy is exposed to strongly alkaline solutions. Like intergranular corrosion, caustic embrittlement starts and proceeds along the boundaries of metal grains. However, unlike intergranular corrosion, the metal must be mechanically stressed simultaneously with exposure to the strongly alkaline solution. Intergranular corrosion can occur in the absence of any stress.

7. Dezincification

Dezincification or selective leaching occurs when a copper zinc alloy (brass) dissolves and the copper redeposits as metal particles. The zinc portion of the brass may be carried away as soluble zinc corrosion product or deposited as insoluble zinc compounds. Dezincification can occur uniformly or locally. Uniform dezincification or layer type dezincification occurs in brasses exposed to acid conditions. Localized or plug type dezincification may be found in brasses exposed to alkaline, neutral, or slightly acid conditions.

High temperatures and low velocity of the corrosive environment accelerate dezincification. Brasses containing 85 percent or more copper are highly resistant to dezincification. Arsenic, antimony, and phosphorus additions to the brass act as powerful retardants to dezincification and are standard additions to inhibited admiralty and aluminum brasses. Exchangers provided with Naval Brass (60 Cu, 39.25 Zn, 0.75 Sn) tubesheets can be particularly affected by dezincification.

Selective leaching also occurs when cast iron is exposed to salt water, dilute acids, and soils. Here metallic iron (ferrite) is leached out and converted into corrosion products; the graphite portion of the cast iron is left behind. This results in increased porosity and loss of density and mechanical strength. Visually the cast iron may show no damage but it can be soft enough to be easily cut with a knife. Graphitization may sometimes be overcome by using refined grades of cast iron, specifically, white cast iron in place of the more commonly used gray cast iron.

8. Corrosion Fatigue

Corrosion fatigue is caused by exposure of metals both to cyclic stresses and a corrosive atmosphere or solution. Corrosion attack in the absence of cyclic stresses would often be superficial and confined to the surface of the metal. However, when the surface layers of the metal are highly stressed, corrosion can penetrate the surface and produce cracking.

Deep steep-walled cracks are characteristics of corrosion fatigue. Exposure of unstressed metal to a corrosive environment is much less likely to lead to premature failure. The stress can cause damage to corrosion films, oxides, and so on, exposing the metal underneath to unrestrained local corrosion. If the film is cracked open by the stress, then corrosion may progress at the cracks. The metal at these cracks, being less accessible to oxygen, becomes anodic to the surrounding metal and consequently is subject to accelerated corrosion.

9. Erosion / Corrosion

Erosion/corrosion occurs when the velocity of the corrosive fluid is sufficiently high to continuously remove the protective scale which normally would remain on the metal surface at lower velocities. Once the scale is removed, new scale starts to form and this in turn is also removed. Metal loss is greater than would be experienced by just direct corrosion. Usually the metal surface in the area of erosion/corrosion is characterized by many smooth impingement pits.

Cavitation corrosion caused by impingement of vacuum cavities can also be classed as a type of erosion/corrosion. Vacuum cavities occur on the tips of pump impellers when there is insufficient net positive suction head or when fluids with dissolved gases are handled.

In some cases, abrasive particles are circulated through equipment without the presence of a corrosive medium. In this case, the deterioration of metal is strictly mechanical erosion. The rate of deterioration then is a function of the hardness of the metal and the particular abrasive particle.

10. Hydrogen Damage

Hydrogen damage is metal deterioration resulting from the presence of, or interaction with, hydrogen. The five types of hydrogen damage are:

Hydrogen blistering.
Hydrogen embrittlement.
Sulfide embrittlement.
Decarburization.
Hydrogen attack.

In all cases the damage occurs within the metal since hydrogen atoms diffuse directly into the metal structure. While hydrogen atoms and ions are small enough to enter into the metallic structure directly, molecular hydrogen, once it is formed, cannot diffuse out from the metal.

Hydrogen attack differs from "corrosion" in that the damage can occur throughout the thickness of the metal without metal loss. It is not practical to provide a "corrosion allowance" to allow for the small amount of attack. In addition, once attack has occurred, the metal cannot be repaired and must be replaced. For these reasons, it is of utmost importance to review all operating conditions and future possible operating conditions before selecting materials for service with hydrogen.

Hydrogen blistering is a phenomenon occurring in mild acids, particularly in the presence of cyanides. The cyanides catalyze the diffusion of atomic hydrogen from the corrosion reaction into carbon steel. The atomic hydrogen collects in the voids and laminations within the steel and forms molecular hydrogen. The molecular hydrogen molecule is too large to diffuse out from the metal and causes an internal pressure buildup inside the voided area with an eventual rupture of the metal. The ruptured area is confined to local areas and appears on the surface of the metal in the form of blisters or fissures.

Hydrogen embrittlement results from exposure of metals to highly corrosive acids when large quantities of molecular and nascent hydrogen collect on the surface of the metal and there is substantial diffusion of hydrogen ions into the metal structure. Although the exact mechanism of hydrogen embrittlement is not well defined, the presence of hydrogen within the metal structure results in a complete deterioration of mechanical properties and embrittlement.

Many metallic materials are subject to spontaneous *sulfide embrittlement* when exposed under stress to environments containing wet hydrogen sulfide. The failure has many names, for example, hydrogen-stress cracking, sulfide corrosion cracking, hard weld cracking. A combination of high-strength metal, stress, and a source of nascent hydrogen (usually from a corrosion reaction) seems to be required to produce the cracking. A specific factor which makes steels susceptible is their hardness, that is, hardness levels in excess of 200–225 Brinell hardness number (BHN) or maximum tensile stress above 90,000 pounds per square inch (psi). The hardness is most critical at highly stressed areas of metal such as at the heat affected zones of welds.

Decarburization occurs in an atmosphere that does not contain carbon but is either oxidizing or reducing. In this case, carbon at the surface combines with oxygen or hydrogen to form carbon monoxide or hydrocarbons. Since these compounds are gases, they leave the steel. Carbon diffuses outward to reduce the concentration gradient set up by the loss of carbon at the surface, and a layer with low carbon content is formed. A layer deficient in carbon results; that layer is softer and weaker than the remainder of the steel with failure being the result.

Hydrogen attack is caused by hydrogen disassociated at temperatures above 204°C (400°F). Methane formation occurs by reaction of hydrogen with the iron carbide (cementite) in the metallic structure. The methane collects in the internal voids, laminations, or other metal discontinuities. Since methane cannot diffuse out of the metal high internal stresses result, ultimately leading to cracking of the metal at the grain boundaries.

Hydrogen attack is prevented by adding to the steel or alloy any of the carbide stabilizing elements. In their order of importance they are: molybdenum, chromium, tungsten, vanadium, titanium, and columbium. All of the austenitic stainless steels, for example, Type 304 (18 Cr, 8 Ni), are resistant because of their high chromium content. The noncarbide forming elements such as nickel and silicon have no effect in preventing hydrogen attack of steel. API Publication 941 (1983) presents detailed information on steels in hydrogen service.

III. CONDUCTING CORROSION TESTS IN PILOT PLANTS

Eight steps are required to complete a corrosion test program in the pilot unit:

Selection of materials and form of material to be tested.
Test specimen selection.
Surface preparation.
Measuring and weighing.
Exposure techniques.
Duration of test.
Cleaning of specimens after exposure.
Calculation of corrosion rate.

Often it is difficult to carry out tests in a pilot unit for a variety of reasons. Some of the major difficulties that can be encountered are given in Table 16-16. While precise corrosion data cannot be obtained in most pilot plant programs, the *qualitative* data that are obtained are satisfactory for most applications.

TABLE 16-16 Difficulties Encountered in Conducting Corrosion Tests in Pilot Plants

Pilot plants tend to utilize highly corrosion-resistant materials.

Emphasis in program is on gathering process data, not testing materials. Corrosion impurities from the test samples can have undesirable process effects.

Small size of equipment leads to difficulties in extrapolation to large-scale unit.

Extreme variation in operating conditions may make corrosion tests less meaningful.

Not all commercial plant operating conditions may be demonstrated in pilot plants.

Short runs in pilot unit can lead to inaccurate corrosion data.

Batch operation or semicontinuous operations of pilot plant can limit the exposure times for obtaining corrosion data.

A. Selection of Materials To Be Tested

Before materials can be selected for testing, the corrodents that are added or formed during processing must be established. See Tables 16-9 and 16-10 for typical corrodents added or formed during processing. Available corrosion

CONDUCTING CORROSION TESTS IN PILOT PLANTS

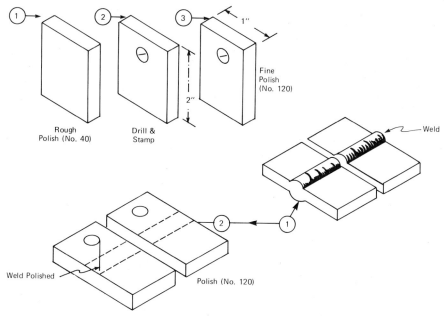

FIGURE 16-2 Standard test pieces for corrosion testing.

data in environments similar to the pilot plant operation must be established and the most likely candidate materials for testing selected. Fontana and Greene (1967, 1978) and NACE Corrosion Data Survey—Metals and Non-Metals [Hamner (1974, 1975)] are excellent starting points for such information. A particularly valuable source of data are equipment suppliers. Potential materials should be considered in order of cost, availability, and fabricability as summarized in Tables 16-3 and 16-4. In addition, it is necessary to test the material in the same physical form that will be used in the commercial plant, for example, cast, wrought, welded, heat treated, stressed, and so on.

B. Test Specimen Selection

Selection of the size and shape of the specimen is quite important. Generally, thin, flat specimens ranging from $\frac{1}{16}$ to $\frac{1}{4}$ in. are desirable. In Figures 16-2 to 16-7 are shown a number of test samples proposed for insertion into pilot plant facilities by Landrum (1961). Specimens are suggested for standard testing, stress corrosion, crevice corrosion, galvanic corrosion, and corrosion under heat transfer conditions.

A special test block assembly in which the specimen is fitted in between a special grooved teflon block is illustrated in Figure 16-8. This sample is recommended by the Materials Technology Institute (1980) for crevice attack corrosion studies.

640 EVALUATING MATERIALS OF CONSTRUCTION IN PILOT PLANT CORROSION TESTS

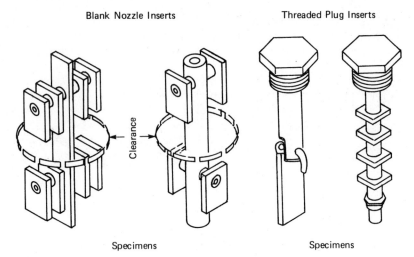

FIGURE 16-3 Specimen racks for limited access.

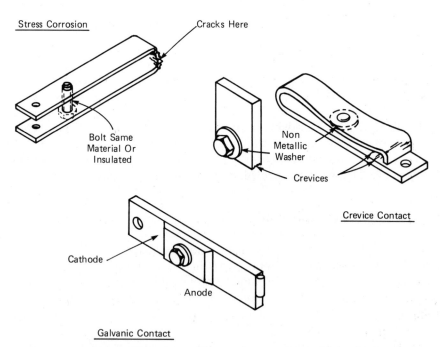

FIGURE 16-4 Specimens for stress corrosion, crevice corrosion and galvanic corrosion tests.

CONDUCTING CORROSION TESTS IN PILOT PLANTS 641

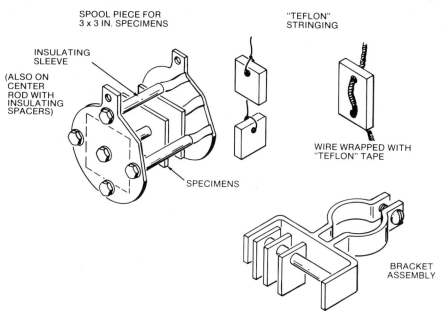

FIGURE 16-5 Specimen racks for easily accessible equipment.

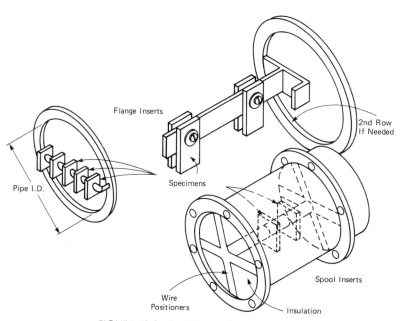

FIGURE 16-6 Specimen racks for pipe lines.

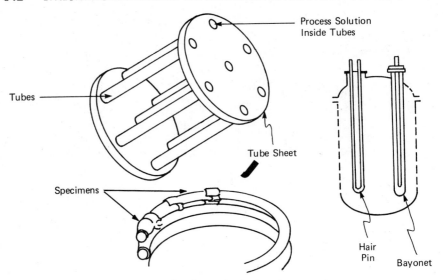

FIGURE 16-7 Specimens for heat exchange surfaces.

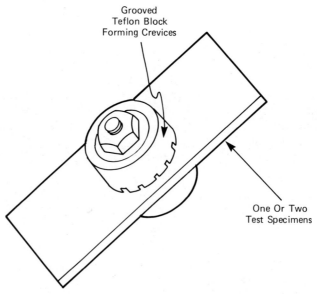

FIGURE 16-8 Crevice block assembly with specimen.

C. Surface Preparation, Measurement, and Weighing

Proper uniform surface preparation, measurement, and weighing of specimens before and after testing is essential, particularly for the short time exposures likely to be encountered in pilot unit testing. For short runs, particularly those of two weeks duration or less, precise and accurate measurements are required.

Important guidelines for pilot unit testing are:

The surface of the specimen should be the same as that which will exist in commercial equipment. However, this is not always possible. Therefore, all surfaces should be abraded to provide a uniform finish removing all sharp edges and burrs. The final abrading of the surface can be completed with either wet No. 80 or dry No. 120 grit abrasive paper. Following the abrasion treatment the samples should be cleaned with a magnesium oxide paste or detergent solution to remove residual dirt and grease. Specimens should then be rinsed in water, washed with acetone, air dried, and placed in a dessicator until weighed. It is critical that all surfaces have equivalent roughnesses. Soft metals may have to be scrubbed with pumice rather than abraded.

Specimens should be carefully measured prior to testing to determine surface area to an accuracy of 0.1 in.^{2}. Area measurements are less precise than weight measurements; therefore, the area measurement will determine the potential accuracy of calculated corrosion rates. For short term tests, specimens with essentially identical dimensions, surface areas, and surface roughnesses should be used.

The dried specimens should be weighed to an accuracy of at least $\pm$ 0.2 mg and then stored in a dessicator until they are used.

Additional details on the preparation of samples are given by Fontana and Greene (1967, 1978) and the Materials Technology Institute (1980).

D. Exposure Techniques and Test Duration

Important considerations regarding exposure technique are summarized in Table 16-17. Proper selection of the time the sample will be exposed is critical. Misleading results can occur when the exposure time is not properly considered. Periods of exposure can be particularly important in batch processes where cyclic exposure conditions are experienced. The test specimen should undergo all of the batch conditions including any shutdowns. Exposure of the sample should be made for two or more complete cycles of operation to ensure that an equivalent of two weeks exposure time is obtained.

Normally initial corrosion rates are high since corrosion products will not be present at the start of testing. Once corrosion products form, they will be somewhat protective; a drop of corrosion rate from the initial rates will result. Therefore, it will be necessary to conduct the tests of sufficient duration to

TABLE 16-17 Considerations in Sample Exposure

Corrodent should have easy access to the specimens.

Supports should not fail during the test period.

Specimens should be insulated or isolated electrically from contact with other metal unless galvanic action is to be studied.

Specimens should be properly positioned to study either effects of complete immersion, partial immersion, or vapor phase.

Specimens should be as readily accessible as possible.

---

compensate for this effect. Usually a two-week test is considered acceptable but a one-month test is preferred. Fontana and Greene (1967, 1978) have suggested the following rough rule for checking corrosion results:

$$\text{duration of test (hr)} = \frac{2000}{\text{expected corrosion rate (MPY)}}$$

E. Examination and Cleaning of Specimens After Exposure

Cleaning of specimens after exposure is extremely critical; improper cleaning is the main source of error in determining corrosion rate. Before proceeding with any cleaning, a careful visual examination of the samples should be made. This examination can give important indications about the causes and mechanisms of the observed corrosion process. If necessary, corrosion films can be further analyzed to determine their exact composition.

In Table 16-18 essential considerations in visual observation and cleaning of test specimens are described. ASTM Standard G1 (1981) provides additional information on the recommended practice for preparing, cleaning, and evaluating corrosion test specimens. In addition, Fontana and Greene (1967, 1978) have useful information, particularly in regard to sample cleaning techniques after exposure.

The measurement of the depths of even extremely shallow pits is important. Pit depth should be measured to an accuracy of 1.0 mil using a micrometer or depth gauge according to the procedure given in Section 6.2.3 of ASTM Standard G46-76 (1983). Alternately, and particularly for shallow pits, the microscopical procedure given in Section 6.2.4 of ASTM G46-76 (1983) can be used.

The corrosion rate is calculated by using the formula given in Table 16-13. Inco's publication, *Corrosion: Processes–Factors–Testing* (1956) and many handbooks give summaries of the densities of common materials of construction. A particularly worthwhile and inexpensive reference book on materials engineering is the *NACE Corrosion Engineer's Reference Book* by R. S. Treseder.

CONDUCTING CORROSION TESTS IN PILOT PLANTS

TABLE 16-18 Examination and Cleaning of Specimens After Exposure

Visual observation can reveal important information on the cause and mechanism of the observed corrosion.

Corrosion products should be classified as loose, or readily removable, as tight or adherent, and as protective or nonprotective.

Heavy loose deposits suggest pitting may be present underneath—this should be checked.

Pitting depth is as important as corrosion rate.

Total cleaning of sample to remove all corrosion products is essential.

Mechanical cleaning-scraping, brushing, scrubbing with abrasives, sandblasting, rubber stopper scrubbing.

Chemical and solvents cleaning.

Electrolytic cleaning—involves use of an impressed current and special chemical reagent to remove corrosion deposits.

F. Evaluation of Nonmetallic Materials

Since a corrosion rate cannot be calculated for nonmetallic materials, an evaluation of nonmetallic materials cannot be quantified. Protective coatings, organic linings, and solid plastics can only be evaluated through visual examination and by checking specific mechanical properties before and after exposure. If softening or swelling occurs and/or there is a significant reduction in some mechanical property, then the nonmetallic material under study is rejected. Usually, an exposure of two weeks to a month is a sufficient test duration. If deterioration is detected in that time, the material is rejected. The tests are strictly "go" or "no go."

The important visual evaluation methods for organic coatings and linings are summarized in Table 16-19. In addition to visual observations, other evaluation techniques include pencil hardness, solvent wipe test, and weight

TABLE 16-19 Visual Observations on Protective Coatings and Linings (Compare Before and After Exposure)

Discoloring
Swelling
Tackiness
Blistering
Shrinkage
Checking
Pinhole
Holidays

TABLE 16-20 Evaluation Methods for Coating Deterioration

Pencil hardness—Compare hardness before and after exposure by scratching coating with lead pencils of various hardnesses. Complete kits for conducting this test are available from suppliers.

Durometer hardness tester[a]—Use for homogeneous materials such as rubber linings and some rigid plastics.

Solvent wipe test—Cotton swab saturated with a suitable solvent (MEK, acetone, etc.) is rubbed vigorously for 50 strokes on coating. Degree of softening and discoloration of cotton is compared before and after exposure.

Weight gain or loss—Compare weight of coated specimens before and after exposure.

[a]ASTM Standard D2240-81 (1981).

gain or loss as shown in Table 16-20. A complete evaluation of coatings and linings must include those items critical for surface preparation, such as surface cleanliness, surface profile, and anchor pattern (surface roughness required to assure good adhesion of coatings) after sandblasting, wet and dry film thickness, and holiday (large pinholes) detection. All of these are important quality control requirements if proper application of protective coatings and linings on metal surfaces is to be obtained. Bryson (1956) reviews quality control requirements that are critical for acceptable coatings and linings applications.

The tests summarized in Tables 16-19 and 16-20 can also be used for solid plastic materials. Usually the Shore Hardness Tester as specified in ASTM D2543-81 (1983) is used for solid plastics and heavy glass/asbestos reinforced thermoset plastics. Evaluation of yield and tensile strength before and after exposure is also often employed.

Glass linings generally fail for reasons other than chemical resistance. Common causes of glass-lined vessel damage are thermal shock, excessive tensile stresses, mechanical damage, and hydrogen attack. Johnson (1964) provides details on failure of glass linings.

Corrosion testing evaluation of glass lined steel equipment is difficult and the manufacturer should be requested to supply necessary data. However, ASTM (1983) does have tests which can be used for evaluation of glass lining. ASTM Test Method C283-54 (1983) for resistance of porcelain enameled utensils to boiling acids and ASTM Test Method C814 (1980) for alkali resistance of porcelain enamels are useful. Both tests determine a weight loss over a specific area; however, it is difficult to assess life expectancy from this data since glass linings are generally only 40 to 80 mils thick. An exception, of course, is when the test shows complete chemical resistance to be present.

Evaluations of cement, concrete, acid brick, and refractories are as difficult as evaluations of glass liners. Applicable ASTM tests are summarized in Table 16-21.

TABLE 16-21 ASTM (1983) Tests for Cement, Concrete, Brick, and Refractories

C133-82a	Cold crushing strength of refractory, brick, and shapes (evaluation made before and after exposure)
C267-82	Chemical resistance of mortars, grouts, and monolithic surfacings
C279-79	Testing of chemical resistant masonry units
C704-76a	Abrasion resistance of refractory material
C768-73	Drip slag testing refractory brick at high temperature

IV. SPECIAL LABORATORY TESTS TO SUPPLEMENT PILOT PLANT DATA

All the corrosion data required to determine the material requirements for a commercial unit *cannot* be obtained from test samples installed in a pilot unit. Therefore, laboratory corrosion tests should be conducted to fill in the "gaps" of corrosion data not provided by the pilot unit. The pilot plant may not adequately duplicate operating conditions such as velocity, heat transfer, and residence time of large-scale plants. Run lengths may also be too short to obtain reliable data. Feed changes, upset operating conditions, and impurities can also have significant effects on corrosion in the large-scale plant. All of these conditions are generally better evaluated in laboratory tests. Laboratory tests also offer greater flexibility and controlability; when used properly they can provide extremely valuable supplemental data. Special forms of corrosion attack such as galvanic corrosion and intergranular attack are better analyzed by special laboratory tests. Material screening is also better done in the laboratory. Wachter and Treseder (1947) have reviewed the applicable laboratory corrosion tests and evaluations.

A. Available Laboratory Tests

There are many special laboratory corrosion testing apparatus and methods available. Apparatus is available not only for running direct corrosion studies, but also for evaluating stress corrosion, the influence of velocity, and heat transfer.

Some of the more important methods and apparatus are:

A typical resin flask operating at atmospheric pressure with capacity from 500 to 5000 ml which can be used to conduct simple immersion tests is described in NACE Standard TM-01-69 (1972, 1983). Also ASTM Test Method G31-72 (1983) provides the recommended practice for laboratory immersion corrosion testing of materials.

Test vessels that can be used to expose corrosion specimens in aqueous solutions at temperatures to 360°C (680°F) are described in NACE Standard TM-01-71 (1971). High-pressure autoclaves for corrosion testing purposes are available from a number of commercial suppliers.

Two-metal gavanic corrosion, crevice corrosion, and intergranular corrosion of sensitized steel tests can be conducted with the laboratory equipment previously described using samples such as shown in Figure 16-4.

A test apparatus for conducting controlled velocity laboratory corrosion tests is described in NACE TM-02-70 (1970). The apparatus is shown in Figure 16-9.

An apparatus for evaluating corrosion of materials under heat transfer conditions which can be assembled with standard laboratory equipment is shown in Figure 16-10.

Evaluating crevice corrosion using the crevice block assembly shown in Figure 16-8 has been described by the Materials Technology Institute (1980). Although the reference describes a procedure using NaCl solutions, it can be used with any solution.

A test for evaluating stress corrosion cracking has also been developed by the Materials Technology Institute (1980). This test utilizes a solution of

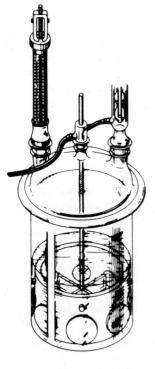
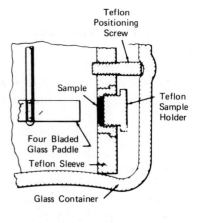

VELOCITY TESTER **VELOCITY TESTER—DETAIL**

FIGURE 16-9 Velocity tester.

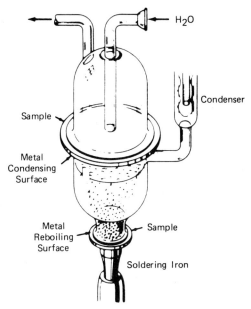

FIGURE 16-10 Heat transfer tester.

sodium chloride that is introduced dropwise onto the test specimens and allowed to evaporate. Any solution can be substituted for the sodium chloride solution.

High-temperature corrosion tests up to 760°C (1400°F) and higher can be carried out in special enclosed chambers within a laboratory furnace. Details of this test method are given in Fontana and Greene (1967, 1978).

B. Rapid Corrosion Testing Techniques

Over the past several years, there has been a new and emerging technology in corrosion measuring instruments that provide extremely rapid determinations of corrosion rate. These instruments are important adjuncts of pilot plant corrosion testing since they record corrosion rates nearly instantaneously without removal of specimen, permitting an evaluation of corrosion rate for each process variable even though the variable may be run for only a few hours. (Valid data with corrosion test specimens require at least two weeks exposure making evaluation of the impact of pilot plant process variables extremely difficult.) The new corrosion probes are available in all conventional construction materials.

There are two types of corrosion measuring instruments available: electrical resistance and linear polarization. Both instruments are suitable for use in the laboratory or directly in the pilot plant. More recently an AC impedance corrosion measuring instrument has been introduced.

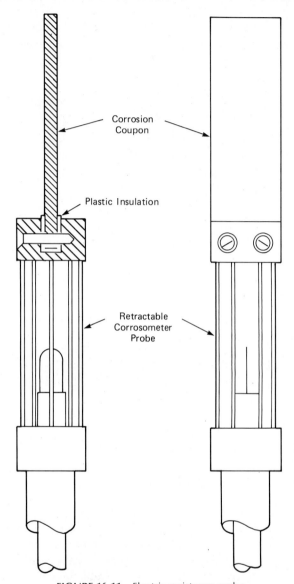

FIGURE 16-11 Electric resistance probe.

An electrical resistance probe fitted with a corrosion coupon is shown in Figure 16-11. This probe correlates the metal loss on a wire or tube element with the change of electrical resistance which is directly related to corrosion rate. The resistance probes require roughly 24-hr exposures or longer to obtain reliable corrosion data and are available from the Rohrback Instruments (1974) under the tradename "Corrosometer."

SPECIAL LABORATORY TESTS TO SUPPLEMENT PILOT PLANT DATA

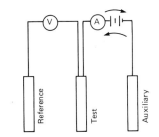

3 electrode system polarized by 10mv anodically and cathodically compared to reference electrode.

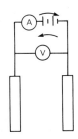

2 electrode systems impress a 20mv difference between the two electrodes. It is assumed that one electrode is shifted 10mv anodically and the other 10mv cathodically. Measurements can be made in either direction.

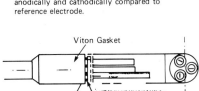

Petrolite 3-Electrode

FIGURE 16-12 Linear polarization probes.

The linear polarization probe, shown schematically in Figure 16-12, measures the current flowing between two electrodes at a fixed voltage. The level of current flow is directly related to the corrosion rate. A third auxiliary electrode can be used to measure the free corrosion potential. These polarization probes measure corrosion rate nearly instantaneously (within a couple of minutes) but require process fluids of suitable conductivity. They are available from Rohrback Instruments (1974) under the tradename "Corrator" and Petrolite Instruments who manufacture "PAIR" [see Shirley (1973)].

The absolute accuracies of the corrosion measurement probe are not as reliable as corrosion test specimens. Corrosion specimens will indicate the true corrosion rate more accurately. However, the corrosion probes are useful for comparing relative corrosion rates. In this manner, it is possible to determine the process variables that provide the lowest corrosion rates, to compare the effectiveness of various inhibitors in the unit, and to develop guidelines about upset operations that can lead to excessive corrosion, and so on.

Corrosion probes should be incorporated in the pilot plant unit. Once the pilot operation stabilizes, arrangements should be made for installation of corrosion specimens to obtain more accurate data on the expected corrosion rates of appropriate candidate construction materials. Moreland and Hines (1978), Henthorne (1971), and Feitler (1970) provide important background information.

V. UNCERTAINTIES IN PILOT PLANT CORROSION TESTING

There are important uncertainties with corrosion testing in pilot plants. The three most important are:

Pilot plant test runs do not accurately reflect long term scaleup conditions.

Exposure time of test samples during pilot plant test runs may be too short to permit extrapolation to long term scaleup conditions.

Variables such as velocity, heat transfer, residence time, feed impurities, temperature, and so on, may be different or may exist in the larger unit but not in the pilot unit.

The key to any successful pilot plant corrosion testing program is to recognize the uncertainties and to arrange supplemental tests, either in the pilot plant or the laboratory, keyed toward resolving the uncertainties. When such supplemental tests are conducted it will be possible to select construction materials that will provide satisfactory life.

In more extreme cases it may not be possible to predict an accurate life for the construction material. However, following the guidelines in this chapter it should almost always be possible to select a material that will provide an acceptable life of at least three to five years. When such a situation exists, a program of corrosion testing in the larger unit itself should be initiated. Valid data can be obtained within three to five months of operations. This allows sufficient time for planning to make supplemental materials substitutions if required.

In new commercial units handling corrosive chemicals, one can schedule a special turnaround after six to twelve months of operation, particularly if uncertainties exist about the performance of the construction materials. During the turnaround a complete inspection of equipment including an accurate measurement of the corrosion rates can be determined and equipment lives predicted. Where the predicted equipment life, based on the inspection, is less than three years, supplemental corrosion testing can be arranged during the turnaround by installing corrosion test racks of potentially more corrosion resistant materials. The test racks can be evaluated at the next turnaround and new, more corrosion-resistant equipment ordered in time for replacing the original equipment.

Onstream corrosion detection methods which utilize ultrasonics to measure wall thicknesses of vessels, piping, and equipment are available on a contractual basis. Many companies maintain staff inspectors to carry out these measurements routinely. While ultrasonics measurements are less accurate than actual inspection during turnaround or use of corrosion test racks, they can forewarn of any unusual or unexpected corrosion. When there is a construction materials concern, arrangements should be made to ultrasonically test all new units at strategic locations before startup and after three to six months.

With the programs outlined in this chapter, a successful scaleup from a construction materials standpoint should be achieved.

REFERENCES

API 941, *Steels for Hydrogen Service at Elevated Temperatures and Pressures in Petroleum Refineries and Petrochemical Plants*, 3rd ed., American Petroleum Institute, 1983.

ASTM Standards, 1983 Annual Book of, Volumes 1–15, American Society for Testing Materials, Philadelphia, 1983.

- C133-82a Standard Test Methods for Cold Crushing Strength and Modules of Rupture of Refractory Bricks and Shapes, **15** (1), 73
- C267082 Test Method for Chemical Resistance of Mortars, Grouts and Monolithic Surfacings, **4** (5), 160
- C279-79 Standard Specification for Chemical Resistant Masonry Units, **4** (5), 170
- C283-54 Test for Resistance of Procelain Enameled Utensils to Boiling Acid, **15** (2), 103
- C614-74 Standard Test Method for Alkali Resistance of Porcelain Enamels, **15** (2), 307
- C704-76a Test Method for Abrasion Resistance of Refractory Materials at Room Temperature, **15** (1), 307
- C768-73 Practice for Drip Slag Test Refractory Brick at High Temperatures, **15** (1), 358
- D2540-81 Standard Test Method for Rubber Property-Durometer Hardness, **9** (1), 602
- 61-81 Recommended Practice for Preparing, Cleaning and Evaluating Corrosion Test Specimens, **3** (2), 87
- 631-72 Recommended Practice for Laboratory Immersion Corrosion Testing of Metals, **3** (2), 175
- 646-76 Recommended Practice for Examination and Evaluation of Pitting Corrosion, **3** (2), 264

Bryson, H. C., *Paint Faults and Remedies*, Scientific Surveys, London, 1956.

Feitler, H., (Questronics, Los Angeles, CA), "Instantaneous Corrosion Rate Measurement," *Mater. Prot. Performance*, **9** (10), 37–41 (1970).

Fontana, M. G., and Greene, N. D., *Corrosion Engineering*, McGraw-Hill, New York, 1967 (2nd ed., 1978).

Hamner, N., *Corrosion Data Survey: Metals Section*, 5th ed., National Association of Corrosion Engineers, Houston, 1974.

Hamner, N., *Corrosion Data Survey: Non-Metals Section*, National Association of Corrosion Engineers, Houston, 1975.

Henthorne, M., (Carpenter Technol. Corp., Reading, Pa) "Fundamental of Corrosion, Measuring Corrosion in the Process Plant," *Chem. Eng.*, **78** (19), 89–94 (1971).

International Nickel Co., *Corrosion: Processes–Factors–Testing*, New York, 1956.

Johnson, D. B., "Caring for Glass-Lined Equipment," *Chem. Eng.*, **71** (12), 244, 246, 248–249 (1964).

Landrum, R. J., "Evaluation of Structural Materials for Corrosion Resistance," *Metals Eng. Quarterly*, **1** (2), 45–57 (1961).

Los Angeles Rubber Group Inc., *1979 Yearbook*, (TLARGI), Los Angeles.

Materials Technology Institute, *Guidelines for Control of Stress Corrosion Cracking of Nickel-Bearing Stainless Steels and Nickel Base Alloys*, MTI Manual #1, Columbus, 1979.

Materials Technology Institute, *Guideline Information on Newer Wrought Iron and Nickel Base Corrosion Resistant Alloys, Phase 1: Corrosion Test Methods*, MTI Manual #3, Columbus, 1980.

Moreland, P. J., and Hines, J. G., "Corrosion Monitoring—Select the Right System," *Hydrocarbon Processing*, **57** (11), 251–255 (1978).

National Association of Corrosion Engineers, *Method of Conduction Controlled Velocity Laboratory Corrosion Tests*, NACE Standard TM-02-70, Houston, 1970.

National Association of Corrosion Engineers, *NACE Book of Standard—1971–72*, Houston.

National Association of Corrosion Engineers, *Autoclave Corrosion Testing of Metals in High Temperature Water*, NACE Standard TM-01-71, Houston, 1971.

National Association of Corrosion Engineers, *Test Method—Laboratory Corrosion Testing of Metals for Process Industries*, NACE Standard TM-01-69, Houston, 1972, Rev. 1983.

Rohrback Instruments, "Principle for Operation of Corrosometers," Bulletin 886A, Sante Fe Springs, 1974.

Shirley, W. L., "Practical Applications of Instantaneous Corrosion Rate Measurements," Unpublished paper (1973).

Treseder, R. S., *NACE Corrosion Engineer's Reference Book*, National Association of Corrosion Engineers, Houston, 1980.

Wachter, A., and Treseder, R.S., "Corrosion Testing, Evaluation of Metals for Process Equipment," *Chem. Eng. Progress*, **43**(6), 315–326 (1947).

17

GAINING EXPERIENCE THROUGH PILOT PLANTS AND DEMONSTRATION UNITS

F. G. AERSTIN, L. A. ROBBINS, AND A. J. VOGEL

I.	Major Issues in Pilot Plant Studies	656
	A. Risk Factors Due to Uncertainty	656
	B. Cost of Pilot Units	656
	C. Timing	657
	D. Fundamental or Empirical Data	658
	E. Producing Products for Market Tests	659
	F. Study of Recycle Streams	659
	G. Safety Considerations	660
II.	Fundamental Considerations in Pilot Plant Programs	662
	A. Objectives of the Program	662
	B. Vendor Tests	664
	C. Show Tubes	666
	D. Integrated Pilot Plants	666
	E. Demonstration Plants	667
III.	Predicting Commercial Performance	671
IV.	Rules of Thumb in Pilot Plants	672
V.	Newer Developments in Pilot Plants	674
	References	675

I. MAJOR ISSUES IN PILOT PLANT STUDIES

Pilot plants are defined as "equipment assembled to generate design data for a large plant and/or material for market testing." Pilot plants vary greatly in size and complexity; for example, as simple as a 500-mL reactor or as large and complex as a 100-t/day coal liquefaction demonstration plant. Some pilot plants are built to test the feasibility of a completely new process, so a "totally integrated pilot plant" may be chosen which includes all recycle streams. Sometimes a pilot plant is built to test a specific unit operation where fundamental data may be lacking, such as a pilot plant for studying a heterogeneous reaction. Regardless of the size or nature of the equipment or process being studied, the fundamental purpose of a pilot plant is to reduce risks and errors in market evaluation and in the design, startup, and operation of full-scale production plants.

The evaluation of risk factors, costs, timing, generation of data, studying recycle streams, and safety are discussed in Section I. The objectives to be obtained with a pilot plant and the strategies used in these cases are discussed in Section II. The predictability of scaleup for a number of unit operations is discussed in Section III, and then a number of guidelines and helpful hints are presented in Section IV. In Section V some of the newer developments in pilot planting such as computer control, microprocessors, and equipment modularization are discussed briefly.

Throughout this chapter examples are provided to illustrate the concepts under discussion. These examples are all based on actual experiences.

A. Risk Factors Due to Uncertainty

In every project the question comes up—Is it necessary to pilot plant this process? In general if the process or product is a new one, the wise decision is yes, build a pilot plant. Most companies have not had much success in building production units directly from laboratory data. From experience we know that certain risks are incurred if the pilot stage is bypassed in the development of a process. These risks may be associated with an inability to predict the performance or value of a new product in the marketplace. Or, the risks may be associated with our inability to predict the performance of equipment in the plant based on first principles or laboratory data. In addition, there are risks due to our oversight or lack of knowledge concerning all of the impurities or by-products that may concentrate in recycle streams or waste streams in a process. Designing a full scale process from pilot plant data by no means eliminates all risk, but it does reduce the risk and minimizes the chances of a total failure.

B. Cost of Pilot Units

Pilot plants are expensive. When the decision is made to build a pilot plant, it is not unusual for a company to commit several million dollars to build and

MAJOR ISSUES IN PILOT PLANT STUDIES

operate the facility. Equipment costs can vary but, normally, instruments and controllers can be as much as 50 percent of the pilot plant cost. This is because the pilot plant will generally require the same number of instruments as the full scale plant; in addition, more instruments and controllers may be necessary to get the data required for scaleup studies. Operating costs are usually high because it is necessary to staff a pilot plant with engineers and operators. After the pilot plant is built and before it is started up, a good research program should be designed so that the necessary data can be obtained in the minimum time to keep the costs down. The normal life of the pilot plant varies from process to process and many times its purpose may change after the "scaleup" data are obtained. For example, it may be decided to use the pilot plant first to generate data, then later to make product for market testing.

Usually the payoff comes when the full scale plant is built and started up. Normal startup costs for a new process may run as high as 30 percent of the direct fixed capital. By building a pilot plant and learning how to operate the process, the startup costs of the first production unit can be reduced to 5-10 percent of the direct fixed capital, according to studies done by Robbins (1979a).

One approach often used to reduce pilot plant costs is to build miniplants as discussed elsewhere in this chapter. It is also possible to reduce costs by building only a part of the process. However, this approach normally prevents study of recycle streams. Yet another approach would be to build a batch pilot plant where many operations can be carried out in the same equipment. This can be a good approach, but if the production plant is continuous, side reactions that were not observed in the batch equipment can cause problems. Also, it is very difficult to study recycle streams in batch equipment.

C. Timing

When to build a pilot plant is a critical decision. In many cases when a new product with unique qualities is being developed, there is tremendous pressure asserted by the sales and marketing groups for material to sell. And since it normally takes three to five years to engineer and build a production facility, the temptation is there to bypass the pilot plant. The risk still exists that there could be high startup costs, but it is sometimes more profitable to take the chance to get into production as soon as possible.

One approach that has been taken is to design the production plant with the best data available and at the same time design and build a pilot plant that is scaled down from the production plant. Since the time to build a pilot plant is only six months to two years, the scaled version of the production plant can be started up and the "bugs" ironed out of the process before the production plant is ready to start up. Of course, there are risks involved in this approach. In an extreme case, the pilot plant will show that the process as designed will not make the desired product, or that the yields are unacceptable. In cases like these, parts of the production unit will have to be scrapped and redesigned based on the latest data. This approach to scaleup gives project and construction managers nightmares since construction costs and schedules become hard to control when there are major field changes.

For example, a product was being developed that would be used as a thixotropic agent in paints and resins. Historically, a fluid bed reactor was used for the final processing step. During scaleup studies it was suggested that a coil reactor would be less expensive for a production plant. A few tests showed that a coil reactor would give the desired conversion, and the production plant was so designed. Several months later, testing of product from the pilot plant showed that the material from the coil reactor did not give the desired thixotropic properties in resins and paints. Although the chemical produced in both reactors had been analyzed as identical with the available physical tests, each had a completely different behavior as a thixotropic agent. Needless to say the coil reactor was scrapped and a new reactor was designed for the production plant.

If the scaled down production plant or pilot plant had not been built, this problem would have not shown up until the production plant was in operation.

D. Fundamental or Empirical Data

One of the major considerations in deciding whether to pilot plant a process is in establishing what the approach to scaleup should be. A determination must be made as to whether the approach will be based on fundamental data or if an empirical approach will be taken. This will dictate both the size and the type of pilot plant that is to be designed. Kline et al. (1974) have classified the general techniques for scaleup in order of increasing difficulty as:

Full scale tests (*no scaleup*): This occurs where the cost of developing data for scaleup is high compared with the cost of a full scale test and where basic scaleup technology is minimal. Such a situation occurs often in solids handling. Full scale tests are discussed in the section on vendor testing.

Modular or linear scaleup: Here, one builds a "show tube," such as one tube of a tubular reactor. All experimental work is done on a small-scale tube; the full size reactor is simply a multiple of that small-scale tube.

Known scaleup correlations: Many correlations are available for the prediction of the performance of large-scale equipment from limited data. Correlations are particularly applicable in the areas of heat transfer and distillation processes with only limited additional practical experience being needed. This scaleup approach can include all the known available correlations. One does not generally build an intermediate size heat exchanger to obtain scaleup data on heat transfer. Acceptable estimates for the full scale unit can generally be made without building a pilot plant unit.

Fundamental approach: This should always be considered first since it can yield the best understanding of the process. This makes extrapolation and, therefore, scaleup, both quick and reliable. To scale up a simple order homogeneous reaction, we can obtain kinetic rate constants most conveniently using a minibatch reactor. The rate constants can then be used to develop a reactor of any size whether for a batch or continuous process. Since the purely fundamental approach is relatively costly and time consuming, it is often not practical. Successful scaleup studies are often based on some combination of the fundamental and other approaches.

Empirical approach (step-by-step scaleup): An example of the empirical approach is the step-by-step scaleup of a batch polymerizer from a 10- to 1000-L to perhaps a 50,000-L reactor.

The distinction between the approaches is not a sharp one and a multiplicity of approaches is often used in a scaleup program.

There is a gray area between fundamental and empirical approaches. Consider a heterogeneous reaction in which diffusion and mixing play important roles. Experimentation may yield a fractional order rate equation. If we scale up with these equations, there certainly is a risk because the approach is empirical. However, if we go back to the laboratory and do some more hard work in modeling the reactions, the diffusional steps can be separated out from the reaction steps so well that we get first or second order kinetics. This would be a fundamental approach. Although the diffusion coefficients may have been empirically determined, successful scaleup is virtually assured.

E. Producing Products for Market Tests

A pilot plant may be built just to fill market development requirements for a new product. In this case, the decision to pilot plant is independent of process considerations since the only real objective of the market development plant is to generate material for market tests. Design and operation of market development plants is discussed later in this chapter.

F. Study of Recycle Streams

In many processes a pilot plant is justified just for the study of recycle streams. Materials are usually recycled in a process to improve yields and reduce both pollution and operating costs. However, by returning materials from the back end of the process to the front end there is always a risk of contamination. How impurities affect a reaction can have a tremendous effect on a process.

Many times the impurities in the recycle streams are not even detected in the laboratory. It is only after a pilot plant or production plant has been operating for some time that impurities show up in sufficient quantities to cause problems. Besides having a detrimental effect on the catalysts, some impurities can also be shock sensitive, corrosive, or toxic. Pilot plant engineers must always be on the lookout for unknown impurities building up when a process is operated for the first time.

Usually it is not difficult to remove impurities once they have been isolated from process streams and identified. For example, a process was redesigned to improve yields by raising the reaction temperature from 100 to 130°C. After several months of operation in the production unit some problems were encountered in the final product formulations. Although the reaction had been studied in both the laboratory and pilot plant an integrated pilot plant had never been operated. The production plant had been built without a complete knowledge of the possible effect of the recycle streams.

An investigation in the plant showed that a by product was being formed at the new reactor temperature to the extent of about 5–10 ppm of this impurity. However, the impurity was soluble in one of the solvents and remained in the process building up to about 20 percent in a solvent stream. At this point the distribution coefficient of the by product was such that the impurity started to show up in the product. Once the cause of the product problem was identified, it was relatively simple to install a solvent recovery column to remove the impurity.

Here, everyone involved was lucky because this impurity was not dangerous. The lesson to be learned from this is to pilot all changes in reactor conditions; researchers should always be on the lookout for possible side reactions.

G. Safety Considerations

The pilot plant, in many ways, represents a greater challenge for safe design and operation than either laboratory studies or operation of a full-scale plant. The larger scale of the pilot plant compared to laboratory setups means potential chemical hazards present a greater risk. In addition, exposure to emissions in the pilot plant is greater than in the laboratory where hazardous chemical reactions are carefully isolated.

Since the pilot plant is an experimental unit, risks are greater than in a full-scale plant developed with proven technology. Experience has shown that the most valuable safety program is based on a thorough evaluation of the process, preparation of a suitable operating information package, and an inspection of the pilot plant before startup of process studies. A good safety program starts with dedication to safety from top management down to the personnel working at the bench. In "A Comprehensive Pilot Plant Safety Program" by Lowenstein (1979), the author points out:

> By definition we deal with unknowns in pilot plants; potential hazards which may have been overlooked in the laboratory can literally become death traps in pilot operation. Since our first responsibility to our people and our management is a safe operation, it is of vital importance that we correctly identify and assess potential hazards and then try to design the process to make it safe. Continuing dialogs with research chemists, detailed knowledge of the physical and chemical characteristics of the reaction under investigation, and safety checklists are the foundations of

MAJOR ISSUES IN PILOT PLANT STUDIES 661

safe pilot plant operation. A realistic safety awareness program which includes the often overlooked categories of good housekeeping, proper safety supervision, enforcement of safety regulations and continuous education, together with management's honest dedication to safety then rounds out the package necessary to assure safe scaleup and pilot plant operation.

An excellent tool for hazard evaluation, which has recently been developed, is the accelerating rate calorimeter. Townsend and Tou (1980) discuss this calorimeter and the theoretical aspects of thermal hazard evaluation. The technique is unique in providing information concerning temperature and pressure of reactions taking place under adiabatic conditions. For an nth order reaction, mathematical modeling of a kinetic event is possible; the information obtained can be used for engineering considerations to prevent a runaway reaction.

Other methods that can be used to elucidate chemical reactivity include differential scanning calorimetry and drop weight shock sensitivity tests. Whatever methods are used, they should be done before scaleup from the laboratory to the pilot plant is started.

An operating information package is the key to safe operating practices in a pilot plant. The contents of a thorough operating package are:

A list or chart indicating what minimum protective equipment would be required for each pilot plant operation.

A list of the chemicals that will be handled, including toxicity, reactivity, corrosiveness, and other hazards for each chemical.

A diagram illustrating location of safety equipment such as safety and eye showers, fire extinguishers, fire hoses, emergency air packs, and siren switches.

Instructions on normal operation of the pilot plant.

Procedures for emergency shutdown of the pilot plant.

The startup inspection represents a final checkpoint before permission is given to initiate the pilot plant work. Usually, the startup inspection will be carried out by a team consisting of technical and managerial personnel from the laboratory, pilot plant, and engineering groups.

The duration of the pilot plant startup inspection depends on the complexity of the pilot plant program. For a miniplant operation, a 2-hr review may suffice. In contrast, for a large-scale, multistep demonstration plant, the startup inspection may take several days to complete. To avoid unnecessary delays, scheduling of the startup of new pilot plants should include contingency plans in case some problems are uncovered during the safety program.

A pilot plant to produce 10,000 kg/month of a new monomer involved three reactions and three distillations; all the operations were to be operated on a continuous basis. The first phase of the safety program prior to startup was a comprehensive presentation of the process chemistry before a Reactive Chemicals Review Committee. In this review, held two months before startup, emphasis was placed on the hazards due to chemical reactivity. A detailed hazard evaluation of the process was presented including thermal and shock sensitivity data on the composite process streams, as well as the raw materials and products. Since the product was a monomer there was concern about the possibility of runaway polymerizations in the process. The final distillation step from which purified monomer was taken overhead was considered to have the greatest potential for polymerization.

Normal operating conditions in the column were:

Top: 110°C, 25 mm Hg absolute pressure.
Bottom: 120°C, 50 mm Hg absolute pressure.

Thermal stability data on the purified monomer indicated a sharp exotherm beginning at 130°C with a resulting peak temperature of 270°C. Although the column's bottom temperature was at least 10° below the temperature at which polymerization would begin in an appreciable rate, an upset in distillation conditions, as might result from a sudden loss of vacuum, could result in a hazardous situation.

Design of the distillation column had already included instrumentation to provide emergency cooling when either a bottoms temperature greater than 125°C, or a column pressure greater than 80 mm Hg absolute occurred. This was considered only the first line of defense. However, a second defense was added—automatic addition to the top of the column of a small amount of inhibitor to prevent or drastically slow down polymerization should it occur. Because the Reactive Chemicals Review had been scheduled well in advance of startup, there was ample time without delay to install the inhibitor system.

II. FUNDAMENTAL CONSIDERATIONS IN PILOT PLANT PROGRAMS

A. Objectives of the Program

The objectives and scope of the pilot plant must be clearly established before a program is undertaken. The objectives can be any of the following:

Generate market development quantities of a new product.
Evaluate the feasibility of a new process.
Generate design data for a commercial plant.
Demonstrate a new process.

FUNDAMENTAL CONSIDERATIONS IN PILOT PLANT PROGRAMS

Only after the objectives of the pilot plant program have been established can an appropriate strategy be selected.

If the objective of the pilot plant is to generate market development quantities of a new product, then a "jumbo lab" strategy may be the most appropriate. The jumbo lab strategy involves:

Study of the laboratory preparation and procedures.
Scaleup of the laboratory concepts to a pilot plant.

This concept is the most familiar and commonly used technique for the development of new products such as pharmaceuticals, agricultural chemicals, and other specialty chemicals.

A minimum of engineering expertise and time is required to produce the first pilot plant kilograms of the new product. However, the operations are generally high in labor requirements, poor in raw material utilization, excessive in solvent consumption, and high in the generation of waste streams. The development and engineering of the ultimate economic process with a totally integrated recycle stream and acceptable ecology control technique is merely postponed when the jumbo lab concept is used. However, the strategy is considered a reasonable approach where the "product" development time involves field studies or clinical testing of product efficacy. The results of these studies can have the greatest impact on success or failure of the project.

If the objective of a pilot plant is to evaluate the feasibility of new process technology or generate design data for a new process then the "miniplant" strategy may be most appropriate. The fundamental considerations in a miniplant, as outlined by Robbins (1979a), are:

Study in the laboratory of the relevant chemistry, that is, thermodynamic equilibria and kinetic rates.
Develop an integrated plant concept.
Scale down the unit operations to the minimum size where credible scaleup relationships exist.
Maintain critical plant concepts and constraints.
Minimize the equipment and installation costs.
Design for high flexibility and the possibility of rapid response to change in procedures and equipment.
Operate with actual plant streams and design conditions.

This concept is less familiar to many chemists and engineers; however, it can enjoy a high ratio of success to failure if the product is an accepted one and where the success of the project is highly dependent on the economic and ecological viability of the new "process." A high level of engineering expertise is required to utilize the miniplant approach; research time and cost can be kept to a minimum for developing a new process.

B. Vendor Tests

Vendor tests are the most familiar form of miniplanting. Equipment suppliers have miniaturized their equipment to the smallest unit size where reliable design data can still be obtained. Some of the areas where vendor tests are commonly used are: filtration, drying, solids conveying, crystallization, agitation, flaking, prilling, gas–solid separations, and liquid–solid separations. Equipment vendors can be excellent sources of scaleup information since they have years of experience with the particular pieces of equipment they sell.

Many vendors have facilities for studying the fundamentals of the unit operation equipment they sell. In the area of mixing, for example, several vendors have laboratories where they can study the different agitator geometries to result in the type of mixing the customer desires. Based on the laboratory data, a "pilot run" can be made at the mixing laboratory to check the scaleup parameters. If the results of this program are satisfactory, vendors may be willing to guarantee that the production size equipment will perform as they predict. While this reduces the risk for a customer, a money back guarantee on a $200,000 piece of equipment will not meet the startup objectives of an expensive commercial plant if the equipment fails to perform or turns out to be a production bottleneck.

One of the greatest uncertainties in selecting equipment from vendor test data is the possibility that the material used for the tests is not truly representative of what will be made in the commercial plant. Quite often the vendor testing period is short and the risk of what may happen after extended periods of operation, such as fouling of heat transfer surfaces or buildup of degradation products, is not completely understood. Similarly, recycle streams cannot generally be incorporated in a vendor test run. Overcoming these problems requires leasing one of the vendor test units and installing it in the pilot plant or as a side stream in a production plant. Here the equipment can be tested at the actual operating conditions and the data that is generated will be more meaningful for scaleup.

Since equipment vendors expect to sell a large piece of equipment, the test charge is usually low or sometimes the test is free. The rental of test equipment also avoids the capital outlay for a pilot plant unit. Quite often the vendor test equipment can be made available at short notice so the testing can be done quickly. However, equipment vendors are not normally well equipped to handle toxic or highly corrosive chemicals. If such materials are to be tested, the customer must both transmit a complete safety information package to the vendor and conduct training sessions with vendor personnel before any work is started.

When production of a herbicide was being considered as a flaked product, a test was arranged with an equipment vendor. Material and a safety informa-

FUNDAMENTAL CONSIDERATIONS IN PILOT PLANT PROGRAMS

tion package were sent to the vendor's test facilities. However, before the arrival of the pilot plant engineers, the vendor decided to try out some of the material. Unfortunately, adequate venting was not available in the test area and, on heating, some of the impurities in the herbicide degraded to give off a noxious gas resulting in evacuation of the area. When adequate venting was provided, the test was conducted and sufficient data were gathered for scaleup studies. A production plant flaker was installed and operated as expected.

Secrecy is an important consideration when using vendor tests. Where a new chemical is under development, or a new processing step is being considered, there is always a chance that as a result of the vendor tests some information will be passed on to the competition. In many cases it is impossible to prove violation of a secrecy agreement. Again the leased equipment route may provide a better option.

In some cases, full-scale equipment must be tested simply because smaller-size equipment scaledown is impossible. Production of a dry polymer from a slurry was under development at the pilot plant scale. The major goals of the development program were:

Obtaining data for reliable scaleup to a commercial-size facility.

Production of several thousand kilograms of material for product performance studies and initial market development.

The crucial unit operation was the drying step.

Since the polymer was heat sensitive, a good candidate would be a spray dryer, since it provides high heat flux at low residence times. These conditions are achieved when the material to be dried is sprayed as small droplets into a relatively large volume of hot gas. Heat and mass transfer occurs by direct contact of the dispersed droplets with the hot gas.

Discussions with a company specializing in spray drying equipment resulted in a decision to carry out tests at the vendor's laboratory. The experimental unit consisted of a spray chamber of a few meters in diameter, with a centrifugal disk for dispersing the feed to the chamber. Basically, the experimental unit was a "small" commercial-size spray dryer—capable of producing several million kg/yr of dry polymer. Scaledown to a smaller size unit was virtually impossible because of the nature of the critical design variables—for example, the radius and centrifugal speed of disk, and droplet mass velocity. The desired physical characteristics of the product, and particularly its particle size and bulk density, in essence "fixed" the dimensions of the drying chamber.

For this reason, the entire program was carried out using the vendor's experimental unit. This allowed him to offer a firm proposal for the planned commercial-size facility based on reliable data. Of equal importance, development quantities of product were obtained under conditions that would be achieved in the proposed commercial-scale plant.

C. Show Tubes

The "show tube" concept is one part of the miniplant approach to pilot planting. If a process is to be scaled up to a bundle of catalyst tubes, then the show tube can be one full-sized tube with the same dimensions as the ones used on a commercial scale. Scaleup is then simply linear, adding more tubes in a bundle or using multiple tube bundles. The approach of taking a characteristic slice from a commercial process into the pilot plant has an extremely high success rate and credibility when scaling up to the commercial plant. Attempts to preserve prespecified L/D ratios of geometric similitude have probably been the greatest deterrent to using the show tube concept more widely.

The success rate of the show tube or "characteristic slice" concept is high since most, if not all, of the parameters are kept the same as in the large-scale plant. The concept is not limited to catalytic processes; falling film evaporator tubes, or a characteristic slice from a fluidized bed with identical velocities and bed height as in a commercial plant also lend themselves to a show tube approach.

In the scaleup of a reactor to make an inorganic chemical, two gases were reacted on a catalyst. The reaction was exothermic and equilibrium considerations favored low temperatures. It was suggested that a tubular catalyst bed should be used with cooling to remove the heat of reaction to be supplied in the shell. The pilot plant reactor was designed with three tubes each 2 in. in diameter and 10 ft long. The use of more than one tube allowed tube interaction to be studied. The unit performed as expected; a production reactor was designed with 900 tubes of the same dimensions and operated successfully.

D. Integrated Pilot Plants

If an objective of the pilot plant is to demonstrate a new process while producing significant quantities of product, then a "totally integrated pilot plant" may be the most appropriate. This concept is commonly used for the development of processes for production of a large volume commodity product where the plants will involve multimillion dollar investments. The key is to design, construct, and operate a "totally integrated pilot plant" including all of the recycle streams and ecology control systems. Material of construction, maintenance, operating labor, instrumentation, process control, raw material handling, and waste disposal can be tested in a totally integrated pilot plant.

The large research expense for a totally integrated pilot plant is justified when the risks associated with the final design are greater than the cost of the totally integrated pilot plant. If the risk associated with starting up a new 50 million dollar plant is 15 million dollars since it is a new technology plant, a 5 million dollar totally integrated pilot plant could be justified. Many of the small plants that were built in the 1950s may in fact have served as the "totally

FUNDAMENTAL CONSIDERATIONS IN PILOT PLANT PROGRAMS

integrated pilot plants" for the large world sized single train plants of the 1970s.

E. Demonstration Plants

The objective of operating a demonstration plant is generally to gain experience on a new process, a new product, or both at once. A demonstration plant differs from a miniplant in scale of operation, design considerations, and operating philosophy. Unlike the miniplant, which is based on a scaledown, a demonstration plant can approach the size of a full-scale production plant. If it is the first full-scale plant of its type, it is often called a prototype or protoplant.

A demonstration plant also differs from a show tube that involves a characteristic slice since it is an integrated unit; all components of the process have been designed for a specific throughput or capacity. As a result of the integrated design concept and relatively large scale of operation, the demonstration plant is relatively expensive to construct and operate. For this reason, demonstration plants are built only when the unknowns in process scaleup or demands in the marketplace preclude selection of other pilot plant alternatives.

In planning construction of a demonstration plant, the same design considerations as those used in construction of a full scale plant apply. In most cases, data for scaleup of equipment to demonstration plant size will have been obtained from one of the smaller units discussed previously. The critical consideration in the design of the demonstration plant is placing emphasis on those portions of the plant which involved process uncertainties. That is, one must carefully design those parts of the process which precluded scaleup to the full-sized commercial plant to begin with.

Often a reactor system presents the greatest degree of uncertainty, particularly if it involves a heterogeneous reaction where diffusion and mixing can have a major role in determining overall reaction rate. In such cases, it is important to install operational flexibility into the reactor system. For example, stirred reactors should be equipped with variable speed agitators; tubular reactors should be designed so that reactor modules can be easily added or removed. Auxiliary systems, such as heat transfer systems and fluid handling equipment must be designed with sufficient spare capacity so as not to unduly limit the operating range of the reactor.

When the feasibility of a process cannot be proven on the miniplant scale, a demonstration plant may be built to provide a first-hand experience with specific operation. This is often the case with solids handling processes consisting of filters, grinders, conveyors, and solid dryers. The minimum size of a demonstration unit is then determined by the minimum size of a particular process unit operation or piece of equipment. Where miniplants concentrate on process fundamentals such as the chemistry, kinetics, and thermodynamics of the process, a demonstration plant is built to a scale which is practical for studying the engineering and mechanical aspects of the process. In coal liquefaction, the process chemistry can be proved in a miniplant operation on a scale of say, 10 kg/hr. However, the uncertainties of scaleup of the liquid/solids separation equipment, since it is mechanical in nature, results in demonstration units of at least 10 t/day being considered.

Once the size is determined, all the process steps must be built to about the same scale to preserve a truly integrated operation. Since the demonstration plant will represent a relatively large expenditure both in capital investment and operating expenses, often there is pressure to reduce the size of the less critical units or even to eliminate a process step. Such an approach destroys the integrated concept of the demonstration plant and is seldom justified by the results. The funds saved in the demonstration plants by these cost cutting maneuvers is usually plowed back into the full-scale plant to make up for uncertainties in the process technology that should have been eliminated by the demonstration unit.

In addition to the need for flexibility and integrated design, it must be remembered that the demonstration plant is a data gathering device. Therefore, it must be equipped with adequate instrumentation and sample gathering devices so that all aspects of the process can be confirmed. In a chlorination process, in the main reaction, chemical "A" was chlorinated to make compound "B" with HCl evolving as a by-product. The steps were as follows:

Drying raw material by distillation to prevent water from causing corrosion in process equipment.

Catalyst mixing with dry raw material to make the chlorination more selective.

Chlorination.

Degassing or vacuum distillation to remove dissolved HCl and other gases.

Product distillation under vacuum to remove "tars."

HCl purification.

The final product had been made for several years in a different process and much was known about processing this chemical. Therefore, it was decided that only the catalyst mixing and the reaction steps should be pilot planted.

After several months of pilot plant work, two continuous stirred reactors in series were selected as an ideal system for the new commercial plant. During the design of the plant it was slowly realized actual data were lacking for other parts of the process. However, because of time pressures and the high cost of building a pilot plant, it was decided to take the risk and continue with design. Gradually, people involved in the project understood that this first commercial plant would be the demonstration plant. Only after it was built would many parts be tied together for the first time.

During startup the plant suffered severe corrosion from water in the degassing column and vacuum system. After much research and frustration, it

FUNDAMENTAL CONSIDERATIONS IN PILOT PLANT PROGRAMS 669

was found that the dry raw material, when stored under dry air, was oxidized and formed water. The phenomenon had never been seen in the older process, since the old system was never truly dry. Operations personnel had learned how to reduce the corrosion by using more expensive glass-lined equipment. When the purge gas was changed to nitrogen, the corrosion problem was solved. However, the vacuum system had to be replaced and the plant was several months late coming onstream.

Other minor problems resulted in the plant not running at capacity for several more months. A vendor's low-temperature refrigeration unit did not perform as expected; the mechanical seals in the vacuum pumps failed repeatedly; the catalyst feeder plugged several times; and water leaks from condensers were a constant problem. In retrospect, it was not appropriate to bypass pilot plant studies of the drying and degassing steps. Better yet, an integrated pilot plant would have brought out most of the problems that occurred during startup. The problems with the vendor-designed refrigeration unit would have been prevented if the equipment calculations had been reviewed in detail. As a result of the long startup time, any benefits of bypassing a pilot plant were completely lost. However, as a footnote, within one year after the first plant was in operation, a second plant was brought onstream and reached capacity within 30 days. All the lessons learned in the first plant (a demonstration plant) had been put to good use in the second unit.

In some pilot plants there can be a number of objectives to be met. For example, in the design and operation of a market development plant the primary goal may be to produce a product with the desired specification and at the required production rate to meet market developmental requirements. Here, obtaining process data is secondary to getting information about the acceptability of the new product. The market development plant can be a large-scale, integrated unit such as a demonstration plant, or far simpler unit. So long as it fulfills the market developmental needs it can be a one-kettle operation.

If, in addition to the product data, considerable process scaleup data is required, it is very important before building a market development plant to consider how both the process data and the manufacture of the product will be accomplished. Several options are available, First, one may obtain the process scaleup data in a miniplant while producing the required product in a separate market development facility. This option has the following advantages:

Two separate facilities allow product and process needs to be met at their separate optimum operation scales.

Potential timing conflicts are eliminated.

Moreover, as the market is developed, the capacity of the production unit can be increased without any adverse effect on the process development program. However, substantially greater resources are often required to maintain two separate facilities. Also, in a multistep process, if raw materials to support the market development requirement are not available, then they must also be produced in the market development plant. This increases both the complexity and capital requirements for the facility.

One can miniplant only a portion of the process and expand the role of the program in the market development plant to demonstrate the low risk portions of the process. Alternatively, one can build a demonstration plant with the dual objective of obtaining both process scaleup and product requirements in a single facility. Here one gains increased confidence through study of the process at a large scale of operation as well as confidence that the product quality will be the same in the commercial plant.

A new herbicide under development required eight chemical reactions for its synthesis. The first five steps leading to a key intermediate K were under extensive study in the laboratory.

$$A \xrightarrow{\text{five steps}} K \qquad (17\text{-}1)$$

For the commercial plant this reaction sequence was envisioned as a continuous process with many separations and recycles.

The last three steps leading to the product P had been studied at a scale of several liters. These steps were envisioned as a series of batch reactions with little recycle of unreacted materials.

$$K \xrightarrow{\text{three steps}} P \qquad (17\text{-}2)$$

Intermediate K could be purchased in quantities up to several metric tons at a price commensurate with developmental quantities, about \$25/kg. Field testing requirements required hundred-kilogram quantities the first year increasing to 20,000 kg in the fifth year of the marketing development. Assuming all programs have an economic justification based on cost, volume, and price projections, let us consider the viable approaches for scaleup to fulfill both product and process requirements. There are several options:

Separate and complete miniplant and market development facilities.

A complete demonstration plant to answer both process and product questions.

Miniplant a portion of the process; build a market development plant for the remainder.

The first option involves a high degree of uncertainty because of the five-step reaction. This would result in considerable time being needed for the design, construction, and startup of a complete market development facility. Similar considerations apply to the second option.

The third option seems best. An integrated miniplant can be designed for the five-step reaction sequence to carry out the extensive process development program leading to the production intermediate K. At the same time, a relatively well-defined three-step sequence starting with purchased K can be scale up to a "kettle" type operation designed to produce thousands of kilograms of P. Enough K can be purchased to start the market development program. After the process of producing K has been studied in the miniplant, the miniplant can be run for a period of several months to produce quantities of K to try out in the market development plant. In this way, a check on both the process and product using "homemade" K can be achieved.

III. PREDICTING COMMERCIAL PERFORMANCE

The prediction of process performance in pilot plants and production plants has been a part of the art and science of chemical engineering for many decades. The goal is to model and predict what will happen while conducting the minimum of experiments. The goal of a plant operator, however, is to make the process produce the product he wants. Pilot plant activities require a combination of skills for predicting, selecting, and designing pilot plant equipment. This must be followed by building and operating the equipment to test the predictions as well as making changes where the predictions were not correct.

Each segment of a unit operation in a pilot plant may have a different degree of predictability for design from first principles. For example, heat transfer can be very predictable for the design of heat exchangers such as preheaters for liquid feed or condensers and reboilers for distillation columns. However, the heat transfer coefficient for the jacket on a polymerization kettle for a new polymer may not be readily predictable if fouling occurs from polymer buildup on the walls.

Plate efficiencies for distillation of "hydrocarbons" are predictable to some degree from first principles such as relative volatility, viscosity, surface tension, and so on. However, the efficiencies or height equivalent to theoretical plate in a packed column may not be predictable. Even in distillation columns, highly nonideal solutions such as priority pollutants in water, highly viscous solutions, foaming solutions, or immiscible mixtures of liquids and solids may require experimental data before commercial performance can be predicted. A high level of skill and experience is required for deciding whether fundamental data alone are required or whether performance data from a pilot plant are mandatory.

The miniplant concept has been developed to predict the performance of large-scale equipment from data generated by low-cost experiments in small equipment, rather than relying solely on fundamental data. For plate-type distillation columns, the accepted miniplant procedure has been to use vacuum-jacketed and silvered glass Oldershaw (1941) columns to both measure plate efficiencies and predict the performance of large-diameter plate columns. Small 2-in.-diameter columns packed with $\frac{1}{4}$-in. ceramic Intalox saddles are recommended by Eckert (1979) for predicting foaming characteristics in packed distillation towers. Karr (1980) developed correlations for generating data for scaleup of his liquid–liquid extraction column from 1-in.-diameter test units. Robbins (1979b) presents an analysis of flooding throughput versus agitation intensity in a 1-in.-diameter Karr column. This permits prediction from Karr data of the performance of other liquid–liquid extraction units.

Handling of slurries in pilot plants and miniplants will predictably cause operating problems from plugging. However, often a concept that is unique to the miniplant but which would not be used in the plant can be employed. For example, a time-pulsed ball valve on the bottom of a mixed suspension crystallizer may be used in the miniplant. A throttled diaphragm valve would be used for the same purpose in a plant crystallizer. For temperature control in the miniplant a time proportioning on–off of control is desirable and keeps capital costs low.

Electric heating coils can be time pulsed for heating and a solenoid valve in the cooling water line can be time pulsed for cooling. In this way, there is no need to use proportioning flow controllers or current-to-pressure (I/P) converters in a miniplant. However, additional instrumentation cost can be justified in a large pilot plant installation. Where instrumentation and control scheme requirements are quite predictable on a plant scale they need not be a critical factor in the pilot plant. However, the fouling of a heat transfer surface in a crystallizer or polymerizer can be a critical constraint. Similarly, in a distillation column the adiabatic condition of a commercial column is a critical constraint that should be met as closely as possible in small distillation column if the performance of large distillation columns is to be adequately predicted.

IV. RULES OF THUMB IN PILOT PLANTS

During the conceptualization stages of a pilot plant, during the design phase, and later during the troubleshooting, rules of thumb are quite valuable. There are a number of general guidelines that experienced pilot plant professionals use to start a point of questioning that may lead to a reasonable approach to the solution of a problem. Some of these rules of thumb are common; others not so common. A number of important rules of thumb we have found useful are:

General

Reaction rates may double every 10°C.
Vapor pressures may double every 20°C.
Liquid and vapor handling are easier than solids handling.

RULES OF THUMB IN PILOT PLANTS

Recycle streams usually build up with impurities which can lead to surprising problems.

A "boiling" reactor or adiabatic reactor is usually easier to scale up than a jacket cooled reactor.

Small volume production favors multipurpose use of equipment; large production favors single use of each item.

Use digital (on–off) control in small equipment to save cost.

Backmixing necessitates longer length to get the same results.

Heat Transfer

Heat transfer coefficients in jacketed reactors may run 40 Btu/hr-ft^2-°F for metal- and 30 Btu/hr-ft^2-°F for glass-lined equipment.

Conservative heat transfer coefficients for condensers may be 95 Btu/hr-ft^2-°F and 200 Btu/hr-ft^2-°F for thermosiphon reboilers.

Separations

Countercurrent multistage equals high efficiency in separations.

Column diameters should be at least eight times the packing size to minimize large voids near the wall.

Column length and throughput per cross-sectional area should be held constant to get identical results.

First pass guess at the boilup vapor rate in a distillation column may be 1000 lb/hr-ft^2.

Use distillation if the relative volatility is greater than 1.2.

If two liquids are immiscible the infinite dilution activity coefficient is greater than 8.

Lower temperatures generally give high relative volatilities in distillation except for alcohols in water.

10 percent salt in water usually doubles the activity coefficient of a dissolved organic.

1000 ppm or less dissolved organic is essentially infinite dilution.

The mass transfer unit is a better mathematical model than the theoretical stage when the stripping factor or absorption factor is high.

A stripping factor or absorption factor of 5 is about as good as infinity.

Liquid distribution is the problem with a packed tower in 80 percent of the cases.

First guess at the residence time required in a decanter may be 20 min or an upflow of 0.5 gpm/ft^2.

Total flow through a liquid–liquid extraction column may be 15 gpm/ft^2 with a density difference of 0.2 g/cm^3.

Solid particles tend to collect at liquid–liquid interfaces.

The relative separation factor in a liquid–liquid extraction should be at least 5.

Liquid–liquid extraction column volume can be reduced with mechanical agitation if the interfacial tension is high.

Dissolving 2 to 20 percent organic solute generally reduces interfacial tension.

Freezing points may be suppressed 1°C for every 1.5 mol% impurity present.

A ratio of impurity concentration between a solid and liquid phase greater than 0.2 is probably due to solid solution.

High viscosity often leads to supercooling in crystallization.

The Ultimate Rule

A simple solution to the problem is elegant.

Rules of thumb for deciding what unit operations need to be pilot planted are given in Table 17-1.

V. NEWER DEVELOPMENT IN PILOT PLANTS

Since one of the primary purposes of a pilot plant is to generate data for scaleup, computers are gradually playing a larger role in process control, data logging, on-line analysis, and report generation. Jongenburger et al. (1981) review recent developments in this area. Computers are also being used more widely to load, control, and dump batch reactors to remove deviations from human elements in the control scheme. Results are more reproducible. Computers can also provide a safe degree of control because programs can be written to respond to emergencies *much* faster than a human operator. Microprocessors, as discussed by Taylor et al. (1981), are now emerging as highly cost effective in controlling miniplant and laboratory equipment.

The pilot plant is an ideal place to test a new computer control scheme or new control hardware since a failure of these new components should not have the same consequences in the pilot plant as it could in a production facility. The ultimate dream may be coming closer to reality where a pilot plant engineer could use a computer to run the process, collect the data, and write the necessary reports. However, the reality of designing and constructing the pilot plant as well as obtaining raw materials and handling the products and by-products still has not succumbed to computer control.

In the realm of equipment procurement and construction, the use of modular construction for standarized shop fabrication and reuse of pilot plant equipment provides an approach for cost reduction. Several companies are

TABLE 17-1 Rules of Thumb for Deciding What Unit Operations Require Pilot Plant Testing

Operation	Pilot Plant Required?	Comments
Distillation	Usually not. Sometimes needed to get tray efficiency or HETP data.	Skipping pilot plant can sometimes produce surprises, for example, foaming.
Fluid Flow	Usually not for single phase. Almost always for two phase flow.	Some polymer systems are also very difficult to predict.
Reactors	Frequently.	Scaleup from lab often justified for homogeneous systems, seldom for heterogeneous.
Polymerizers	Almost always.	
Evaporators, reboilers, heat exchangers, coolers, condensers	Usually not unless there is a possibility of fouling.	
Dryers	Yes—almost always.	Usually done using vendor equipment.
Solids handling	Almost always.	Usually done using vendor equipment.
Crystallization	Almost always.	Many times done with vendor equipment.
Extraction	Almost always.	Should pilot in similar type of equipment proposed for production plant.

already building standard unit operation modules such as furnaces, reactors, distillation columns, and extraction columns. These modules can be assembled for a process and after the data are collected, the modules can be cleaned up, disassembled, and reused for another process. Pilot plants will always be needed but they can be made more efficient and timely.

REFERENCES

Ackerman, G. D., Huling, G. P., and Metzger, K. J., "The Effects and Benefits of Pilot Plant Automation," *Chem. Eng. Prog.*, **67** (4), 50–53 (4 Apr. 1971).

Bunning, D. L., and Foster, R. D., "Piloting an Operating Plant," *Chem. Eng. Prog.*, **77** (8), 50–54 (Aug. 1981).

Cinadr, B. F., Curley, J. K., and Schooley, A. T., "Miniplant Design and Use," *Chem. Eng.*, **78**, 62–76 (25 Jan. 1971).

Eckert, J. S., "Design of Packed Columns," Section 1.7, *Schweitzer's Handbook of Separation Techniques for Chemical Engineers*, McGraw-Hill, New York, 1979.

Henthorne, M., "Measuring Corrosion in the Process Plant," *Chem. Eng.*, **78**, 89–94 (23 Aug. 1971).

Johnson, D. B., "Caring for Glass-Lined Equipment," *Chem. Eng.*, **71**, 244, 246, 248–249 (8 June 1964).

Jones, A. R., and Hopkins, D. J., "Controlling Health Hazards in Pilot Plant Operations," *Chem. Eng. Progress*, **62** (12), 59–67 (Dec. 1966).

Jongenburger, H. S., Dicke, W. F., Kahn, H. A., and Shah, S. S., "Putting a Data Logging System to Work," *Chem. Eng. Prog.*, **77** (8), 44–49 (Aug. 1981).

Karr, A. E., "Design, Scaleup and Applications of the Reciprocating Plate Extraction Column," *Separ. Sci. Tech.*, **15** (4), 877–905 (1980).

Katell, S., "Justifying Pilot Plant Operations," *Chem. Eng. Prog.*, **69** (4), 55–56 (Apr. 1973).

Kline, P. E., Vogel, A. J., Young, A. E., Townsend, D. I., Moyer, M. P., and Aerstin, F. G., "Guidelines for Process Scaleup," *Chem. Eng. Prog.*, **70** (10), 67–70 (10 Oct. 1974).

Lowenstein, J. G., "Pilot Plant Safety Program," *Chem. Eng. Prog.*, **75** (9), 51–59 (Sept. 1979).

Moreland, P. J., and Hines, J. G., "Corrosion Monitoring—Select the Right System," *Hydrocarbon Processing*, **57** (11), 251–255 (Nov. 1978).

Muller, K. A., Hamilton, L. W., and Johnston, W. F., "Specialized Pilot Plant Equipment," *Chem. Eng. Prog.*, **59** (6), 33–37 (June 1963).

Ohsol, E. O., "What Does It Cost to Pilot a Process?" *Chem. Eng. Prog.*, **69** (4), 17–20 (Apr. 1973).

Oldershaw, C. F., "Perforated Plate Columns for Analytical Batch Distillations," *Ind. Eng. Chem. Anal. Ed.*, **13**, 265–268 (1941).

Prugh, R. W., "Preferred Pilot Plant Control Practices," *Chem. Eng. Prog.*, **70** (11), 61–64 (Nov. 1974).

Robbins, L. A., "The Miniplant Concept," *Chem. Eng. Prog.*, **9**, 45–48 (1979a).

Robbins, L. A., Liquid-Liquid Extraction," Section 1.9, *Schweitzer's Handbook of Separation Techniques for Chemical Engineers*, McGraw-Hill, New York, 1979.

Shields, S. E., and Vander Klay, A. P., "Amoco's New Generation of Pilot Plants," *Chem. Eng. Prog.*, **73** (9), 73–79 (Sept. 1977).

Taylor, J. H., Johnson, J. H., Barton, W. J., and Wang, R., "An Information System for the 80's," *Chem. Eng. Prog.*, **77** (8), 41–43 (Aug. 1981).

Townsend, D. I., and Tou, J. C., "Thermal Hazard Evaluation by an Accelerating Rate Calorimeter," *Thermochim. AGTA (Thagas)*, **37** (1), 1–30 (1980).

18

SCALEUP: OVERVIEW, CLOSING REMARKS, AND CAUTIONS

G. ASTARITA

I.	Is There a Royal Road to Scaleup?	678
II.	Experiments	681
III.	Modeling	683
IV.	Dimensional Parameters in Constitutive Equations	686
V.	Scaleup in Retrospect	690
References		690

Once King Ptolemy of Egypt asked Euclid whether he could make his geometry simpler, so that Ptolemy himself could learn it with less effort. Euclid's answer was: "There is no Royal road to geometry."

I. IS THERE A ROYAL ROAD TO SCALEUP?

The definition of scaleup given in Chapter 1 is:

> The successful startup and operation of a commercial size unit whose design and operating procedures are *in part* based upon experimentation and demonstration at a smaller scale of operation.

Like all defintions of some area of knowledge, not all people will agree with the one given above. Likewise, no definition of chemical engineering would be agreed upon by all practitioners of the profession. However, whatever the definitions of scaleup and of chemical engineering may be in anybody's mind, they are likely to largely overlap each other. Indeed, to a very significant extent, scaleup *is* chemical engineering.

One may ask oneself why do we feel the need to speak of scaleup as distinguished from chemical engineering. One possible answer to that question is that perhaps there *is* a Royal road to scaleup. Somehow, we chemical engineers (without necessarily even knowing it) may secretly hope that, although we may not have developed it yet, a scaleup *algorithm* exists, the use of which would allow us to *rigorously* predict the behavior of a large-scale unit from measurements made on a small-scale model of it. Should we be able to develop such an algorithm, the goal of "successful startup and operation of a commercial size unit" could be achieved without any need of a deep understanding of the physical and chemical phenomena that are taking place.

Unfortunately, it is by now known that no such algorithm can be developed except for degenerately simple problems. The scaleup algorithm is, in fact, well known, and has been well known for a long time. Complete geometrical similarity should exist between the small-scale and the large-scale unit, and all the relevant dimensionless groups should have the same value for the two units. There are, however, three serious catches with the algorithm.

The first catch is that, if one tries to satisfy all the constraints imposed by the algorithm, often it turns out that they can be satisfied only if the linear dimensions of the two units are the same—a conclusion that obviously nullifies the usefulness of the algorithm. This point has been known for a long time. The classical example comes from hydraulic engineering; one cannot keep both the Reynolds number and the Froude number constant and yet change the linear dimensions—unless one uses a liquid different from water on the small scale. And even if one is willing to do so (which hydraulic engineers generally are not), the liquid one would need to use on the small scale may require properties which no real liquid possesses (like, for example, a density 20 times less than that of water). There are subtle, and in a sense funny, variations of this result, such as the case discussed in Chapter 5, where the conclusion is reached that the catalyst used on the small scale should be much more active than the one used on the large scale—and, of course, should such a catalyst exist, one would want to use it on the large scale!

The second catch is that, in fact, one can never really achieve complete geometrical similarity. For example, how could one scale up (or down) completely the geometry of a catalyst particle? And, even if one could, the pores in the small particle would be of such a small diameter that the mode of diffusion through them might change to Knudsen diffusion—and one would need to scale down the size of molecules to avoid that! Variations of this kind of difficulty are easy to find whenever some possibly relevant linear dimension (like the size of gas bubbles in a sparged reactor) are not independently controllable. The ultimate form of the geometrical impossibility is the following one. While it is in principle possible to have the same length-to-diameter ratio on two different scales, it is patently impossible to have the same surface-to-volume ratio. Since that ratio is certainly relevant for heat transfer, the conclusion is reached that full geometrical similarity at two different scales is impossible. (Strictly speaking, since the surface-to-volume ratio is dimensional, an appropriate dimensionless group containing it could be held constant to avoid this problem. However, this turns out to be very difficult to do.)

The third catch of the algorithm is related to the fact that the "relevant" dimensionless groups are involved. In order to convince oneself about which are the *relevant* dimensionless groups, one needs to understand quite thoroughly the physico-chemical phenomena that take place. There is no Royal road to the identification of the relevance of dimensionless groups. For any given process, if one understands it enough, one can come to the conclusion that some of the dimensionless groups involved are irrelevant—and *then* apply the algorithm to the simpler problem resulting from such a conclusion; but doing so requires a deep understanding. One can certainly predict the pressure drop in a large pipe from measurements made on a small one, because we know that the Reynolds number is the only relevant dimensionless group (though making measurements in a small pipe at Reynolds numbers as large as those prevailing in a very big pipe is not easy). But it takes a lot of physical understanding to convince oneself that indeed the Reynolds number is the only relevant dimensionless group. Getting one's students in fluid mechanics to understand why gravity forces do not play any role in pipe flow is not easy, as anyone who has ever taught fluid mechanics knows all too well. And, of course, should they play any role, then the Reynolds number would not be the only relevant dimensionless group.

One, therefore, comes to the conclusion that, like for geometry, there is no Royal road to scaleup. The only road possible is a republican one, where one does need to understand the phenomena involved. This has been stressed again and again in this book, a large fraction of which is really not dedicated to scaleup as such, but to a review of the state of the art of our understanding of the physical and chemical phenomena taking place in any given process. If any reader of this book has come all the way to here, it is quite likely that he or she has received the impression that it is really a book on chemical engineering.

What, then, is scaleup as distinguished from chemical engineering? It is not a Royal road to the goals of scaleup which avoids the need to understand typical of chemical engineering. It is, however, a *viewpoint*; a sort of guideline of what a chemical engineer should always keep in mind when doing his work. A chemical engineer should always keep in mind that whatever he or she is doing, the ultimate goal should be "the successful startup and operation of a commercial size unit whose design and operating procedures are *in part* based upon experimentation and demonstration at a smaller scale of operation." The authors of the several chapters in this book have certainly kept this in mind while writing their contributions, though at times parts of the chapters may read like research reports. But, of course, chemical engineering research is significant when it is aimed at making the "part based on experimentation at a smaller scale of operation" larger and larger.

Thus, one comes to the conclusion that the intricacies of scaleup methodology are not any less or any different from those of chemical engineering. This conclusion is also borne out by the fact that this book was written by several authors: the scope of scaleup methodology is so wide-ranging that its discussion requires the concerted efforts of several people, each one discussing the part which falls into his or her area of expertise.

There is, however, a drawback in this breaking up of scaleup methodology into a number of component parts—scaleup of chemical reactors (Chapters 4–7), countercurrent separation processes (Chapters 12 & 13), mixers (Chapter 8), and so on. The various chapters of this book are concerned with these component parts of scaleup, and as our understanding of the phenomena involved in each area progresses, so does the reliability of the scaleup procedure for each unit.

However, if one reads carefully the definition of scaleup given in Chapter 1, one realizes that the ultimate goal is to scale up the whole process. The commercial process will indeed consist of a number of units (reactors, distillation towers, etc.) linked together in a complicated network, and maybe one does know how to scaleup each individual unit. But the commercial process is *more* than the sum of its parts, and unexpected bugs may develop which are related to the fact that good scaleup of all the units of a process does not necessarily imply that the process as a whole has been successfully scaled up.

The three examples of unexpected problems in the commercial size unit, which were discussed in Chapter 1, are all in a sense examples of this dichotomy between *unit* scaleup and *process* scaleup. The two problems of the explosive limits and of materials storage bring back the geometrical trick of the surface-to-volume ratio: in both cases the large size unit has inferior heat transfer characteristics, and this causes the problems to arise. One could argue that these are unit scaleup problems, not process scaleup ones. However, one does not usually regard storage of materials as a unit operation. It is simply taken for granted on the laboratory scale, and nobody worries about storage scaleup; so that the ensuing problem arises at the commercial size process level.

The best example is the one of water leaking into a stream in a heat exchanger, and "killing" the catalyst in a reactor. Here both the heat ex-

EXPERIMENTS   681

changer and the reactor may have been scaled up perfectly: the first one does exchange heat as it was supposed to do, and the second one *would* give the right conversion and selectivity if fed with a dehydrated stream (as was the case in the laboratory). However, the process that incorporates the two units does not work—a perfect example of the fact that the process is more than the sum of its component parts.

It is interesting to observe that process scaleup problems arise in connection with areas where the available fundamental understanding is comparatively poor. The water leak example falls in the area of catalyst poisoning by trace impurities; other common areas where process scaleup problems arise are corrosion, foaming, and so on. These are indeed areas where the fundamental understanding is rather poor: for only one of them (corrosion) is there a chapter in this book and one may notice that Chapter 16 approaches the problem by considering directly pilot plant tests, that is, tests made on as large a scale as possible without making them on the commercial size unit directly!

II. EXPERIMENTS

The scaleup methodologies that have been discussed in the preceding chapters range over a broad spectrum that spans the conceptual distance between a mockup and a mathematical model. As was discussed in Chapter 2, even mathematical models, no matter how firmly rooted in physico-chemical understanding, actually describe a conceptual model of the real process rather than the real process itself. In this sense, there is a certain amount of "mockup-ness" in every mathematical model. Moreover, every mathematical model includes a lot of parameters, the values of which need to be *measured*, so that experimentation is not ruled out by a "mathematical modeling" approach—though of course the nature of the experimentation is changed drastically.

At the other extreme, even the most crude mockups, such as the ones made of transparent plastic, so that one can just *look* at what happens, are of no use at all without a modicum of theoretical understanding on which to base the *interpretation* of whatever one observes when the mockup is operated. This shows modeling is not ruled out by a "mockup" approach.

In fact, modeling and experimentation are intimately interconnected, and any scaleup methodology incorporates both. In this section, attention is focused on the experimental part of the methodology, and the mockup-to-model spectrum is analyzed in the light of the following question: what are the experiments that should be made?

A good chemical engineer should always strive to be as close to the model end of the spectrum as is allowed by the complexity of the process to be scaled up. After all, the Royal road, which does not exist, is nothing else but a reliable mockup. It has been reiterated over and over again in the preceding chapters, that one should be as close as possible to the model end. Suffice it to quote from Chapter 1:

> Scaleup studies involve modeling relevant phenomena, not the study of miniaturized commercial systems.

However, the mockup does represent our secret hope that there *is* a Royal road, and there is often a tendency in all of us to go closer to the mockup extreme than we really need to. In industrial research and development studies when kinetic data are needed for a reaction which is foreseen as going to rather high conversions in the commercial unit, the data are often obtained at rather low space velocities; that is, at total conversion levels comparable to those foreseen for the industrial unit. This is the Royal road syndrome: the right way to do it is of course to work at very high space velocities, so that the reactor is effectively a differential, rather than a integral one. Of course, the reaction kinetics may be different at low and at high conversions, so one does need data at high conversion. But such data should be collected by running at high space velocities a reactor fed with a feed which is already highly converted to products.

The tendency to move toward the mockup extreme even when one does not need to do so is not restricted to industrial environments; academics also secretly hope that there is a Royal road. An example of this is the chemical mockup of a countercurrent absorber where bench scale reactions take place, which was discussed in Chapter 6. It is possible to *calculate* what an absorber of that type will do, as has recently been demonstrated by Joshi et al. (1981). Therefore, one should not use the mockup: it is a problem where one can be closer to the modeling extreme than a bench scale mockup. (This is not meant as a criticism of the mockup procedure, which was developed well before the calculation procedure became available.)

The example discussed above is interesting in another respect. The *algorithm* for the bench scale mockup assumes that the process is isothermal; there is no heat balance equation in it. So does also the model discussed by Joshi et al. (1981). However, real absorbers using chemical solvents are almost never isothermal; and the extension of the model to include heat effects is comparatively trivial. The bench scale mockup will not do for nonisothermal systems, because one gets faced with the surface-to-volume ratio problem. Note that the commercial unit is likely to be nearly adiabatic, and the bench scale unit to be nearly isothermal.

One may wonder why academics also have a tendency to slip toward the mockup side of the spectrum; after all, they should know better. One of the reasons may be the experience of academics who do consulting work with industry. It is very hard to convince industrial practitioners to do it by measuring significant physical parameters and developing a reliable model, rather than using some variation of the mockup idea; in other words, to model the relevant phenomena, not to miniaturize the commercial system. Also, even academics may need answers on a timetable, for example, to satisfy a sponsor interested in relevant work. They may also seek broad perspective on a problem before breaking it down into fundamental parts. But these are not the only reasons; academics are not immune to the Royal road syndrome. In whatever may be their own area of expertise, many secretly hope "I can do it," secretly hope that there *is* a Royal road. Many people who live in a republic would still rather like to be kings.

III. MODELING

The development of a fundamentally based mathematical model of any given process is made up of *three* different elements: balance equations, constitutive equations, and choice of the system to be considered. These three elements have different features, and they are strictly interlocked with each other.

The choice of the system to be considered is essentially a choice of the *scale of description*. The scale that is usually regarded as the appropriate one for fundamental modeling is "a neighborhood of a point in a continuum." As an example, the description of flow phenomena on this scale results in the formulation of the Navier–Strokes equations. This scale will be called the continuum scale in the following.

It is important, however, to realize that the continuum scale is not the smallest one on which one could choose to set up the description. On the continuum scale, one is allowed to handle such quantities as the concentration or the temperature *at a point*. This would not be allowed at the (plausible) lower scale of molecular size: a single molecule is either component A or B (so that the concept of concentration is lost at this lower level) and it moves at one particular velocity, (so that the temperature concept is also lost). And of course, the molecular scale is not the smallest possible one either: the atomic and subatomic scales come immediately to mind.

The choice of the scale is dictated by a balance between two conflicting requirements. On the one side, we know that, the smaller the scale, the more accurate is the description. On the other side, one is really interested in the accurate description of some macroscopically observable quantities, and one would like to develop models that yield directly the values of such macroscopic quantities.

Once a scale of description has been chosen, whatever phenomena take place at all underlying smaller scales have hopefully been "chunked" successfully. A fluid mechanicist is happy with the Navier–Stokes equations; he believes that, whatever may be the molecular-level mechanism which is responsible for the fact that the viscosity of water is 1 cP, the only relevant bit of information needed is indeed the value of the viscosity. That one can measure in a viscometer, and just forget about molecules.

The considerations above lead to the question of the distinction between balance equations and constitutive equations. As far as the balance equations are concerned, the scale of description is irrelevant: mass, energy, momentum, and so on, are never created and never destroyed. That is true whatever the scale of description (some trouble arises with mass and energy at the subatomic level, but that is not a concern in the present context). The form of the balance equations will of course be different at different scales of description, but the *principles* of conservation which is invoked in writing them down is the same at all scales.

Constitutive equations are a completely different matter. Viscosity is not even defined at the molecular level, but it enters the constitutive equation for a Newtonian fluid at the continuum level—in fact, its value supposedly subsumes whatever of the molecular scale phenomena are relevant to a continuum scale description. However, as one moves to progressively larger scales of description, the reliability of the constitutive equations becomes worse and worse, as discussed below.

Consider first the case of viscosity. The kinetic theory of gases, as developed by Maxwell in the second half of the nineteenth century, is a molecular scale description. It handles viscous dissipation in gases as well as relaxation phenomena. Stresses do relax in finite times in a Maxwellian gas—though admittedly the times of relaxation are exceedingly short at all but unrealistically low pressures. Newton's linear law of friction, which is a continuum scale constitutive equation, does not allow for relaxation phenomena—in other words, its use is equivalent to the assumption that stresses relax instantaneously. In moving from the molecular scale description of Maxwellian gases to the continuum scale description of Newton's law of friction, one has lost *some* of the features of the actual behavior of gases.

Lest one may think that this is an "academic" argument with no "engineering" interest, I will discuss two examples which show the engineering relevance of the argument. Gases do indeed relax very rapidly, and therefore, from an engineering viewpoint, one may as well describe them as if they relaxed instantaneously. However, this simplified description does not work with polymers; Newton's linear law of friction does not apply to polymers. In this case, the molecular scale is just too large (polymers are very long molecules) and the jump from molecular scale to an all-encompassing continuum scale viscosity leads to serious trouble. It is interesting that a constitutive equation which is often used to describe polymers at the continuum level is the so-called Maxwell model, and it is indeed a variation of the equation deduced by Maxwell in his kinetic theory of gases!

The other example is the breakdown of Fick's law of diffusion (which is of course a continuum level constitutive equation) in the very fine pores of a catalyst. Here one has a case where the relevant scale (the pore diameter) is so small that a continuum description will not do, and one has to go one step down and describe the phenomenon at the molecular scale (Knudsen diffusion is a molecular scale constitutive equation).

In spite of the two examples given above, it is in fact generally true that the continuum scale is acceptable for fundamental mathematical modeling. Newton's linear law of friction, Fick's law, Fourier's law, and so on, have been

MODELING

tested again and again and we belive that they do give an accurate constitutive description at the continuum level. In fact, we believe it so strongly that, when referring to a constitutive equation for the rate of chemical reaction at the continuum level we call it the "intrinsic" or "true" kinetics of the reaction.

However, description at the continuum level leads to hopelessly complex equations, which can only be solved for very special cases. A class of such special cases, laminar flows, is discussed in Chapter 11. The author there correctly states that laminar flows are difficult to analyze correctly. Indeed they are so difficult that, for problems at a slightly higher level of complexity, a route other than solution of the continuum level description equations must be found.

At this point, engineers perform a new step of "conceptual scaleup": the scale of description is not the three-dimensional continuum one, but some larger scale. A "chunking" process needs to be made on the continuum level phenomena, and one needs to jump to some new description. This is often called "lumping parameters."

How does one do the chunking? An interesting example (or rather, a series of examples) can be found in Chapter 5. The first model is called the one-dimensional pseudohomogeneous model. Here everything over any given cross section has been chunked together, so that the problem collapses from three-dimensional to one-dimensional. Moreover, the two phases (gas and catalyst) have been chunked together into a "pseudohomogeneous" one. Finally, the assumption of steady state has chunked the time variable, and only the axial coordinate has been left unchunked, so that ordinary differential equations are obtained. [Had one chunked also the axial coordinate, one would have obtained the algebraic equations describing a Continuous Flow Stirred Tank Reactor (CSTR)—the scale of description would have been the whole reactor!]

Now this first model is nice and its equations are easy to solve: that is the advantage of chunking. However, what about the constitutive equations? One would need one for the *average* rate of reaction as a function of the *average* composition and the *average* temperatures (all averages being on the "chunked" parts). These "chunked" constitutive equations are less reliable than the continuum scale ones, and so is the resulting model.

Perhaps one has done more chunking than reasonable, and indeed the next model discussed is the one-dimensional heterogeneous one: gas and catalyst get unchunked again. However, it is a curious form of unchunking: the resulting model equations correspond to the assumptions that the gas and catalyst phases simultaneously fill the same space—which of course at the catalyst pellet dimension scale they do not. Additional unchunkings occur with the more sophisticated two-dimensional models. However, the unchunking never goes back all the way to the continuum scale: the "coexistence at a point" of catalyst and gas is a feature of all the models discussed in Chapter 5. The scale of description is at some ill-defined level somewhat larger than a catalyst particle. Whatever happens at the lower scale (inside the particle) is hopefully subsumed into the value of an effectiveness factor for the catalyst pellets. (Incidentally, calculations of an effectiveness factor are based on models which regard the catalyst as a homogeneous phase, so that their scale of description is somewhat larger than the pore size. Hopefully the underlying continuum scale phenomena are subsumed into an appropriate value of the effective diffusivity).

Chunking of this type occurs over and over again, and it can be found in almost every chapter of this book. The scale of description may be as small as the active site on a catalyst (see Chapter 3), a molecular scale, and it may be as large as the whole mixing tank (Chapter 9) or an entire tray of a distillation column (Chapter 12). And it is generally true that, the larger the scale of description, the less reliable the constitutive equations. The best example of this can be found in Chapter 10, where the statement is made that "Physical properties of fluid beds change as bed diameter and depth are increased"; in particular, the bed density is discussed. Now density is a constitutive property, very well defined at the continuum level. The density of a ton of water is the same as that of a gallon of water. However, when modeling a fluid bed, one is forced into chunking, and the description takes place at a scale where the property of density is not so well defined any more—its value depends on bed depth and diameter!

To summarize: as the scale of description goes down, the constitutive equations become more reliable, and the resulting equations less tractable. One needs to strike a balance: do as much chunking as required to obtain a tractable system of equations, but no more. Put in other words, loose as much reliability as needed to have a working model, but not any more than that.

This spectrum of scales of description, which goes all the way from the very large (the whole reactor) to the very small (the molecular scale) is reminiscent of the mockup-to-fundamental understanding spectrum. Indeed, the former is the spectrum of modes of modeling, the latter the spectrum of modes of making experiments. In both cases, one should be as close to one end (small-scale, fundamental understanding) as possible. And in both cases there is a temptation to move to the "Royal road" end: mockups in experiments, a lot of chunking in the models. But there is no Royal road to scaleup modeling any more than there is a Royal road to scaleup experimentation.

IV. DIMENSIONAL PARAMETERS IN CONSTITUTIVE EQUATIONS

There is a subtle point concerning the number of dimensional parameters appearing in a constitutive equation which has been discussed by Astarita (1974, 1977, 1979, 1983) for the specific case of the constitutive equations for Newtonian fluids. The point, however, is of a general nature, and it is of importance whenever constitutive equations are nonlinear, and therefore in particular for chemical kinetics. This point is discussed below for the simple case where the constitutive equation assigns a relationship between a single

DIMENSIONAL PARAMETERS IN CONSTITUTIVE EQUATIONS

constitutive quantity y (in the non-Newtonian fluid case, the stress) and a single state variable x (in the same case, the rate of strain):

$$y = f(x) \tag{18-1}$$

The discussion is easily generalized to the case where there are several state variables $x_1, x_2, \ldots$.

Equation 18-1, as such, is simply a statement that the value of y is believed to be uniquely determined by the value of x. The question to be addressed is: what is the minimum number of dimensional parameters which will appear in some explicit form of the function $f(\cdot)$? Since the number of dimensionless groups increases as the number of dimensional parameters is increased, it is obvious that scaleup by similarity will be easier the lower the number of dimensional parameters.

If both x and y are dimensional (and do not have the same dimensions), two dimensional parameters will in general be needed. Indeed, let X and Y be two parameters having the dimensions of x and y, respectively. Then one can write, instead of Equation 18-1, the following equation:

$$\frac{y}{Y} = F\left(\frac{x}{X}\right) \tag{18-2}$$

Since $F(\cdot)$ is now a transformation of one dimensionless quantity into another, it is clearly invariant under a change of units.

There is, however, a special case where only *one* dimensional parameter is needed, and that is when $f(x)$ can be expressed as a power law:

$$y = Kx^m \tag{18-3}$$

Equation 18-3 does contain two parameters, K and m, but only one of them, K, is dimensional.

It is important to observe that the linear case ($m = 1$) is a special case of the power-law form. Many constitutive equations are linear, and therefore contain only one parameter (this will be discussed in more detail below). However, it is seldom realized that, among the infinitely many possible nonlinear forms of the function $f(\cdot)$, the power law is dimensionally peculiar. The power law nonlinearity is (from the dimensional viewpoint) a mild one, because it does not increase the number of dimensional parameters beyond that corresponding to the linear case.

An example drawn from Chapter 3 is of interest here. The discussion concerning the Langmuir–Hinshelwood–Houghen–Watson (LHHW) formulation and Temkhin–Pyzhev rate equations in ammonia synthesis is about the constitutive equation for the rate of reaction; the state variables are the partial pressures of H_2, N_2, and NH_3. The LHHW form, Equation 3-51, contains three dimensional parameters, the Temkhin–Pyzhev form, Equation 3-57, only two. The reason is, or course, that in the latter form the state variables are aggregated into power-law chunks. In fact, there is more than meets the eye in the transformation leading from Equation 3-51 to Equation 3-57: the number of dimensional parameters has been decreased, and scaleup by similarity is easier if based on Equation 3-57. The price to pay is the usual one: less reliability. The LHHW rate equation, being more firmly rooted in our understanding of the true mechanism of reaction, is more reliable.

Incidentally, the point above is related to the question of chunking. It is well known [Aris (1968)] that, if one measures the global rate of simultaneous reaction of a whole group of homogeneous components in a mixture, the apparent "order" of the reaction (the value of m) is different from that of any of the component reactions. If the latter are all first order, the global rate will appear to be of an order close to 2. If one uses a second order global rate equation, one is fooling oneself into believing that there is only one dimensional parameter, when indeed there are many (the kinetic constants of the component reactions). Change the initial distribution of concentrations of the reacting components, and the apparent global kinetics will change. A chunked constitutive equation is less reliable than an unchunked one.

Now turn attention back to the fact that many continuum-level constitutive equations are linear: Newton's law of friction, Fick's law of diffusion, Fourier's law of heat conduction, and so on. It is indeed fortunate that most, if not all materials do obey these linear constitutive equations, so that the relevant dimensionless groups are rather few. This simplicity does not hold in chemistry, where simple first order kinetics is the exception rather than the rule (and, as the discussion in the preceding paragraph shows, even when the kinetics are linear at the unchunked stage, they become nonlinear when one chunks).

Now what do engineers do when they are confronted by a nonlinear relationship between two variables (say when some data plotted on a linear scale are not correlated by a straight line)? They like straight lines, and so they plot their data on a log–log scale, where it is well known that almost invariably one can draw a straight line through the data if they do not span a very wide range. And of course a straight line in a log–log plot is a power law, and so this procedure forces a dimensionally peculiar form of nonlinearity, which may in fact not be justified. The funny thing is that, if on a linear scale the data are well correlated by a straight line with a nonzero intercept, people are happy to write Equation 18-1 in the form

$$y = a + bx \qquad (18\text{-}4)$$

which contains two dimensional parameters!

Data correlated by a straight line in a log–log plot could equally well be fitted by a straight line in some other nonlinear plot, and in that case a different, nonpeculiar kind of nonlinearity would be inferred. Indeed, the argument in Chapter 3 shows this quite clearly, when the following "mathematical equivalency" is invoked:

$$Cx^n \sim \frac{bx}{1 + bx}, \quad 0 < n < 1 \qquad (18\text{-}5)$$

In Equation 18-5, if X are the dimensions of x, then X^{-1} are the dimensions of b and X^{-n} those of C. The two sides of Equation 18-5 appear to have the same number of dimensional parameters: one. However, consider the following two statements

$$y = ACx^n \qquad (18\text{-}6)$$

$$y = A\frac{bx}{1 + bx} \qquad (18\text{-}7)$$

which the "mathematical equivalency" would lead one to believe are equivalent. In Equations 18-6 and 18-7, A has the dimensions of y. Now in Equation 18-6 there really is only *one* dimensional parameter, the product AC; but in Equation 18-7 there are two, A and b, which cannot be collapsed into one. From the dimensional viewpoint, Equations 18-6 and 18-7 are not equivalent at all!

One may ask oneself why linear constitutive equations, which are a special case of the power law, so often turn out to hold true. Is perhaps a similar "magic" to be expected also for power law equations with $m \neq 1$? The answer to that question is unfortunately no. The reason why linear forms work is related to the fact that $f(x)$ in Equation 18-1 is likely to be expandable in a Taylor series

$$f(x) = a + bx + cx^2 + \ldots \qquad (18\text{-}8)$$

and that, generally, appropriate choice of the definition of x and y will make $a = 0$. It follows that, as long as x is, in some sense, "small" (in fact, smaller than b/c), $f(x)$ can be approximated with its linear form bx. However, by the same token, as soon as nonlinearity sets in (x grows larger than b/c), $f(x)$ will *not* be a power law!

It may be argued that a power-law equation does not result in any real scaleup advantage, since in addition to all other dimensionless groups m also should be held constant. However, if one fools oneself into believing that m is a true constitutive parameter, holding it constant would be simply accomplished by using the same system on two different scales, which one would do anyway (particularly if chemical reactions are involved). However, since m is not a true intrinsic parameter, its apparent value may turn out to be different at the two scales!

To summarize: when constitutive equations are such that aggregation of variables into power-law groups occurs, the number of dimensional parameters appearing in the equations is decreased. If the aggregation is related to some well-understood physical fact (like the mechanism of a chemical reaction, or a thermodynamic constraint), the decrease is a real one, and scaleup by similarity is indeed made easier. If, however, the aggregation is the result of some variation of the "plot on log–log scale" procedure, the decrease is a red herring and should be viewed with suspicion. It is interesting to note that lack of such suspicion has led occasionally to gross misconceptions [Astarita (1977, 1983)].

V. SCALEUP IN RETROSPECT

The dilemma on whether there is or there is not a Royal road to scaleup is a very old one. Leonardo da Vinci wrote: "Vitruvius says that small models are of no avail for ascertaining the behavior of big ones. I here propose to show that this conclusion is a false one." It is a false one; but it takes Leonardo da Vinci to show that it is.

REFERENCES

Aris, R., *Arch. Ratl. Mech. Anal.*, **27**, 356 (1968).
Astarita, G., *Chem. Eng. Sci.*, **29**, 1973 (1974).
Astarita, G., *J. Non-Newt. Fluid Mech.*, **2**, 343 (1977).
Astarita, G., *J. Non-Newt. Fluid Mech.*, **5**, 285 (1979).
Astarita, G., *J. Non-Newt. Fluid Mech.*, **13**, 223 (1983).
Joshi, S. V., Astarita, G., and Savage, D. W., *Chem. Eng. Prog. Symp. Series*, **77**, 63 (1981).

INDEX

Absorbers, mass transfer efficiency, 530
Absorption, 463, 510, 530
  enhancement, 231
  factors, 465
  multicomponent, 467
  phase equilibria, 463
  rates, 215
  in reactive solutions, 268
Absorption column tray hydraulics, 469
Absorption acid gas, 109, 269
Absorption reaction process, 238
Absorption reaction regimes, 212
Absorption stage calculations, 465
Accelerating rate calorimeter, 661
Adiabatic reactor, 168
Adiabatic temperature rise, 168
Agglomeration of particles, 398
Agitation, 69
Air emissions, 601
Alper-Danckwerts method, 213, 214
Ammonia synthesis, 85, 86, 104
Analytical distribution functions, 290
Apparent bed thermal conductivity, 365
Arranged packings characteristics, 516
Arrhenius equation, 109
Attrition of particles, 397
Autothermal reactors, 181
Average shear rates, 325
Axial diffusion, 408
Axial dispersion, 295, 297
Axial dispersion model, washout function, 296
Axial gas-phase dispersion coefficient E_G, 242

Basis for commercial plant, 31
Batch, mass balances, 120
Batch reactor conversion, 123
Batch reactors, 118, 255, 270
Bench scale mockup, 682
Benzene, sulfonation of, 142, 149, 151, 157
Boundary conditions, 53
Bubble behavior, fluidized beds, 354, 355

Bubble column, 217, 236
  Fischer–Tropsch slurry process, 243
  gas holdup, 221
  model simplifications, 235
Bubble column flow regimes, 217
Bubble column reactor, 234
  dynamic variables, 227
Bubble contacting limitations, 351
Bubble slug flow, 358

Cake compressibility, 590
Cake dewatering, 554
Cake filters, 574, 587
Cake washing, 553
Calculated risks, 3
Carman–Kozeny equation, 563
Catalyst deactivation, 171, 193
Catalyst performance, 7
Cavity flows, 421
  heat transfer, 422
Cell models, 187
Centrifugal sedimentation, 562
Centrifuges, 566, 573
Change, rate of, *see* Rate of change
Characteristic particle size, 556
Chemical reaction, 108
  absorption, 108
  liquid extraction, 108
Chunking, 685
Churn-turbulent flow, 219, 231
Churn turbulent regime, 218
Circular tubes, 297
Clarification, 577
Clarifiers, 566
Classification, 46, 54
Clean Air Act, 601
Cleaning of corrosion specimens, 644
Clean Water Act, 604
Closed systems, 288
Coating deterioration, 646
Cocurrent flow, 172

691

Cocurrent upflow, 173
Coking, 193
Column type extractors, 483
Commercial extractors, 487
Commercial grid design, 380
Commercial plant, basis for, 31
Commercial reactors, 258
Complete segregation, 283
Complete segregation yields, 285
Compressibility of filter cakes, 590
Computer code verification, 41
Concentration gradients, 184
Conductivity, effective, Λ_{er}, 187
Consecutive coking, 193
Constitutive equations, 684, 686
Construction materials, 626
Contactor height, 509
Continuous stirred tank, 130, 131
Continuous stirred tank reactors, 119, 257
   dynamic analysis, 140
Corrodents, 628, 629
Corrosion:
   forms of, 630, 632
   preventing, 629
   rapid, tests, 649
   stress, 634
   uniform, 630
Corrosion allowances, 631
Corrosion data, 622
Corrosion fatigue, 635
Corrosion measuring instruments, 649
Corrosion rate, 631
Corrosion sample exposure, 644
Corrosion specimen racks, 640, 641
Corrosion specimens, 642
   cleaning, 644
   preparation, 643
   selection, 639
Corrosion tests, 638
   duration, 643
   laboratory, 647
   in pilot plants, 638
   programs, 623, 624
Cost of pilot units, 656
Countercurrent extraction, 478, 534
Countercurrent flow, 172
Countercurrent contacting, 513
Countercurrent mass transfer, 506
Countercurrent processing, 505
Crack formation, 634
Crevice corrosion concentration cell, 632
Critical settling flux, 573
CSTRs, 265
   in series, 257
Cumulative distribution function, 278

Cyclone, 382, 384
   efficiency, 382
   scaleup factor, 384
Cyclone separator, 383

Danckwerts boundary conditions, 229
Danckwerts–Gillham–Alper method, 213, 214
Darcy's law, 564
Decarburization, 637
Deep bed filters, 571, 575, 589
Degree of segregation, 302
Delta distribution, 280, 284, 298
Demonstration plant, 667
Department of Transportation (DOT), 596
Description, scale of, 683
Designing mixers, 327
Detoxification, 617
Development program, 25
Dewatering screens, 574, 587
Dezincification, 635
Differential distribution function, 278
Differential mass transfer operations, 505
Differential scanning calorimeter, 661
Diffusion coefficients, 439
Dimensional analysis, 62
Dimensional parameters, 686
Dimensionless equation, 63
Dimensionless groups, 64, 66, 67, 316, 679
Discrete variables, 42
Dispersed air flotation, 568
Dispersion coefficients, 245
Dispersion models, 49
Disposal alternatives, 616
Distillation, 435
   infinite stages, 440
Distillation column efficiency, 450
Distillation energy reduction, 493
Distillation hydraulics, 442
Distillation liquid-phase transfer units, 453
Distillation minimum reflux, 442
Distillation problems classification, 441
Distillation stage calculations, 439
Distillation total reflux, 440
Distillation tray, 457
   bubble caps, 448
   efficiency prediction, 461
   entrainment, 458
   froth height, 448
   liquid mixing, 456
   point efficiency, 452
   spray, 447
Distributed parameters, 42, 419
Distribution functions, 278
Distribution moments, 278
Distribution variance, 279

INDEX

693

DOT, see Department of Transportation (DOT)
Drop weight shock sensitivity tests, 661
Dumped packings characteristics, 515
Dynamic analysis, continuous stirred tank reactors, 140

Effective conductivity, Λ_{er}, 187
Effective diffusion model, 189
Effective diffusivity, D_{er}, 186, 189
Effect of temperature, 92
Elastomeric construction materials, 627
Electrical resistance probes, 650
Electrolytic flotation, 569
Empirical models, 56, 59
Empirical transfer functions, 57
Energy balance, 134, 156
   adiabatic reactor, 135
   batch reactor, 136
   continuous stirred tank, 138
   isothermal reactor, 134
   nonisothermal reactors, 119
   tubular reactor, 138
Energy considerations, separations, 493
Engineering correlations, 8
Entrainment, 359, 380
Entrainment rate, 380
Entrainment transport height, 381
Environmental data, 615
Environmental impact, 33, 616
Environmental issues, 595
Environmental permit applications, 600
Environmental Protection Agency (EPA), 596
Environmental regulations, 600, 601
Environmental responsibilities of federal agencies, 599
Environmental sinks, 596
EPA, see Environmental Protection Agency (EPA)
Equations of change, 410
Equilibrium curve, Γ_e, 169
Erosion corrosion, 636
Erosion of particles, 398
Esterification reaction, 121
Evaluation of nonmetallic materials, 645
Experimental residence time data model parameters, 287
Exploratory research, 25
Explosive limits, 2
Exponential distribution, 280, 281, 300, 302
Extraction, 469, 474
   extract phase, 471
   heavy phases, 470
   light phase, 470
   mass transfer relationships, 489
   minimum solvent rate, 475
   multicomponent systems, 478
   raffinate phase, 471
   scaleup, 492
   solvent selection, 479
Extraction column, 536
Extraction column efficiency sieve tray, 491
Extraction devices, 481
   application areas, 488
   classification, 486
Extraction notation, 470
Extraction phase equilibrium, 472
   plait point, 473
Extraction stage determination, 477
Extractors:
   column type, 483
   commercial, 487
   packing material, 534
Extruder, 425

Fabricability, 625
FDA, see Food and Drug Administration (FDA)
Federal agencies, environmental responsibilities of, 599
Fick's law, 435, 684
Filterability tests, 575
Filtration centrifuges, 589
Fixed-bed reactor, 126, 171
   models, 179
Flannery agreement toxic pollutants, 606
Flocculation, 558, 559, 590
Flooding, 445
Flotation, 568, 574, 586
Flow regime, 218, 246
Flow through packed beds, 563
Flow transition, 219, 221
Flow velocity variation, 187
Fluid bed:
   baffles, 401
   bubble coalescence, 372
   bubble size, 371
   contacting model, 389
   "empirical" models, 399
   "fundamental" models, 399
   gas mixing, 375
   grid design, 377
   grid pressure drop, 379
   grids, 378
   heat transfer, 350
   mixing rates, 375
   particle size balances, 396
   solids mixing, 375
Fluid bed reactor:
   scaleup, 401
   scaleup debits, 352
Fluid coking, 385, 386

INDEX

Fluid-fluid reactions, 202, 230
Fluid-fluid reactors, 246
Fluid-fluid systems, 108
Fluid hydroforming, 387
Fluidization, 348
Fluidized beds, 349
  bubble velocity, 355
  heat transfer, 365
  heat transfer coefficients, 366
  slug flow, 369
Fluidized bed density, 370
Fluidized bed expansion, 356, 357
Fluidized bed reactors, 204
Fluidized bed terminal velocity, 361
Fluid reactor crossflow unit, 394
Fluid shear rates, 314
Fluid-solid reaction systems, 102
Food and Drug Administration (FDA), 596
Forms of corrosion, 630
Fractional tubularity model, 290
Freedom of information act, 598
Froth contacting, 455
Fugitive emissions, 614
Full scale tests, 14, 658

Galvanic corrosion, 632
Galvanic series, 632
Gas bubble exchange, 359
Gas exchange rate, 394
Gas flow through bubbles, 358
Gas holdup, 221, 222, 232, 246
Gas holdup axial variation, 229
Gas-liquid bubble column, 204, 228
Gas-liquid interfacial area a, 211
Gas-liquid reaction-film theory, 209
Gas-liquid reactor, 210
  backmixing, 228
Gas-liquid-solid slurry, 204
Gas phase dispersion, 225, 233
Gas phase mass transfer coefficient, 512
Gas phase mixing (P_{eG}), 238
Gas side mass transfer coefficient, k_G, 212, 224
Gas-solid reactors, 262
Gas-solids contacting, 361
Gas Sparger, 217
Gas-to-particle heat transfer, 365
Geometrical similarity, 679
Gibbs-Duhem thermodynamic relationship, 437
Global rate, 688
Grade efficiency curves, 552, 583
Gravity settling, 559, 565
Gravity settling tanks, 572, 576
Grid hole velocity, 380

Hazardous designations, 607
Hazardous pollutant control, 605
Hazardous waste, 612
Heat and material balance, 31
Heat transfer, 246
  axial dispersion coefficient, 228
Heat transfer coefficient, 184
Heat transfer, packed beds, 187
Height equivalent to a theoretical plate (HETP), 511
Height of a gas-phase transfer unit, 508, 509
Heterogeneous models, 179
Heterogeneous reactions, 102
High-temperature corrosion tests, 649
Hindered settling, 560
Homogeneous reactions, 99
Homogeneous reactors, 118, 119
Homogeneous reactor scaleup, 142
Hot spots, 194
Hougen-Watson, 104, 106
Hydraulic similarity:
  blending, 318
  heat transfer, 317
Hydrocyclones, 566, 567, 573, 578, 580
  dimensional analysis, 578, 581
Hydrodynamic differences, 14
Hydrodynamic mixing, 234
Hydrodynamic models, 202
Hydrodynamic parameters, 210
Hydrodynamic similarity, 230
Hydrogen attack, 637
Hydrogen blistering, 637
Hydrogen damage, 636
Hydrogen embrittlement, 637

Ideal mixing, 290
Impeller fluid mechanics, 312
Impeller mechanical specifications, 343
Impeller power, 317
  measurements, 331
Impeller power/volume scaleup, 329
Impeller process specification, 343
Impeller pumping capacity, 312
Impeller radial velocities, 325
Impeller scaleup factors, 322
Impeller shear rate, 338
  relationships, 319
Impeller speed, 330, 334
Impeller velocity head, 312
Impeller vortex, 335
Impurities, 2, 163
Instability, 271
Instrumentation, 32
Instrument specifications, 33

INDEX

Integrated pilot plants, 666
Interfacial area, 213, 222, 223, 246, 450
Interfacial mass transfer, 208
Intergranular corrosion, 634
Intermediate residence time distribution, 303

K_a, 88, 89
K_b, 88, 89
K_f, 88, 89, 90
K_{oG}, 511
K_p, 89
K_v, 90, 91
K_y, 91
Kinetic modeling, catalytic cracking, 192
Kinetic resistances, 210
Kinetics:
  controlled regime, 208
  sulfonation reactions, 143
Kinetics static mixer, 294
Known scaleup, 15
Known scaleup correlations, 658

Laboratory corrosion tests, 647
Laminar flow, stirred tanks, 426
Laminar flow heat exchangers, 413
Laminar flow isothermal, 412
Laminar flow models, 293
Laminar flow reactor, 129, 409
Laminar flow scaleup, 407
Laminar flow systems, residence time distributions, 294, 299
Langmuir-Hinshelwood, 104, 106
Langmuir-Hinshelwood-Houghen-Watson, 687
Life of construction materials, 652
Linear polarization probes, 651
Line/block flow diagrams, 20
Liquid distribution, packed columns, 518
Liquid-liquid extraction data, 474
Liquid phase axial dispersion coefficients, 225
Liquid phase dispersion, 233
Liquid phase mass transfer coefficient, 512
Liquid side mass transfer coefficient, k_L, 212
Liquid side mass transfer coefficients, $(k_L a)$, 207
Liquid-solid mass transfer, 224
Long tube procedure, 572
Lumped parameter, 42

Macroscale mixing, 315
Macroscopic balances, 52
Macroscopic model, 51
Major equipment outline drawings, 32
Maldistribution of fluid, 194
Mass axial dispersion coefficients, 228

Mass balances, 120
  semibatch reactors, 119
Mass transfer, two-resistance model, 453
Mass transfer coefficients, 213, 489
Mass transfer efficiency, 450
  absorbers, 530
Materials selection criteria, 624
Maximum contaminant levels, 605
Maximum gradient balances, 50
Maximum mixedness, 284
Maximum mixedness yields, 285
Maximum shear rates, 325
Maximum stable bubble size, 374
Mean residence time, t, 277
Merrill and Hamrin criterion, 415
Micromixing, 283, 301
Microscale mixing, 315
Minimum fluidization velocity, 348, 353
Minimum size pilot plant, 328
Miniplant, 657, 663, 670
Mixed feed streams, 304
Mixer power, 340
Mixing, 230
  dynamic similarity, 315
  scaleup of, 310
Mixing models, 202
Mixing tank fluid mechanics, 313
Mockups, 5, 17, 163, 176
Model:
  building, 71
  dispersion, 47, 48
  microscopic description, 47
  nonsteady state, 43
  parameters, estimation, 41
  simplification, 39
  solution, 54
  validation, 70
  verification, 41
  see also Models
Modeling, 35, 36, 42, 683
  complex feedstocks kinetics, 191
  detailed design phase, 38
  evaluation phase, 39
  fixed beds, 178
  of fluid bed reactors, 362
  practical aspects, 69
  preliminary design phase, 38
  problems, 39
  reactors, 205
Models, 16, 42, 46, 54
  classification, 55
  classification transport of phenomena, 46
  continuous, 44
  discrete variables, 44

Models (*Continued*)
  distributed, 44
  empirical, 45
  frequency domain, 55
  linear, 43
  lumped parameters, 44
  nonlinear, 43
  physicochemical principles, 45
  population balance, 45
  steady state, 43
  transport phenomena, 45
  see also Model
Modular scaleup, 15, 658
Molecular level mixing, 283
Motionless mixers, 294, 416, 419
  blending applications, 417
  heat transfer, 418
  laminar flow reactors, 419
Moving wall devices, 421
Multibed adiabatic reactor, 169, 170
Multiphase reactors, 4, 173, 188
Multiple fluid phases, 172
Multiple of steady states, 186
Multiple steady states, 141, 271
Multitubular reactor, 169
Murphee efficiency, 459

National Ambient Air Quality Standards, 602
Neutralization, 617
New source performance standards, 603
Nonhomogeneous reactions, 305
Nonideal flow, stirred tank reactor, 162
Nonisothermal reactions, 305
Nonlinear model, 42
Nonmetallic construction materials, 628
Nonmetallic materials, evaluation of, 645
Nonsteady state, 42
Normalized distributions, 279
Normalized grade efficiency, 384
Nuclear Regulatory Agency (NRA), 596

Occupational Health and Safety Administration, 596
Open systems, 288, 289
Orthokinetic flocculation, 558
Overall pumping capacity, 330

Packed bed absorber, 202, 204
Packed column, 514
  flooding, 521
  HETP, 523
  liquid distribution, 519
  mass transfer efficiency, 522
  pressure drop, 519, 520
  scaleup, 528
  theoretical stages, 526
  transfer units, 526
Packed contactor, 514
Packed extraction columns, 481, 543
  distributors, 537
  efficiency, 540
  flooding, 538
  scaleup, 544
Packing materials, 517
Packing sizes, 513
Parallel coking, 193
Particle density, 557
Particle diffusivity, 393
Particle settling, 590
Particle shape, 557
Particles:
  agglomeration of, 398
  attrition of, 397
  characteristic size of, 556
  erosion of, 398
Penetration theory, 454
Perikinetic flocculation, 558
Phase behavior, 94
Phase contacting, 513
Phase equilibrium, 433, 437
Phosgene, synthesis of, 100
Pilot plant programs, 662
Pilot plants, 9, 11, 669
  integrated, 666
  minimum size, 328
  new developments, 674
  rules of thumb, 672, 675
  safety considerations, 660
  startup inspection, 661
  studies, 656
Pilot units:
  cost of, 656
  timing of, 657
Piping and line specifications, 33
Piston flow, 290, 291
Pitting, 631
Plot plan, 32
Plug flow reactor, 129
Preliminary analysis, 36
Pressure drop limits, 171
Pressure filter equipment, 570
Pretreatment, 590
  of suspensions, 555
Preventing corrosion, 629
Problem definition, 36
Problem definition phase, 37
Problem formulation, 36
Process data estimation, 206
Process design, scope of, 31
Process development, 25

INDEX 697

Process documentation, 26
Process flow sheet, 19, 31
Process piping and instrumentation diagrams, 31
Process rate, 79, 80, 98
Process research, 25
Process safety, 32
Process scaleup, 680
Process simulator, 462
Process synthesis, 254
Process system, 23
Process variables, 41
Project description, 31
Pseudohomogeneous, 245
Pseudohomogeneous behavior, 210
Pseudohomogeneous model, 179, 180, 183
Pulsed extractor, 481
Pumping capacity, 314, 336

Q_e curve, 170
Q_m curve, 170

Radial concentration gradients, 187
Radial diffusion, 408
Radial temperature gradients, 187
Rapid corrosion tests, 649
Rate concept, 79
Rate controlling step, 107
Rate data, 82
    correlation, 97
Rate equation, 104, 106
Rate of change, 79, 80, 81, 83
Rate of reaction, 80, 98
Reacting flows, 415
Reaction mechanism, 100
Reaction plane, 208
Reaction system, 112
Reactor, 170
    autothermal, 181
    batch, 118, 255, 270
    commercial, 258
    fixed-bed, 126, 171
    gas-liquid, 210
    laminar flow, 129, 409
    multibed adiabatic, 169, 170
    multiphase, 4, 173, 188
    multitubular, 169
    plug flow, 129
    recycle, 257
    semibatch, 133
    tubular, 124, 125, 128, 186, 257, 268
    two-dimensional models, 186
Reactor design structure, 4
Reactor dynamics, 271
Reactor instability, 164
Reactor models, 229

Reactor phases, 258
Reactor ratings, 267
Reactor scaleup, 17, 68, 78, 112
Reactor selection, 254, 263
    cost minimization, 263
    example, 261
    flexibility, 263
    product selectivity, 264
Reactor types, 259, 262
Reciprocating plate extractor, 481, 484
Recycle reactors, 257
Recycle streams, 659
Reentrainment, 591
Regulation, 596
Reinforced thermoset plastics construction materials, 627
Relative volatility, 437
Replica, 35
Residence time distributions, 277
    closed systems, 286
    scaleup, 298
Right to know, 618
Root Mean Square Velocity Fluctuation, 316, 320
Rotating disk contactor, 485

Scaledown, 10, 19, 272
Scale of description, 683
Scaleup, 28, 67, 112, 678
    approaches, 14
    definition, 3
    empirical approach, 15, 659
    fundamental approach, 15
    methodology, 680, 681
    problems, 8
Scaleup experience, 11
Scaleup of mixing, 310
Scaleup ratio, 6, 7
Scaleup studies, 10, 682
Scaling, 443
Scheibel column extractor, 485
Scope of process design, 31
Sedimenting centrifuges, 583
Segregation, degree of, 302
Semibatch reaction, 149
Semibatch reactors, 133, 142, 256, 270
Separation criteria, 436
Separation factor, 434
Shear rate, 315, 316, 321, 322, 326, 328, 336, 337
    average, 325
    maximum, 325
Shear stress, 335
Shell chlorine process, 392
Show tubes, 666

Sieve bend, 571, 575
Sieve tray column, scaleup, 443
Sieve tray extractor, 485
  efficiency, 490
  interfacial area of froths, 451
Sigma concept, 586
Sigma factors, 585, 586
Similarity, 15, 16, 173, 175
Single bubble rise velocity, 354
Size distribution, 556
Slow reaction, 208
Slow reaction regime, 233, 239
Sludge blanket clarifiers, 573
Slug flow regime, 218
Slug-gas exchange, 359
Slug shape factor, 363
Slurry-wall heat transfer coefficient, 226
Solid, hazardous discharges, 607
Solid phase dispersion coefficient, 225
Solids, eddy diffusivities, 376
Solid separation efficiencies, 551, 553
Solid settling velocity, 578
Solids mixing, 351
Solubility, 95
Special distribution functions, 280
Specific cake resistance, 587
Spray contacting, 455
Spray drying, 665
Stable bubble size, 373
Stagewide processing, 432
Stanton number, 237
Startup inspection, 661
Steady state, 42
Step-by-step scaleup, 659
Stokes' law, 590
Stress corrosion, 634
Stripper, mass transfer efficiency, 530, 532
Stripping, 463, 510, 530
  multicomponent, 467
Stripping column, tray hydraulics, 469
Stripping factors, 465
Stripping stage calculations, 465
Structure of reactor design, 4
Sulfonation of benzene, 142, 149, 151, 157
Surface filters, cake filtration, 569
Surge damping, 281
Synthesis of phosgene, 100

Tanks in series, 292
  residence time distribution, 292
Temkhin-Pyzhev rate equations, 111, 687

Temperature, effect of, 92
Temperature profiles, 182
Temperature unstable materials, 2
Terminal velocity, 359
Theoretical stages, 436
Thermal control, 265
Thermodynamic consistency, 111
Thermodynamics, 84, 85, 87, 92, 94, 110
Thermoplastic construction materials, 627
Thickener, 565, 566, 572, 576, 577
Three phase reactor, 202, 243
Time distribution, first order reactions, 282
Timing of pilot units, 657
Tracer experiments, 288
Transfer function, 56
Transfer units, 436
Transition, bubble-churn-turbulent flow, 231
Transport, 55
Transport disengaging height, 381
Transport equations, 53
Transport parameters, 234
Transport phenomena, 46
Transport properties, 230
Transport resistances, 210
Tray columns design, 444
Tray liquid residence, 449
Trickle flow regime, 189
t transfer, axial dispersion coefficient, 228
Tubular plug flow, 124
Tubular reactor, 124, 125, 128, 186, 257, 268
Turbulent flow, Newtonian fluids, 302
Two-dimensional models, reactors, 186
Two phase gas/liquid flow, 5
Typical hazardous pollutants, 603

Unacceptable waste products, 613
Uncertainty, 28
Uniform corrosion, 630
Units, 509
Unit scaleup, 680
Unmixed feed streams, 304
Unsteady state devices, 270
Used metals, 625
Utilities requirements, 33
Utility piping and instrumentation, 32

Van Laar vapor-liquid equilibria, 438
Velocity fluctuations, 323, 325
Velocity head, 314
Velocity profile, 314
Vendor tests, 664

INDEX

Viscosity, 333
Volumetric mass transfer coefficients, $k_L a$, 223

Wall heat transfer coefficient, α_w, 187
Washout determination, 286, 287
Washout function, 278, 291
Waste destruction, 617
Water discharges, 604

Water quality standards[a], 604
Weldability, 625

Yield, 285
Y-X equilibrium curve, 438

Zone settling, 561